Randall D. Knight • Brian Jones • Stuart Field

College Physics
A Strategic Approach
Volume 1

Second Custom Edition for University of Minnesota Twin Cities

Taken from:
College Physics: A Strategic Approach, Third Edition
by Randall D. Knight, Brian Jones, and Stuart Field

Cover Art: Courtesy of Photodisc/Getty Images.

Taken from:

College Physics: A Strategic Approach, Third Edition
by Randall D. Knight, Brian Jones, and Stuart Field
Copyright © 2015 by Pearson Education, Inc.
Published by Addison-Wesley
Boston, Massachusetts 02116

This special edition published in cooperation with Pearson Learning Solutions.

Pearson Learning Solutions, 501 Boylston Street, Suite 900, Boston, MA 02116
A Pearson Education Company
www.pearsoned.com

Printed in the United States of America

000200010271905587

LA

ISBN 10: 1-269-91836-2
ISBN 13: 978-1-269-91836-7

About the Authors

Randy Knight taught introductory physics for 32 years at Ohio State University and California Polytechnic University, where he is Professor Emeritus of Physics. Randy received a Ph.D. in physics from the University of California, Berkeley and was a post-doctoral fellow at the Harvard-Smithsonian Center for Astrophysics before joining the faculty at Ohio State University. It was at Ohio that he began to learn about the research in physics education that, many years later, led to *Five Easy Lessons: Strategies for Successful Physics Teaching, Physics for Scientists and Engineers: A Strategic Approach,* and now to this book. Randy's research interests are in the fields of laser spectroscopy and environmental science. When he's not in front of a computer, you can find Randy hiking, sea kayaking, playing the piano, or spending time with his wife Sally and their six cats.

Brian Jones has won several teaching awards at Colorado State University during his 25 years teaching in the Department of Physics. His teaching focus in recent years has been the College Physics class, including writing problems for the MCAT exam and helping students review for this test. In 2011, Brian was awarded the Robert A. Millikan Medal of the American Association of Physics Teachers for his work as director of the Little Shop of Physics, a hands-on science outreach program. He is actively exploring the effectiveness of methods of informal science education and how to extend these lessons to the college classroom. Brian has been invited to give workshops on techniques of science instruction throughout the United States and in Belize, Chile, Ethiopia, Azerbaijan, Mexico and Slovenia. Brian and his wife Carol have dozens of fruit trees and bushes in their yard, including an apple tree that was propagated from a tree in Isaac Newton's garden.

Stuart Field has been interested in science and technology his whole life. While in school he built telescopes, electronic circuits, and computers. After attending Stanford University, he earned a Ph.D. at the University of Chicago, where he studied the properties of materials at ultralow temperatures. After completing a postdoctoral position at the Massachusetts Institute of Technology, he held a faculty position at the University of Michigan. Currently at Colorado State University, Stuart teaches a variety of physics courses, including algebra-based introductory physics, and was an early and enthusiastic adopter of Knight's *Physics for Scientists and Engineers.* Stuart maintains an active research program in the area of superconductivity. Stuart enjoys Colorado's great outdoors, where he is an avid mountain biker; he also plays in local ice hockey leagues.

Preface to the Instructor

In 2006, we published *College Physics: A Strategic Approach,* a new algebra-based physics textbook for students majoring in the biological and life sciences, architecture, natural resources, and other disciplines. As the first such book built from the ground up on research into how students can more effectively learn physics, it quickly gained widespread critical acclaim from professors and students alike. For the second edition, and now for this third edition, we have continued to build on the research-proven instructional techniques introduced in the first edition and the extensive feedback from thousands of users to take student learning even further.

Objectives

Our primary goals in writing *College Physics: A Strategic Approach* have been:

- To provide students with a textbook that's a more manageable size, less encyclopedic in its coverage, and better designed for learning.
- To integrate proven techniques from physics education research into the classroom in a way that accommodates a range of teaching and learning styles.
- To help students develop both quantitative reasoning skills and solid conceptual understanding, with special focus on concepts well documented to cause learning difficulties.
- To help students develop problem-solving skills and confidence in a systematic manner using explicit and consistent tactics and strategies.
- To motivate students by integrating real-world examples relevant to their majors—especially from biology, sports, medicine, the animal world—and that build upon their everyday experiences.
- To utilize proven techniques of visual instruction and design from educational research and cognitive psychology that improve student learning and retention and address a range of learner styles.

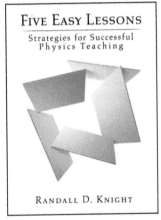

FIVE EASY LESSONS

Strategies for Successful
Physics Teaching

RANDALL D. KNIGHT

A more complete explanation of these goals and the rationale behind them can be found in Randy Knight's paperback book, *Five Easy Lessons: Strategies for Successful Physics Teaching.* Please request a copy from your local Pearson sales representative if it would be of interest to you (ISBN 978-0-805-38702-5).

What's New to This Edition

In revising the book for this third edition, we have renewed our basic focus on students and how they learn. We've considered extensive feedback from scores of instructors and thousands of students, including our student advisory panel, in order to enhance and improve the text, figures, and the end-of-chapter problems. Changes include the following:

- More focused **Chapter Previews** provide a brief, visual, and non-technical preview, proven to help students organize their thinking and improve their understanding of the upcoming material.
- New **Synthesis boxes** bring together key concepts, principles, and equations in order to highlight connections and differences.
- New **Concept Check figures** encourage students to actively engage with key or complex figures by asking them to reason with a related Stop To Think question.
- Additional **Stop To Think questions** provide students with more crucial practice and concept checks as they go through the chapters.

- New in-line **Looking Back pointers** encourage students to review important material from earlier chapters. These are given right at the moment they are needed, rather than at the start of the chapter (where they are often overlooked).
- New **Problem-Solving Strategy Overviews** give students the "big picture" of the strategy before delving into details, just as the chapter previews give the "big picture" of the chapter.
- **Streamlined text and figures** tighten and focus the presentation to be more closely matched to student needs. We've scrutinized every figure, caption, discussion, and photo in order to enhance their clarity and focus their role.
- Expanded use of **annotated equations** helps students decipher what they "say," and what the variables and units are.
- Increased emphasis on **critical thinking and reasoning,** both in worked examples and end-of-chapter problems, promotes these key skills. These skills are especially important for those taking the MCAT exam.
- Expanded use of **realistic and real-world data** ensures students can make sense of answers that are grounded in the real world. Our examples and problems use real numbers and real data, and test different types of reasoning using equations, ratios, and graphs.
- **Enhanced end-of-chapter problems,** based on the wealth of data from MasteringPhysics, student advisory panel input, and a rigorous blind-solving and accuracy cross-checking process, optimize clarity, utility, and variety. We've added problems based on real-world and biomedical situations and problems that expand the range of reasoning skills students need to use in the solution.

We have made many small changes to the flow of the text throughout, streamlining derivations and discussions, providing more explanation for complex concepts and situations, and reordering and reorganizing material so that each section and each chapter has a clearer focus. There are small changes on nearly every page. The more significant content changes include:

- The circular motion material in Chapters 3, 6 and 7 has been reworked for a more natural progression of topics. Acceleration in circular motion is now introduced in Chapter 3, frequency and period are now introduced in Chapter 6, while angular position and angular velocity are now in Chapter 7. The treatment of circular motion in Chapter 3 emphasizes the use of vectors to understand the nature of centripetal acceleration. In Chapter 6, the focus is on dynamics, and in Chapter 7, we extend these ideas to rotational motion.
- The discussion of the law of conservation of energy in Section 10.6 has been updated to provide a more logical and coherent flow from the most general form of the law to more specialized versions for isolated systems and then to systems with only mechanical energy.
- The material in Chapter 11 making the microscopic connection between thermal energy and temperature for an ideal gas has been moved to Chapter 12, where it fits better with the atomic model of an ideal gas presented there.
- Minor topics that have been removed to focus the presentation include antinodal lines for sound waves in Chapter 16, maximum intensity of a diffraction grating's bright fringes in Chapter 17, exposure in Chapter 19, and elevation graphs in Chapter 21.
- The start of Chapter 21 has been revised to clarify the origin of electric potential energy by making a more concrete connection between electric potential energy and more familiar potential energies, such as gravitational and elastic potential energy.
- The treatment of electromagnetic waves in Chapter 25 was streamlined to focus on the nature of the waves, the meaning of polarization, and the application of these ideas to real-world situations.
- Chapters 29 and 30 have been significantly streamlined, improving the overall flow and removing some extraneous details so that students can better focus on the physics.

We know that students increasingly rely on sources of information beyond the text, and instructors are looking for quality resources that prepare students for engagement in lecture. The text will always be the central focus, but we are adding additional media elements closely tied to the text that will enhance student understanding. In the Technology Update to the Second Edition, we added Class Videos, Video Tutor Solutions, and Video Tutor Demonstrations. In the Third Edition, we are adding an exciting new supplement, **Prelecture Videos,** short videos with author Brian Jones that introduce the topics of each chapter with accompanying assessment questions.

Textbook Organization

College Physics: A Strategic Approach is a 30-chapter text intended for use in a two-semester course. The textbook is divided into seven parts: Part I: *Force and Motion*, Part II: *Conservation Laws*, Part III: *Properties of Matter*, Part IV: *Oscillations and Waves*, Part V: *Optics*, Part VI: *Electricity and Magnetism*, and Part VII: *Modern Physics*.

Part I covers Newton's laws and their applications. The coverage of two fundamental conserved quantities, momentum and energy, is in Part II, for two reasons. First, the way that problems are solved using conservation laws—comparing an *after* situation to a *before* situation—differs fundamentally from the problem-solving strategies used in Newtonian dynamics. Second, the concept of energy has a significance far beyond mechanical (kinetic and potential) energies. In particular, the key idea in thermodynamics is energy, and moving from the study of energy in Part II into thermal physics in Part III allows the uninterrupted development of this important idea.

Optics (Part V) is covered directly after oscillations and waves (Part IV), but *before* electricity and magnetism (Part VI). Further, we treat wave optics before ray optics. Our motivations for this organization are twofold. First, wave optics is largely just an extension of the general ideas of waves; in a more traditional organization, students will have forgotten much of what they learned about waves by the time they get to wave optics. Second, optics as it is presented in introductory physics makes no use of the properties of electromagnetic fields. The documented difficulties that students have with optics are difficulties with waves, not difficulties with electricity and magnetism. There's little reason other than historical tradition to delay optics. However, the optics chapters are easily deferred until after Part VI for instructors who prefer that ordering of topics.

The Student Workbook

A key component of *College Physics: A Strategic Approach* is the accompanying *Student Workbook*. The workbook bridges the gap between textbook and homework problems by providing students the opportunity to learn and practice skills prior to using those skills in quantitative end-of-chapter problems, much as a musician practices technique separately from performance pieces. The workbook exercises, which are keyed to each section of the textbook, focus on developing specific skills, ranging from identifying forces and drawing free-body diagrams to interpreting field diagrams.

The workbook exercises, which are generally qualitative and/or graphical, draw heavily upon the physics education research literature. The exercises deal with issues known to cause student difficulties and employ techniques that have proven to be effective at overcoming those difficulties. **New to the third edition workbook** are *jeopardy problems* that ask students to work backwards from equations to physical situations, enhancing their understanding and critical thinking skills. The workbook exercises can be used in-class as part of an active-learning teaching strategy, in recitation sections, or as assigned homework. More information about effective use of the *Student Workbook* can be found in the *Instructor's Guide*.

Available versions: Volume 1 (ISBN 978-0-321-90886-5): Chapters 1–16, and Volume 2 (978-0-321-90887-2): Chapters 17–30. A package of both volumes is also available (ISBN 978-0-321-90724-0).

■ **Complete edition,** with MasteringPhysics® and Student Workbook (ISBN 978-0-321-90255-9): Chapters 1–30.

■ **Books a la Carte edition,** with MasteringPhysics® and Student Workbook (ISBN 978-0-321-90882-7)

■ **Complete edition,** without MasteringPhysics® (ISBN 978-0-321-87972-1): Chapters 1–30.

■ **Volume 1** without MasteringPhysics® (ISBN 978-0-321-90877-3): Chapters 1–16.

■ **Volume 2** without MasteringPhysics® (ISBN 978-0-321-90878-0): Chapters 17–30.

Split your text the way you split your course! Log on to www.pearsoncustomlibrary.com and create your own splits of *College Physics: A Strategic Approach,* 3e, including *your* choice of chapters.

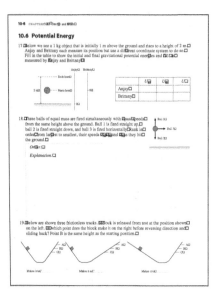

Instructor Supplements

NOTE ► For convenience, most instructor supplements can be downloaded from the "Instructor Resources" area of MasteringPhysics®. ◄

■ (MP) **MasteringPhysics**® is a powerful, yet simple, online homework, tutorial, and assessment system designed to improve student learning and results. Students benefit from wrong-answer specific feedback, hints, and a huge variety of educationally effective content while unrivalled gradebook diagnostics allow an instructor to pinpoint the weaknesses and misconceptions of their class.

NSF-sponsored published research (and subsequent studies) show that MasteringPhysics has dramatic educational results. MasteringPhysics allows instructors to build wide-ranging homework assignments of just the right difficulty and length and provides them with efficient tools to analyze in unprecedented detail both class trends and the work of any student.

■ The cross-platform **Instructor's Resource DVD** (ISBN 978-0-321-90725-7) provides invaluable and easy-to-use resources for your class, organized by textbook chapter. The **Instructor's Solutions Manual** and the **Instructor's Guide** are provided in PDF format and as editable Word files. Comprehensive **Lecture Slides** (with embedded classroom response system **"Clicker" Questions**) are provided in PowerPoint, as well as high-quality versions of all the **Prelecture Videos.** In addition, all figures, photos, tables, previews, and summaries from the textbook are given in JPEG format. All Problem-Solving Strategies, Math Relationships Boxes, Tactics Boxes, and Key Equations are provided in editable Word and JPEG format.

■ The **Instructor's Solutions Manual,** written by Professor Larry Smith, Snow College, provides *complete* solutions to all the end-of-chapter questions and problems. All solutions follow the Prepare/Solve/Assess problem-solving strategy used in the textbook for quantitative problems, and Reason/Assess strategy for qualitative ones. The solutions are available by chapter in Word and PDF format, are included on the *Instructor's Resource DVD,* and can also be downloaded from the *Instructor Resource Center* (www.pearsonhighered.com/educator).

■ The **Instructor's Guide for *College Physics: A Strategic Approach,*** a comprehensive and highly acclaimed resource, provides chapter-by-chapter creative ideas and teaching tips for using *College Physics: A Strategic Approach* in your class. In addition, it contains an extensive review of what has been learned from physics education research, and provides guidelines for using active-learning techniques in your classroom. Instructor Guide chapters are provided in Word and PDF format, are included on the *Instructor's Resource DVD,* and can also be downloaded from the *Instructor Resource Center* (www.pearsonhighered.com/educator).

■ The **Test Bank** contains 4,000 high-quality problems, with a range of multiple-choice and regular homework-type questions. Test files are provided in both TestGen® (an easy-to-use, fully networkable program for creating and editing quizzes and exams) and Word format, and can also be downloaded from www.pearsonhighered.com/educator. The Test Bank problems are also assignable via MasteringPhysics.

Student Supplements

■ (MP) **MasteringPhysics**® is a powerful, yet simple, online homework, tutorial, and assessment system designed to improve student learning and results. Students benefit from wrong-answer specific feedback, hints, and a huge variety of educationally effective content while unrivalled gradebook diagnostics allow an instructor to pinpoint the weaknesses and misconceptions of their class. The individualized, 24/7 Socratic tutoring is recommended by 9 out of 10 students to their peers as the most effective and time-efficient way to study.

■ The **Student Workbook** (Volume 1: Chapters 1–16 (ISBN 978-0-321-90886-5), Volume 2: Chapters 17–30 (978-0-321-90887-2), or a package of both volumes (ISBN 978-0-321-90724-0)) is a key component of *College Physics: A Strategic Approach.* The workbook bridges the gap between textbook and homework problems by providing students the opportunity to learn and practice skills prior to using those skills in quantitative end-of-chapter problems, much as a musician practices technique separately from performance pieces.

■ **Pearson eText** is available through MasteringPhysics, either automatically when MasteringPhysics is packaged with new books, or available as a purchased upgrade online. Allowing students access to the text wherever they have access to the Internet, Pearson eText comprises the full text, including figures that can be enlarged for better viewing. Within eText, students are also able to pop up definitions and terms to help with vocabulary and the reading of the material. Students can also take notes in eText using the annotation feature at the top of each page.

■ Over 140 **Video Tutors** about relevant demonstrations or problem-solving strategies play directly on a smartphone or tablet via Class Video scannable QR codes in the printed book.

These interactive videos are also viewable via links within the Pearson eText and the Study Area of MasteringPhysics.

■ **ActivPhysics Online**™ applets and applet-based tutorials, developed by education pioneers Professors Alan Van Heuvelen and Paul D'Alessandris, are available in the Study Area of MasteringPhysics. Also provided are over 70 **PhET Simulations** from the University of Colorado.

■ **Pearson Tutor Services** (www.pearsontutorservices.com) Each student's subscription to MasteringPhysics also contains complimentary access to Pearson Tutor Services, powered by Smarthinking, Inc. By logging in with their MasteringPhysics ID and password, they will be connected to highly qualified e-instructors™ who provide additional, interactive online tutoring on the major concepts of physics. Some restrictions apply; offer subject to change.

■ The **Student Solutions Manuals, Chapters 1–16** (ISBN 978-0-321-90884-1) and **Chapters 17–30** (ISBN 978-0-321-90885-8), written by Professor Larry Smith, Snow College, provide *detailed* solutions to more than half of the odd-numbered end-of-chapter problems. Following the problem-solving strategy presented in the text, thorough solutions are provided to carefully illustrate both the qualitative (Reason/Assess) and quantitative (Prepare/Solve/Assess) steps in the problem-solving process.

Acknowledgments

We have relied upon conversations with and, especially, the written publications of many members of the physics education community. Those who may recognize their influence include Arnold Arons, Uri Ganiel, Fred Goldberg, Ibrahim Halloun, David Hestenes, Leonard Jossem, Jill Larkin, Priscilla Laws, John Mallinckrodt, Lillian McDermott and members of the Physics Education Research Group at the University of Washington, Edward "Joe" Redish, Fred Reif, John Rigden, Rachel Scherr, Bruce Sherwood, David Sokoloff, Ronald Thornton, Sheila Tobias, and Alan Van Heuleven.

We are very grateful to Larry Smith for the difficult task of writing the *Instructor Solutions Manual*; to Scott Nutter for writing out the Student Workbook answers; to Wayne Anderson, Jim Andrews, Nancy Beverly, David Cole, Karim Diff, Jim Dove, Marty Gelfand, Kathy Harper, Charlie Hibbard, Robert Lutz, Matt Moelter, Kandiah Manivannan, Ken Robinson, and Cindy Schwarz-Rachmilowitz for their contributions to the end-of-chapter questions and problems; to Wayne again for helping with the Test Bank questions; and to Steven Vogel for his careful review of the biological content of many chapters and for helpful suggestions.

We especially want to thank our editor Becky Ruden, development editor Alice Houston, project manager Martha Steele, and all the other staff at Pearson for their enthusiasm and hard work on this project. Having a diverse author team is one of the strengths of this book, but it has meant that we rely a great deal on Becky to help us keep to a single focus, on Martha to be certain that one of us attends to all details, and on Alice's tireless efforts and keen editorial eye as she helps us synthesize our visions into a coherent whole.

Rose Kernan and the team at Nesbitt Graphics/Cenveo, copy editor Carol Reitz, and photo researcher Eric Schrader get much credit for making this complex project all come together. In addition to the reviewers and classroom testers listed below, who gave invaluable feedback, we are particularly grateful to Charlie Hibbard for his close scrutiny of every word, symbol, number, and figure.

Randy Knight: I would like to thank my Cal Poly colleagues, especially Matt Moelter, for many valuable conversations and suggestions. I am endlessly grateful to my wife Sally for her love, encouragement, and patience, and to our many cats for nothing in particular other than being cats.

Brian Jones: I would like to thank my fellow AAPT and PIRA members for their insight and ideas, the creative students and colleagues who are my partners in the Little Shop of Physics, the students in my College Physics classes who help me become a better teacher, and, most of all, my wife Carol, my best friend and gentlest editor, whose love makes the journey worthwhile.

Stuart Field: I would like to thank my wife Julie and my children, Sam and Ellen, for their love, support, and encouragement.

Reviewers and Classroom Testers

Special thanks go to our third edition review panel: Taner Edis, Marty Gelfand, Jason Harlow, Charlie Hibbard, Jeff Loats, Amy Pope, and Bruce Schumm.

David Aaron, *South Dakota State University*
Susmita Acharya, *Cardinal Stritch University*
Ugur Akgun, *University of Iowa*
Ralph Alexander, *University of Missouri—Rolla*
Kyle Altmann, *Elon University*
Donald Anderson, *Ivy Tech*

Michael Anderson, *University of California—San Diego*
Steve Anderson, *Montana Tech*
James Andrews, *Youngstown State University*
Charles Ardary, *Edmond Community College*
Charles Bacon, *Ferris State University*
John Barry, *Houston Community College*

David H. Berman, *University of Northern Iowa*
Phillippe Binder, *University of Hawaii—Hilo*
Jeff Bodart, *Chipola College*
James Boger, *Flathead Valley Community College*
Richard Bone, *Florida International University*
James Borgardt, *Juniata College*
Daniela Bortoletto, *Purdue University*
Don Bowen, *Stephen F. Austin State University*
Asa Bradley, *Spokane Falls Community College*
Elena Brewer, *SUNY at Buffalo*
Dieter Brill, *University of Maryland*
Hauke Busch, *Augusta State University*
Kapila Castoldi, *Oakland University*
Raymond Chastain, *Louisiana State University*
Michael Cherney, *Creighton University*
Lee Chow, *University of Central Florida*
Song Chung, *William Paterson University*
Alice Churukian, *Concordia College*
Christopher M. Coffin, *Oregon State University*
John S. Colton, *Brigham Young University*
Kristi Concannon, *Kings College*
Teman Cooke, *Georgia Perimeter College at Lawrenceville*
Daniel J. Costantino, *The Pennsylvania State University*
Jesse Cude, *Hartnell College*
Melissa H. Dancy, *University of North Carolina at Charlotte*
Loretta Dauwe, *University of Michigan—Flint*
Mark Davenport, *San Antonio College*
Chad Davies, *Gordon College*
Lawrence Day, *Utica College*
Carlos Delgado, *Community College of Southern Nevada*
David Donovan, *Northern Michigan University*
James Dove, *Metropolitan State University of Denver*
Archana Dubey, *University of Central Florida*
Andrew Duffy, *Boston University*
Taner Edis, *Truman State University*
Ralph Edwards, *Lurleen B. Wallace Community College*
Steve Ellis, *University of Kentucky*
Paula Engelhardt, *Tennessee Technical University*
Davene Eryes, *North Seattle Community College*
Gerard Fasel, *Pepperdine University*
Luciano Fleischfresser, *OSSM Autry Tech*
Cynthia Galovich, *University of Northern Colorado*
Bertram Gamory, *Monroe Community College*
Sambandamurthy Ganapathy, *SUNY at Buffalo*
Delena Gatch, *Georgia Southern University*
Richard Gelderman, *Western Kentucky University*
Martin Gelfand, *Colorado State University*
Terry Golding, *University of North Texas*
Robert Gramer, *Lake City Community College*
William Gregg, *Louisiana State University*
Paul Gresser, *University of Maryland*
Robert Hagood, *Washtenaw Community College*
Jason Harlow, *University of Toronto*
Heath Hatch, *University of Massachusetts*
Carl Hayn, *Santa Clara University*

James Heath, *Austin Community College*
Zvonko Hlousek, *California State University Long Beach*
Greg Hood, *Tidewater Community College*
Sebastian Hui, *Florence-Darlington Technical College*
Eric Hudson, *The Pennsylvania State University*
Joey Huston, *Michigan State University*
David Iadevaia, *Pima Community College—East Campus*
Fred Jarka, *Stark State College*
Ana Jofre, *University of North Carolina—Charlotte*
Daniel Jones, *Georgia Tech*
Erik Jensen, *Chemeketa Community College*
Todd Kalisik, *Northern Illinois University*
Ju H. Kim, *University of North Dakota*
Armen Kocharian, *California State University Northridge*
J. M. Kowalski, *University of North Texas*
Laird Kramer, *Florida International University*
Christopher Kulp, *Eastern Kentucky University*
Richard Kurtz, *Louisiana State University*
Kenneth Lande, *University of Pennsylvania*
Tiffany Landry, *Folsom Lake College*
Todd Leif, *Cloud County Community College*
John Levin, *University of Tennessee—Knoxville*
John Lindberg, *Seattle Pacific University*
Jeff Loats, *Metropolitan State University of Denver*
Rafael López-Mobilia, *The University of Texas at San Antonio*
Robert W. Lutz, *Drake University*
Lloyd Makorowitz, *SUNY Farmingdale*
Colleen Marlow, *Rhode Island College*
Eric Martell, *Millikin University*
Mark Masters, *Indiana University—Purdue*
John McClain, *Temple College*
Denise Meeks, *Pima Community College*
Henry Merrill, *Fox Valley Technical College*
Mike Meyer, *Michigan Technological University*
Karie Meyers, *Pima Community College*
Tobias Moleski, *Nashville State Tech*
April Moore, *North Harris College*
Gary Morris, *Rice University*
Krishna Mukherjee, *Slippery Rock University*
Charley Myles, *Texas Tech University*
Meredith Newby, *Clemson University*
David Nice, *Bryn Mawr*
Fred Olness, *Southern Methodist University*
Charles Oliver Overstreet, *San Antonio College*
Paige Ouzts, *Lander University*
Russell Palma, *Minnesota State University—Mankato*
Richard Panek, *Florida Gulf Coast University*
Joshua Phiri, *Florence-Darling Technical College*
Iulia Podariu, *University of Nebraska at Omaha*
David Potter, *Austin Community College*
Promod Pratap, *University of North Carolina—Greensboro*
Michael Pravica, *University of Nevada, Las Vegas*
Earl Prohofsky, *Purdue University*
Marilyn Rands, *Lawrence Technological University*

Andrew Rex, *University of Puget Sound*
Andrew Richter, *Valparaiso University*
William Robinson, *North Carolina State University*
Phyliss Salmons, *Embry—Riddle Aeronautical University*
Michael Schaab, *Maine Maritime Academy*
Bruce Schumm, *University of California, Santa Cruz*
Mizuho Schwalm, *University of Minnesota Crookston*
Cindy Schwarz, *Vassar College*
Natalia Semushkhina, *Shippensburg University*
Khazhgery (Jerry) Shakov, *Tulane University*
Kathy Shan, *University of Toledo*
Anwar Sheikh, *Colorado Mesa University*
Bart Sheinberg, *Houston Community College*
Marllin Simon, *Auburn University*
Kenneth Smith, *Pennsylvania State University*
Michael Smutko, *Northwestern University*
Jon Son, *Boston University*
Noel Stanton, *Kansas State University*
Donna Stokes, *University of Houston*
Chuck Stone, *North Carolina A&T*
Chun Fu Su, *Mississippi State University*
Jeffrey Sudol, *West Chester University*

Scott Thompson, *Georgia Gwinnett College*
William Tireman, *Northern Michigan University*
Negussie Tirfessa, *Manchester Community College*
Rajive Tiwari, *Belmont Abbey College*
Herman Trivilino, *College of the Mainland*
Dmitri Tsybychev, *Stony Brook University*
Douglas Tussey, *Pennsylvania State University*
Stephen Van Hook, *Pennsylvania State University*
Manuel Valera, *Slippery Rocky University*
Christos Valiotis, *Antelope Valley College*
James Vesenka, *University of New England*
Stamatis Vokos, *Seattle Pacific University*
James Wanliss, *Embry—Riddle Aeronautical University*
Henry Weigel, *Arapahoe Community College*
Luc T. Wille, *Florida Atlantic University*
Courtney Willis, *University of Northern Colorado*
Katherine Wu, *University of Tampa*
Ali Yazdi, *Jefferson State Community College*
David Young, *Louisiana State University*
Hsiao-Ling Zhou, *Georgia State University*
Todd Zimmerman, *University of Wisconsin—Stout*
Ulrich Zurcher, *Cleveland State University*

Student Advisory Board for the Third Edition

Nathalia Alzate, *Auburn University*
Hannah Chapman, *Colorado State University*
Rachel Eckert, *California Polytechnic State University—San Luis Obispo*
Emily Garban, *Colorado State University*
Tyrel Heckendorf, *Georgia State University*
Alex Keifer, *California Polytechnic State University—San Luis Obispo*
Isaac Moore, *The Pennsylvania State University*
John Peter Polites, *Georgia State University*
Blair Porterfield, *The Pennsylvania State University*

Rebecca Rogers, *Auburn University*
Rashawn D. Simmons, *Georgia State University*
Mary-Catherin Skoulos, *Stony Brook University*
Brittany Swiderski, *Stony Brook University*
Jenna Tustin, *Colorado State University*
Aaron Vermeersch, *Michigan State University*
Philip E. Weinberg, *Michigan State University*
Keith Wood, *Auburn University*
Timothy Yuan, *Stony Brook University*
Andrew Zilavy, *Colorado State University*

Preface to the Student

The most incomprehensible thing about the universe is that it is comprehensible.
—Albert Einstein

If you are taking a course for which this book is assigned, you probably aren't a physics major or an engineering major. It's likely that you aren't majoring in a physical science. So why are you taking physics?

It's almost certain that you are taking physics because you are majoring in a discipline that requires it. Someone, somewhere, has decided that it's important for you to take this course. And they are right. There is a lot you can learn from physics, even if you don't plan to be a physicist. We regularly hear from doctors, physical therapists, biologists and others that physics was one of the most interesting and valuable courses they took in college.

So, what can you expect to learn in this course? Let's start by talking about what physics is. Physics is a way of thinking about the physical aspects of nature. Physics is not about "facts." It's far more focused on discovering *relationships* between facts and the *patterns* that exist in nature than on learning facts for their own sake. Our emphasis will be on thinking and reasoning. We are going to look for patterns and relationships in nature, develop the logic that relates different ideas, and search for the reasons *why* things happen as they do.

The concepts and techniques you will learn will have a wide application. In this text we have a special emphasis on applying physics to understanding the living world. You'll use your understanding of charges and electric potential to analyze the electric signal produced when your heart beats. You'll learn how sharks can detect this signal to locate prey and, further, how and why this electric sensitivity seems to allow hammerhead sharks to detect magnetic fields, aiding navigation in the open ocean.

Like any subject, physics is best learned by doing. "Doing physics" in this course means solving problems, applying what you have learned to answer questions at the end of the chapter. When you are given a homework assignment, you may find yourself tempted to simply solve the problems by thumbing through the text looking for a formula that seems like it will work. This isn't how to do physics; if it was, whoever required you to take this course wouldn't bother. The folks who designed your major want you to learn to *reason*, not to "plug and chug." Whatever you end up studying or doing for a career, this ability will serve you well.

How do you learn to reason in this way? There's no single strategy for studying physics that will work for all students, but we can make some suggestions that will certainly help:

- **Read each chapter *before* it is discussed in class.** Class attendance is much more effective if you have prepared.
- **Participate actively in class.** Take notes, ask and answer questions, take part in discussion groups. There is ample scientific evidence that *active participation* is far more effective for learning science than is passive listening.
- **After class, go back for a careful rereading of the chapter.** In your second reading, pay close attention to the details and the worked examples. Look for the *logic* behind each example, not just at what formula is being used.
- **Apply what you have learned to the homework problems at the end of each chapter.** By following the techniques of the worked examples, applying the tactics and problem-solving strategies, you'll learn how to apply the knowledge you are gaining.
- **Form a study group with two or three classmates.** There's good evidence that students who study regularly with a group do better than the rugged individualists who try to go it alone.

And we have one final suggestion. As you read the book, take part in class, and work through problems, step back every now and then to appreciate the big picture. You are going to study topics that range from motions in the solar system to the electrical signals in the nervous system that let you tell your hand to turn the pages of this book. It's a remarkable breadth of topics and techniques that is based on a very compact set of organizing principles.

Now, let's get down to work.

Studying for and Taking the MCAT Exam

If you are taking the College Physics course, there's a good chance that you are majoring in the biological sciences. There's also a good chance that you are preparing for a career in the health professions, and so might well be required to take the Medical College Admission Test, the MCAT exam.

The *Chemical and Physical Foundations of Biological Systems* section of the MCAT assesses your understanding of the concepts of this course by testing your ability to apply these concepts to living systems. You will be expected to use what you've learned to analyze situations you've never seen before, making simplified but realistic models of the world. Your reasoning skills will be just as important as your understanding of the universal laws of physics.

Structure of the MCAT Exam

Most of the test consists of a series of passages of technical information followed by a series of questions based on each passage, much like the passage problems at the end of each chapter in this book. Some details:

- **The passages and the questions are *always* integrated.** Understanding the passage and answering the questions will require you to use knowledge from several different areas of physics.
- **Passages will generally be about topics for which you do not have detailed knowledge.** But, if you read carefully, you'll see that the treatment of the passage is based on information you should know well.
- **The test assumes a basic level of background knowledge.** You'll need to have facility with central themes and major concepts, but you won't need detailed knowledge of any particular topic. Such detailed information, if needed, will be provided in the passage.
- **You can't use calculators on the test, so any math that you do will be reasonably simple.** Quickly estimating an answer with ratio reasoning or a knowledge of the scale of physical quantities will be a useful skill.
- **The answers to the questions are all designed to be plausible.** You can't generally weed out the "bad" answers with a quick inspection.
- **The test is given online.** Practicing with MasteringPhysics will help you get used to this format.

Preparing for the Test

Because you have used this book as a tool for learning physics, you should use it as a tool for reviewing for the MCAT exam.

Several of the key features of the book will be useful for this, including some that were explicitly designed with the MCAT exam in mind.

As you review the chapters:

- Start with the *Chapter Previews,* which provide a "big picture" overview of the content. What are the major themes of each chapter?
- Look for the *Synthesis* boxes that bring together key concepts and equations. These show connections and highlight differences that you should understand and be ready to apply.
- Go through each chapter and review the *Stop to Think* exercises. These are a good way to test your understanding of the key concepts and techniques.
- Each chapter closes with a passage problem that is designed to be "MCAT-exam-like." They'll give you good practice with the "read a passage, answer questions" structure of the MCAT exam.

The passage problems are a good tool, but the passages usually don't integrate topics that span several chapters—a key feature of the MCAT exam. For integrated passages and problems, turn to the *Part Summaries:*

- For each Part Summary, read the *One Step Beyond* passage and answer the associated questions.
- After this, read the passages and answer the questions that end each Part Summary section. These passages and associated problems are—by design—very similar to the passages and questions you'll see on the actual MCAT exam.

Taking the Test: Reading the Passage

As you read each passage, you'll need to interpret the information presented and connect it with concepts you are familiar with, translating it into a form that makes sense based on your background.

The next page shows a passage that was written to very closely match the style and substance of an actual MCAT passage. Blue annotations highlight connections you should make as you read. The passage describes a situation (the mechanics and energetics of sled dogs) that you probably haven't seen before. But the basic physics (friction, energy conversion) are principles that you are familiar with, principles that you have seen applied to related situations. When you read the passage, think about the underlying physics concepts and how they apply to this case.

Translating the Passage
As you read the passage, do some translation.
Connect the scenario to examples you've seen before,
translate given information into forms you are
familiar with, think about the basic physical
principles that apply.

Passage X

For travel over snow, a sled with runners that slide
on snow is the best way to get around. Snow is
slippery, but there is still friction between runners
and the ground; the forward force required to pull a
sled at a constant speed might be 1/6 of the sled's
weight.

As you read this part of the passage, think about the forces involved:
For a sled moving at a constant speed, there is no net force. The
downward weight force is equal to the upward normal force; the
forward pulling force must be equal to the friction force, which is
acting opposite the sled's motion. There are many problems like this
in Chapter 5.

The pulling force might well come from a dog. In a
typical sled, the rope that the dog uses to pull
attaches at a slight angle, as in Figure 1. The pulling
force is the horizontal component of the tension in
the rope.

Part of translating is converting given information into a more usual
or more useful form. This is really a statement about the coefficient
of kinetic friction.

The force applied to the sled is the tension force in the rope, which is
shown at an angle. The horizontal component is the pulling force;
you're told this. There is a vertical component of the force as well.

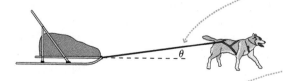

Figure 1

In the data given here, and the description given above, the sled
moves at a constant speed—there is no mention of acceleration
anywhere in this passage. In such cases, the net force is zero and the
kinetic energy of the sled isn't changing.

Sled dogs have great aerobic capacity; a 40 kg dog
can provide output power to pull with a 60 N force
at 2.2 m/s for hours. The output power is related to
force and velocity by $P = F \cdot v$, so they can pull
lighter loads at higher speeds.

Notice that the key equation relating power, force and velocity is
given to you. That's to be expected. Any specific information,
including equations, constants and other such details, will generally
be given in the passage. The MCAT is a test of reasoning, not recall.

The concepts of metabolic energy and energy output are treated in
Chapter 11. The details here match those in the chapter (as they
should!); this corresponds to an efficiency of 25%. 400 J of energy
is used by the body; 25% of this, 100 J, is the energy output. This
means that 300 J is exhausted as heat.

Doing 100 J of work means that a dog must
expend 400 J of metabolic energy. The difference
must be exhausted as heat; given the excellent
insulation provided by a dog's fur, this is mostly
via evaporation as it pants. At a typical body
temperature, the evaporation of 1.0 l of water carries
away 240,000 J, so this is an effective means of
cooling.

Chapter 12 discusses means of heat transfer: conduction,
convection, radiation, evaporation. This paragraph gives biological
details about dogs that you can interpret as follows: A dog's fur limits
transfer by conduction, convection and radiation; evaporation of
water by a panting dog must take up the slack.

The specific data for energy required to evaporate water is given. If
you need such information to answer questions, it will almost
certainly be provided. As we noted above, this is a test of
reasoning, not recall.

FIGURE MCAT-EXAM.1 Interpreting a passage.

Taking the Test: Answering the Questions

The passages on the MCAT exam seem complicated at first, but, as we've seen, they are about basic concepts and central themes that you know well. The same is true of the questions; they aren't as difficult as they may seem at first. As with the passage, you should start by translating the questions, identifying the physical concepts that apply in each case. You then proceed by reasoning, determining the solution to the question, using your understanding of these basic concepts. The practical suggestions below are followed by a detailed overview of the solutions to the questions based on the passage on the previous page.

You Can Answer the Questions in Any Order

The questions test a range of skills and have a range of difficulties. Many questions will involve simple reading comprehension; these are usually quite straightforward. Some require sophisticated reasoning and (slightly) complex mathematical manipulations. Start with the easy ones, ones that you can quickly solve. Save the more complex ones for later, and skip them if time is short.

Take Steps to Simplify or Eliminate Calculations

You won't be allowed to use a calculator on the exam, so any math that you do will be reasonably straightforward. To rapidly converge on a correct answer choice, there are some important "shortcuts" that you can take.

- **Use ratio reasoning.** What's the relationship between the variables involved in a question? You can use this to deduce the answer with only a very simple calculation, as we've seen many times in the book. For instance, suppose you are asked the following question:

  A model rocket is powered by chemical fuel. A student launches a rocket with a small engine containing 1.0 g of combustible fuel. The rocket reaches a speed of 10 m/s. The student then launches the rocket again, using an engine with 4.0 g of fuel. If all other parameters of the launch are kept the same, what final speed would you expect for this second trial?

  This is an energy conversion problem: Chemical energy of the fuel is converted to kinetic energy of the rocket. Kinetic energy is related to the speed by $K = \frac{1}{2}mv^2$. The chemical energy—and thus the kinetic energy—in the second trial is increased by a factor of 4. Since $K \sim v^2$, the speed must increase by a factor of 2, to 20 m/s.

- **Simplify calculations by liberally rounding numbers.** You can round off numbers to make calculations more straightforward. Your final result will probably be close enough to choose the correct answer from the list given. For instance, suppose you are asked the following question:

  A ball moving at 2.0 m/s rolls off edge of table that's 1.2 m high. How far from the edge of the table does the ball land?

  A. 2 m   B. 1.5 m   C. 1 m   D. 0.5 m

We know that the vertical motion of the ball is free fall; so the vertical distance fallen by the ball in a time Δt is $\Delta y = -\frac{1}{2}gt^2$. The time to fall 1.2 m is $\Delta t = \sqrt{2(1.2 \text{ m})/g}$. Rather than complete this calculation, we estimate the results as follows: $\Delta t = \sqrt{2.4/9.8} \approx \sqrt{1/4} = 1/2 = 0.5$ s During this free fall time, the horizontal motion is constant at 2.0 m/s, so we expect the ball to land about 1 m away. Our quick calculation shows us that the correct answer is choice C—no other answer is close.

- **For calculations using values in scientific notation, compute either the first digits or the exponents, not both.** In some cases, a quick calculation can tell you the correct leading digit, and that's all you need to figure out the correct answer. In other cases, you'll find possible answers with the same leading digit but very different exponents or decimal places. In this case, all you need is a simple order of magnitude estimate to decide on the right result.

- **Where possible, use your knowledge of the expected scale of physical quantities to quickly determine the correct answer.** For instance, suppose a question asks you to find the photon energy for green light of wavelength 550 nm. Visible light has photon energies of about 2 eV, or about 3×10^{-19} J, and that might be enough information to allow you to pick out the correct answer with no calculation.

- **Beware of "distractors", answers that you'll get if you make common mistakes.** For example, Question 4 on the next page is about energy conversion. The dog is keeping the sled in motion, so it's common for students to say that the dog is converting chemical energy in its body into kinetic energy. However, the kinetic energy isn't changing. The two answer choices that involve kinetic energy are common, but incorrect, choices. Be aware that the questions are constructed to bring out such misconceptions and that these tempting, but wrong, answer choices will be provided.

One Final Tip: Look at the Big Picture

The MCAT exam tests your ability to look at a technical passage about which you have some background knowledge and quickly get a sense of what it is saying, enough to answer questions about it. Keep this big picture in mind:

- **Don't get bogged down in technical details of the particular situation.** Focus on the basic physics.
- **Don't spend too much time on any one question.** If one question is taking too much time, make an educated guess and move on.
- **Don't get confused by details of notation or terminology.** For instance, different people use different symbols for physical variables; in this text we use the symbol K for kinetic energy; others use E_K.

Finally, don't forget the most important aspect of success on the MCAT exam: The best way to prepare for this or any test is simply to understand the subject. As you prepare for the test, focus your energy on reviewing and refining your knowledge of central topics and techniques, and practice applying your knowledge by solving problems like you'll see on the actual MCAT.

Translating

Look at the questions and think about the physics principles that apply, how they connect to concepts you know and understand.

Tips

• Numerical choices are presented in order; that's the usual practice on the test. Estimate the size of the answer, and think about where it falls.

• For questions with sentences as choices, decide on the solution before you look at the choices; this will save time reading.

Reasoning

Think about the question and the range of possible answers, and converge to a solution with as few steps as possible—time is limited!

This is a question about the size of the friction force. You are told that it takes a force that's about 1/6 of the sled's weight to pull it forward on snow. You can estimate the friction coefficient from this information.

1. What is the approximate coefficient of kinetic friction for a sled on snow?

  A. 0.35
  B. 0.25
  C. 0.15
  D. 0.05

For an object on level ground, the normal force equals the weight force. If the sled is moving at a constant speed, the pulling force equals the friction force. This implies that $\mu = f_k/n = f_{pull}/w = 1/6$. Two of the answer choices convert easily to fractions: $0.25 = 1/4$; $0.05 = 1/20$. 1/6 is between these, so C must be our choice. (Indeed, $1/6 = 0.167$, so 0.15 is pretty close.)

If the speed is constant, there is no net force. We are told that the pulling force is the horizontal component of the tension force, not the tension force itself. Because there is no net force, this horizontal component is equal to the friction force, which is directed backward. So this is really a question about the friction force.

2. If a rope pulls at an angle, as in Figure 1, how will this affect the pulling force necessary to keep the sled moving at a constant speed?

  A. This will reduce the pulling force.
  B. This will not change the pulling force.
  C. This will increase the pulling force.
  D. It will increase or decrease the pulling force, depending on angle.

A vertical component of the tension force will reduce the normal force, reducing the friction force—and thus the pulling force.

We assume that the output power is the same for the two cases—this is implied in the passage.

3. A dog pulls a 40 kg sled at a maximum speed of 2 m/s. What is the maximum speed for an 80 kg sled?

  A. 2 m/s
  B. 1.5 m/s
  C. 1.0 m/s
  D. 0.5 m/s

Doubling the weight doubles the normal force, which doubles the friction force. This will double the necessary pulling force as well. Given the expression for power given in the passage, this means the maximum speed will be halved.

This is a question about energy transformation. For such questions, think about changes. What forms of energy are *changing*? We know that thermal energy is part of the picture because some of the chemical energy is converted to thermal energy in the dog's body.

4. As a dog pulls a sled at constant speed, chemical energy in the dog's body is converted to

  A. kinetic energy
  B. thermal energy
  C. kinetic energy and thermal energy
  D. kinetic energy and potential energy

Choice B is correct, but A and C are clever distractors. It's tempting to choose an answer that includes kinetic energy. The sled is in motion, after all! But don't be swayed. The kinetic energy isn't changing, and friction to the sled converts any energy the dog supplies into thermal energy.

Increasing speed increases power, as the passage told us. But the energy to pull the sled is not the *power*, it's the *work*, and we know that the work is $W = F\Delta x$. This is a question about work and energy, not about power.

5. A dog pulls a sled for a distance of 1.0 km at a speed of 1 m/s, requiring an energy output of 60,000 J. If the dog pulls the sled at 2 m/s, the necessary energy is

  A. 240,000 J
  B. 120,000 J
  C. 60,000 J
  D. 30,000 J

Doubling the speed doubles the power, but it doesn't change the force; that's fixed by friction. The distance is the same as well, and so is the work done, the energy required. Since the speed doubles, it's tempting to think the energy doubles, though. This "obvious" but incorrect solution is one of the choices—expect such situations on the actual MCAT.

The passage tells us that the dog uses 400 J of metabolic energy to do 100 J of work. 300 J, or 75%, must be exhausted to the environment. We can assume the same efficiency here.

6. A dog uses 100,000 J of metabolic energy pulling a sled. How much energy must the dog exhaust by panting?

  A. 100,000 J
  B. 75,000 J
  C. 50,000 J
  D. 25,000 J

If 75% of the energy must be exhausted to the environment, that's 75,000 J.

FIGURE MCAT-EXAM.2 Answering the questions for the passage of Figure MCAT-EXAM.1.

Real-World Applications

Applications of biological or medical interest are marked BIO in the list below, including MCAT-style Passage Problems. Other end-of-chapter problems of biological or medical interest are marked BIO in the chapter.

Detailed Contents

PART I Force and Motion

PART II Conservation Laws

Force and Motion

The cheetah is the fastest land animal, able to run at speeds exceeding 60 miles per hour. Nonetheless, the rabbit has an advantage in this chase. It can *change* its motion more quickly and will likely escape. How can you tell, by looking at the picture, that the cheetah is changing its motion?

Why Things Change

Each of the seven parts of this text opens with an overview that gives you a look ahead, a glimpse of where your journey will take you in the next few chapters. It's easy to lose sight of the big picture while you're busy negotiating the terrain of each chapter. In Part I, the big picture is, in a word, *change*.

Simple observations of the world around you show that most things change. Some changes, such as aging, are biological. Others, such as sugar dissolving in your coffee, are chemical. We will look at changes that involve *motion* of one form or another—running and jumping, throwing balls, lifting weights.

There are two big questions we must tackle to study how things change by moving:

- **How do we describe motion?** How should we measure or characterize the motion if we want to analyze it mathematically?
- **How do we explain motion?** Why do objects have the particular motion they do? Why, when you toss a ball upward, does it go up and then come back down rather than keep going up? What are the "laws of nature" that allow us to predict an object's motion?

Two key concepts that will help answer these questions are *force* (the "cause") and *acceleration* (the "effect"). Our basic tools will be three laws of motion worked out by Isaac Newton. Newton's laws relate force to acceleration, and we will use them to explain and explore a wide range of problems. As we learn to solve problems dealing with motion, we will learn basic techniques that we can apply in all the parts of this text.

Simplifying Models

Reality is extremely complicated. We would never be able to develop a science if we had to keep track of every detail of every situation. Suppose we analyze the tossing of a ball. Is it necessary to analyze the way the atoms in the ball are connected? Do we need to analyze what you ate for breakfast and the biochemistry of how that was translated into muscle power? These are interesting questions, of course. But if our task is to understand the motion of the ball, we need to simplify!

We can do a perfectly fine analysis if we treat the ball as a round solid and your hand as another solid that exerts a force on the ball. This is a *model* of the situation. A model is a simplified description of reality—much as a model airplane is a simplified version of a real airplane—that is used to reduce the complexity of a problem to the point where it can be analyzed and understood.

Model building is a major part of the strategy that we will develop for solving problems in all parts of the text. We will introduce different models in different parts. We will pay close attention to where simplifying assumptions are being made, and why. Learning *how* to simplify a situation is the essence of successful modeling— and successful problem solving.

1 Representing Motion

As this falcon moves in a graceful arc through the air, the direction of its motion and the distance between each of its positions and the next are constantly changing. What language should we use to describe this motion?

LOOKING AHEAD »

Goal: To introduce the fundamental concepts of motion and to review related basic mathematical principles.

Chapter Preview

Each chapter starts with a preview outlining the major themes and what you'll be learning for each theme.

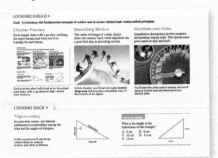

Each preview also looks back at an important past topic, with a question to help refresh your memory.

Describing Motion

This series of images of a skier clearly shows his motion. Such visual depictions are a good first step in describing motion.

In this chapter, you'll learn to make **motion diagrams** that provide a simplified view of the motion of an object.

Numbers and Units

Quantitative descriptions involve numbers, and numbers require units. This speedometer gives speed in mph and km/h.

You'll learn the units used in science, and you'll learn to convert between these and more familiar units.

LOOKING BACK «

Trigonometry

In a previous course, you learned mathematical relationships among the sides and the angles of triangles.

In this course you'll use these relationships to analyze motion and other problems.

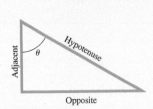

STOP TO THINK

What is the length of the hypotenuse of this triangle?

A. 6 cm   B. 8 cm
C. 10 cm   D. 12 cm
E. 14 cm

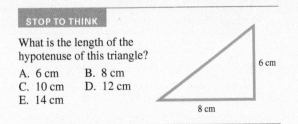

1.1 Motion: A First Look

The concept of motion is a theme that will appear in one form or another throughout this entire text. You have a well-developed intuition about motion based on your experiences, but you'll see that some of the most important aspects of motion can be rather subtle. We need to develop some tools to help us explain and understand motion, so rather than jumping immediately into a lot of mathematics and calculations, this first chapter focuses on visualizing motion and becoming familiar with the concepts needed to describe a moving object.

One key difference between physics and other sciences is how we set up and solve problems. We'll often use a two-step process to solve motion problems. The first step is to develop a simplified *representation* of the motion so that key elements stand out. For example, the photo of the falcon at the start of the chapter allows us to observe its position at many successive times. It is precisely by considering this sort of picture of motion that we will begin our study of this topic. The second step is to analyze the motion with the language of mathematics. The process of putting numbers on nature is often the most challenging aspect of the problems you will solve. In this chapter, we will explore the steps in this process as we introduce the basic concepts of motion.

Types of Motion

As a starting point, let's define **motion** as the change of an object's position or orientation with time. Examples of motion are easy to list. Bicycles, baseballs, cars, airplanes, and rockets are all objects that move. The path along which an object moves, which might be a straight line or might be curved, is called the object's **trajectory.**

FIGURE 1.1 Four basic types of motion.

Straight-line motion

Circular motion

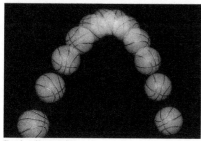
Projectile motion

Rotational motion

FIGURE 1.1 shows four basic types of motion that we will study in this text. In this chapter, we will start with the first type of motion in the figure, motion along a straight line. In later chapters, we will learn about circular motion, which is the motion of an object along a circular path; projectile motion, the motion of an object through the air; and rotational motion, the spinning of an object about an axis.

Making a Motion Diagram

An easy way to study motion is to record a video of a moving object with a stationary camera. A video camera takes images at a fixed rate, typically 30 images every second. Each separate image is called a *frame*. As an example, FIGURE 1.2 shows several frames from a video of a car going past, with the camera in a fixed position. Not surprisingly, the car is in a different position in each frame.

FIGURE 1.2 Several frames from the video of a car.

FIGURE 1.3 A motion diagram of the car shows all the frames simultaneously.

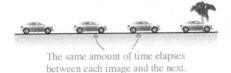

The same amount of time elapses between each image and the next.

Suppose we now edit the video by layering the frames on top of each other and then look at the final result. We end up with the picture in FIGURE 1.3. This composite image, showing an object's positions at several *equally spaced instants of time,* is called a **motion diagram.** As simple as motion diagrams seem, they will turn out to be powerful tools for analyzing motion.

Now let's take our camera out into the world and make some motion diagrams. The following table illustrates how a motion diagram shows important features of different kinds of motion.

Examples of motion diagrams

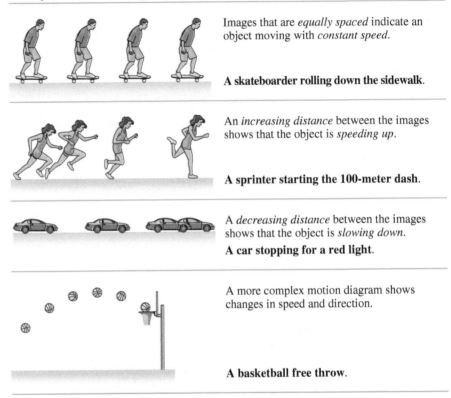

Images that are *equally spaced* indicate an object moving with *constant speed.*

A skateboarder rolling down the sidewalk.

An *increasing distance* between the images shows that the object is *speeding up.*

A sprinter starting the 100-meter dash.

A *decreasing distance* between the images shows that the object is *slowing down.*
A car stopping for a red light.

A more complex motion diagram shows changes in speed and direction.

A basketball free throw.

We have defined several concepts (constant speed, speeding up, and slowing down) in terms of how the moving object appears in a motion diagram. These are called **operational definitions,** meaning that the concepts are defined in terms of a particular procedure or operation. For example, we could answer the question Is the airplane speeding up? by checking whether the images in the plane's motion diagram are getting farther apart. Many of the concepts in physics will be introduced as operational definitions. This reminds us that physics is an experimental science.

STOP TO THINK 1.1   Which car is going faster, A or B? Assume there are equal intervals of time between the frames of both videos.

Car A

Car B

NOTE ▶ Each chapter in this text has several *Stop to Think* questions. These questions are designed to see if you've understood the basic ideas that have just been presented. The answers are given at the end of the chapter, but you should make a serious effort to think about these questions before turning to the answers. ◀

The Particle Model

For many objects, the motion of the object *as a whole* is not influenced by the details of the object's size and shape. To describe the object's motion, all we really need to keep track of is the motion of a single point: You could imagine looking at the motion of a dot painted on the side of the object.

In fact, for the purposes of analyzing the motion, we can often consider the object *as if* it were just a single point. We can also treat the object *as if* all of its mass were concentrated into this single point. An object that can be represented as a mass at a single point in space is called a **particle.**

If we treat an object as a particle, we can represent the object in each frame of a motion diagram as a simple dot. FIGURE 1.4 shows how much simpler motion diagrams appear when the object is represented as a particle. Note that the dots have been numbered 0, 1, 2, . . . to tell the sequence in which the frames were exposed. These diagrams still convey a complete understanding of the object's motion.

Treating an object as a particle is, of course, a simplification of reality. Such a simplification is called a **model.** Models allow us to focus on the important aspects of a phenomenon by excluding those aspects that play only a minor role. The **particle model** of motion is a simplification in which we treat a moving object as if all of its mass were concentrated at a single point. Using the particle model may allow us to see connections that are very important but that are obscured or lost by examining all the parts of an extended, real object. Consider the motion of the two objects shown in FIGURE 1.5. These two very different objects have exactly the same motion diagram. As we will see, all objects falling under the influence of gravity move in exactly the same manner if no other forces act. The simplification of the particle model has revealed something about the physics that underlies both of these situations.

FIGURE 1.4 Simplifying a motion diagram using the particle model.

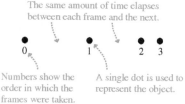

(a) Motion diagram of a car stopping

(b) Same motion diagram using the particle model

The same amount of time elapses between each frame and the next.

0   1   2 3

Numbers show the order in which the frames were taken.

A single dot is used to represent the object.

FIGURE 1.5 The particle model for two falling objects.

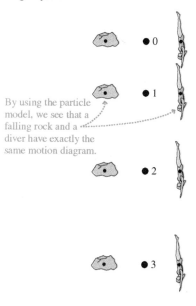

By using the particle model, we see that a falling rock and a diver have exactly the same motion diagram.

STOP TO THINK 1.2 Three motion diagrams are shown. Which is a dust particle settling to the floor at constant speed, which is a ball dropped from the roof of a building, and which is a descending rocket slowing to make a soft landing on Mars?

A.	B.	C.
0 ●	0 ●	0 ●
1 ●		
2 ●	1 ●	
		1 ●
3 ●	2 ●	
		2 ●
4 ●	3 ●	
	4 ●	3 ●
		4 ●
5 ●	5 ●	5 ●

1.2 Position and Time: Putting Numbers on Nature

To develop our understanding of motion further, we need to be able to make quantitative measurements: We need to use numbers. As we analyze a motion diagram, it is useful to know where the object is (its *position*) and when the object was at that position (the *time*). We'll start by considering the motion of an object that can move only along a straight line. Examples of this **one-dimensional** or "1-D" motion are a bicyclist moving along the road, a train moving on a long straight track, and an elevator moving up and down a shaft.

Position and Coordinate Systems

Suppose you are driving along a long, straight country road, as in FIGURE 1.6, and your friend calls and asks where you are. You might reply that you are 4 miles east of the post office, and your friend would then know just where you were.

FIGURE 1.6 Describing your position.

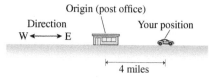

Origin (post office)

Direction   Your position

W ◄——► E

4 miles

This gauge's vertical scale measures the depth of snow when it falls. It has a natural origin at the level of the road.

Your location at a particular instant in time (when your friend phoned) is called your **position**. Notice that to know your position along the road, your friend needed three pieces of information. First, you had to give her a reference point (the post office) from which all distances are to be measured. We call this fixed reference point the **origin**. Second, she needed to know how far you were from that reference point or origin—in this case, 4 miles. Finally, she needed to know which side of the origin you were on: You could be 4 miles to the west of it or 4 miles to the east.

We will need these same three pieces of information in order to specify any object's position along a line. We first choose our origin, from which we measure the position of the object. The position of the origin is arbitrary, and we are free to place it where we like. Usually, however, there are certain points (such as the well-known post office) that are more convenient choices than others.

In order to specify how far our object is from the origin, we lay down an imaginary axis along the line of the object's motion. Like a ruler, this axis is marked off in equally spaced divisions of distance, perhaps in inches, meters, or miles, depending on the problem at hand. We place the zero mark of this ruler at the origin, allowing us to locate the position of our object by reading the ruler mark where the object is.

Finally, we need to be able to specify which side of the origin our object is on. To do this, we imagine the axis extending from one side of the origin with increasing positive markings; on the other side, the axis is marked with increasing *negative* numbers. By reporting the position as either a positive or a negative number, we know on what side of the origin the object is.

These elements—an origin and an axis marked in both the positive and negative directions—can be used to unambiguously locate the position of an object. We call this a **coordinate system.** We will use coordinate systems throughout this text, and we will soon develop coordinate systems that can be used to describe the positions of objects moving in more complex ways than just along a line. FIGURE 1.7 shows a coordinate system that can be used to locate various objects along the country road discussed earlier.

Although our coordinate system works well for describing the positions of objects located along the axis, our notation is somewhat cumbersome. We need to keep saying things like "the car is at position $+4$ miles." A better notation, and one that will become particularly important when we study motion in two dimensions, is to use a symbol such as x or y to represent the position along the axis. Then we can say "the cow is at $x = -5$ miles." The symbol that represents a position along an axis is called a **coordinate**. The introduction of symbols to represent positions (and, later, velocities and accelerations) also allows us to work with these quantities mathematically.

FIGURE 1.8 shows how we would set up a coordinate system for a sprinter running a 50 meter race (we use the standard symbol "m" for meters). For horizontal motion like this we usually use the coordinate x to represent the position.

FIGURE 1.7 The coordinate system used to describe objects along a country road.

The post office defines the zero, or origin, of the coordinate system.

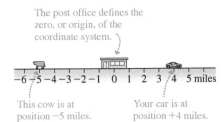

This cow is at position -5 miles.

Your car is at position $+4$ miles.

FIGURE 1.9 Examples of one-dimensional motion.

For vertical motion, we'll use the coordinate y.

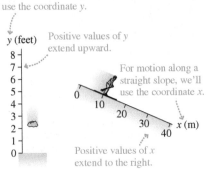

Positive values of y extend upward.

For motion along a straight slope, we'll use the coordinate x.

Positive values of x extend to the right.

FIGURE 1.8 A coordinate system for a 50 meter race.

This is the symbol, or coordinate, used to represent positions along the axis.

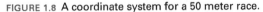
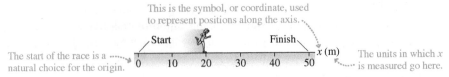

The start of the race is a natural choice for the origin.

The units in which x is measured go here.

Motion along a straight line need not be horizontal. As shown in FIGURE 1.9, a rock falling vertically downward and a skier skiing down a straight slope are also examples of straight-line or one-dimensional motion.

Time

The pictures in Figure 1.9 show the position of an object at just one instant of time. But a full motion diagram represents how an object moves as time progresses. So far, we have labeled the dots in a motion diagram by the numbers 0, 1, 2, . . . to indicate the order in which the frames were exposed. But to fully describe the motion, we need to indicate the *time,* as read off a clock or a stopwatch, at which each frame of a video was made. This is important, as we can see from the motion diagram of a stopping car in FIGURE 1.10. If the frames were taken 1 second apart, this motion diagram shows a leisurely stop; if 1/10 of a second apart, it represents a screeching halt.

For a complete motion diagram, we thus need to label each frame with its corresponding time (symbol t) as read off a clock. But when should we start the clock? Which frame should be labeled $t = 0$? This choice is much like choosing the origin $x = 0$ of a coordinate system: You can pick any arbitrary point in the motion and label it "$t = 0$ seconds." This is simply the instant you decide to start your clock or stopwatch, so it is the origin of your time coordinate. A video frame labeled "$t = 4$ seconds" means it was taken 4 seconds after you started your clock. We typically choose $t = 0$ to represent the "beginning" of a problem, but the object may have been moving before then.

To illustrate, FIGURE 1.11 shows the motion diagram for a car moving at a constant speed and then braking to a halt. Two possible choices for the frame labeled $t = 0$ seconds are shown; our choice depends on what part of the motion we're interested in. Each successive position of the car is then labeled with the clock reading in seconds (abbreviated by the symbol "s").

Changes in Position and Displacement

Now that we've seen how to measure position and time, let's return to the problem of motion. To describe motion we'll need to measure the *changes* in position that occur with time. Consider the following:

> Sam is standing 50 feet (ft) east of the corner of 12th Street and Vine. He then walks to a second point 150 ft east of Vine. What is Sam's change of position?

FIGURE 1.12 shows Sam's motion on a map. We've placed a coordinate system on the map, using the coordinate x. We are free to place the origin of our coordinate system wherever we wish, so we have placed it at the intersection. Sam's initial position is then at $x_i = 50$ ft. The positive value for x_i tells us that Sam is east of the origin.

> NOTE ▶ We will label special values of x or y with subscripts. The value at the start of a problem is usually labeled with a subscript "i," for *initial,* and the value at the end is labeled with a subscript "f," for *final.* For cases having several special values, we will usually use subscripts "1," "2," and so on. ◄

Sam's final position is $x_f = 150$ ft, indicating that he is 150 ft east of the origin. You can see that Sam has changed position, and a *change* of position is called a **displacement.** His displacement is the distance labeled Δx in Figure 1.12. The Greek letter delta (Δ) is used in math and science to indicate the *change* in a quantity. Thus Δx indicates a change in the position x.

> NOTE ▶ Δx is a *single* symbol. You cannot cancel out or remove the Δ in algebraic operations. ◄

To get from the 50 ft mark to the 150 ft mark, Sam clearly had to walk 100 ft, so the change in his position—his displacement—is 100 ft. We can think about displacement in a more general way, however. Displacement is the *difference* between a final position x_f and an initial position x_i. Thus we can write

$$\Delta x = x_f - x_i = 150 \text{ ft} - 50 \text{ ft} = 100 \text{ ft}$$

> NOTE ▶ A general principle, used throughout this text, is that the change in any quantity is the final value of the quantity minus its initial value. ◄

FIGURE 1.10 Is this a leisurely stop or a screeching halt?

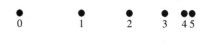

FIGURE 1.11 The motion diagram of a car that travels at constant speed and then brakes to a halt.

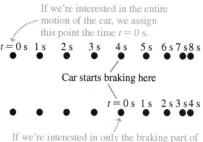

FIGURE 1.12 Sam undergoes a displacement Δx from position x_i to position x_f.

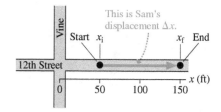

The size and the direction of the displacement both matter. Roy Riegels (pursued above by teammate Benny Lom) found this out in dramatic fashion in the 1928 Rose Bowl when he recovered a fumble and ran 69 yards—toward his own team's end zone. An impressive distance, but in the wrong direction!

FIGURE 1.13 A displacement is a signed quantity. Here Δx is a negative number.

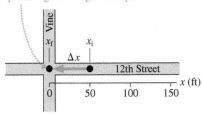

A final position to the left of the initial position gives a negative displacement.

Displacement is a *signed quantity;* that is, it can be either positive or negative. If, as shown in **FIGURE 1.13**, Sam's final position x_f had been at the origin instead of the 150 ft mark, his displacement would have been

$$\Delta x = x_f - x_i = 0 \text{ ft} - 50 \text{ ft} = -50 \text{ ft}$$

The negative sign tells us that he moved to the *left* along the x-axis, or 50 ft *west*.

Change in Time

A displacement is a change in position. In order to quantify motion, we'll need to also consider changes in *time,* which we call **time intervals.** We've seen how we can label each frame of a motion diagram with a specific time, as determined by our stopwatch. **FIGURE 1.14** shows the motion diagram of a bicycle moving at a constant speed, with the times of the measured points indicated.

The displacement between the initial position x_i and the final position x_f is

$$\Delta x = x_f - x_i = 120 \text{ ft} - 0 \text{ ft} = 120 \text{ ft}$$

Similarly, we define the time interval between these two points to be

$$\Delta t = t_f - t_i = 6 \text{ s} - 0 \text{ s} = 6 \text{ s}$$

FIGURE 1.14 The motion diagram of a bicycle moving to the right at a constant speed.

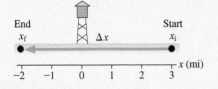

A time interval Δt measures the elapsed time as an object moves from an initial position x_i at time t_i to a final position x_f at time t_f. Note that, unlike Δx, Δt is always positive because t_f is always greater than t_i.

EXAMPLE 1.1 **How long a ride?**

Carol is enjoying a bicycle ride on a country road that runs east-west past a water tower. Define a coordinate system so that increasing x means moving east. At noon, Carol is 3 miles (mi) east of the water tower. A half-hour later, she is 2 mi west of the water tower. What is her displacement during that half-hour?

PREPARE Although it may seem like overkill for such a simple problem, you should start by making a drawing, like the one in **FIGURE 1.15**, with the x-axis along the road. Distances are measured with respect

FIGURE 1.15 A drawing of Carol's motion.

to the water tower, so it is a natural origin for the coordinate system. Once the coordinate system is established, we can show Carol's initial and final positions and her displacement between the two.

SOLVE We've specified values for Carol's initial and final positions in our drawing. We can thus compute her displacement:

$$\Delta x = x_f - x_i = (-2 \text{ mi}) - (3 \text{ mi}) = -5 \text{ mi}$$

ASSESS Once we've completed the solution to the problem, we need to go back to see if it makes sense. Carol is moving to the west, so we expect her displacement to be negative—and it is. We can see from our drawing in Figure 1.15 that she has moved 5 miles from her starting position, so our answer seems reasonable. As part of the Assess step, we also check our answers to see if they make physical sense. Carol travels 5 miles in a half-hour, quite a reasonable pace for a cyclist.

NOTE ▶ All of the numerical examples in the text are worked out with the same three-step process: Prepare, Solve, Assess. It's tempting to cut corners, especially for the simple problems in these early chapters, but you should take the time to do all of these steps now, to practice your problem-solving technique. We'll have more to say about our general problem-solving strategy in Chapter 2. ◀

STOP TO THINK 1.3 Sarah starts at a positive position along the x-axis. She then undergoes a negative displacement. Her final position

A. Is positive.   B. Is negative.   C. Could be either positive or negative.

1.3 Velocity

We all have an intuitive sense of whether something is moving very fast or just cruising slowly along. To make this intuitive idea more precise, let's start by examining the motion diagrams of some objects moving along a straight line at a *constant* speed, objects that are neither speeding up nor slowing down. This motion at a constant speed is called **uniform motion.** As we saw for the skateboarder in Section 1.1, for an object in uniform motion, successive frames of the motion diagram are *equally spaced*, so the object's displacement Δx is the same between successive frames.

To see how an object's displacement between successive frames is related to its speed, consider the motion diagrams of a bicycle and a car, traveling along the same street as shown in FIGURE 1.16. Clearly the car is moving faster than the bicycle: In any 1 second time interval, the car undergoes a displacement $\Delta x = 40$ ft, while the bicycle's displacement is only 20 ft.

The greater the distance traveled by an object in a given time interval, the greater its speed. This idea leads us to define the speed of an object as

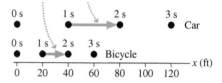

FIGURE 1.16 Motion diagrams for a car and a bicycle.

$$\text{speed} = \frac{\text{distance traveled in a given time interval}}{\text{time interval}} \qquad (1.1)$$

Speed of an object in uniform motion

For the bicycle, this equation gives

$$\text{speed} = \frac{20 \text{ ft}}{1 \text{ s}} = 20 \,\frac{\text{ft}}{\text{s}}$$

while for the car we have

$$\text{speed} = \frac{40 \text{ ft}}{1 \text{ s}} = 40 \,\frac{\text{ft}}{\text{s}}$$

The speed of the car is twice that of the bicycle, which seems reasonable.

NOTE ▶ The division gives units that are a fraction: ft/s. This is read as "feet per second," just like the more familiar "miles per hour." ◀

To fully characterize the motion of an object, we must specify not only the object's speed but also the *direction* in which it is moving. FIGURE 1.17 shows the motion diagrams of two bicycles traveling at 20 ft/s. The two bicycles have the same speed, but something about their motion is different—the *direction* of their motion.

The "distance traveled" in Equation 1.1 doesn't capture any information about the direction of travel. But we've seen that the *displacement* of an object does contain this information. We can introduce a new quantity, the **velocity,** as

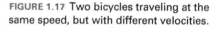
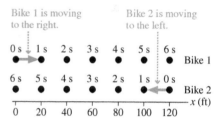

FIGURE 1.17 Two bicycles traveling at the same speed, but with different velocities.

$$\text{velocity} = \frac{\text{displacement}}{\text{time interval}} = \frac{\Delta x}{\Delta t} \qquad (1.2)$$

Velocity of a moving object

The velocity of bicycle 1 in Figure 1.17, computed using the 1 second time interval between the $t = 2$ s and $t = 3$ s positions, is

$$v = \frac{\Delta x}{\Delta t} = \frac{x_3 - x_2}{3 \text{ s} - 2 \text{ s}} = \frac{60 \text{ ft} - 40 \text{ ft}}{1 \text{ s}} = +20 \,\frac{\text{ft}}{\text{s}}$$

while the velocity for bicycle 2, during the same time interval, is

$$v = \frac{\Delta x}{\Delta t} = \frac{x_3 - x_2}{3 \text{ s} - 2 \text{ s}} = \frac{60 \text{ ft} - 80 \text{ ft}}{1 \text{ s}} = -20 \,\frac{\text{ft}}{\text{s}}$$

NOTE ▶ We have used x_2 for the position at time $t = 2$ seconds and x_3 for the position at time $t = 3$ seconds. The subscripts serve the same role as before—identifying particular positions—but in this case the positions are identified by the time at which each position is reached. ◀

The two velocities have opposite signs because the bicycles are traveling in opposite directions. **Speed measures only how fast an object moves, but velocity tells us both an object's speed** *and its direction.* In this text, we'll use a positive velocity to indicate motion to the right or, for vertical motion, upward. We'll use negative velocity for an object moving to the left, or downward.

NOTE ▶ Learning to distinguish between speed, which is always a positive number, and velocity, which can be either positive or negative, is one of the most important tasks in the analysis of motion. ◀

The velocity as defined by Equation 1.2 is actually what is called the *average* velocity. On average, over each 1 s interval bicycle 1 moves 20 ft, but we don't know if it was moving at exactly the same speed at every moment during this time interval. In Chapter 2, we'll develop the idea of *instantaneous* velocity, the velocity of an object at a particular instant in time. Since our goal in this chapter is to *visualize* motion with motion diagrams, we'll somewhat blur the distinction between average and instantaneous quantities, refining these definitions in Chapter 2.

EXAMPLE 1.2   Finding the speed of a seabird

Albatrosses are seabirds that spend most of their lives flying over the ocean looking for food. With a stiff tailwind, an albatross can fly at high speeds. Satellite data on one particularly speedy albatross showed it 60 miles east of its roost at 3:00 PM and then, at 3:15 PM, 80 miles east of its roost. What was its velocity?

PREPARE The statement of the problem provides us with a natural coordinate system: We can measure distances with respect to the roost, with distances to the east as positive. With this coordinate system, the motion of the albatross appears as in FIGURE 1.18.

FIGURE 1.18 The motion of an albatross at sea.

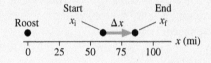

The motion takes place between 3:00 and 3:15, a time interval of 15 minutes, or 0.25 hour.

SOLVE We know the initial and final positions, and we know the time interval, so we can calculate the velocity:

$$v = \frac{\Delta x}{\Delta t} = \frac{x_f - x_i}{0.25 \text{ h}} = \frac{20 \text{ mi}}{0.25 \text{ h}} = 80 \text{ mph}$$

ASSESS The velocity is positive, which makes sense because Figure 1.18 shows that the motion is to the right. A speed of 80 mph is certainly fast, but the problem said it was a "particularly speedy" albatross, so our answer seems reasonable. (Indeed, albatrosses have been observed to fly at such speeds in the very fast winds of the Southern Ocean. This problem is based on real observations, as will be our general practice in this text.)

The "Per" in Meters Per Second

The units for speed and velocity are a unit of distance (feet, meters, miles) divided by a unit of time (seconds, hours). Thus we could measure velocity in units of m/s or mph, pronounced "meters *per* second" and "miles *per* hour." The word "per" will often arise in physics when we consider the ratio of two quantities. What do we mean, exactly, by "per"?

If a car moves with a speed of 23 m/s, we mean that it travels 23 meters *for each* second of elapsed time. The word "per" thus associates the number of units in the numerator (23 m) with *one* unit of the denominator (1 s). We'll see many other examples of this idea as the text progresses. You may already know a bit about *density;* you can look up the density of gold and you'll find that it is 19.3 g/cm³ ("grams *per* cubic centimeter"). This means that there are 19.3 grams of gold *for each* cubic centimeter of the metal.

STOP TO THINK 1.4   Jane starts from her house to take a stroll in her neighborhood. After walking for 2 hours at a steady pace, she has walked 4 miles and is 2 miles from home. For this time interval, what was her speed?

A. 4 mph   B. 3 mph   C. 2 mph   D. 1 mph

1.4 A Sense of Scale: Significant Figures, Scientific Notation, and Units

Physics attempts to explain the natural world, from the very small to the exceedingly large. And in order to understand our world, we need to be able to *measure* quantities both minuscule and enormous. A properly reported measurement has three elements. First, we can measure our quantity with only a certain precision. To make this precision clear, we need to make sure that we report our measurement with the correct number of *significant figures.*

Second, writing down the really big and small numbers that often come up in physics can be awkward. To avoid writing all those zeros, scientists use *scientific notation* to express numbers both big and small.

Finally, we need to choose an agreed-upon set of *units* for the quantity. For speed, common units include meters per second and miles per hour. For mass, the kilogram is the most commonly used unit. Every physical quantity that we can measure has an associated set of units.

Measurements and Significant Figures

When we measure any quantity, such as the length of a bone or the weight of a specimen, we can do so with only a certain *precision*. The digital calipers in FIGURE 1.19 can make a measurement to within ± 0.01 mm, so they have a precision of 0.01 mm. If you made the measurement with a ruler, you probably couldn't do better than about ± 1 mm, so the precision of the ruler is about 1 mm. The precision of a measurement can also be affected by the skill or judgment of the person performing the measurement. A stopwatch might have a precision of 0.001 s, but, due to your reaction time, your measurement of the time of a sprinter would be much less precise.

It is important that your measurement be reported in a way that reflects its actual precision. Suppose you use a ruler to measure the length of a particular frog. You judge that you can make this measurement with a precision of about 1 mm, or 0.1 cm. In this case, the frog's length should be reported as, say, 6.2 cm. We interpret this to mean that the actual value falls between 6.15 cm and 6.25 cm and thus rounds to 6.2 cm. Reporting the frog's length as simply 6 cm is saying less than you know; you are withholding information. On the other hand, to report the number as 6.213 cm is wrong. Any person reviewing your work would interpret the number 6.213 cm as meaning that the actual length falls between 6.2125 cm and 6.2135 cm, thus rounding to 6.213 cm. In this case, you are claiming to have knowledge and information that you do not really possess.

The way to state your knowledge precisely is through the proper use of **significant figures.** You can think of a significant figure as a digit that is reliably known. A measurement such as 6.2 cm has *two* significant figures, the 6 and the 2. The next decimal place—the hundredths—is not reliably known and is thus not a significant figure. Similarly, a time measurement of 34.62 s has four significant figures, implying that the 2 in the hundredths place is reliably known.

When we perform a calculation such as adding or multiplying two or more measured numbers, we can't claim more accuracy for the result than was present in the initial measurements. Determining the proper number of significant figures is straightforward, but there are a few definite rules to follow. We will often spell out such technical details in what we call a "Tactics Box." A Tactics Box is designed to teach you particular skills and techniques. Each Tactics Box will include the ✐ icon to designate exercises in the *Student Workbook* that you can use to practice these skills.

FIGURE 1.19 The precision of a measurement depends on the instrument used to make it.

These calipers have a precision of 0.01 mm.

Walter Davis's best long jump on this day was reported as 8.24 m. This implies that the actual length of the jump was between 8.235 m and 8.245 m, a spread of only 0.01 m, which is 1 cm. Does this claimed accuracy seem reasonable?

❶ When you multiply or divide several numbers, or when you take roots, the number of significant figures in the answer should match the number of significant figures of the *least* precisely known number used in the calculation:

Three significant figures

$$3.73 \times 5.7 = 21$$

Two significant figures

Answer should have the *lower* of the two, or two significant figures.

❷ When you add or subtract several numbers, the number of decimal places in the answer should match the *smallest* number of decimal places of any number used in the calculation:

$$\begin{array}{r} 18.54 \\ +106.6 \\ \hline 125.1 \end{array}$$ — Two decimal places
— One decimal place

Answer should have the *lower* of the two, or one decimal place.

❸ **Exact numbers** have no uncertainty and, when used in calculations, do not change the number of significant figures of measured numbers. Examples of exact numbers are π and the number 2 in the relation $d = 2r$ between a circle's diameter and radius.

There is one notable exception to these rules:

■ It is acceptable to keep one or two extra digits during *intermediate* steps of a calculation to minimize round-off errors in the calculation. But the *final* answer must be reported with the proper number of significant figures.

Exercise 15

EXAMPLE 1.3   **Measuring the velocity of a car**

To measure the velocity of a car, clocks A and B are set up at two points along the road, as shown in FIGURE 1.20. Clock A is precise to 0.01 s, while clock B is precise to only 0.1 s. The distance between these two clocks is carefully measured to be 124.5 m. The two clocks are automatically started when the car passes a trigger in the road; each clock stops automatically when the car passes that clock. After the car has passed both clocks, clock A is found to read $t_A = 1.22$ s, and clock B to read $t_B = 4.5$ s. The time from the less-precise clock B is correctly reported with fewer significant figures than that from A. What is the velocity of the car, and how should it be reported with the correct number of significant figures?

FIGURE 1.20 Measuring the velocity of a car.

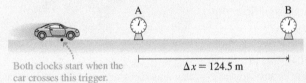

Both clocks start when the car crosses this trigger.

$\Delta x = 124.5$ m

PREPARE To calculate the velocity, we need the displacement Δx and the time interval Δt as the car moves between the two clocks. The displacement is given as $\Delta x = 124.5$ m; we can calculate the time interval as the difference between the two measured times.

SOLVE The time interval is:

This number has one decimal place.   This number has two decimal places.

$$\Delta t = t_B - t_A = (4.5 \text{ s}) - (1.22 \text{ s}) = 3.3 \text{ s}$$

By rule 2 of Tactics Box 1.1, the result should have *one* decimal place.

We can now calculate the velocity with the displacement and the time interval:

The displacement has four significant figures.

$$v = \frac{\Delta x}{\Delta t} = \frac{124.5 \text{ m}}{3.3 \text{ s}} = 38 \text{ m/s}$$

The time interval has two significant figures.   By rule 1 of Tactics Box 1.1, the result should have *two* significant figures.

ASSESS Our final value has two significant figures. Suppose you had been hired to measure the speed of a car this way, and you reported 37.72 m/s. It would be reasonable for someone looking at your result to assume that the measurements you used to arrive at this value were correct to four significant figures and thus that you had measured time to the nearest 0.001 second. Our correct result of 38 m/s has all of the accuracy that you can claim, but no more!

Scientific Notation

It's easy to write down measurements of ordinary-sized objects: Your height might be 1.72 meters, the weight of an apple 0.34 pound. But the radius of a hydrogen atom is 0.000 000 000 053 m, and the distance to the moon is 384,000,000 m. Keeping track of all those zeros is quite cumbersome.

Beyond requiring you to deal with all the zeros, writing quantities this way makes it unclear how many significant figures are involved. In the distance to the moon given above, how many of those digits are significant? Three? Four? All nine?

Writing numbers using scientific notation avoids both these problems. A value in scientific notation is a number with one digit to the left of the decimal point and zero or more to the right of it, multiplied by a power of ten. This solves the problem of all the zeros and makes the number of significant figures immediately apparent. In scientific notation, writing the distance to the sun as 1.50×10^{11} m implies that three digits are significant; writing it as 1.5×10^{11} m implies that only two digits are.

Even for smaller values, scientific notation can clarify the number of significant figures. Suppose a distance is reported as 1200 m. How many significant figures does this measurement have? It's ambiguous, but using scientific notation can remove any ambiguity. If this distance is known to within 1 m, we can write it as 1.200×10^3 m, showing that all four digits are significant; if it is accurate to only 100 m or so, we can report it as 1.2×10^3 m, indicating two significant figures.

TACTICS
BOX 1.2 **Using scientific notation** (MP)

To convert a number into scientific notation:

❶ For a number greater than 10, move the decimal point to the left until only one digit remains to the left of the decimal point. The remaining number is then multiplied by 10 to a power; this power is given by the number of spaces the decimal point was moved. Here we convert the diameter of the earth to scientific notation:

We move the decimal point until there is only   Since we moved the decimal point
one digit to its left, counting the number of steps.   6 steps, the power of ten is 6.

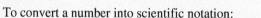

$$6\,370\,000 \text{ m} = 6.37 \times 10^6 \text{ m}$$

The number of digits here equals
the number of significant figures.

❷ For a number less than 1, move the decimal point to the right until it passes the first digit that isn't a zero. The remaining number is then multiplied by 10 to a negative power; the power is given by the number of spaces the decimal point was moved. For the diameter of a red blood cell we have:

We move the decimal point until it passes the first   Since we moved the decimal point
digit that is not a zero, counting the number of steps.   6 steps, the power of ten is −6.

$$0.000\,007\,5 \text{ m} = 7.5 \times 10^{-6} \text{ m}$$

The number of digits here equals
the number of significant figures.

Exercise 16 🖉

Proper use of significant figures is part of the "culture" of science. We will frequently emphasize these "cultural issues" because you must learn to speak the same language as the natives if you wish to communicate effectively!

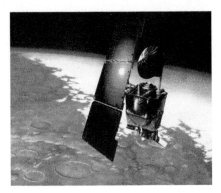

◄ **The importance of units** In 1999, the $125 million Mars Climate Orbiter burned up in the Martian atmosphere instead of entering a safe orbit from which it could perform observations. The problem was faulty units! An engineering team had provided critical data on spacecraft performance in English units, but the navigation team assumed these data were in metric units. As a consequence, the navigation team had the spacecraft fly too close to the planet, and it burned up in the atmosphere.

Units

As we have seen, in order to measure a quantity we need to give it a numerical value. But a measurement is more than just a number—it requires a *unit* to be given. You can't go to the deli and ask for "three quarters of cheese." You need to use a unit—here, one of weight, such as pounds—in addition to the number.

In your daily life, you probably use the English system of units, in which distances are measured in inches, feet, and miles. These units are well adapted for daily life, but they are rarely used in scientific work. Given that science is an international discipline, it is also important to have a system of units that is recognized around the world. For these reasons, scientists use a system of units called *le Système Internationale d'Unités,* commonly referred to as **SI units.** We often refer to these as *metric units* because the meter is the basic standard of length.

The three basic SI quantities, shown in Table 1.1, are time, length (or distance), and mass. Other quantities needed to understand motion can be expressed as combinations of these basic units. For example, speed and velocity are expressed in meters per second or m/s. This combination is a ratio of the length unit (the meter) to the time unit (the second).

TABLE 1.1 Common SI units

Quantity	Unit	Abbreviation
time	second	s
length	meter	m
mass	kilogram	kg

Using Prefixes

We will have many occasions to use lengths, times, and masses that are either much less or much greater than the standards of 1 meter, 1 second, and 1 kilogram. We will do so by using *prefixes* to denote various powers of ten. For instance, the prefix "kilo" (abbreviation k) denotes 10^3, or a factor of 1000. Thus 1 km equals 1000 m, 1 MW equals 10^6 watts, and 1 μV equals 10^{-6} V. Table 1.2 lists the common prefixes that will be used frequently throughout this text. A more extensive list of prefixes is shown at the front of the text.

Although prefixes make it easier to talk about quantities, the proper SI units are meters, seconds, and kilograms. Quantities given with prefixed units are usually converted to base SI units before any calculations are done. Thus 23.0 cm should be converted to 0.230 m before starting calculations. The exception is the kilogram, which is already the base SI unit.

TABLE 1.2 Common prefixes

Prefix	Abbreviation	Power of 10
mega-	M	10^6
kilo-	k	10^3
centi-	c	10^{-2}
milli-	m	10^{-3}
micro-	μ	10^{-6}
nano-	n	10^{-9}

Unit Conversions

Although SI units are our standard, we cannot entirely forget that the United States still uses English units. Even after repeated exposure to metric units in classes, most of us "think" in English units. Thus it remains important to be able to convert back and forth between SI units and English units. Table 1.3 shows some frequently used conversions that will come in handy.

One effective method of performing unit conversions begins by noticing that since, for example, 1 mi = 1.609 km, the ratio of these two distances—*including their units*—is equal to 1, so that

$$\frac{1 \text{ mi}}{1.609 \text{ km}} = \frac{1.609 \text{ km}}{1 \text{ mi}} = 1$$

A ratio of values equal to 1 is called a **conversion factor.** The following Tactics Box shows how to make a unit conversion.

TABLE 1.3 Useful unit conversions

1 inch (in) = 2.54 cm
1 foot (ft) = 0.305 m
1 mile (mi) = 1.609 km
1 mile per hour (mph) = 0.447 m/s
1 m = 39.37 in
1 km = 0.621 mi
1 m/s = 2.24 mph

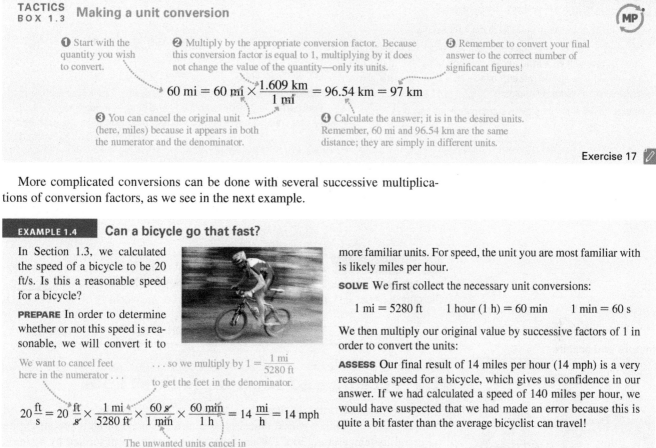

TACTICS
BOX 1.3 **Making a unit conversion**

❶ Start with the quantity you wish to convert.

❷ Multiply by the appropriate conversion factor. Because this conversion factor is equal to 1, multiplying by it does not change the value of the quantity—only its units.

❺ Remember to convert your final answer to the correct number of significant figures!

$$60 \text{ mi} = 60 \text{ mi} \times \frac{1.609 \text{ km}}{1 \text{ mi}} = 96.54 \text{ km} = 97 \text{ km}$$

❸ You can cancel the original unit (here, miles) because it appears in both the numerator and the denominator.

❹ Calculate the answer; it is in the desired units. Remember, 60 mi and 96.54 km are the same distance; they are simply in different units.

Exercise 17

More complicated conversions can be done with several successive multiplications of conversion factors, as we see in the next example.

EXAMPLE 1.4 **Can a bicycle go that fast?**

In Section 1.3, we calculated the speed of a bicycle to be 20 ft/s. Is this a reasonable speed for a bicycle?

PREPARE In order to determine whether or not this speed is reasonable, we will convert it to

more familiar units. For speed, the unit you are most familiar with is likely miles per hour.

SOLVE We first collect the necessary unit conversions:

1 mi = 5280 ft   1 hour (1 h) = 60 min   1 min = 60 s

We then multiply our original value by successive factors of 1 in order to convert the units:

We want to cancel feet here in the numerator . . .   . . . so we multiply by $1 = \frac{1 \text{ mi}}{5280 \text{ ft}}$ to get the feet in the denominator.

$$20 \frac{\text{ft}}{\text{s}} = 20 \frac{\text{ft}}{\text{s}} \times \frac{1 \text{ mi}}{5280 \text{ ft}} \times \frac{60 \text{ s}}{1 \text{ min}} \times \frac{60 \text{ min}}{1 \text{ h}} = 14 \frac{\text{mi}}{\text{h}} = 14 \text{ mph}$$

The unwanted units cancel in pairs, as indicated by the colors.

ASSESS Our final result of 14 miles per hour (14 mph) is a very reasonable speed for a bicycle, which gives us confidence in our answer. If we had calculated a speed of 140 miles per hour, we would have suspected that we had made an error because this is quite a bit faster than the average bicyclist can travel!

▶The man has a mass of 70 kg. What is the mass of the elephant standing next to him? By thinking about the relative dimensions of the two, you can make a reasonable one-significant-figure *estimate*.

Estimation

When scientists and engineers first approach a problem, they may do a quick measurement or calculation to establish the rough physical scale involved. This will help establish the procedures that should be used to make a more accurate measurement—or the estimate may well be all that is needed.

Suppose you see a rock fall off a cliff and would like to know how fast it was going when it hit the ground. By doing a mental comparison with the speeds of familiar objects, such as cars and bicycles, you might judge that the rock was traveling at about 20 mph. This is a one-significant-figure estimate. With some luck, you can probably distinguish 20 mph from either 10 mph or 30 mph, but you certainly cannot distinguish 20 mph from 21 mph just from a visual appearance. A one-significant-figure estimate or calculation, such as this estimate of speed, is called an **order-of-magnitude estimate.** An order-of-magnitude estimate is indicated by the symbol ~, which indicates even less precision than the "approximately equal" symbol ≈. You would report your estimate of the speed of the falling rock as $v \sim 20$ mph.

It's a useful skill to make reliable order-of-magnitude estimates on the basis of known information (or information found on the Internet), simple reasoning, and common sense. It may help to convert from SI units to more familiar units to make such estimates. You can also do this to assess problem solutions given in SI units. Table 1.4 lists some approximate conversion factors to apply in such cases.

TABLE 1.4 Some approximate conversion factors

Quantity	SI unit	Approximate conversion
Mass	kg	1 kg ≈ 2 lb
Length	m	1 m ≈ 3 ft
	cm	3 cm ≈ 1 in
	km	5 km ≈ 3 mi
Speed	m/s	1 m/s ≈ 2 mph
	km/h	10 km/h ≈ 6 mph

EXAMPLE 1.5   **How fast do you walk?**

Estimate how fast you walk, in meters per second.

PREPARE In order to compute speed, we need a distance and a time. If you walked a mile to campus, how long would this take? You'd probably say 30 minutes or so—half an hour. Let's use this rough number in our estimate.

SOLVE Given this estimate, we compute your speed as

$$\text{speed} = \frac{\text{distance}}{\text{time}} \sim \frac{1 \text{ mile}}{1/2 \text{ hour}} = 2 \frac{\text{mi}}{\text{h}}$$

But we want the speed in meters per second. Since our calculation is only an estimate, we use an approximate conversion factor from Table 1.4:

$$1 \frac{\text{mi}}{\text{h}} \sim 0.5 \frac{\text{m}}{\text{s}}$$

This gives an approximate walking speed of 1 m/s.

ASSESS Is this a reasonable value? Let's do another estimate. Your stride is probably about 1 yard long—about 1 meter. And you take about one step per second; next time you are walking, you can count and see. So a walking speed of 1 meter per second sounds pretty reasonable.

STOP TO THINK 1.5   Rank in order, from the most to the fewest, the number of significant figures in the following numbers. For example, if B has more than C, C has the same number as A, and A has more than D, give your answer as B > C = A > D.

A. 0.43   B. 0.0052   C. 0.430   D. 4.321×10^{-10}

Vectors and scalars

Scalars

Time, temperature and weight are all *scalar* quantities. To specify your weight, the temperature outside, or the current time, you only need a single number.

Vectors

The velocity of the race car is a *vector*. To fully specify a velocity, we need to give its magnitude (e.g., 120 mph) *and* its direction (e.g., west).

The force with which the boy pushes on his friend is another example of a vector. To completely specify this force, we must know not only how hard he pushes (the magnitude) but also in which direction.

1.5 Vectors and Motion: A First Look

Many physical quantities, such as time, temperature, and weight, can be described completely by a number with a unit. For example, the mass of an object might be 6 kg and its temperature 30°C. When a physical quantity is described by a single number (with a unit), we call it a **scalar quantity**. A scalar can be positive, negative, or zero.

Many other quantities, however, have a directional quality and cannot be described by a single number. To describe the motion of a car, for example, you must specify not only how fast it is moving, but also the *direction* in which it is moving. A **vector quantity** is a quantity that has both a *size* (How far? or How fast?) and a *direction* (Which way?). The size or length of a vector is called its **magnitude**. The magnitude of a vector can be positive or zero, but it cannot be negative.

We **graphically represent a vector as an** *arrow,* as illustrated for the velocity and force vectors. The arrow is drawn to point in the direction of the vector quantity, and the *length* of the arrow is proportional to the magnitude of the vector quantity.

When we want to represent a vector quantity with a *symbol*, we need somehow to indicate that the symbol is for a vector rather than for a scalar. We do this by drawing an arrow over the letter that represents the quantity. Thus $\vec{r}$ and $\vec{A}$ are symbols for vectors, whereas r and A, without the arrows, are symbols for scalars. In handwritten work you *must* draw arrows over all symbols that represent vectors. This may seem strange until you get used to it, but it is very important because we will often use both r and $\vec{r}$, or both A and $\vec{A}$ in the same problem, and they mean different things!

NOTE ▶ The arrow over the symbol always points to the right, regardless of which direction the actual vector points. Thus we write $\vec{r}$ or $\vec{A}$, never $\overleftarrow{r}$ or $\overleftarrow{A}$. ◀

Displacement Vectors

For motion along a line, we found in Section 1.2 that the displacement is a quantity that specifies not only how *far* an object moves but also the *direction*—to the left or to the right—that the object moves. Since displacement is a quantity that has both a magnitude (How far?) and a direction, it can be represented by a vector, the **displacement vector**. FIGURE 1.21 shows the displacement vector for Sam's trip that we discussed earlier. We've simply drawn an arrow—the vector—from his initial to his final position and assigned it the symbol $\vec{d}_S$. Because $\vec{d}_S$ has both a magnitude and a direction, it is convenient to write Sam's displacement as $\vec{d}_S = (100\text{ ft, east})$. The first value in the parentheses is the magnitude of the vector (i.e., the size of the displacement), and the second value specifies its direction.

Also shown in Figure 1.21 is the displacement vector $\vec{d}_J$ for Jane, who started on 12th Street and ended up on Vine. As with Sam, we draw her displacement vector as an arrow from her initial to her final position. In this case, $\vec{d}_J = 100\text{ ft},$ 30° east of north).

Jane's trip illustrates an important point about displacement vectors. Jane started her trip on 12th Street and ended up on Vine, leading to the displacement vector shown. But to get from her initial to her final position, she needn't have walked along the straight-line path denoted by $\vec{d}_J$. If she walked east along 12th Street to the intersection and then headed north on Vine, her displacement would still be the vector shown. **An object's displacement vector is drawn from the object's initial position to its final position, regardless of the actual path followed between these two points**.

Vector Addition

Let's consider one more trip for the peripatetic Sam. In FIGURE 1.22, he starts at the intersection and walks east 50 ft; then he walks 100 ft to the northeast through a vacant lot. His displacement vectors for the two legs of his trip are labeled $\vec{d}_1$ and $\vec{d}_2$ in the figure.

Sam's trip consists of two legs that can be represented by the two vectors $\vec{d}_1$ and $\vec{d}_2$, but we can represent his trip as a whole, from his initial starting position to his overall final position, with the *net* displacement vector labeled $\vec{d}_{net}$. Sam's net displacement is in a sense the *sum* of the two displacements that made it up, so we can write

$$\vec{d}_{net} = \vec{d}_1 + \vec{d}_2$$

Sam's net displacement thus requires the *addition* of two vectors, but vector addition obeys different rules from the addition of two scalar quantities. The directions of the two vectors, as well as their magnitudes, must be taken into account. Sam's trip suggests that we can add vectors together by putting the "tail" of one vector at the tip of the other. This idea, which is reasonable for displacement vectors, in fact is how *any* two vectors are added. Tactics Box 1.4 shows how to add two vectors $\vec{A}$ and $\vec{B}$.

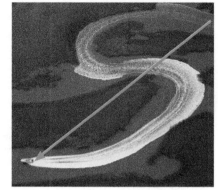

The boat's displacement is the straight-line connection from its initial to its final position.

FIGURE 1.21 Two displacement vectors.

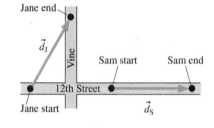

FIGURE 1.22 Sam undergoes two displacements.

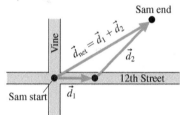

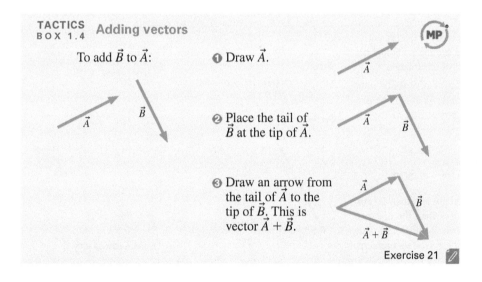

TACTICS BOX 1.4 Adding vectors

To add $\vec{B}$ to $\vec{A}$:

❶ Draw $\vec{A}$.

❷ Place the tail of $\vec{B}$ at the tip of $\vec{A}$.

❸ Draw an arrow from the tail of $\vec{A}$ to the tip of $\vec{B}$. This is vector $\vec{A} + \vec{B}$.

Exercise 21

Vectors and Trigonometry

When we need to add displacements or other vectors in more than one dimension, we'll end up computing lengths and angles of triangles. This is the job of trigonometry. FIGURE 1.23 reviews the basic ideas of trigonometry.

FIGURE 1.23 Relating sides and angles of triangles using trigonometry.

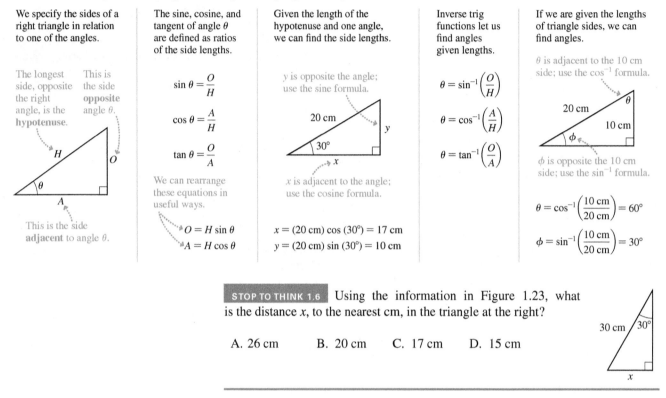

We specify the sides of a right triangle in relation to one of the angles.

The longest side, opposite the right angle, is the **hypotenuse**.

This is the side **opposite** angle θ.

This is the side **adjacent** to angle θ.

The sine, cosine, and tangent of angle θ are defined as ratios of the side lengths.

$$\sin \theta = \frac{O}{H}$$

$$\cos \theta = \frac{A}{H}$$

$$\tan \theta = \frac{O}{A}$$

We can rearrange these equations in useful ways.

$$O = H \sin \theta$$
$$A = H \cos \theta$$

Given the length of the hypotenuse and one angle, we can find the side lengths.

y is opposite the angle; use the sine formula.

x is adjacent to the angle; use the cosine formula.

$x = (20 \text{ cm}) \cos (30°) = 17 \text{ cm}$
$y = (20 \text{ cm}) \sin (30°) = 10 \text{ cm}$

Inverse trig functions let us find angles given lengths.

$$\theta = \sin^{-1}\left(\frac{O}{H}\right)$$

$$\theta = \cos^{-1}\left(\frac{A}{H}\right)$$

$$\theta = \tan^{-1}\left(\frac{O}{A}\right)$$

If we are given the lengths of triangle sides, we can find angles.

θ is adjacent to the 10 cm side; use the $\cos^{-1}$ formula.

ϕ is opposite the 10 cm side; use the $\sin^{-1}$ formula.

$$\theta = \cos^{-1}\left(\frac{10 \text{ cm}}{20 \text{ cm}}\right) = 60°$$

$$\phi = \sin^{-1}\left(\frac{10 \text{ cm}}{20 \text{ cm}}\right) = 30°$$

STOP TO THINK 1.6 Using the information in Figure 1.23, what is the distance x, to the nearest cm, in the triangle at the right?

A. 26 cm   B. 20 cm   C. 17 cm   D. 15 cm

EXAMPLE 1.6   **How far north and east?**

Suppose Alex is navigating using a compass. She starts walking at an angle 60° north of east and walks a total of 100 m. How far north is she from her starting point? How far east?

PREPARE A sketch of Alex's motion is shown in FIGURE 1.24a. We've shown north and east as they would be on a map, and we've noted Alex's displacement as a vector, giving its magnitude and direction. FIGURE 1.24b shows a triangle with this displacement as the hypotenuse. Alex's distance north of her starting point and her distance east of her starting point are the sides of this triangle.

SOLVE The sine and cosine functions are ratios of sides of right triangles, as we saw above. With the 60° angle as noted, the distance north of the starting point is the opposite side of the triangle; the distance east is the adjacent side. Thus:

distance north of start = (100 m) sin (60°) = 87 m

distance east of start = (100 m) cos (60°) = 50 m

ASSESS Both of the distances we calculated are less than 100 m, as they must be, and the distance east is less than the distance north, as our diagram in Figure 1.24b shows it should be. Our answers seem reasonable. In finding the solution to this problem,

we "broke down" the displacement into two different distances, one north and one east. This hints at the idea of the *components* of a vector, something we'll explore in the next chapter.

FIGURE 1.24 An analysis of Alex's motion.

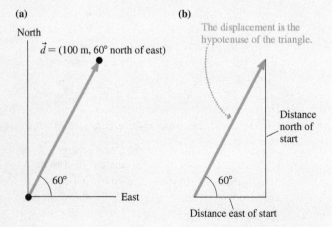

(a)

$\vec{d} = (100 \text{ m}, 60°$ north of east)

North

60°

East

(b)

The displacement is the hypotenuse of the triangle.

Distance north of start

Distance east of start

60°

EXAMPLE 1.7   **How far away is Anna?**

Anna walks 90 m due east and then 50 m due north. What is her displacement from her starting point?

PREPARE Let's start with the sketch in FIGURE 1.25a. We set up a coordinate system with Anna's original position as the origin, and then we drew her two subsequent motions as the two displacement vectors $\vec{d}_1$ and $\vec{d}_2$.

FIGURE 1.25 Analyzing Anna's motion.

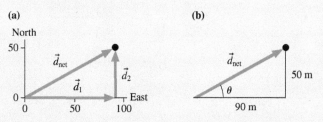

SOLVE We drew the two vector displacements with the tail of one vector starting at the head of the previous one—exactly what is needed to form a vector sum. The vector $\vec{d}_{net}$ in Figure 1.25a is the vector sum of the successive displacements and thus represents Anna's net displacement from the origin.

Anna's distance from the origin is the length of this vector $\vec{d}_{net}$. FIGURE 1.25b shows that this vector is the hypotenuse of a right triangle with sides 50 m (because Anna walked 50 m north)

and 90 m (because she walked 90 m east). We can compute the magnitude of this vector, her net displacement, using the Pythagorean theorem (the square of the length of the hypotenuse of a triangle is equal to the sum of the squares of the lengths of the sides):

$$d_{net}^2 = (50 \text{ m})^2 + (90 \text{ m})^2$$
$$d_{net} = \sqrt{(50 \text{ m})^2 + (90 \text{ m})^2} = 103 \text{ m} \approx 100 \text{ m}$$

We have rounded off to the appropriate number of significant figures, giving us 100 m for the magnitude of the displacement vector. How about the direction? Figure 1.25b identifies the angle that gives the angle north of east of Anna's displacement. In the right triangle, 50 m is the opposite side and 90 m is the adjacent side, so the angle is given by

$$\theta = \tan^{-1}\left(\frac{50 \text{ m}}{90 \text{ m}}\right) = \tan^{-1}\left(\frac{5}{9}\right) = 29°$$

Putting it all together, we get a net displacement of

$$\vec{d}_{net} = (100 \text{ m}, 29° \text{ north of east})$$

ASSESS We can use our drawing to assess our result. If the two sides of the triangle are 50 m and 90 m, a length of 100 m for the hypotenuse seems about right. The angle is certainly less than 45°, but not too much less, so 29° seems reasonable.

Velocity Vectors

We've seen that a basic quantity describing the motion of an object is its velocity. Velocity is a vector quantity because its specification involves how fast an object is moving (its speed) and also the direction in which the object is moving. We thus represent the velocity of an object by a **velocity vector** $\vec{v}$ that points in the direction of the object's motion, and whose magnitude is the object's speed.

FIGURE 1.26a shows the motion diagram of a car accelerating from rest. We've drawn vectors showing the car's displacement between successive positions in the motion diagram. To draw the velocity vectors, first note that the direction of the displacement vector is the direction of motion between successive points in the motion diagram. The velocity of an object also points in the direction of motion, so the velocity vector points in the same direction as its displacement vector. Next, note that the magnitude of the velocity vector—How fast?—is the object's speed. Higher speeds imply greater displacements, so the length of the velocity vector should be proportional to the length of the displacement vector between successive points on a motion diagram. All this means that the vectors connecting each dot of a motion diagram to the next, which we have labeled as displacement vectors, could equally well be identified as velocity vectors, as shown in FIGURE 1.26b. **From now on, we'll show and label velocity vectors on motion diagrams rather than displacement vectors.**

NOTE ▶ The velocity vectors shown in Figure 1.26b are actually *average* velocity vectors. Because the velocity is steadily increasing, it's a bit less than this average at the start of each time interval, and a bit more at the end. In Chapter 2 we'll refine these ideas as we develop the idea of instantaneous velocity. ◀

FIGURE 1.26 The motion diagram for a car starting from rest.

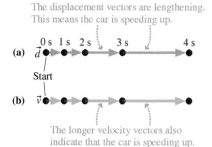

The displacement vectors are lengthening. This means the car is speeding up.

The longer velocity vectors also indicate that the car is speeding up.

EXAMPLE 1.8   **Drawing a ball's motion diagram**

Jake hits a ball at a 60° angle from the horizontal. It is caught by Jim. Draw a motion diagram of the ball that shows velocity vectors rather than displacement vectors.

PREPARE This example is typical of how many problems in science and engineering are worded. The problem does not give a clear statement of where the motion begins or ends. Are we interested in the motion of the ball only during the time it is in the air between Jake and Jim? What about the motion *as* Jake hits it (ball rapidly speeding up) or *as* Jim catches it (ball rapidly slowing down)? Should we include Jim dropping the ball after he catches it? The point is that *you* will often be called on to make a *reasonable interpretation* of a problem statement. In this problem, the details of hitting and catching the ball are complex. The motion of the ball through the air is easier to describe, and it's a motion you might expect to learn about in a physics class. So our *interpretation* is that the motion diagram should start as the ball leaves Jake's bat (ball already moving) and should end the instant it touches Jim's hand (ball still moving). We will model the ball as a particle.

SOLVE With this interpretation in mind, FIGURE 1.27 shows the motion diagram of the ball. Notice how, in contrast to the car of Figure 1.26, the ball is already moving as the motion diagram movie begins. As before, the velocity vectors are shown by con-

necting the dots with arrows. You can see that the velocity vectors get shorter (ball slowing down), get longer (ball speeding up), and change direction. Each $\vec{v}$ is different, so this is *not* constant-velocity motion.

FIGURE 1.27 The motion diagram of a ball traveling from Jake to Jim.

ASSESS We haven't learned enough to make a detailed analysis of the motion of the ball, but it's still worthwhile to do a quick assessment. Does our diagram make sense? Think about the velocity of the ball—we show it moving upward at the start and downward at the end. This does match what happens when you toss a ball back and forth, so our answer seems reasonable.

STOP TO THINK 1.7 $\vec{P}$ and $\vec{Q}$ are two vectors of equal length but different direction. Which vector shows the sum $\vec{P} + \vec{Q}$?

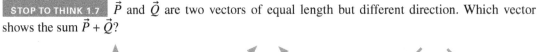

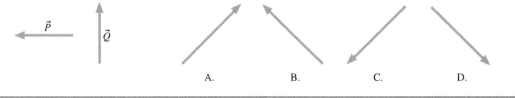

1.6 Where Do We Go from Here?

This first chapter has been an introduction to some of the fundamental ideas about motion and some of the basic techniques that you will use in the rest of the course. You have seen some examples of how to make *models* of a physical situation, thereby focusing on the essential elements of the situation. You have learned some practical ideas, such as how to convert quantities from one kind of units to another. The rest of this text—and the rest of your course—will extend these themes.

In each chapter of this text, you'll learn both new principles and more tools and techniques. As you proceed, you'll find that each new chapter depends on those that preceded it. The principles and the problem-solving strategies you learned in this chapter will still be needed in Chapter 30.

We'll give you some assistance integrating new ideas with the material of previous chapters. When you start a chapter, the **chapter preview** will let you know which topics are especially important to review. And the last element in each chapter will be an **integrated example** that brings together the principles and techniques you have just learned with those you learned previously. The integrated nature of

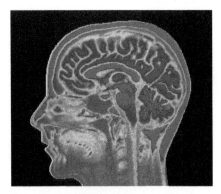

BIO Chapter 28 ends with an integrated example that explores the basic physics of magnetic resonance imaging (MRI), explaining how the interaction of magnetic fields with the nuclei of atoms in the body can be used to create an image of the body's interior.

these examples will also be a helpful reminder that the problems of the real world are similarly complex, and solving such problems requires you to do just this kind of integration.

Our first integrated example is reasonably straightforward because there's not much to integrate yet. The examples in future chapters will be much richer.

INTEGRATED EXAMPLE 1.9   **A goose gets its bearings**

Migrating geese determine direction using many different tools: by noting local landmarks, by following rivers and roads, and by using the position of the sun in the sky. When the weather is overcast so that they can't use the sun's position to get their bearings, geese may start their day's flight in the wrong direction. **FIGURE 1.28** shows the path of a Canada goose that flew in a straight line for some time before making a corrective right-angle turn. One hour after beginning, the goose made a rest stop on a lake due east of its original position.

FIGURE 1.28 Trajectory of a misdirected goose.

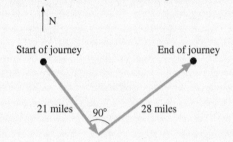

a. How much extra distance did the goose travel due to its initial error in flight direction? That is, how much farther did it fly than if it had simply flown directly to its final position on the lake?
b. What was the flight speed of the goose?
c. A typical flight speed for a migrating goose is 80 km/h. Given this, does your result seem reasonable?

PREPARE Figure 1.28 shows the trajectory of the goose, but it's worthwhile to redraw Figure 1.28 and note the displacement from the start to the end of the journey, the shortest distance the goose could have flown. (The examples in the chapter to this point have used professionally rendered drawings, but these are much more careful and detailed than you are likely to make. **FIGURE 1.29** shows a drawing that is more typical of what you might actually do when working problems yourself.) Drawing and labeling the displacement between the starting and ending points in Figure 1.29 show that it is the hypotenuse of a right triangle, so we can use our rules for triangles as we look for a solution.

FIGURE 1.29 A typical student sketch shows the motion and the displacement of the goose.

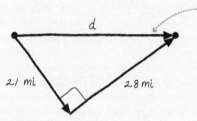

The displacement is the hypotenuse of a right triangle, with the two legs of the journey as the sides.

SOLVE

a. The minimum distance the goose *could* have flown, if it flew straight to the lake, is the hypotenuse of a triangle with sides 21 mi and 28 mi. This straight-line distance is

$$d = \sqrt{(21 \text{ mi})^2 + (28 \text{ mi})^2} = 35 \text{ mi}$$

The actual distance the goose flew is the sum of the distances traveled for the two legs of the journey:

$$\text{distance traveled} = 21 \text{ mi} + 28 \text{ mi} = 49 \text{ mi}$$

The extra distance flown is the difference between the actual distance flown and the straight-line distance—namely, 14 miles.

b. To compute the flight speed, we need to consider the distance that the bird actually flew. The flight speed is the total distance flown divided by the total time of the flight:

$$v = \frac{49 \text{ mi}}{1.0 \text{ h}} = 49 \text{ mi/h}$$

c. To compare our calculated speed with a typical flight speed, we must convert our solution to km/h, rounding off to the correct number of significant digits:

$$49 \frac{\text{mi}}{\text{h}} \times \frac{1.61 \text{ km}}{1.00 \text{ mi}} = 79 \frac{\text{km}}{\text{h}}$$

A calculator will return many more digits, but the original data had only two significant figures, so we report the final result to this accuracy.

ASSESS In this case, an assessment was built into the solution of the problem. The calculated flight speed matches the expected value for a goose, which gives us confidence that our answer is correct. As a further check, our calculated net displacement of 35 mi seems about right for the hypotenuse of the triangle in Figure 1.29.

S U M M A R Y

Goal: To introduce the fundamental concepts of motion and to review related basic mathematical principles.

IMPORTANT CONCEPTS

Motion Diagrams

The particle model represents a moving object as if all its mass were concentrated at a single point. Using this model, we can represent motion with a **motion diagram,** where dots indicate the object's positions at successive times. In a motion diagram, the time interval between successive dots is always the same.

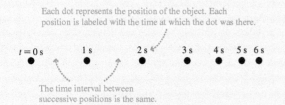

Each dot represents the position of the object. Each position is labeled with the time at which the dot was there.

$t = 0$ s   1 s   2 s   3 s   4 s   5 s   6 s

The time interval between successive positions is the same.

Scalars and Vectors

Scalar quantities have only a magnitude and can be represented by a single number. Temperature, time, and mass are scalars.

A vector is a quantity described by both a magnitude and a direction. Velocity and displacement are vectors.

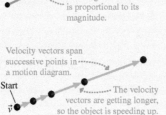

Direction

$\vec{A}$

The length of a vector is proportional to its magnitude.

Velocity vectors can be drawn on a motion diagram by connecting successive points with a vector.

Velocity vectors span successive points in a motion diagram.

Start

$\vec{v}$

The velocity vectors are getting longer, so the object is speeding up.

Describing Motion

Position locates an object with respect to a chosen coordinate system. It is described by a **coordinate.**

The *coordinate* is the variable used to describe the position.

x (mi)

$-6\ -5\ -4\ -3\ -2\ -1\ \ 0\ \ 1\ \ 2\ \ 3\ \ 4\ \ 5$

This cow is at $x = -5$ miles.   This car is at $x = +4$ miles.

A change in position is called a **displacement.** For motion along a line, a displacement is a signed quantity. The displacement from x_i to x_f is $\Delta x = x_f - x_i$.

Time is measured from a particular instant to which we assign $t = 0$. A **time interval** is the elapsed time between two specific instants t_i and t_f. It is given by $\Delta t = t_f - t_i$.

Velocity is the ratio of the displacement of an object to the time interval during which this displacement occurs:

$$v = \frac{\Delta x}{\Delta t}$$

Units

Every measurement of a quantity must include a unit.

The standard system of units used in science is the SI system. Common SI units include:

• Length: meters (m)

• Time: seconds (s)

• Mass: kilograms (kg)

APPLICATIONS

Working with Numbers

In scientific notation, a number is expressed as a decimal number between 1 and 10 multiplied by a power of ten. In scientific notation, the diameter of the earth is 1.27×10^7 m.

A prefix can be used before a unit to indicate a multiple of 10 or 1/10. Thus we can write the diameter of the earth as 12,700 km, where the k in km denotes 1000.

We can perform a unit conversion to convert the diameter of the earth to a different unit, such as miles. We do so by multiplying by a conversion factor equal to 1, such as 1 = 1 mi/1.61 km.

Significant figures are reliably known digits. The number of significant figures for:

• **Multiplication, division, and powers** is set by the value with the fewest significant figures.

• **Addition and subtraction** is set by the value with the smallest number of decimal places.

An order-of-magnitude estimate is an estimate that has an accuracy of about one significant figure. Such estimates are usually made using rough numbers from everyday experience.

Problem difficulty is labeled as | (straightforward) to |||| (challenging). Problems labeled BIO are of biological or medical interest.

 For assigned homework and other learning materials, go to MasteringPhysics®

 Scan this QR code to launch a Video Tutor Solution that will help you solve problems for this chapter.

QUESTIONS

Conceptual Questions

1. a. Write a paragraph describing the *particle model*. What is it, and why is it important?
   b. Give two examples of situations, different from those described in the text, for which the particle model is appropriate.
   c. Give an example of a situation, different from those described in the text, for which it would be inappropriate.

2. A softball player slides into second base. Use the particle model to draw a motion diagram of the player from the time he begins to slide until he reaches the base. Number the dots in order, starting with zero.

3. A car travels to the left at a steady speed for a few seconds, then brakes for a stop sign. Use the particle model to draw a motion diagram of the car for the entire motion described here. Number the dots in order, starting with zero.

4. A ball is dropped from the roof of a tall building and students in a physics class are asked to sketch a motion diagram for this situation. A student submits the diagram shown in Figure Q1.4. Is the diagram correct? Explain.

   • 0
   • 1
   • 2
   • 3
   • 4
   FIGURE Q1.4

5. Write a sentence or two describing the difference between position and displacement. Give one example of each.

6. Give an example of a trip you might take in your car for which the distance traveled as measured on your car's odometer is not equal to the displacement between your initial and final positions.

7. Write a sentence or two describing the difference between speed and velocity. Give one example of each.

8. The motion of a skateboard along a horizontal axis is observed for 5 s. The initial position of the skateboard is negative with respect to a chosen origin, and its velocity throughout the 5 s is also negative. At the end of the observation time, is the skateboard closer to or farther from the origin than initially? Explain.

9. You are standing on a straight stretch of road and watching the motion of a bicycle; you choose your position as the origin. At one instant, the position of the bicycle is negative and its velocity is positive. Is the bicycle getting closer to you or farther away? Explain.

10. Two friends watch a jogger complete a 400 m lap around the track in 100 s. One of the friends states, "The jogger's velocity was 4 m/s during this lap." The second friend objects, saying, "No, the jogger's speed was 4 m/s." Who is correct? Justify your answer.

11. A softball player hits the ball and starts running toward first base. Draw a motion diagram, using the particle model, showing her velocity vectors during the first few seconds of her run.

12. A child is sledding on a smooth, level patch of snow. She encounters a rocky patch and slows to a stop. Draw a motion diagram, using the particle model, showing her velocity vectors.

13. A skydiver jumps out of an airplane. Her speed steadily increases until she deploys her parachute, at which point her speed quickly decreases. She subsequently falls to earth at a constant rate, stopping when she lands on the ground. Draw a motion diagram, using the particle model, that shows her position at successive times and includes velocity vectors.

14. Your roommate drops a tennis ball from a third-story balcony. It hits the sidewalk and bounces as high as the second story. Draw a motion diagram, using the particle model, showing the ball's velocity vectors from the time it is released until it reaches the maximum height on its bounce.

15. A car is driving north at a steady speed. It makes a gradual 90° left turn without losing speed, then continues driving to the west. Draw a motion diagram, using the particle model, showing the car's velocity vectors as seen from a helicopter hovering over the highway.

16. A toy car rolls down a ramp, then across a smooth, horizontal floor. Draw a motion diagram, using the particle model, showing the car's velocity vectors.

17. Density is the ratio of an object's mass to its volume. Would you expect density to be a vector or a scalar quantity? Explain.

Multiple-Choice Questions

18. | A student walks 1.0 mi west and then 1.0 mi north. Afterward, how far is she from her starting point?
    A. 1.0 mi   B. 1.4 mi   C. 1.6 mi   D. 2.0 mi

19. | You throw a rock upward. The rock is moving upward, but it is slowing down. If we define the ground as the origin, the position of the rock is _____ and the velocity of the rock is _____.
    A. positive, positive   B. positive, negative
    C. negative, positive   D. negative, negative

20. | Which of the following motions could be described by the motion diagram of Figure Q1.20?
    A. A hockey puck sliding across smooth ice.
    B. A cyclist braking to a stop.
    C. A sprinter starting a race.
    D. A ball bouncing off a wall.

FIGURE Q1.20   5 4   3   2   1   0
                     •• •   •   •   •

21. | Which of the following motions is described by the motion diagram of Figure Q1.21?
    A. An ice skater gliding across the ice.
    B. An airplane braking to a stop after landing.
    C. A car pulling away from a stop sign.
    D. A pool ball bouncing off a cushion and reversing direction.

FIGURE Q1.21   0 1   2   3   4   5
                     •• •   •   •   •

22. | A bird flies 3.0 km due west and then 2.0 km due north. What is the magnitude of the bird's displacement?
    A. 2.0 km   B. 3.0 km   C. 3.6 km   D. 5.0 km

23. ‖   Weddell seals make holes in sea ice so that they can swim
BIO   down to forage on the ocean floor below. Measurements for one seal showed that it dived straight down from such an opening, reaching a depth of 0.30 km in a time of 5.0 min. What was the speed of the diving seal?
    A. 0.60 m/s   B. 1.0 m/s   C. 1.6 m/s   D. 6.0 m/s
    E. 10 m/s

24. ‖   A bird flies 3.0 km due west and then 2.0 km due north. Another bird flies 2.0 km due west and 3.0 km due north. What is the angle between the net displacement vectors for the two birds?
    A. 23°   B. 34°   C. 56°   D. 90°

25. | A woman walks briskly at 2.00 m/s. How much time will it take her to walk one mile?
    A. 8.30 min   B. 13.4 min   C. 21.7 min   D. 30.0 min

26. | Compute 3.24 m + 0.532 m to the correct number of significant figures.
    A. 3.7 m   B. 3.77 m   C. 3.772 m   D. 3.7720 m

27. | A rectangle has length 3.24 m and height 0.532 m. To the correct number of significant figures, what is its area?
    A. 1.72 m^2 B. 1.723 m^2
    C. 1.7236 m^2 D. 1.72368 m^2

28. | The earth formed 4.57×10^9 years ago. What is this time in seconds?
    A. 1.67×10^{12} s   B. 4.01×10^{13} s
    C. 2.40×10^{15} s   D. 1.44×10^{17} s

29. ‖| An object's average density ρ is defined as the ratio of its mass to its volume: $\rho = M/V$. The earth's mass is 5.94×10^{24} kg, and its volume is 1.08×10^{12} km^3. What is the earth's average density?
    A. 5.50×10^3 kg/m^3 B. 5.50×10^6 kg/m^3
    C. 5.50×10^9 kg/m^3 D. 5.50×10^{12} kg/m^3

PROBLEMS

Section 1.1 Motion: A First Look

1. | A car skids to a halt to avoid hitting an object in the road. Draw a motion diagram of the car from the time the skid begins until the instant the car stops.

2. | A man rides a bike along a straight road for 5 min, then has a flat tire. He stops for 5 min to repair the flat, but then realizes he cannot fix it. He continues his journey by walking the rest of the way, which takes him another 10 min. Use the particle model to draw a motion diagram of the man for the entire motion described here. Number the dots in order, starting with zero.

3. | A jogger running east at a steady pace suddenly develops a cramp. He is lucky: A westbound bus is sitting at a bus stop just ahead. He gets on the bus and enjoys a quick ride home. Use the particle model to draw a motion diagram of the jogger for the entire motion described here. Number the dots in order, starting with zero.

Section 1.2 Position and Time: Putting Numbers on Nature

4. | Figure P1.4 shows Sue along the straight-line path between her home and the cinema. What is Sue's position x if
    a. Her home is the origin?
    b. The cinema is the origin?

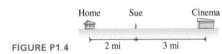

FIGURE P1.4

5. | Keira starts at position $x = 23$ m along a coordinate axis. She then undergoes a displacement of -45 m. What is her final position?

6. | A car travels along a straight east-west road. A coordinate system is established on the road, with x increasing to the east. The car ends up 14 mi west of the origin, which is defined as the intersection with Mulberry Road. If the car's displacement was -23 mi, what side of Mulberry Road did the car start on? How far from the intersection was the car at the start?

7. | Foraging bees often move in straight lines away from and
BIO   toward their hives. Suppose a bee starts at its hive and flies 500 m due east, then flies 400 m west, then 700 m east. How far is the bee from the hive?

Section 1.3 Velocity

8. | A security guard walks at a steady pace, traveling 110 m in one trip around the perimeter of a building. It takes him 240 s to make this trip. What is his speed?

9. ‖   List the following items in order of decreasing speed, from greatest to least: (i) A wind-up toy car that moves 0.15 m in 2.5 s. (ii) A soccer ball that rolls 2.3 m in 0.55 s. (iii) A bicycle that travels 0.60 m in 0.075 s. (iv) A cat that runs 8.0 m in 2.0 s.

10. ‖   Figure P1.10 shows the motion diagram for a horse galloping in one direction along a straight path. Not every dot is labeled, but the dots are at equally spaced instants of time. What is the horse's velocity
    a. During the first ten seconds of its gallop?
    b. During the interval from 30 s to 40 s?
    c. During the interval from 50 s to 70 s?

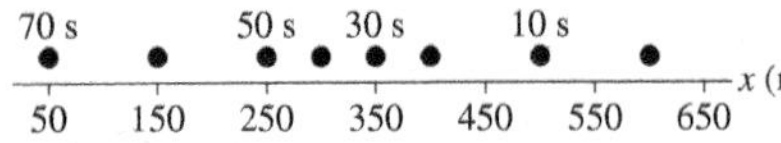

```
70 s        50 s  30 s      10 s
  ●      ●    ●  ●  ●      ●      ●
                                      → x (m)
FIGURE P1.10    50   150   250   350   450   550   650
```

11. ‖   It takes Harry 35 s to walk from $x = -12$ m to $x = -47$ m. What is his velocity?

12. | A dog trots from $x = -12$ m to $x = 3$ m in 10 s. What is its velocity?

13. | A ball rolling along a straight line with velocity 0.35 m/s goes from $x = 2.1$ m to $x = 7.3$ m. How much time does this take?

Section 1.4 A Sense of Scale: Significant Figures, Scientific Notation, and Units

14. ‖   Convert the following to SI units:
    a. 9.12 μs   b. 3.42 km
    c. 44 cm/ms   d. 80 km/h

15. | Convert the following to SI units:
    a. 8.0 in   b. 66 ft/s   c. 60 mph

16. | Convert the following to SI units:
    a. 1.0 hour   b. 1.0 day   c. 1.0 year

17. ‖ How many significant figures does each of the following numbers have?
    a. 6.21   b. 62.1   c. 0.620   d. 0.062
18. | How many significant figures does each of the following numbers have?
    a. 0.621   b. 0.006200   c. 1.0621   d. 6.21×10^3
19. | Compute the following numbers to three significant figures.
    a. 33.3×25.4 b. $33.3 - 25.4$
    c. $\sqrt{33.3}$ d. $333.3 \div 25.4$
20. ‖ BIO If you make multiple measurements of your height, you are likely to find that the results vary by nearly half an inch in either direction due to measurement error and actual variations in height. You are slightly shorter in the evening, after gravity has compressed and reshaped your spine over the course of a day. One measurement of a man's height is 6 feet and 1 inch. Express his height in meters, using the appropriate number of significant figures.

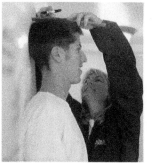

21. | The Empire State Building has a height of 1250 ft. Express this height in meters, giving your result in scientific notation with three significant figures.
22. ‖ BIO Blades of grass grow from the bottom, so, as growth occurs, the top of the blade moves upward. During the summer, when your lawn is growing quickly, estimate this speed, in m/s. Make this estimate from your experience noting, for instance, how often you mow the lawn and what length you trim. Express your result in scientific notation.
23. ‖ BIO Estimate the average speed, in m/s, with which the hair on your head grows. Make this estimate from your own experience noting, for instance, how often you cut your hair and how much you trim. Express your result in scientific notation.

Section 1.5 Vectors and Motion: A First Look

24. | Carol and Robin share a house. To get to work, Carol walks north 2.0 km while Robin drives west 7.5 km. How far apart are their workplaces?
25. | Loveland, Colorado, is 18 km due south of Fort Collins and 31 km due west of Greeley. What is the distance between Fort Collins and Greeley?
26. | Joe and Max shake hands and say goodbye. Joe walks east 0.55 km to a coffee shop, and Max flags a cab and rides north 3.25 km to a bookstore. How far apart are their destinations?
27. ‖ A city has streets laid out in a square grid, with each block 135 m long. If you drive north for three blocks, then west for two blocks, how far are you from your starting point?
28. ‖ A butterfly flies from the top of a tree in the center of a garden to rest on top of a red flower at the garden's edge. The tree is 8.0 m taller than the flower, and the garden is 12 m wide. Determine the magnitude of the butterfly's displacement.
29. ‖ A garden has a circular path of radius 50 m. John starts at the easternmost point on this path, then walks counterclockwise around the path until he is at its southernmost point. What is John's displacement? Use the (magnitude, direction) notation for your answer.

30. ‖ A circular test track for cars in England has a circumference of 3.2 km. A car travels around the track from the southernmost point to the northernmost point.
    a. What distance does the car travel?
    b. What is the car's displacement from its original position?
31. | BIO Migrating geese tend to travel at approximately constant speed, flying in segments that are straight lines. A goose flies 32 km south, then turns to fly 20 km west. Afterward, how far is the goose from its original position?
32. ‖ BIO Black vultures excel at gliding flight; they can move long distances through the air without flapping their wings while undergoing only a modest drop in height. A vulture in a typical glide in still air moves along a path tipped 3.5° below the horizontal. If the vulture moves a horizontal distance of 100 m, how much height does it lose?
33. | A hiker walks 25° north of east for 200 m. How far north and how far east is he from his starting position?
34. | A hiker is climbing a steep 10° slope. Her pedometer shows that she has walked 1500 m along the slope. How much elevation has she gained?
35. ‖‖‖ A ball on a porch rolls 60 cm to the porch's edge, drops 40 cm, continues rolling on the grass, and eventually stops 80 cm from the porch's edge. What is the magnitude of the ball's net displacement, in centimeters?
36. ‖ A kicker punts a football from the very center of the field to the sideline 43 yards downfield. What is the net displacement of the ball? (A football field is 53 yards wide.)

Problems 37 and 38 relate to the gliding flight of flying squirrels. These squirrels glide from tree to tree at a constant speed, moving in a straight line tipped below the vertical and steadily losing altitude as they move forward. Short and long glides have different profiles.

37. ‖ BIO A squirrel completing a short glide travels in a straight line tipped 40° below the horizontal. The squirrel starts 9.0 m above the ground on one tree and glides to a second tree that is a horizontal distance of 3.5 m away.
    a. What is the length of the squirrel's glide path?
    b. What is the squirrel's height above the ground when it lands?
38. ‖ BIO A squirrel in a typical long glide covers a horizontal distance of 16 m while losing 8.0 m of elevation. During this glide,
    a. What is the angle of the squirrel's path below the horizontal?
    b. What is the total distance covered by the squirrel?

General Problems

Problems 39 through 45 are motion problems similar to those you will learn to solve in Chapter 2. For now, simply *interpret* the problem by drawing a motion diagram showing the object's position and its velocity vectors. **Do *not* solve these problems** or do any mathematics.

39. ‖ BIO In a typical greyhound race, a dog accelerates to a speed of 20 m/s over a distance of 30 m. It then maintains this speed. What would be a greyhound's time in the 100 m dash?
40. ‖ Billy drops a watermelon from the top of a three-story building, 10 m above the sidewalk. How fast is the watermelon going when it hits?

41. ‖ Sam is recklessly driving 60 mph in a 30 mph speed zone when he suddenly sees the police. He steps on the brakes and slows to 30 mph in three seconds, looking nonchalant as he passes the officer. How far does he travel while braking?

42. ‖ A speed skater moving across frictionless ice at 8.0 m/s hits a 5.0-m-wide patch of rough ice. She slows steadily, then continues on at 6.0 m/s. What is her acceleration on the rough ice?

43. ‖ The giant eland, an African antelope, is an exceptional
BIO jumper, able to leap 1.5 m off the ground. To jump this high, with what speed must the eland leave the ground?

44. ‖ A ball rolls along a smooth horizontal floor at 10 m/s, then starts up a 20° ramp. How high does it go before rolling back down?

45. ‖ A motorist is traveling at 20 m/s. He is 60 m from a stop light when he sees it turn yellow. His reaction time, before stepping on the brake, is 0.50 s. What steady deceleration while braking will bring him to a stop right at the light?

Problems 46 through 50 show a motion diagram. For each of these problems, write a one or two sentence "story" about a *real object* that has this motion diagram. Your stories should talk about people or objects by name and say what they are doing. Problems 39 through 45 are examples of motion short stories.

46. |

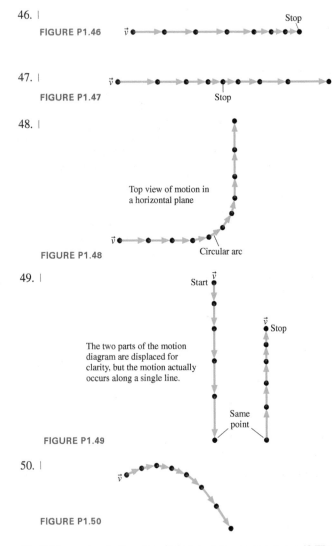

FIGURE P1.46

47. |

FIGURE P1.47

48. |

FIGURE P1.48

49. |

FIGURE P1.49

50. |

FIGURE P1.50

51. ‖‖‖ How many inches does light travel in one nanosecond? The speed of light is 3.0×10^8 m/s.

52. | Joseph watches the roadside mile markers during a long car trip on an interstate highway. He notices that at 10:45 AM they are passing a marker labeled 101, and at 11:00 AM the car reaches marker 119. What is the car's speed, in mph?

53. ‖ Alberta is going to have dinner at her grandmother's house, but she is running a bit behind schedule. As she gets onto the highway, she knows that she must exit the highway within 45 min if she is not going to arrive late. Her exit is 32 mi away. What is the slowest speed at which she could drive and still arrive in time? Express your answer in miles per hour.

54. ‖ The end of Hubbard Glacier in Alaska advances by an average of 105 feet per year. What is the speed of advance of the glacier in m/s?

55. | The earth completes a circular orbit around the sun in one year. The orbit has a radius of 93,000,000 miles. What is the speed of the earth around the sun in m/s? Report your result using scientific notation.

56. ‖‖ Shannon decides to check the accuracy of her speedometer. She adjusts her speed to read exactly 70 mph on her speedometer and holds this steady, measuring the time between successive mile markers separated by exactly 1.00 mile. If she measures a time of 54 s, is her speedometer accurate? If not, is the speed it shows too high or too low?

57. ‖ The Nardo ring is a circular test track for cars. It has a circumference of 12.5 km. Cars travel around the track at a constant speed of 100 km/h. A car starts at the easternmost point of the ring and drives for 15 minutes at this speed.
    a. What distance, in km, does the car travel?
    b. What is the magnitude of the car's displacement, in km, from its initial position?
    c. What is the speed of the car in m/s?

58. ‖ Motor neurons in mammals transmit signals from the brain to
BIO skeletal muscles at approximately 25 m/s. Estimate how much time in ms $(10^{-3}$ s) it will take for a signal to get from your brain to your hand.

59. ‖‖ Satellite data taken several times per hour on a particular
BIO albatross showed travel of 1200 km over a time of 1.4 days.
    a. Given these data, what was the bird's average speed in mph?
    b. Data on the bird's position were recorded only intermittently. Explain how this means that the bird's actual average speed was higher than what you calculated in part a.

60. ‖ The bacterium *Escherichia*
BIO *coli* (or *E. coli*) is a single-celled organism that lives in the gut of healthy humans and animals. Its body shape can be modeled as a 2-μm-long cylinder with a 1 μm diameter, and it has a mass of 1×10^{-12} g. Its chromosome consists of

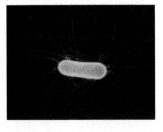

a single double-stranded chain of DNA 700 times longer than its body length. The bacterium moves at a constant speed of 20 μm/s, though not always in the same direction. Answer the following questions about *E. coli* using SI units (unless specifically requested otherwise) and correct significant figures.
    a. What is its length?
    b. Diameter?
    c. Mass?
    d. What is the length of its DNA, in millimeters?
    e. If the organism were to move along a straight path, how many meters would it travel in one day?

61. ‖ The bacterium *Escherichia*
BIO *coli* (or *E. coli*) is a single-
celled organism that lives in
the gut of healthy humans and
animals. When grown in a uniform medium rich in salts and
amino acids, it swims along
zig-zag paths at a constant
speed changing direction at
varying time intervals. Figure

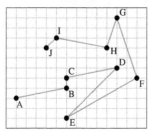

FIGURE P1.61

P1.61 shows the positions of an *E. coli* as it moves from point
A to point J. Each segment of the motion can be identified by
two letters, such as segment BC. During which segments, if any,
does the bacterium have the same
    a. Displacement?   b. Speed?   c. Velocity?

62. ‖ The sun is 30° above the horizon. It makes a 52-m-long
shadow of a tall tree. How high is the tree?

63. ‖ Weddell seals foraging in open water dive toward the ocean
BIO bottom by swimming forward in a straight-line path tipped
below the horizontal. The tracking data for one seal showed
it taking 4.0 min to descend 360 m below the surface while
moving 920 m horizontally.
    a. What was the angle of the seal's path below the horizontal?
    b. What distance did the seal cover in making this dive?
    c. What was the seal's speed, in m/s?

64. ‖ A large passenger aircraft accelerates down the runway for a
distance of 3000 m before leaving the ground. It then climbs at a
steady 3.0° angle. After the plane has traveled 3000 m along this
new trajectory, (a) how high is it, and (b) how far horizontally is
it, from its initial position?

65. ‖ Whale sharks swim forward while ascending or descend-
BIO ing. They swim along a straight-line path at a shallow angle
as they move from the surface to deep water or from the
depths to the surface. In one recorded dive, a shark started
50 m below the surface and swam at 0.85 m/s along a path
tipped at a 13° angle above the horizontal until reaching the
surface.
    a. What was the horizontal distance between the shark's start-
    ing and ending positions?
    b. What was the total distance that the shark swam?
    c. How much time did this motion take?

66. ‖ Starting from its nest, an eagle flies at constant speed for 3.0 min
due east, then 4.0 min due north. From there the eagle flies directly
to its nest at the same speed. How long is the eagle in the air?

67. ‖ John walks 1.00 km north, then turns right and walks 1.00 km
east. His speed is 1.50 m/s during the entire stroll.
    a. What is the magnitude of his displacement, from beginning
    to end?
    b. If Jane starts at the same time and place as John, but walks in
    a straight line to the endpoint of John's stroll, at what speed
    should she walk to arrive at the endpoint just when John does?

MCAT-Style Passage Problems

Growth Speed

The images of trees in Figure P1.68 come from a catalog advertising
fast-growing trees. If we mark the position of the top of the tree in
the successive years, as shown in the graph in the figure, we obtain
a motion diagram much like ones we have seen for other kinds of
motion. The motion isn't steady, of course. In some months the tree
grows rapidly; in other months, quite slowly. We can see, though, that
the average speed of growth is fairly constant for the first few years.

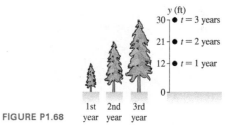

FIGURE P1.68

68. ‖ What is the tree's speed of growth, in feet per year, from
$t = 1$ yr to $t = 3$ yr?
    A. 12 ft/yr   B. 9 ft/yr   C. 6 ft/yr   D. 3 ft/yr

69. ‖ What is this speed in m/s?
    A. 9×10^{-8} m/s   B. 3×10^{-9} m/s
    C. 5×10^{-6} m/s   D. 2×10^{-6} m/s

70. ‖ At the end of year 3, a rope is tied to the very top of the tree to
steady it. This rope is staked into the ground 15 feet away from
the tree. What angle does the rope make with the ground?
    A. 63°   B. 60°   C. 30°   D. 27°

STOP TO THINK ANSWERS

Chapter Preview Stop to Think: C. The sides of a right triangle are
related by the Pythagorean theorem. The length of the hypotenuse is
thus $\sqrt{(6 \text{ cm})^2 + (8 \text{ cm})^2} = 10$ cm. Note that this triangle is a version of a 3-4-5 right triangle; the lengths of the sides are in this ratio.

Stop to Think 1.1: B. The images of B are farther apart, so B travels
a greater distance than does A during the same intervals of time.

Stop to Think 1.2: A. Dropped ball. **B.** Dust particle. **C.** Descending
rocket.

Stop to Think 1.3: C. Depending on her initial positive position and
how far she moves in the negative direction, she could end up on
either side of the origin.

Stop to Think 1.4: C. Her speed is given by Equation 1.1. Her speed
is the distance traveled (4 miles) divided by the time interval (2
hours), or 2 mph.

Stop to Think 1.5: D > C > B = A.

Stop to Think 1.6: D. x is the length of the side opposite the 30°
angle, so $x = (30 \text{ cm}) \sin 30° = 15$ cm.

Stop to Think 1.7: B. The vector sum is found by placing the tail of
one vector at the head of the other.

2 Motion in One Dimension

A horse can run at 35 mph, much faster than a human. And yet, surprisingly, a man can win a race against a horse if the length of the course is right. When, and how, can a man outrun a horse?

LOOKING AHEAD »

Goal: To describe and analyze linear motion.

Uniform Motion

Successive images of the rider are the same distance apart, so the velocity is constant. This is **uniform motion.**

You'll learn to describe motion in terms of quantities such as distance and velocity, an important first step in analyzing motion.

Acceleration

A cheetah is capable of very high speeds but, more importantly, it is capable of a rapid *change* in speed—a large **acceleration.**

You'll use the concept of acceleration to solve problems of changing velocity, such as races, or predators chasing prey.

Free Fall

When you toss a coin, the motion—both going up and coming down—is determined by gravity alone. We call this **free fall.**

How long does it take the coin to go up and come back down? This is the type of free-fall problem you'll learn to solve.

LOOKING BACK «

Motion Diagrams

As you saw in Section 1.5, a good first step in analyzing motion is to draw a motion diagram, marking the position of an object in subsequent times.

In this chapter, you'll learn to create motion diagrams for different types of motion along a line. Drawing pictures like this is a good starting point for solving problems.

STOP TO THINK

A bicycle is moving to the left with increasing speed. Which of the following motion diagrams illustrates this motion?

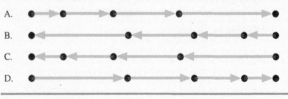

2.1 Describing Motion

The name for the mathematical description of motion is **kinematics,** from the Greek word *kinema,* meaning "movement." You know this word through its English variation *cinema*—motion pictures!

Representing Position

As we saw in Chapter 1, kinematic variables such as position and velocity are measured with respect to a coordinate system, an axis that *you* impose on a system. We will use an *x*-axis to analyze both horizontal motion and motion on a ramp; a *y*-axis will be used for vertical motion. We will adopt the convention that the positive end of an *x*-axis is to the right and the positive end of a *y*-axis is up. This convention is illustrated in **FIGURE 2.1**.

> NOTE ▶ The conventions illustrated in Figure 2.1 aren't absolute. In most cases, we are free to define the coordinate system, and doing so in this standardized way makes sense. In some cases, though, we'll want to make a different choice. ◀

FIGURE 2.2 The motion diagram of a student walking to school and a coordinate axis for making measurements.

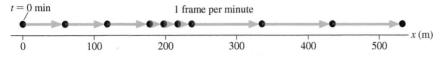

Now, let's look at a practical problem. **FIGURE 2.2** is a motion diagram of a straightforward situation: a student walking to school. She is moving horizontally, so we use the variable *x* to describe her motion. We have set the origin of the coordinate system, *x* = 0, at her starting position, and we measure her position in meters. We have included velocity vectors connecting successive positions on the motion diagram, as we saw we could do in Chapter 1. The motion diagram shows that she leaves home at a time we choose to call *t* = 0 min, and then makes steady progress for a while. Beginning at *t* = 3 min there is a period in which the distance traveled during each time interval becomes shorter—perhaps she slowed down to speak with a friend. Then, at *t* = 6 min, the distances traveled within each interval are longer—perhaps, realizing she is running late, she begins walking more quickly.

Every dot in the motion diagram of Figure 2.2 represents the student's position at a particular time. For example, the student is at position *x* = 120 m at *t* = 2 min. Table 2.1 lists her position for every point in the motion diagram.

The motion diagram of Figure 2.2 is one way to represent the student's motion. Presenting the data as in Table 2.1 is a second way to represent this motion. A third way to represent the motion is to use the data to make a graph. **FIGURE 2.3** is a graph of the positions of the student at different times; we say it is a graph of *x* versus *t* for the student.

> NOTE ▶ A graph of "*a* versus *b*" means that *a* is graphed on the vertical axis and *b* on the horizontal axis. ◀

We can flesh out the graph of Figure 2.3, though. We can assume that the student moved *continuously* through all intervening points of space, so we can represent her motion as a continuous curve that passes through the measured points, as shown in **FIGURE 2.4**. Such a continuous curve that shows an object's position as a function of time is called a **position-versus-time graph** or, sometimes, just a *position graph*.

> NOTE ▶ A graph is *not* a "picture" of the motion. The student is walking along a straight line, but the graph itself is not a straight line. Further, we've graphed her position on the vertical axis even though her motion is horizontal. A graph is an *abstract representation* of motion. ◀

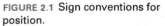

FIGURE 2.1 Sign conventions for position.

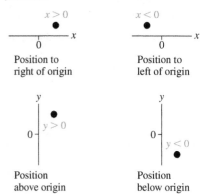

TABLE 2.1 Measured positions of a student walking to school

Time t (min)	Position x (m)	Time t (min)	Position x (m)
0	0	5	220
1	60	6	240
2	120	7	340
3	180	8	440
4	200	9	540

FIGURE 2.3 A graph of the student's motion.

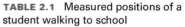

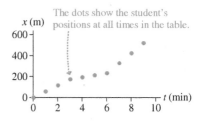

FIGURE 2.4 Extending the graph of Figure 2.3 to a position-versus-time graph.

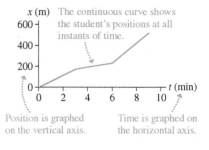

CONCEPTUAL EXAMPLE 2.1 Interpreting a car's position-versus-time graph

The graph in **FIGURE 2.5** represents the motion of a car along a straight road. Describe (in words) the motion of the car.

FIGURE 2.5 Position-versus-time graph for the car.

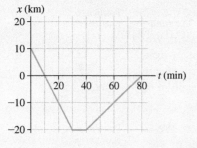

FIGURE 2.6 Looking at the position-versus-time graph in detail.

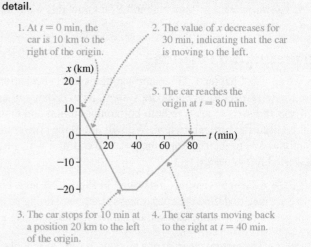

1. At $t = 0$ min, the car is 10 km to the right of the origin.

2. The value of x decreases for 30 min, indicating that the car is moving to the left.

5. The car reaches the origin at $t = 80$ min.

3. The car stops for 10 min at a position 20 km to the left of the origin.

4. The car starts moving back to the right at $t = 40$ min.

REASON The vertical axis in Figure 2.5 is labeled "x (km)"; position is measured in kilometers. Our convention for motion along the x-axis given in Figure 2.1 tells us that x increases as the car moves to the right and x decreases as the car moves to the left. The graph thus shows that the car travels to the left for 30 minutes, stops for 10 minutes, then travels to the right for 40 minutes. It ends up 10 km to the left of where it began. **FIGURE 2.6** gives a full explanation of the reasoning.

ASSESS The car travels to the left for 30 minutes and to the right for 40 minutes. Nonetheless, it ends up to the left of where it started. This means that the car was moving faster when it was moving to the left than when it was moving to the right. We can deduce this fact from the graph as well, as we will see in the next section.

Representing Velocity

FIGURE 2.7 Sign conventions for velocity.

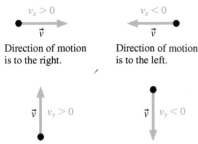

Direction of motion is to the right.

Direction of motion is to the left.

Direction of motion is up.

Direction of motion is down.

Velocity is a vector; it has both a magnitude and a direction. When we draw a velocity vector on a diagram, we use an arrow labeled with the symbol $\vec{v}$ to represent the magnitude and the direction. For motion in one dimension, vectors are restricted to point only "forward" or "backward" for horizontal motion (or "up" or "down" for vertical motion). This restriction lets us simplify our notation for vectors in one dimension. When we solve problems for motion along an x-axis, we will represent velocity with the symbol v_x. As with positions, we will have a usual sign convention for velocities, although we are free to make other choices. v_x will be positive or negative, corresponding to motion to the right or the left, as shown in **FIGURE 2.7**. For motion along a y-axis, we will use the symbol v_y to represent the velocity. The sign conventions are also illustrated in Figure 2.7. We will use the symbol v, with no subscript, to represent the speed of an object. **Speed is the *magnitude* of the velocity vector** and is always positive.

For motion along a line, the definition of velocity from ◀SECTION 1.3 can be written as

$$v_x = \frac{\Delta x}{\Delta t} \tag{2.1}$$

This agrees with the sign conventions in Figure 2.7. If Δx is positive, x is increasing, the object is moving to the right, and Equation 2.1 gives a positive value for velocity. If Δx is negative, x is decreasing, the object is moving to the left, and Equation 2.1 gives a negative value for velocity.

Equation 2.1 is the first of many kinematic equations we'll see in this chapter. We'll often specify equations in terms of the coordinate x, but if the motion is vertical, in which case we use the coordinate y, the equations can be easily adapted. For example, Equation 2.1 for motion along a vertical axis becomes

$$v_y = \frac{\Delta y}{\Delta t} \tag{2.2}$$

From Position to Velocity

Let's take another look at the motion diagram of the student walking to school. As we see in FIGURE 2.8, where we have repeated the motion diagram of Figure 2.2, her motion has three clearly defined phases. In each phase her speed is constant (because the velocity vectors have the same length) but the speed varies from phase to phase.

FIGURE 2.8 Revisiting the motion diagram of the student walking to school.

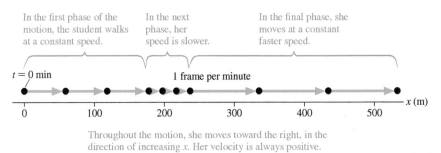

In the first phase of the motion, the student walks at a constant speed.

In the next phase, her speed is slower.

In the final phase, she moves at a constant faster speed.

$t = 0$ min   1 frame per minute

x (m)

Throughout the motion, she moves toward the right, in the direction of increasing x. Her velocity is always positive.

Her motion has three different phases. Similarly, the position-versus-time graph redrawn in FIGURE 2.9a has three clearly defined segments with three different slopes. We can see that there's a relationship between her speed and the slope of the graph: **A faster speed corresponds to a steeper slope.**

The correspondence is actually deeper than this. Let's look at the slope of the third segment of the position-versus-time graph, as shown in FIGURE 2.9b. The slope of a graph is defined as the ratio of the "rise," the vertical change, to the "run," the horizontal change. For the segment of the graph shown, the slope is

$$\text{slope of graph} = \frac{\text{rise}}{\text{run}} = \frac{\Delta x}{\Delta t}$$

This ratio has a physical meaning—it's the velocity, exactly as we defined it in Equation 2.1. We've shown this correspondence for one particular graph, but it is a general principle: **The slope of an object's position-versus-time graph is the object's velocity at that point in the motion.** This principle also holds for negative slopes, which correspond to negative velocities. We can associate the slope of a position-versus-time graph, a *geometrical* quantity, with velocity, a *physical* quantity.

FIGURE 2.9 Revisiting the graph of the motion of the student walking to school.

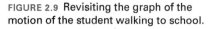

(a)

When the student slows down, the slope of the graph decreases.

When she speeds up, the slope increases.

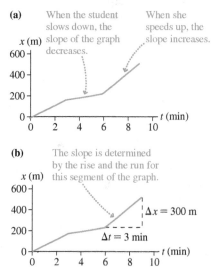

(b)

The slope is determined by the rise and the run for this segment of the graph.

$\Delta x = 300$ m

$\Delta t = 3$ min

| TACTICS BOX 2.1 | Interpreting position-versus-time graphs | (MP) |

Information about motion can be obtained from position-versus-time graphs as follows:

❶ Determine an object's *position* at time t by reading the graph at that instant of time.

❷ Determine the object's *velocity* at time t by finding the slope of the position graph at that point. Steeper slopes correspond to faster speeds.

❸ Determine the *direction of motion* by noting the sign of the slope. Positive slopes correspond to positive velocities and, hence, to motion to the right (or up). Negative slopes correspond to negative velocities and, hence, to motion to the left (or down).

Exercises 2,3 ✐

NOTE ▶ The slope is a ratio of intervals, $\Delta x/\Delta t$, not a ratio of coordinates; that is, the slope is *not* simply x/t. ◀

FIGURE 2.10 Deducing the velocity-versus-time graph from the position-versus-time graph.

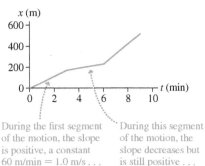

During the first segment of the motion, the slope is positive, a constant 60 m/min = 1.0 m/s . . .

During this segment of the motion, the slope decreases but is still positive . . .

. . . so the velocity is positive, a constant 1.0 m/s.

. . . so the velocity is positive, but with a smaller magnitude.

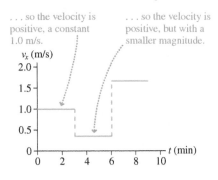

NOTE ▶ We are distinguishing between the actual slope and the *physically meaningful* slope. If you were to use a ruler to measure the rise and the run of the graph, you could compute the actual slope of the line as drawn on the page. That is not the slope we are referring to when we equate the velocity with the slope of the line. Instead, we find the *physically meaningful* slope by measuring the rise and run using the scales along the axes. The "rise" Δx is some number of meters; the "run" Δt is some number of seconds. The physically meaningful rise and run include units, and the ratio of these units gives the units of the slope. ◀

We can now use the approach of Tactics Box 2.1 to analyze the student's position-versus-time graph that we saw in Figure 2.4. We can determine her velocity during the first phase of her motion by measuring the slope of the line:

$$v_x = \text{slope} = \frac{\Delta x}{\Delta t} = \frac{180 \text{ m}}{3 \text{ min}} = 60 \frac{\text{m}}{\text{min}} \times \frac{1 \text{ min}}{60 \text{ s}} = 1.0 \text{ m/s}$$

In completing this calculation, we've converted to more usual units for speed, m/s. During this phase of the motion, her velocity is constant, so a graph of velocity versus time appears as a horizontal line at 1.0 m/s, as shown in FIGURE 2.10. We can do similar calculations to show that her velocity during the second phase of her motion is +0.33 m/s, and then increases to +1.7 m/s during the final phase. We combine this information to create the **velocity-versus-time graph** shown in Figure 2.10.

An inspection of the velocity-versus-time graph shows that it matches our understanding of the student's motion: There are three phases of the motion, each with constant speed. In each phase, the velocity is positive because she is always moving to the right. The second phase is slow (low velocity) and the third phase fast (high velocity). All of this can be clearly seen on the velocity-versus-time graph, which is yet another way to represent her motion.

NOTE ▶ The velocity-versus-time graph in Figure 2.10 includes vertical segments in which the velocity changes instantaneously. Such rapid changes are an idealization; it actually takes a small amount of time to change velocity. ◀

EXAMPLE 2.2   **Analyzing a car's position graph**

FIGURE 2.11 gives the position-versus-time graph of a car.
a. Draw the car's velocity-versus-time graph.
b. Describe the car's motion in words.

FIGURE 2.11 The position-versus-time graph of a car.

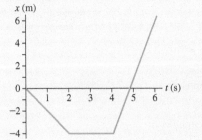

PREPARE Figure 2.11 is a graphical representation of the motion. The car's position-versus-time graph is a sequence of three straight lines. Each of these straight lines represents uniform motion at a constant velocity. We can determine the car's velocity during each interval of time by measuring the slope of the line.

SOLVE

a. From $t = 0$ s to $t = 2$ s ($\Delta t = 2$ s) the car's displacement is $\Delta x = -4 \text{ m} - 0 \text{ m} = -4$ m. The velocity during this interval is

$$v_x = \frac{\Delta x}{\Delta t} = \frac{-4 \text{ m}}{2 \text{ s}} = -2 \text{ m/s}$$

The car's position does not change from $t = 2$ s to $t = 4$ s ($\Delta x = 0$ m), so $v_x = 0$ m/s. Finally, the displacement

between $t = 4$ s and $t = 6$ s ($\Delta t = 2$ s) is $\Delta x = 10$ m. Thus the velocity during this interval is

$$v_x = \frac{10 \text{ m}}{2 \text{ s}} = 5 \text{ m/s}$$

These velocities are represented graphically in FIGURE 2.12.

FIGURE 2.12 The velocity-versus-time graph for the car.

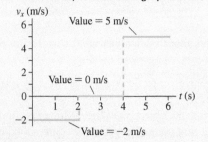

b. The velocity-versus-time graph of Figure 2.12 shows the motion in a way that we can describe in a straightforward manner: The car backs up for 2 s at 2 m/s, sits at rest for 2 s, then drives forward at 5 m/s for 2 s.

ASSESS Notice that the velocity graph and the position graph look completely different. They should! The value of the velocity graph at any instant of time equals the *slope* of the position graph. Since the position graph is made up of segments of constant slope, the velocity graph should be made up of segments of constant *value*, as it is. This gives us confidence that the graph we have drawn is correct.

From Velocity to Position

We've now seen how to move between different representations of uniform motion. There's one last issue to address: If you have a graph of velocity versus time, how can you determine the position graph?

Suppose you leave a lecture hall and begin walking toward your next class, which is down the hall to the west. You then realize that you left your textbook (which you always bring to class with you!) at your seat. You turn around and run back to the lecture hall to retrieve it. A velocity-versus-time graph for this motion appears as the top graph in **FIGURE 2.13**. There are two clear phases to the motion: walking away from class (velocity +1.0 m/s) and running back (velocity −3.0 m/s.) How can we deduce your position-versus-time graph?

As before, we can analyze the graph segment by segment. This process is shown in Figure 2.13, in which the upper velocity-versus-time graph is used to deduce the lower position-versus-time graph. For each of the two segments of the motion, the sign of the velocity tells us whether the slope of the graph is positive or negative. The magnitude of the velocity tells how steep the slope is. The final result makes sense: It shows 15 seconds of slowly increasing position (walking away) and then 5 seconds of rapidly decreasing position (running back.) And you end up back where you started.

There's one important detail that we didn't talk about in the preceding paragraph: How did we know that the position graph started at $x = 0$ m? The velocity graph tells us the *slope* of the position graph, but it doesn't tell us where the position graph should start. Although you're free to select any point you choose as the origin of the coordinate system, here it seems reasonable to set $x = 0$ m at your starting point in the lecture hall; as you walk away, your position increases.

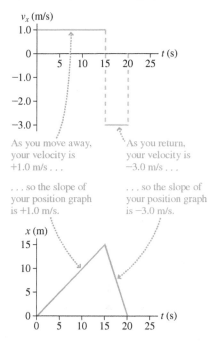

FIGURE 2.13 Deducing a position graph from a velocity-versus-time graph.

As you move away, your velocity is +1.0 m/s . . .

. . . so the slope of your position graph is +1.0 m/s.

As you return, your velocity is −3.0 m/s . . .

. . . so the slope of your position graph is −3.0 m/s.

STOP TO THINK 2.1   Which position-versus-time graph best describes the motion diagram at left?

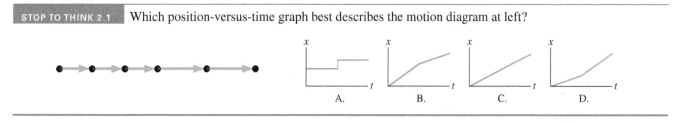

A.   B.   C.   D.

2.2 Uniform Motion

If you drive your car on a straight road at a perfectly steady 60 miles per hour (mph), you will cover 60 mi during the first hour, another 60 mi during the second hour, yet another 60 mi during the third hour, and so on. This is an example of what we call *uniform motion*. **Straight-line motion in which equal displacements occur during any successive equal-time intervals is called uniform motion or constant-velocity motion.**

> **NOTE** ▶ The qualifier "any" is important. If during each hour you drive 120 mph for 30 min and stop for 30 min, you will cover 60 mi during each successive 1 hour interval. But you will *not* have equal displacements during successive 30 min intervals, so this motion is not uniform. ◀

FIGURE 2.14 shows a motion diagram and a graph for an object in uniform motion. Notice that the position-versus-time graph for uniform motion is a straight line. This follows from the requirement that all values of Δx corresponding to the same value of Δt be equal. In fact, an alternative definition of uniform motion is: **An object's motion is uniform if and only if its position-versus-time graph is a straight line.**

Equations of Uniform Motion

An object is in uniform motion along the *x*-axis with the linear position-versus-time graph shown in **FIGURE 2.15** on the next page. Recall from Chapter 1 that we denote the object's initial position as x_i at time t_i. The term "initial" refers to the starting point of

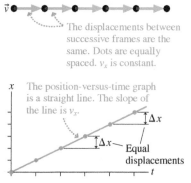

FIGURE 2.14 Motion diagram and position-versus-time graph for uniform motion.

Uniform motion

The displacements between successive frames are the same. Dots are equally spaced. v_x is constant.

The position-versus-time graph is a straight line. The slope of the line is v_x.

Equal displacements

FIGURE 2.15 Position-versus-time graph for an object in uniform motion.

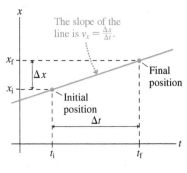

our analysis or the starting point in a problem. The object may or may not have been in motion prior to t_i. We use the term "final" for the ending point of our analysis or the ending point of a problem, and denote the object's final position x_f at the time t_f. As we've seen, the object's velocity v_x along the x-axis can be determined by finding the slope of the graph:

$$v_x = \frac{\text{rise}}{\text{run}} = \frac{\Delta x}{\Delta t} = \frac{x_f - x_i}{t_f - t_i} \qquad (2.3)$$

Equation 2.3 can be rearranged to give

$$x_f = x_i + v_x \, \Delta t \qquad (2.4)$$

Position equation for an object in uniform motion (v_x is constant)

where $\Delta t = t_f - t_i$ is the interval of time in which the object moves from position x_i to position x_f. Equation 2.4 applies to any time interval Δt during which the velocity is constant. We can also write this in terms of the object's displacement, $\Delta x = x_f - x_i$:

$$\Delta x = v_x \, \Delta t \qquad (2.5)$$

The velocity of an object in uniform motion tells us the amount by which its position changes during each second. An object with a velocity of 20 m/s *changes* its position by 20 m during every second of motion: by 20 m during the first second of its motion, by another 20 m during the next second, and so on. We say that position is changing at the *rate* of 20 m/s. If the object starts at $x_i = 10$ m, it will be at $x = 30$ m after 1 s of motion and at $x = 50$ m after 2 s of motion. Thinking of velocity like this will help you develop an intuitive understanding of the connection between velocity and position.

Physics may seem densely populated with equations, but most equations follow a few basic forms. The mathematical form of Equation 2.5 is a type that we will see again: The displacement Δx is *proportional* to the time interval Δt.

Mathematical forms These three figures show graphs of a mathematical equation, the kinetic energy of a moving object versus its speed, and the potential energy of a spring versus the displacement of the end of the spring. All three graphs have the same overall appearance. The three expressions differ in their variables, but all three equations have the same **mathematical form**. There are only a handful of different mathematical forms that we'll use in this text. As we meet each form for the first time, we will give an overview. When you see it again, we'll insert an icon that refers back to the overview so that you can remind yourself of the key details.

Proportional relationships

We say that y is **proportional** to x if they are related by an equation of this form:

$$y = Cx$$

y is proportional to x

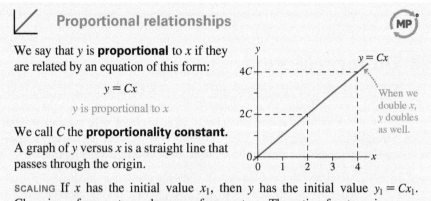

We call C the **proportionality constant.** A graph of y versus x is a straight line that passes through the origin.

SCALING If x has the initial value x_1, then y has the initial value $y_1 = Cx_1$. Changing x from x_1 to x_2 changes y from y_1 to y_2 The ratio of y_2 to y_1 is

$$\frac{y_2}{y_1} = \frac{Cx_2}{Cx_1} = \frac{x_2}{x_1}$$

The ratio of y_2 to y_1 is exactly the same as the ratio of x_2 to x_1. If y is proportional to x, which is often written $y \propto x$, then x and y change by the same factor:

■ If you double x, you double y.
■ If you decrease x by a factor of 3, you decrease y by a factor of 3.

If two variables have a proportional relationship, we can draw important conclusions from ratios without knowing the value of the proportionality constant C. We can often solve problems in a very straightforward manner by looking at such ratios. This is an important skill called *ratio reasoning*.

EXAMPLE 2.3 **If a train leaves Cleveland at 2:00 ...**

A train is moving due west at a constant speed. A passenger notes that it takes 10 minutes to travel 12 km. How long will it take the train to travel 60 km?

PREPARE For an object in uniform motion, Equation 2.5 shows that the distance traveled Δx is proportional to the time interval Δt, so this is a good problem to solve using ratio reasoning.

SOLVE We are comparing two cases: the time to travel 12 km and the time to travel 60 km. Because Δx is proportional to Δt, the ratio of the times will be equal to the ratio of the distances. The ratio of the distances is

$$\frac{\Delta x_2}{\Delta x_1} = \frac{60 \text{ km}}{12 \text{ km}} = 5$$

This is equal to the ratio of the times:

$$\frac{\Delta t_2}{\Delta t_1} = 5$$

$\Delta t_2 = $ time to travel 60 km $= 5\Delta t_1 = 5 \times (10 \text{ min})$
$= 50 \text{ min}$

It takes 10 minutes to travel 12 km, so it will take 50 minutes—5 times as long—to travel 60 km.

ASSESS For an object in steady motion, it makes sense that 5 times the distance requires 5 times the time. We can see that using ratio reasoning is a straightforward way to solve this problem. We don't need to know the proportionality constant (in this case, the velocity); we just used ratios of distances and times.

From Velocity to Position, One More Time

We've seen that we can deduce an object's velocity by measuring the slope of its position graph. Conversely, if we have a velocity graph, we can say something about position—not by looking at the slope of the graph, but by looking at what we call the *area under the graph*. Let's look at an example.

Suppose a car is in uniform motion at 12 m/s. How far does it travel—that is, what is its displacement—during the time interval between $t = 1.0$ s and $t = 3.0$ s?

Equation 2.5, $\Delta x = v_x \Delta t$, describes the displacement mathematically; for a graphical interpretation, consider the graph of velocity versus time in FIGURE 2.16. In the figure, we've shaded a rectangle whose height is the velocity v_x and whose base is the time interval Δt. The area of this rectangle is $v_x \Delta t$. Looking at Equation 2.5, we see that the quantity is also equal to the displacement of the car. The area of this rectangle is the area between the axis and the line representing the velocity; we call it the "area under the graph." We see that **the displacement Δx is equal to the area under the velocity graph during interval Δt.**

Whether we use Equation 2.5 or the area under the graph to compute the displacement, we get the same result:

$$\Delta x = v_x \Delta t = (12 \text{ m/s})(2.0 \text{ s}) = 24 \text{ m}$$

Although we've shown that the displacement is the area under the graph only for uniform motion, where the velocity is constant, we'll soon see that this result applies to any one-dimensional motion.

> **NOTE** ▸ Wait a minute! The displacement $\Delta x = x_f - x_i$ is a length. How can a length equal an area? Recall that earlier, when we found that the velocity is the slope of the position graph, we made a distinction between the *actual* slope and the *physically meaningful* slope? The same distinction applies here. The velocity graph does indeed bound a certain area on the page. That is the actual area, but it is *not* the area to which we are referring. Once again, we need to measure the quantities we are using, v_x and Δt, by referring to the scales on the axes. Δt is some number of seconds, while v_x is some number of meters per second. When these are multiplied together, the *physically meaningful* area has units of meters, appropriate for a displacement. ◂

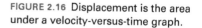

FIGURE 2.16 Displacement is the area under a velocity-versus-time graph.

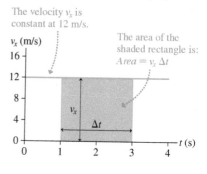

STOP TO THINK 2.2 Four objects move with the velocity-versus-time graphs shown. Which object has the largest displacement between $t = 0$ s and $t = 2$ s?

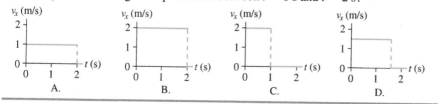

2.3 Instantaneous Velocity

Drag racers move with rapidly changing velocity.

The objects we've studied so far have moved with a constant, unchanging velocity or, like the car in Example 2.1, changed abruptly from one constant velocity to another. This is not very realistic. Real moving objects speed up and slow down, *changing* their velocity. In a race, a drag racer begins at rest but, 1 second later, is moving at over 25 miles per hour!

For one-dimensional motion, an object changing its velocity is either speeding up or slowing down. When you drive your car, as you speed up or slow down—changing your velocity—a glance at your speedometer tells you how fast you're going *at that instant*. An object's velocity—a speed *and* a direction—at a specific *instant* of time t is called the object's **instantaneous velocity.**

But what does it mean to have a velocity "at an instant"? An instantaneous velocity of magnitude 60 mph means that the rate at which your car's position is changing—at that exact instant—is such that it would travel a distance of 60 miles in 1 hour *if* it continued at that rate without change. If *just for an instant* your car matches the velocity of another car driving at a steady 60 mph, then your instantaneous velocity is 60 mph. **From now on, the word "velocity" will always mean instantaneous velocity.**

FIGURE 2.17 Position-versus-time graph for a drag racer.

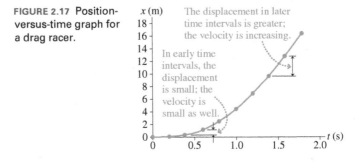

For uniform motion, an object's position-versus-time graph is a straight line and the object's velocity is the slope of that line. In contrast, FIGURE 2.17 shows that the position-versus-time graph for a drag racer is a *curved* line. The displacement Δx during equal intervals of time gets greater as the car speeds up. Even so, we can use the slope of the position graph to measure the car's velocity. We can say that

instantaneous velocity v_x at time t = slope of position graph at time t (2.6)

But how do we determine the slope of a curved line at a particular point?

Finding the instantaneous velocity

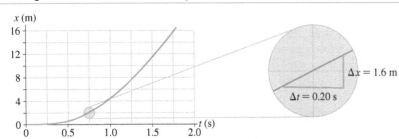

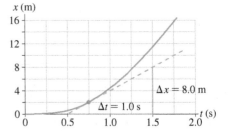

If the velocity changes, the position graph is a curved line. But we can compute a slope at a point by considering a small segment of the graph. Let's look at the motion in a very small time interval right around $t = 0.75$ s. This is highlighted with a circle, and we show a closeup in the next graph at the right.

In this magnified segment of the position graph, the curve isn't apparent. It appears to be a line segment. We can find the slope by calculating the rise over the run, just as before:

$$v_x = (1.6 \text{ m})/(0.20 \text{ s}) = 8.0 \text{ m/s}$$

This is the slope at $t = 0.75$ s and thus the velocity at this instant of time.

Graphically, the slope of the curve at a point is the same as the slope of a straight line drawn *tangent* to the curve at that point. Calculating rise over run for the tangent line, we get

$$v_x = (8.0 \text{ m})/(1.0 \text{ s}) = 8.0 \text{ m/s}$$

This is the same value we obtained from the close-up view. **The slope of the tangent line is the instantaneous velocity at that instant of time.**

CONCEPTUAL EXAMPLE 2.4 **Analyzing a hockey player's position graph**

A hockey player moves in a straight line along the length of the ice in a game. We measure position from the center of the rink. **FIGURE 2.18** shows a position-versus-time graph for his motion.

a. Sketch an approximate velocity-versus-time graph.
b. At which point or points is the player moving the fastest?
c. Is the player ever at rest? If so, at which point or points?

FIGURE 2.18 The position-versus-time graph for a hockey player.

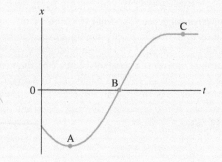

FIGURE 2.19 Finding a velocity graph from a position graph.

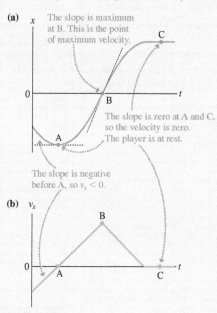

REASON a. The velocity at a particular instant of time is the slope of a tangent line to the position-versus-time graph at that time. We can move point-by-point along the position-versus-time graph, noting the slope of the tangent at each point to find the velocity at that point.

Initially, to the left of point A, the slope is negative and thus the velocity is negative (i.e., the player is moving to the left). But the slope decreases as the curve flattens out, and by the time the graph gets to point A, the slope is zero. The slope then increases to a maximum value at point B, decreases back to zero a little before point C, and remains at zero thereafter. This reasoning process is outlined in **FIGURE 2.19a**, and **FIGURE 2.19b** shows the approximate velocity-versus-time graph that results.

The other questions were answered during the construction of the graph:

b. The player moves the fastest at point B where the slope of the position graph is the steepest.

c. If the player is at rest, $v_x = 0$. Graphically, this occurs at points where the tangent line to the position-versus-time graph is horizontal and thus has zero slope. Figure 2.19 shows that the slope is zero at points A and C. At point A, the velocity is only instantaneously zero—the player is reversing direction, changing from moving to the left to moving to the right. At point C, he has stopped moving and stays at rest.

ASSESS The best way to check our work is to look at different segments of the motion and see if the velocity and position graphs match. Until point A, x is decreasing. The player is moving to the left, so the velocity should be negative, which our graph shows. Between points A and C, x is increasing, so the velocity should be positive, which is also a feature of our graph. The steepest slope is at point B, so this should be the high point of our velocity graph, as it is.

FIGURE 2.20 shows a typical velocity-versus-time graph for a lion speeding up to pursue prey. Even though the speed varies, we can still use the graph to determine how far the lion moves during the time interval t_i to t_f. For uniform motion we showed that the displacement Δx is the area under the velocity-versus-time graph during the time interval. But there was nothing special about the type of motion: We can generalize this idea to the case of an object whose velocity varies. If we draw a velocity graph for the motion, the object's displacement is given by

$$x_f - x_i = \text{area under the velocity graph between } t_i \text{ and } t_f \qquad (2.7)$$

The area under the graph in Figure 2.20 tells us how far the lion ran during this segment of the chase.

In many cases, as in the next example, the area under the graph will be a simple shape whose area we can easily compute. If the shape is complex, however, we can approximate the area using a number of simpler shapes that closely match it.

FIGURE 2.20 Velocity-versus-time graph for a lion pursuing prey.

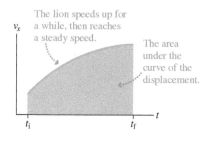

EXAMPLE 2.5   **The displacement during a rapid start**

FIGURE 2.21 shows the velocity-versus-time graph of a car pulling away from a stop. How far does the car move during the first 3.0 s?

PREPARE Figure 2.21 is a graphical representation of the motion. The question How far? indicates that we need to find a displacement Δx rather than a position x. According to Equation 2.7, the car's displacement $\Delta x = x_f - x_i$ between $t = 0$ s and $t = 3$ s is the area under the curve from $t = 0$ s to $t = 3$ s.

FIGURE 2.21 Velocity-versus-time graph for the car of Example 2.5.

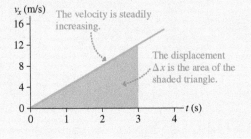

SOLVE The curve in this case is an angled line, so the area is that of a triangle:

$$\Delta x = \text{area of triangle between } t = 0 \text{ s and } t = 3 \text{ s}$$
$$= \tfrac{1}{2} \times \text{base} \times \text{height} = \tfrac{1}{2} \times 3 \text{ s} \times 12 \text{ m/s} = 18 \text{ m}$$

The car moves 18 m during the first 3 seconds as its velocity changes from 0 to 12 m/s.

ASSESS The physically meaningful area is a product of s and m/s, so Δx has the proper units of m. Let's check the numbers to see if they make physical sense. The final velocity, 12 m/s, is about 25 mph. Pulling away from a stop, you might expect to reach this speed in about 3 s—at least if you have a reasonably sporty vehicle! If the car had moved at a constant 12 m/s (the final velocity) during these 3 s, the distance would be 36 m. The actual distance traveled during the 3 s is 18 m—half of 36 m. This makes sense, as the velocity was 0 m/s at the start of the problem and increased steadily to 12 m/s.

STOP TO THINK 2.3   Which velocity-versus-time graph goes with the position-versus-time graph on the left?

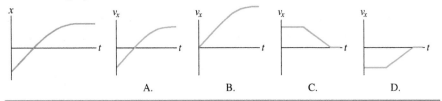

2.4 Acceleration

The goal of this chapter is to describe motion. We've seen that velocity describes the rate at which an object changes position. We need one more motion concept to complete the description, one that will describe an object whose velocity is changing.

As an example, let's look at a frequently quoted measurement of car performance, the time it takes the car to go from 0 to 60 mph. Table 2.2 shows this time for two different cars, a sporty Corvette and a compact Sonic with a much more modest engine.

Let's look at motion diagrams for the Corvette and the Sonic in FIGURE 2.22. We can see two important facts about the motion. First, the lengths of the velocity vectors are increasing, showing that the speeds are increasing. Second, the velocity vectors for the Corvette are increasing in length more rapidly than those of the Sonic. The quantity we seek is one that measures how rapidly an object's velocity vectors change in length.

TABLE 2.2 Performance data for vehicles

Vehicle	Time to go from 0 to 60 mph
2011 Chevy Corvette	3.6 s
2012 Chevy Sonic	9.0 s

FIGURE 2.22 Motion diagrams for the Corvette and Sonic.

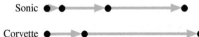

When we wanted to measure changes in position, the ratio $\Delta x/\Delta t$ was useful. This ratio, which we defined as the velocity, is the *rate of change of position*. Similarly, we can measure how rapidly an object's velocity changes with the ratio $\Delta v_x/\Delta t$. Given our experience with velocity, we can say a couple of things about this new ratio:

- The ratio $\Delta v_x/\Delta t$ is the *rate of change of velocity*.
- The ratio $\Delta v_x/\Delta t$ is the *slope of a velocity-versus-time graph*.

We will define this ratio as the **acceleration,** for which we use the symbol a_x:

$$a_x = \frac{\Delta v_x}{\Delta t} \qquad (2.8)$$

Definition of acceleration as the rate of change of velocity

Similarly, $a_y = \Delta v_y/\Delta t$ for vertical motion.

As an example, let's calculate the acceleration for the Corvette and the Sonic. For both, the initial velocity $(v_x)_i$ is zero and the final velocity $(v_x)_f$ is 60 mph. Thus the *change* in velocity is $\Delta v_x = 60$ mph. In m/s, our SI unit of velocity, $\Delta v_x = 27$ m/s.

Now we can use Equation 2.8 to compute acceleration. Let's start with the Corvette, which speeds up to 27 m/s in $\Delta t = 3.6$ s:

$$a_{\text{Corvette}x} = \frac{\Delta v_x}{\Delta t} = \frac{27 \text{ m/s}}{3.6 \text{ s}} = 7.5 \frac{\text{m/s}}{\text{s}}$$

Here's the meaning of this final figure: Every second, the Corvette's velocity changes by 7.5 m/s. In the first second of motion, the Corvette's velocity increases by 7.5 m/s; in the next second, it increases by another 7.5 m/s, and so on. After 1 second, the velocity is 7.5 m/s; after 2 seconds, it is 15 m/s. We thus interpret the units as 7.5 meters per second, per second—7.5 (m/s)/s.

The Sonic's acceleration is

$$a_{\text{Sonic}x} = \frac{\Delta v_x}{\Delta t} = \frac{27 \text{ m/s}}{9.0 \text{ s}} = 3.0 \frac{\text{m/s}}{\text{s}}$$

In each second, the Sonic changes its speed by 3.0 m/s. This is only 2/5 the acceleration of the Corvette! The reason the Corvette is capable of greater acceleration has to do with what *causes* the motion. We will explore the reasons for acceleration in Chapter 4. For now, we will simply note that the Corvette is capable of much greater acceleration, something you would have suspected.

NOTE ▶ It is customary to abbreviate the acceleration units (m/s)/s as m/s^2, which we say as "meters per second squared." For example, the Sonic has an acceleration of 3.0 m/s^2. When you use this notation, keep in mind its *meaning* as "(meters per second) per second." ◄

Cushion kinematics When a car hits an obstacle head-on, the damage to the car and its occupants can be reduced by making the acceleration as small as possible. As we can see from Equation 2.8, acceleration can be reduced by making the *time* for a change in velocity as long as possible. This is the purpose of the yellow crash cushion barrels you may have seen in work zones on highways—to lengthen the time of a collision with a barrier.

EXAMPLE 2.6   Animal acceleration BIO

Lions, like most predators, are capable of very rapid starts. From rest, a lion can sustain an acceleration of 9.5 m/s^2 for up to one second. How much time does it take a lion to go from rest to a typical recreational runner's top speed of 10 mph?

PREPARE We can start by converting to SI units. The speed the lion must reach is

$$v_f = 10 \text{ mph} \times \frac{0.45 \text{ m/s}}{1.0 \text{ mph}} = 4.5 \text{ m/s}$$

The lion can accelerate at 9.5 m/s^2, changing its speed by 9.5 m/s per second, for only 1.0 s—long enough to reach 9.5 m/s. It will take the lion less than 1.0 s to reach 4.5 m/s, so we can use $a_x = 9.5 \text{ m/s}^2$ in our solution.

SOLVE We know the acceleration and the desired change in velocity, so we can rearrange Equation 2.8 to find the time:

$$\Delta t = \frac{\Delta v_x}{a_x} = \frac{4.5 \text{ m/s}}{9.5 \text{ m/s}^2} = 0.47 \text{ s}$$

ASSESS The lion changes its speed by 9.5 meters per second in one second. So it's reasonable (if a bit intimidating) that it will reach 4.5 m/s in just under half a second.

Representing Acceleration

Let's use the values we have computed for acceleration to make a table of velocities for the Corvette and the Sonic we considered earlier. Table 2.3 uses the idea that the Sonic's velocity increases by 3.0 m/s every second while the Corvette's velocity increases by 7.5 m/s every second. The data in Table 2.3 are the basis for the velocity-versus-time graphs in FIGURE 2.23. As you can see, an object undergoing constant acceleration has a straight-line velocity graph.

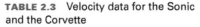

TABLE 2.3   Velocity data for the Sonic and the Corvette

Time (s)	Velocity of Sonic (m/s)	Velocity of Corvette (m/s)
0	0	0
1	3.0	7.5
2	6.0	15.0
3	9.0	22.5
4	12.0	30.0

FIGURE 2.23   Velocity-versus-time graphs for the two cars.

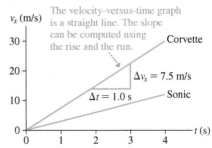

The slope of either of these lines—the rise over run—is $\Delta v_x / \Delta t$. Comparing this with Equation 2.8, we see that the equation for the slope is the same as that for the acceleration. That is, **an object's acceleration is the slope of its velocity-versus-time graph:**

$$\text{acceleration } a_x \text{ at time } t = \text{ slope of velocity graph at time } t \qquad (2.9)$$

The Sonic has a smaller acceleration, so its velocity graph has a smaller slope.

CONCEPTUAL EXAMPLE 2.7   **Analyzing a car's velocity graph**

FIGURE 2.24a is a graph of velocity versus time for a car. Sketch a graph of the car's acceleration versus time.

REASON The graph can be divided into three sections:

■ An initial segment, in which the velocity increases at a steady rate

■ A middle segment, in which the velocity is constant

■ A final segment, in which the velocity decreases at a steady rate

In each section, the acceleration is the slope of the velocity-versus-time graph. Thus the initial segment has constant, positive acceleration, the middle segment has zero acceleration, and the final segment has constant, *negative* acceleration. The acceleration graph appears in FIGURE 2.24b.

ASSESS This process is analogous to finding a velocity graph from the slope of a position graph. The middle segment having zero acceleration does *not* mean that the velocity is zero. The velocity is constant, which means it is *not changing* and thus the car is not accelerating. The car does accelerate during the initial and final segments. The magnitude of the acceleration is a measure of how quickly the velocity is changing. How about the sign? This is an issue we will address in the next section.

FIGURE 2.24   Finding an acceleration graph from a velocity graph.

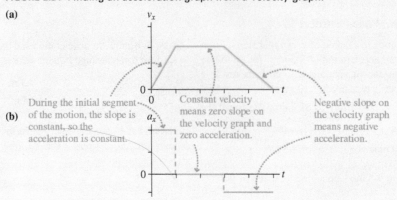

The Sign of the Acceleration

It's a natural tendency to think that a positive value of a_x or a_y describes an object that is speeding up while a negative value describes an object that is slowing down. Unfortunately, this simple interpretation *does not work.*

Because an object can move right or left (or, equivalently, up or down) while either speeding up or slowing down, there are four situations to consider. FIGURE 2.25 shows a motion diagram and a velocity graph for each of these situations. Acceleration, like velocity, is a vector quantity. The acceleration vector points in the same direction as the velocity vector for an object that is speeding up and opposite to the velocity vector for an object that is slowing down.

As we've seen, an object's acceleration is the slope of its velocity graph, so a positive slope implies a positive acceleration and a negative slope implies a negative acceleration. We've adopted the convention for horizontal motion that motion to the right corresponds to increasing x and a positive velocity v_x. With this convention, an object that speeds up as it moves to the right has a positive acceleration, but an object that speeds up as it moves to the left (negative v_x) has a negative acceleration.

FIGURE 2.25 Determining the sign of the acceleration.

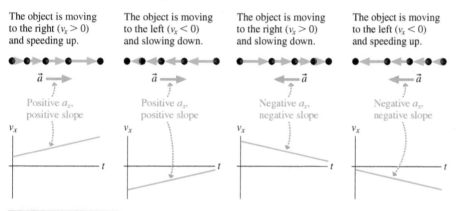

STOP TO THINK 2.4 An elevator is moving downward. It is slowing down as it approaches the ground floor. Adapt the information in Figure 2.25 to determine which of the following velocity graphs best represents the motion of the elevator.

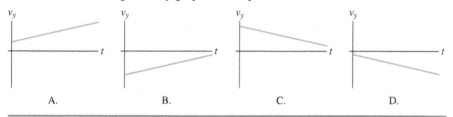

Ultimately, whether or not an object that is slowing down has a negative acceleration depends on whether the object is moving to the right or to the left, or whether it is moving up or down. This is admittedly a bit more complex than thinking that negative acceleration always means slowing down, but our definition of acceleration as the slope of the velocity graph forces us to pay careful attention to the sign of the acceleration.

STOP TO THINK 2.5 A particle moves with the velocity-versus-time graph shown here. At which labeled point is the magnitude of the acceleration the greatest?

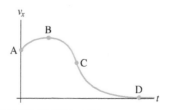

FIGURE 2.26 The red dots show the positions of the top of the Saturn V rocket at equally spaced intervals of time during liftoff.

2.5 Motion with Constant Acceleration

For uniform motion—motion with constant velocity—we found in Equation 2.3 a simple relationship between position and time. It's no surprise that there are also simple relationships that connect the various kinematic variables in constant-acceleration motion. We will start with a concrete example, the launch of a Saturn V rocket like the one that carried the Apollo astronauts to the moon in the 1960s and 1970s. FIGURE 2.26 shows one frame from a video of a rocket lifting off the launch pad. The red dots show the positions of the top of the rocket at equally spaced intervals of time in earlier frames of the video. This is a motion diagram for the rocket, and we can see that the velocity is increasing. The graph of velocity versus time in FIGURE 2.27 shows that the velocity is increasing at a fairly constant rate. We can approximate the rocket's motion as constant acceleration.

We can use the slope of the graph in Figure 2.27 to determine the acceleration of the rocket:

$$a_y = \frac{\Delta v_y}{\Delta t} = \frac{27 \text{ m/s}}{1.5 \text{ s}} = 18 \text{ m/s}^2$$

This acceleration is more than double the acceleration of the Corvette, and it goes on for quite a long time—the first phase of the launch lasts over 2 minutes! How fast is the rocket moving at the end of this acceleration, and how far has it traveled? To answer questions like these, we first need to work out some basic kinematic formulas for motion with constant acceleration.

FIGURE 2.27 A graph of the rocket's velocity versus time.

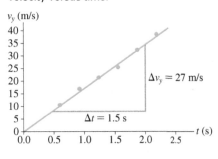

Constant-Acceleration Equations

Consider an object whose acceleration a_x remains constant during the time interval $\Delta t = t_f - t_i$. At the beginning of this interval, the object has initial velocity $(v_x)_i$ and initial position x_i. Note that t_i is often zero, but it need not be. FIGURE 2.28a shows the acceleration-versus-time graph. It is a horizontal line between t_i and t_f, indicating a *constant* acceleration.

The object's velocity is changing because the object is accelerating. We can use the acceleration to find $(v_x)_f$ at a later time t_f. We defined acceleration as

$$a_x = \frac{\Delta v_x}{\Delta t} = \frac{(v_x)_f - (v_x)_i}{\Delta t} \qquad (2.10)$$

FIGURE 2.28 Acceleration and velocity graphs for motion with constant acceleration.

(a) Acceleration

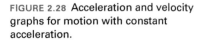

which is rearranged to give

$$(v_x)_f = (v_x)_i + a_x \Delta t \qquad (2.11)$$

Velocity equation for an object with constant acceleration

NOTE ▶ We have expressed this equation for motion along the *x*-axis, but it is a general result that will apply to any axis. ◄

The velocity-versus-time graph for this constant-acceleration motion, shown in FIGURE 2.28b, is a straight line with value $(v_x)_i$ at time t_i and with slope a_x.

We would also like to know the object's position x_f at time t_f. As you learned earlier, the displacement Δx during a time interval Δt is the area under the velocity-versus-time graph. The shaded area in Figure 2.28b can be subdivided into a rectangle of area $(v_x)_i \Delta t$ and a triangle of area $\frac{1}{2}(a_x \Delta t)(\Delta t) = \frac{1}{2}a_x(\Delta t)^2$. Adding these gives

(b) Velocity

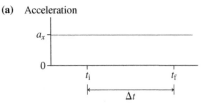

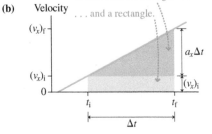

$$x_f = x_i + (v_x)_i \Delta t + \tfrac{1}{2}a_x(\Delta t)^2 \qquad (2.12)$$

Position equation for an object with constant acceleration

where $\Delta t = t_f - t_i$ is the elapsed time. The fact that the time interval Δt appears in the equation as $(\Delta t)^2$ causes the position-versus-time graph for constant-acceleration motion to have a parabolic shape. For the rocket launch of Figure 2.26, a graph of the position of the top of the rocket versus time appears as in FIGURE 2.29.

Equations 2.11 and 2.12 are two of the basic kinematic equations for motion with constant acceleration. They allow us to predict an object's position and velocity at a future instant of time. We need one more equation to complete our set, a direct relationship between displacement and velocity. To derive this relationship, we first use Equation 2.11 to write $\Delta t = ((v_x)_f - (v_x)_i)/a_x$. We can substitute this into Equation 2.12 to obtain

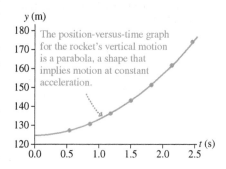

FIGURE 2.29 Position-versus-time graph for the Saturn V rocket launch.

$$(v_x)_f^2 = (v_x)_i^2 + 2a_x \, \Delta x \qquad (2.13)$$

Relating velocity and displacement for constant-acceleration motion

In Equation 2.13 $\Delta x = x_f - x_i$ is the *displacement* (not the distance!). Notice that Equation 2.13 does not require knowing the time interval Δt. This is an important equation in problems where you're not given information about times.

At this point, it's worthwhile to summarize the relationships among kinematic variables that we've seen. This will help you solve problems by gathering together the most important information that you'll use in your solutions. But, more important, gathering this information together allows you to compare graphs, equations, and details from different parts of the chapter in one place. This will help you make important connections. The emphasis is on *synthesis*—hence the title of this box. You'll find other such synthesis boxes in most chapters.

SYNTHESIS 2.1 Describing motion in one dimension

We describe motion in terms of position, velocity and acceleration.

For all motion:

Velocity is the rate of change of position, in m/s. $\qquad v_x = \dfrac{\Delta x}{\Delta t}$

Acceleration is the rate of change of velocity, in m/s². $\qquad a_x = \dfrac{\Delta v_x}{\Delta t}$

For uniform motion:
• acceleration is zero
• velocity is constant
• position changes steadily

The velocity is constant, so the slope of the position graph is constant as well.

The slope is v_x.

Final and initial position (m)

$$x_f = x_i + v_x \Delta t$$

Velocity (m/s)   Time interval (s)

For motion with constant acceleration:
• acceleration is steady; it does not change

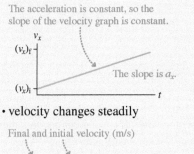

The acceleration is constant, so the slope of the velocity graph is constant.

The slope is a_x.

• velocity changes steadily

Final and initial velocity (m/s)

$$(v_x)_f = (v_x)_i + a_x \Delta t \qquad (1)$$

Acceleration (m/s²)   Time interval (s)

• the position changes as the square of the time interval

The velocity steadily increases, so the slope of the position graph steadily increases.

Final and initial position (m)   Time interval (s)

$$x_f = x_i + (v_x)_i \, \Delta t + \tfrac{1}{2} a_x (\Delta t)^2 \qquad (2)$$

Initial velocity (m/s)   Acceleration (m/s²)

• we can also express the change in velocity in terms of **distance, not time**.

This gives us a third equation, which is useful for many kinematics problems.

$$(v_x)_f^2 = (v_x)_i^2 + 2a_x \Delta x \qquad (3)$$

Acceleration (m/s²)   Change in position (m)

EXAMPLE 2.8   **Coming to a stop**

As you drive in your car at 15 m/s (just a bit under 35 mph), you see a child's ball roll into the street ahead of you. You hit the brakes and stop as quickly as you can. In this case, you come to rest in 1.5 s. How far does your car travel as you brake to a stop?

PREPARE The problem statement gives us a description of motion in words. To help us visualize the situation, FIGURE 2.30 illustrates the key features of the motion with a motion diagram and a velocity graph. The graph is based on the car slowing from 15 m/s to 0 m/s in 1.5 s.

FIGURE 2.30 Motion diagram and velocity graph for a car coming to a stop.

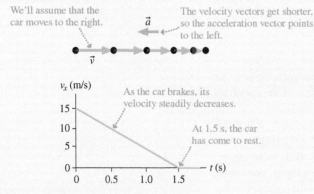

SOLVE We've assumed that your car is moving to the right, so its initial velocity is $(v_x)_i = +15$ m/s. After you come to rest, your final velocity is $(v_x)_f = 0$ m/s. We use the definition of acceleration from Synthesis 2.1:

$$a_x = \frac{\Delta v_x}{\Delta t} = \frac{(v_x)_f - (v_x)_i}{\Delta t} = \frac{0 \text{ m/s} - 15 \text{ m/s}}{1.5 \text{ s}} = -10 \text{ m/s}^2$$

An acceleration of -10 m/s^2 (really -10 m/s per second) means the car slows by 10 m/s every second.

Now that we know the acceleration, we can compute the distance that the car moves as it comes to rest using the second constant acceleration equation in Synthesis 2.1:

$$x_f - x_i = (v_x)_i \Delta t + \tfrac{1}{2}a_x(\Delta t)^2$$
$$= (15 \text{ m/s})(1.5 \text{ s}) + \tfrac{1}{2}(-10 \text{ m/s}^2)(1.5 \text{ s})^2 = 11 \text{ m}$$

ASSESS 11 m is a little over 35 feet. That's a reasonable distance for a quick stop while traveling at about 35 mph. The purpose of the Assess step is not to prove that your solution is correct but to use common sense to recognize answers that are clearly wrong. Had you made a calculation error and ended up with an answer of 1.1 m—less than 4 feet—a moment's reflection should indicate that this couldn't possibly be correct.

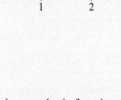

Getting up to speed BIO A bird must have a minimum speed to fly. Generally, the larger the bird, the faster the takeoff speed. Small birds can get moving fast enough to fly with a vigorous jump, but larger birds may need a running start. This swan must accelerate for a long distance in order to achieve the high speed it needs to fly, so it makes a frenzied dash across the frozen surface of a pond. Swans require a long, clear stretch of water or land to become airborne.

As we've noted, for motion at constant acceleration, the position changes as the square of the time interval. If $(v_x)_i = 0$ and $x_i = 0$, the second constant-acceleration equation in Synthesis 2.1 reduces to

$$x_f = \tfrac{1}{2}a_x(\Delta t)^2$$

This is a new mathematical form, one that we will see again and one that we can use as the basis of reasoning to solve problems. This is an example of a *quadratic relationship*.

Quadratic relationships

Two quantities are said to have a **quadratic relationship** if y is proportional to the square of x. We write the mathematical relationship as

$$y = Ax^2$$
y is proportional to x^2

The graph of a quadratic relationship is a parabola.

SCALING If x has the initial value x_1, then y has the initial value $y_1 = A(x_1)^2$. Changing x from x_1 to x_2 changes y from y_1 to y_2. The ratio of y_2 to y_1 is

$$\frac{y_2}{y_1} = \frac{A(x_2)^2}{A(x_1)^2} = \left(\frac{x_2}{x_1}\right)^2$$

The ratio of y_2 to y_1 is the square of the ratio of x_2 to x_1. If y is a quadratic function of x, a change in x by some factor changes y by the square of that factor:

- If you increase x by a factor of 2, you increase y by a factor of $2^2 = 4$.
- If you decrease x by a factor of 3, you decrease y by a factor of $3^2 = 9$.

Generally, we can say that:
Changing x by a factor of c changes y by a factor of c^2.

Exercise 19

EXAMPLE 2.9 Displacement of a drag racer

A drag racer, starting from rest, travels 6.0 m in 1.0 s. Suppose the car continues this acceleration for an additional 4.0 s. How far from the starting line will the car be?

PREPARE We assume that the acceleration is constant, and the initial speed is zero, so the displacement will scale as the square of the time.

SOLVE After 1.0 s, the car has traveled 6.0 m; after another 4.0 s, a total of 5.0 s will have elapsed. The initial elapsed time was 1.0 s, so the elapsed time increases by a factor of 5. The displacement thus increases by a factor of 5^2, or 25. The total displacement is

$$\Delta x = 25(6.0 \text{ m}) = 150 \text{ m}$$

ASSESS This is a big distance in a short time, but drag racing is a fast sport, so our answer makes sense.

STOP TO THINK 2.6 A cyclist is at rest at a traffic light. When the light turns green, he begins accelerating at 1.2 m/s^2. How many seconds after the light turns green does he reach his cruising speed of 6.0 m/s?

A. 1.0 s   B. 2.0 s   C. 3.0 s   D. 4.0 s   E. 5.0 s

2.6 Solving One-Dimensional Motion Problems

The big challenge when solving a physics problem is to translate the words into symbols that can be manipulated, calculated, and graphed. This translation from words to symbols is the heart of problem solving in physics. Ambiguous words and phrases must be clarified, the imprecise must be made precise, and you must arrive at an understanding of exactly what the question is asking.

Problem-Solving Strategy

The first step in solving a seemingly complicated problem is to break it down into a series of smaller steps. In worked examples in the text, we use a problem-solving strategy that consists of three steps: *prepare, solve,* and *assess.* Each of these steps has important elements that you should follow when you solve problems on your own.

PREPARE The Prepare step of a solution is where you identify important elements of the problem and collect information. It's tempting to jump right to the Solve step, but a skilled problem solver will spend the most time on preparation, which includes:

- **Drawing a picture.** This is often the most important part of a problem. The picture lets you model the problem and identify the important elements. As you add information to your picture, the outline of the solution will take shape. For the problems in this chapter, a picture could be a motion diagram or a graph—or perhaps both.
- **Collecting necessary information.** The problem's statement may give you some values of variables. Other information may be implied, or looked up in a table, or estimated or measured.
- **Doing preliminary calculations.** Some calculations, such as unit conversions, are best done in advance.

SOLVE The Solve step of a solution is where you actually do the mathematics or reasoning necessary to arrive at the answer needed. This is the part of the problem-solving strategy that you likely think of as "solving problems." But don't make the mistake of starting here! The Prepare step will help you be certain you understand the problem before you start putting numbers in equations.

ASSESS The Assess step of your solution is very important. Once you have an answer, you should check to see whether it makes sense. Ask yourself:

- **Does my solution answer the question that was asked?** Make sure you have addressed all parts of the question and clearly written down your solutions.
- **Does my answer have the correct units and number of significant figures?**
- **Does the value I computed make physical sense?** In this book all calculations use physically reasonable numbers. If your answer seems unreasonable, go back and check your work.
- **Can I estimate what the answer should be to check my solution?**
- **Does my final solution make sense in the context of the material I am learning?**

Dinner at a distance BIO A chameleon's tongue is a powerful tool for catching prey. Certain species can extend the tongue to a distance of over 1 ft in less than 0.1 s! A study of the kinematics of the motion of the chameleon tongue reveals that the tongue has a period of rapid acceleration followed by a period of constant velocity. This knowledge is a very valuable clue in the analysis of the evolutionary relationships between chameleons and other animals.

The Pictorial Representation

Many physics problems, including one-dimensional motion problems, often have several variables and other pieces of information to keep track of. The best way to tackle such problems is to draw a picture, as we noted when we introduced a general problem-solving strategy. But what kind of picture should you draw?

In this section, we will begin to draw **pictorial representations** as an aid to solving problems. A pictorial representation shows all of the important details that we need to keep track of and will be very important in solving motion problems.

TACTICS BOX 2.2 Drawing a pictorial representation

❶ **Sketch the situation.** Not just any sketch: Show the object at the *beginning* of the motion, at the *end,* and at any point where the character of the motion changes. Very simple drawings are adequate.
❷ **Establish a coordinate system.** Select your axes and origin to match the motion.
❸ **Define symbols.** Use the sketch to define symbols representing quantities such as position, velocity, acceleration, and time. *Every* variable used later in the mathematical solution should be defined on the sketch.

We will generally combine the pictorial representation with a **list of values,** which will include:

■ *Known information.* Make a table of the quantities whose values you can determine from the problem statement or that you can find quickly with simple geometry or unit conversions.
■ *Desired unknowns.* What quantity or quantities will allow you to answer the question?

Exercise 21 ✐

EXAMPLE 2.10 **Drawing a pictorial representation**

Complete a pictorial representation and a list of values for the following problem: A rocket sled accelerates at 50 m/s² for 5 s. What are the total distance traveled and the final velocity?

PREPARE FIGURE 2.31a shows a pictorial representation as drawn by an artist in the style of the figures in this book. This is certainly neater and more artistic than the sketches you will make when solving problems yourself! FIGURE 2.31b shows a sketch like one you might actually do. It's less formal, but it contains all of the important information you need to solve the problem.

NOTE ► Throughout this book we will illustrate select examples with actual hand-drawn figures so that you have them to refer to as you work on your own pictures for homework and practice. ◄

Let's look at how these pictures were constructed. The motion has a clear beginning and end; these are the points sketched. A coordinate system has been chosen with the origin at the starting point. The quantities x, v_x, and t are needed at both points, so these have been defined on the sketch and distinguished by subscripts. The acceleration is associated with an interval between these points. Values for two of these quantities are given in the problem statement. Others, such as $x_i = 0$ m and $t_i = 0$ s, are inferred from our choice of coordinate system. The value $(v_x)_i = 0$ m/s is part of our *interpretation* of the problem. Finally, we identify x_f and $(v_x)_f$ as the quantities that will answer the question. We now understand quite a bit about the problem and would be ready to start a quantitative analysis.

ASSESS We didn't *solve* the problem; that was not our purpose. Constructing a pictorial representation and a list of values is part of a systematic approach to interpreting a problem and getting ready for a mathematical solution.

FIGURE 2.31 Constructing a pictorial representation and a list of values.

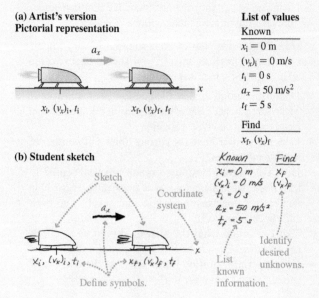

(a) Artist's version
Pictorial representation

List of values

Known
$x_i = 0$ m
$(v_x)_i = 0$ m/s
$t_i = 0$ s
$a_x = 50$ m/s²
$t_f = 5$ s

Find
$x_f, (v_x)_f$

(b) Student sketch

The Visual Overview

The pictorial representation and the list of values are a very good complement to the motion diagram and other ways of looking at a problem that we have seen. As we translate a problem into a form we can solve, we will combine these elements into what we will term a **visual overview.** The visual overview will consist of some or all of the following elements:

- A *motion diagram.* A good strategy for solving a motion problem is to start by drawing a motion diagram.
- A *pictorial representation,* as defined above.
- A *graphical representation.* For motion problems, it is often quite useful to include a graph of position and/or velocity.
- A *list of values.* This list should sum up all of the important values in the problem.

Future chapters will add other elements to this visual overview of the physics.

EXAMPLE 2.11   **Kinematics of a rocket launch**

A Saturn V rocket is launched straight up with a constant acceleration of 18 m/s². After 150 s, how fast is the rocket moving and how far has it traveled?

PREPARE FIGURE 2.32 shows a visual overview of the rocket launch that includes a motion diagram, a pictorial representation, and a list of values. The visual overview shows the whole problem in a nutshell. The motion diagram illustrates the motion of the rocket. The pictorial representation (produced according to Tactics Box 2.2) shows axes, identifies the important points of the motion, and defines variables. Finally, we have included a list of values that gives the known and unknown quantities. In the visual overview we have taken the statement of the problem in words and made it much more precise. The overview contains everything you need to know about the problem.

SOLVE Our first task is to find the final velocity. Our list of values includes the initial velocity, the acceleration, and the time

interval, so we can use the first kinematic equation of Synthesis 2.1 to find the final velocity:

$$(v_y)_f = (v_y)_i + a_y \, \Delta t = 0 \text{ m/s} + (18 \text{ m/s}^2)(150 \text{ s})$$

$$= 2700 \text{ m/s}$$

The distance traveled is found using the second equation in Synthesis 2.1:

$$y_f = y_i + (v_y)_i \, \Delta t + \tfrac{1}{2} a_y (\Delta t)^2$$

$$= 0 \text{ m} + (0 \text{ m/s})(150 \text{ s}) + \tfrac{1}{2}(18 \text{ m/s}^2)(150 \text{ s})^2$$

$$= 2.0 \times 10^5 \text{ m} = 200 \text{ km}$$

ASSESS The acceleration is very large, and it goes on for a long time, so the large final velocity and large distance traveled seem reasonable.

FIGURE 2.32 Visual overview of the rocket launch.

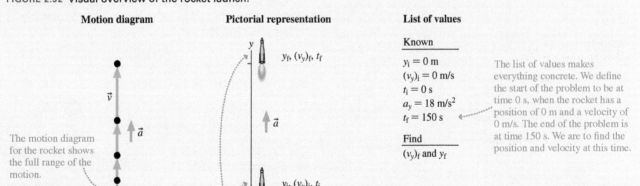

Problem-Solving Strategy for Motion with Constant Acceleration

Earlier in this section, we introduced a general problem-solving strategy. In this and future chapters we will adapt this general strategy to specific types of problems.

PROBLEM-SOLVING STRATEGY 2.1   Motion with constant acceleration

Problems involving constant acceleration—speeding up, slowing down, vertical motion, horizontal motion—can all be treated with the same problem-solving strategy.

PREPARE Draw a visual overview of the problem. This should include a motion diagram, a pictorial representation, and a list of values; a graphical representation may be useful for certain problems.

SOLVE The mathematical solution is based on the three equations in Synthesis 2.1.

- Though the equations are phrased in terms of the variable x, it's customary to use y for motion in the vertical direction.
- Use the equation that best matches what you know and what you need to find. For example, if you know acceleration and time and are looking for a change in velocity, the first equation is the best one to use.
- Uniform motion with constant velocity has $a = 0$.

ASSESS Is your result believable? Does it have proper units? Does it make sense?

Exercise 25 ✐

EXAMPLE 2.12   **Calculating the minimum length of a runway**

A fully loaded Boeing 747 with all engines at full thrust accelerates at 2.6 m/s². Its minimum takeoff speed is 70 m/s. How much time will the plane take to reach its takeoff speed? What minimum length of runway does the plane require for takeoff?

PREPARE The visual overview of **FIGURE 2.33** summarizes the important details of the problem. We set x_i and t_i equal to zero at the starting point of the motion, when the plane is at rest and the acceleration begins. The final point of the motion is when the plane achieves the necessary takeoff speed of 70 m/s. The plane is accelerating to the right, so we will compute the time for the plane to reach a velocity of 70 m/s and the position of the plane at this time, giving us the minimum length of the runway.

FIGURE 2.33 Visual overview for an accelerating plane.

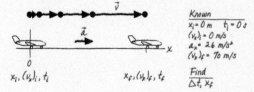

SOLVE First we solve for the time required for the plane to reach takeoff speed. We can use the first equation in Synthesis 2.1 to compute this time:

$$(v_x)_f = (v_x)_i + a_x \, \Delta t$$

$$70 \text{ m/s} = 0 \text{ m/s} + (2.6 \text{ m/s}^2) \, \Delta t$$

$$\Delta t = \frac{70 \text{ m/s}}{2.6 \text{ m/s}^2} = 26.9 \text{ s}$$

We keep an extra significant figure here because we will use this result in the next step of the calculation.

Given the time that the plane takes to reach takeoff speed, we can compute the position of the plane when it reaches this speed using the second equation in Synthesis 2.1:

$$x_f = x_i + (v_x)_i \, \Delta t + \tfrac{1}{2} a_x (\Delta t)^2$$

$$= 0 \text{ m} + (0 \text{ m/s})(26.9 \text{ s}) + \tfrac{1}{2}(2.6 \text{ m/s}^2)(26.9 \text{ s})^2$$

$$= 940 \text{ m}$$

Our final answers are thus that the plane will take 27 s to reach takeoff speed, with a minimum runway length of 940 m.

ASSESS Think about the last time you flew; 27 s seems like a reasonable time for a plane to accelerate on takeoff. Actual runway lengths at major airports are 3000 m or more, a few times greater than the minimum length, because they have to allow for emergency stops during an aborted takeoff. (If we had calculated a distance far greater than 3000 m, we would know we had done something wrong!)

EXAMPLE 2.13 **Finding the braking distance**

A car is traveling at a speed of 30 m/s, a typical highway speed, on wet pavement. The driver sees an obstacle ahead and decides to stop. From this instant, it takes him 0.75 s to begin applying the brakes. Once the brakes are applied, the car experiences an acceleration of -6.0 m/s². How far does the car travel from the instant the driver notices the obstacle until stopping?

PREPARE This problem is more involved than previous problems we have solved, so we will take more care with the visual overview in FIGURE 2.34. In addition to a motion diagram and a pictorial representation, we include a graphical representation. Notice that there are two different phases of the motion: a constant-velocity phase before braking begins, and a steady slowing down once the brakes are applied. We will need to do two different calculations, one for each phase. Consequently, we've used numerical subscripts rather than a simple i and f.

SOLVE From t_1 to t_2 the velocity stays constant at 30 m/s. This is uniform motion, so the position at time t_2 is computed using Equation 2.4:

$$x_2 = x_1 + (v_x)_1(t_2 - t_1) = 0 \text{ m} + (30 \text{ m/s})(0.75 \text{ s})$$
$$= 22.5 \text{ m}$$

At t_2, the velocity begins to decrease at a steady -6.0 m/s² until the car comes to rest at t_3. This time interval can be computed using the first equation in Synthesis 2.1, $(v_x)_3 = (v_x)_2 + a_x \Delta t$:

$$\Delta t = t_3 - t_2 = \frac{(v_x)_3 - (v_x)_2}{a_x} = \frac{0 \text{ m/s} - 30 \text{ m/s}}{-6.0 \text{ m/s}^2} = 5.0 \text{ s}$$

The position at time t_3 is computed using the second equation in Synthesis 2.1. We take point 2 as the initial point and point 3 as the final point for this phase of the motion and use $\Delta t = t_3 - t_2$:

$$x_3 = x_2 + (v_x)_2 \Delta t + \tfrac{1}{2}a_x(\Delta t)^2$$
$$= 22.5 \text{ m} + (30 \text{ m/s})(5.0 \text{ s}) + \tfrac{1}{2}(-6.0 \text{ m/s}^2)(5.0 \text{ s})^2$$
$$= 98 \text{ m}$$

x_3 is the position of the car at the end of the problem—and so the car travels 98 m before coming to rest.

ASSESS The numbers for the reaction time and the acceleration on wet pavement are reasonable ones for an alert driver in a car with good tires. The final distance is quite large—more than the length of a football field.

FIGURE 2.34 Visual overview for a car braking to a stop.

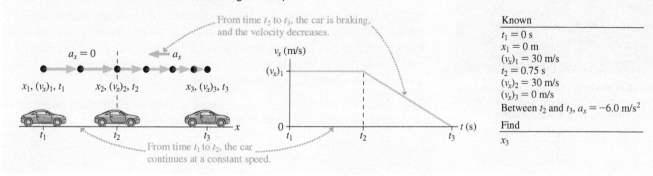

Known
$t_1 = 0$ s
$x_1 = 0$ m
$(v_x)_1 = 30$ m/s
$t_2 = 0.75$ s
$(v_x)_2 = 30$ m/s
$(v_x)_3 = 0$ m/s
Between t_2 and t_3, $a_x = -6.0$ m/s²

Find
x_3

2.7 Free Fall

If you drop a hammer and a feather, you know what will happen. The hammer quickly strikes the ground, and the feather flits and floats and lands some time later. But if you do this experiment on the moon, the result is strikingly different: Both the hammer and the feather experience the exact same acceleration, undergo the exact same motion, and strike the ground at the same time.

The moon lacks an atmosphere, and so objects in motion above its surface experience no air resistance. There is one and only one force that matters—gravity. If an object moves under the influence of gravity only, and no other forces, we call the resulting motion **free fall**. Early investigators concluded, correctly, that **any two objects in free fall, regardless of their mass, have the same acceleration.** Thus, if you drop two objects and they are both in free fall, they hit the ground at the same time.

On the earth, air resistance is a factor. But when you drop a hammer, air resistance is very small, so we make only a slight error in treating the hammer *as if* it were in free fall. Motion with air resistance is a problem we will study in Chapter 5. Until then, we will restrict our attention to situations in which air resistance can be ignored, and we will make the reasonable assumption that falling objects are in free fall.

Free-falling feather Apollo 15 lunar astronaut David Scott performed a classic experiment on the moon, simultaneously dropping a hammer and a feather from the same height. Both hit the ground at the exact same time—something that would not happen in the atmosphere of the earth!

FIGURE 2.35a shows the motion diagram for an object that was released from rest and falls freely. Since the acceleration is the same for all objects, the diagram and graph would be the same for a falling baseball or a falling boulder! FIGURE 2.35b shows the object's velocity graph. The velocity changes at a steady rate. The slope of the velocity-versus-time graph is the free-fall acceleration $a_{\text{free fall}}$.

FIGURE 2.35 Motion of an object in free fall.

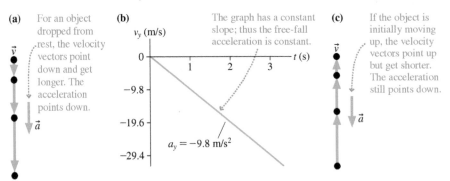

(a) For an object dropped from rest, the velocity vectors point down and get longer. The acceleration points down.

(b) The graph has a constant slope; thus the free-fall acceleration is constant. $a_y = -9.8 \text{ m/s}^2$

(c) If the object is initially moving up, the velocity vectors point up but get shorter. The acceleration still points down.

Some of the children are moving up and some are moving down, but all are in free fall—and so are accelerating downward at 9.8 m/s^2.

Instead of dropping the object, suppose we throw it upward. What happens then? You know that the object will move up and that its speed will decrease as it rises. This is illustrated in the motion diagram of FIGURE 2.35c, which shows a surprising result: Even though the object is moving up, its acceleration still points down. In fact, **the free-fall acceleration always points down,** no matter what direction an object is moving.

NOTE ▶ Despite the name, free fall is not restricted to objects that are literally falling. Any object moving under the influence of gravity only, and no other forces, is in free fall. This includes objects falling straight down, objects that have been tossed or shot straight up, objects in projectile motion (such as a passed football), and, as we will see, satellites in orbit. ◀

The value of the free-fall acceleration varies slightly at different places on the earth, but for the calculations in this book we will use the the following average value:

$$\vec{a}_{\text{free fall}} = (9.80 \text{ m/s}^2, \text{ vertically downward}) \tag{2.14}$$

Standard value for the acceleration of an object in free fall

The magnitude of the **free-fall acceleration** has the special symbol g:

$$g = 9.80 \text{ m/s}^2$$

We will generally work with two significant figures and so will use $g = 9.8 \text{ m/s}^2$. Several points about free fall are worthy of note:

- g, by definition, is *always* positive. **There will never be a problem that uses a negative value for g.**
- The velocity graph in Figure 2.35b has a negative slope. Even though a falling object speeds up, it has *negative* acceleration. Alternatively, notice that the acceleration vector $\vec{a}_{\text{free fall}}$ points down. Thus g is *not* the object's acceleration, simply the magnitude of the acceleration. The one-dimensional acceleration is

$$a_y = a_{\text{free fall}} = -g$$

- Because free fall is motion with constant acceleration, we can use the kinematic equations for constant acceleration with $a_y = -g$.
- g is not called "gravity." Gravity is a force, not an acceleration. g is the *free-fall acceleration*.
- $g = 9.80 \text{ m/s}^2$ only on earth. Other planets have different values of g. You will learn in Chapter 6 how to determine g for other planets.

■ We will sometimes compute acceleration in units of g. An acceleration of 9.8 m/s^2 is an acceleration of $1g$; an acceleration of 19.6 m/s^2 is $2g$. Generally, we can compute

$$\text{acceleration (in units of } g) = \frac{\text{acceleration (in units of m/s}^2)}{9.8 \text{ m/s}^2} \quad (2.15)$$

This allows us to express accelerations in units that have a definite physical reference.

EXAMPLE 2.14   Analyzing a rock's fall

A heavy rock is dropped from rest at the top of a cliff and falls 100 m before hitting the ground. How long does the rock take to fall to the ground, and what is its velocity when it hits?

PREPARE FIGURE 2.36 shows a visual overview with all necessary data. We have placed the origin at the ground, which makes $y_i = 100$ m.

FIGURE 2.36 Visual overview of a falling rock.

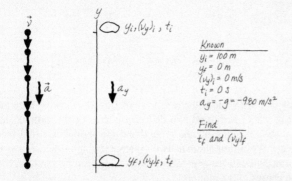

SOLVE Free fall is motion with the specific constant acceleration $a_y = -g$. The first question involves a relation between time and distance, a relation expressed by the second equation in Synthesis 2.1. Using $(v_y)_i = 0$ m/s and $t_i = 0$ s, we find

$$y_f = y_i + (v_y)_i \, \Delta t + \tfrac{1}{2} a_y \, (\Delta t)^2 = y_i - \tfrac{1}{2} g \, (\Delta t)^2 = y_i - \tfrac{1}{2} g t_f^2$$

We can now solve for t_f:

$$t_f = \sqrt{\frac{2(y_i - y_f)}{g}} = \sqrt{\frac{2(100 \text{ m} - 0 \text{ m})}{9.80 \text{ m/s}^2}} = 4.52 \text{ s}$$

Now that we know the fall time, we can use the first kinematic equation to find $(v_y)_f$:

$$(v_y)_f = (v_y)_i - g \, \Delta t = -g t_f = -(9.80 \text{ m/s}^2)(4.52 \text{ s})$$
$$= -44.3 \text{ m/s}$$

ASSESS Are the answers reasonable? Well, 100 m is about 300 feet, which is about the height of a 30-floor building. How long does it take something to fall 30 floors? Four or five seconds seems pretty reasonable. How fast would it be going at the bottom? Using an approximate version of our conversion factor 1 m/s ≈ 2 mph, we find that 44.3 m/s ≈ 90 mph. That also seems like a pretty reasonable speed for something that has fallen 30 floors. Suppose we had made a mistake. If we misplaced a decimal point we could have calculated a speed of 443 m/s, or about 900 mph! This is clearly *not* reasonable. If we had misplaced the decimal point in the other direction, we would have calculated a speed of 4.3 m/s ≈ 9 mph. This is another unreasonable result, because this is slower than a typical bicycling speed.

CONCEPTUAL EXAMPLE 2.15   Analyzing the motion of a ball tossed upward

Draw a motion diagram and a velocity-versus-time graph for a ball tossed straight up in the air from the point that it leaves the hand until just before it is caught.

REASON You know what the motion of the ball looks like: The ball goes up, and then it comes back down again. This complicates the drawing of a motion diagram a bit, as the ball retraces its route as it falls. A literal motion diagram would show the upward motion and downward motion on top of each other, leading to confusion. We can avoid this difficulty by horizontally separating the upward motion and downward motion diagrams. This will not affect our conclusions because it does not change any of the vectors. The motion diagram and velocity-versus-time graph appear as in FIGURE 2.37 on the next page.

ASSESS The highest point in the ball's motion, where it reverses direction, is called a *turning point*. What are the velocity and the acceleration at this point? We can see from the motion diagram that the velocity vectors are pointing upward but getting shorter as the ball

approaches the top. As it starts to fall, the velocity vectors are pointing downward and getting longer. There must be a moment—just an instant as $\vec{v}$ switches from pointing up to pointing down—when the velocity is zero. Indeed, the ball's velocity *is* zero for an instant at the precise top of the motion! We can also see on the velocity graph that there is one instant of time when $v_y = 0$. This is the turning point.

But what about the acceleration at the top? Many people expect the acceleration to be zero at the highest point. But recall that the velocity at the top point is changing—from up to down. If the velocity is changing, there *must* be an acceleration. The slope of the velocity graph at the instant when $v_y = 0$—that is, at the highest point—is no different than at any other point in the motion. The ball is still in free fall with acceleration $a_y = -g$!

Another way to think about this is to note that zero acceleration would mean no change of velocity. When the ball reached zero velocity at the top, it would hang there and not fall if the acceleration were also zero!

Continued

FIGURE 2.37 Motion diagram and velocity graph of a ball tossed straight up in the air.

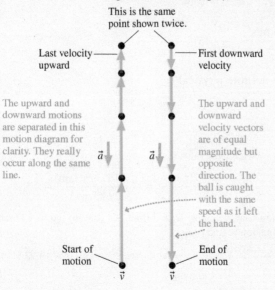

This is the same point shown twice.

Last velocity upward

First downward velocity

The upward and downward motions are separated in this motion diagram for clarity. They really occur along the same line.

The upward and downward velocity vectors are of equal magnitude but opposite direction. The ball is caught with the same speed as it left the hand.

$\vec{a}$ $\vec{a}$

Start of motion

End of motion

$\vec{v}$ $\vec{v}$

The ball starts with a positive (upward) velocity that steadily decreases.

When the ball reaches its highest point, its velocity is instantaneously zero. This is the turning point of the motion.

Now the ball is moving downward. The velocity is negative.

v_y

t

$a_y = -9.8 \text{ m/s}^2$

During the entire motion, the acceleration is that of free fall. The slope of the velocity graph is constant and negative.

EXAMPLE 2.16   **Finding the height of a leap** BIO

A springbok is an antelope found in southern Africa that gets its name from its remarkable jumping ability. When a springbok is startled, it will leap straight up into the air—a maneuver called a "pronk." A springbok goes into a crouch to perform a pronk. It then extends its legs forcefully, accelerating at 35 m/s² for 0.70 m as its legs straighten. Legs fully extended, it leaves the ground and rises into the air.

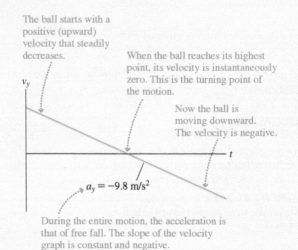

a. At what speed does the springbok leave the ground?
b. How high does it go?

PREPARE We begin with the visual overview shown in FIGURE 2.38, where we've identified two different phases of the motion: the springbok pushing off the ground and the springbok rising into the air. We'll treat these as two separate problems that we solve in turn. We will "re-use" the variables y_i, y_f, $(v_y)_i$, and $(v_y)_f$ for the two phases of the motion.

For the first part of our solution, in Figure 2.38a we choose the origin of the y-axis at the position of the springbok deep in the crouch. The final position is the top extent of the push, at the instant the springbok leaves the ground. We want to find the velocity at this position because that's how fast the springbok is moving as it leaves the ground. Figure 2.38b essentially starts over—we have defined a new vertical axis with its origin at the ground, so the highest point of the springbok's motion is a

FIGURE 2.38 A visual overview of the springbok's leap.

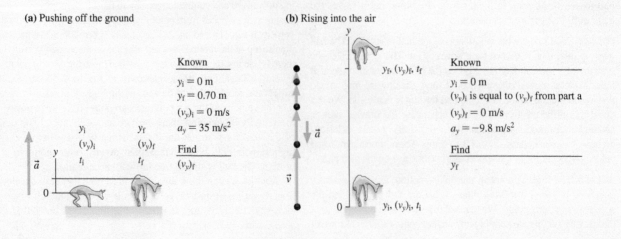

(a) Pushing off the ground

$\vec{a}$ y

y_i y_f
$(v_y)_i$ $(v_y)_f$
t_i t_f

0

Known
$y_i = 0$ m
$y_f = 0.70$ m
$(v_y)_i = 0$ m/s
$a_y = 35$ m/s²

Find
$(v_y)_f$

(b) Rising into the air

y

y_f, $(v_y)_f$, t_f

$\vec{a}$

$\vec{v}$

0   y_i, $(v_y)_i$, t_i

Known
$y_i = 0$ m
$(v_y)_i$ is equal to $(v_y)_f$ from part a
$(v_y)_f = 0$ m/s
$a_y = -9.8$ m/s²

Find
y_f

distance above the ground. The table of values shows the key piece of information for this second part of the problem: The initial velocity for part b is the final velocity from part a.

After the springbok leaves the ground, this is a free-fall problem because the springbok is moving under the influence of gravity only. We want to know the height of the leap, so we are looking for the height at the top point of the motion. This is a turning point of the motion, with the instantaneous velocity equal to zero. Thus y_f, the height of the leap, is the springbok's position at the instant $(v_y)_f = 0$.

SOLVE a. For the first phase, pushing off the ground, we have information about displacement, initial velocity, and acceleration, but we don't know anything about the time interval. The third equation in Synthesis 2.1 is perfect for this type of situation. We can rearrange it to solve for the velocity with which the springbok lifts off the ground:

$$(v_y)_f^2 = (v_y)_i^2 + 2a_y \Delta y = (0 \text{ m/s})^2 + 2(35 \text{ m/s}^2)(0.70 \text{ m}) = 49 \text{ m}^2/\text{s}^2$$

$$(v_y)_f = \sqrt{49 \text{ m}^2/\text{s}^2} = 7.0 \text{ m/s}$$

The springbok leaves the ground with a speed of 7.0 m/s.

b. Now we are ready for the second phase of the motion, the vertical motion after leaving the ground. The third equation in Synthesis 2.1 is again appropriate because again we don't know the time. Because $y_i = 0$, the springbok's displacement is $\Delta y = y_f - y_i = y_f$, the height of the vertical leap. From part a, the initial velocity is $(v_y)_i = 7.0$ m/s, and the final velocity is $(v_y)_f = 0$. This is free-fall motion, with $a_y = -g$; thus

$$(v_y)_f^2 = 0 = (v_y)_i^2 - 2g\Delta y = (v_y)_i^2 - 2gy_f$$

which gives

$$(v_y)_i^2 = 2gy_f$$

Solving for y_f, we get a jump height of

$$y_f = \frac{(7.0 \text{ m/s})^2}{2(9.8 \text{ m/s}^2)} = 2.5 \text{ m}$$

ASSESS 2.5 m is a remarkable leap—a bit over 8 ft—but these animals are known for their jumping ability, so this seems reasonable.

The caption accompanying the photo at the start of the chapter suggested a question about animals and their athletic abilities: Who is the winner in a race between a horse and a man? The surprising answer is "It depends." Specifically, the winner depends on the length of the race.

Some animals are capable of high speed; others are capable of great acceleration. Horses can run much faster than humans, but, when starting from rest, humans are capable of much greater initial acceleration. FIGURE 2.39 shows velocity and position graphs for an elite male sprinter and a thoroughbred racehorse. The horse's maximum velocity is about twice that of the man, but the man's initial acceleration—the slope of the velocity graph at early times—is greater than that of the horse. As the second graph shows, a man could win a *very* short race. For a longer race, the horse's higher maximum velocity will put it in the lead. The men's world-record time for the mile is a bit under 4 min, but a horse can easily run this distance in less than 2 min.

For a race of many miles, another factor comes into play: energy. A very long race is less about velocity and acceleration than about endurance—the ability to continue expending energy for a long time. In such endurance trials, humans often win. We will explore such energy issues in Chapter 11.

STOP TO THINK 2.7 A volcano ejects a chunk of rock straight up at a velocity of $v_y = 30$ m/s. Ignoring air resistance, what will be the velocity v_y of the rock when it falls back into the volcano's crater?

A. > 30 m/s   B. 30 m/s   C. 0 m/s   D. −30 m/s   E. < −30 m/s

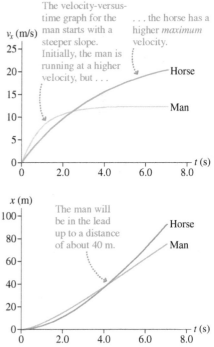

FIGURE 2.39 BIO Velocity-versus-time and position-versus-time graphs for a sprint between a man and a horse.

INTEGRATED EXAMPLE 2.17   **Speed versus endurance** BIO

Cheetahs have the highest top speed of any land animal, but they usually fail in their attempts to catch their prey because their endurance is limited. They can maintain their maximum speed of 30 m/s for only about 15 s before they need to stop.

Thomson's gazelles, their preferred prey, have a lower top speed than cheetahs, but they can maintain this speed for a few minutes. When a cheetah goes after a gazelle, success or failure is a simple matter of kinematics: Is the cheetah's high speed enough to allow it to reach its prey before the cheetah runs out of steam? The following problem uses realistic data for such a chase.

A cheetah has spotted a gazelle. The cheetah leaps into action, reaching its top speed of 30 m/s in a few seconds. At this instant, the gazelle, 160 m from the running cheetah, notices the danger and heads directly away. The gazelle accelerates at 4.5 m/s^2 for 6.0 s, then continues running at a constant speed. After reaching its maximum speed, the cheetah can continue running for only 15 s. Does the cheetah catch the gazelle, or does the gazelle escape?

PREPARE The example asks, "Does the cheetah catch the gazelle?" Our most challenging task is to translate these words into a graphical and mathematical problem that we can solve using the techniques of the chapter.

There are two related problems: the motion of the cheetah and the motion of the gazelle, for which we'll use the subscripts "C" and "G." Let's take our starting time, $t_1 = 0$ s, as the instant that the gazelle notices the cheetah and begins to run. We'll take the position of the cheetah at this instant as the origin of our coordinate system, so $x_{1C} = 0$ m and $x_{1G} = 160$ m—the gazelle is 160 m away when it notices the cheetah. We've used this information to draw the visual overview in FIGURE 2.40, which includes motion diagrams and velocity graphs for the cheetah and the gazelle. The visual overview sums up everything we know about the problem.

With a clear picture of the situation, we can now rephrase the problem this way: Compute the position of the cheetah and the position of the gazelle at $t_3 = 15$ s, the time when the cheetah needs to break off the chase. If $x_{3G} \geq x_{3C}$, then the gazelle stays out in front and escapes. If $x_{3G} < x_{3C}$, the cheetah wins the race—and gets its dinner.

SOLVE The cheetah is in uniform motion for the entire duration of the problem, so we can use Equation 2.4 to solve for its position at $t_3 = 15$ s:

$$x_{3C} = x_{1C} + (v_x)_{1C}\Delta t = 0 \text{ m} + (30 \text{ m/s})(15 \text{ s}) = 450 \text{ m}$$

The gazelle's motion has two phases: one of constant acceleration and then one of constant velocity. We can solve for the position and the velocity at t_2, the end of the first phase, using the first two equations in Synthesis 2.1. Let's find the velocity first:

$$(v_x)_{2G} = (v_x)_{1G} + (a_x)_G\Delta t = 0 \text{ m/s} + (4.5 \text{ m/s}^2)(6.0 \text{ s}) = 27 \text{ m/s}$$

The gazelle's position at t_2 is

$$x_{2G} = x_{1G} + (v_x)_{1G}\Delta t + \tfrac{1}{2}(a_x)_G(\Delta t)^2$$

$$= 160 \text{ m} + 0 + \tfrac{1}{2}(4.5 \text{ m/s}^2)(6.0 \text{ s})^2 = 240 \text{ m}$$

The gazelle has a head start; it begins at $x_{1G} = 160$ m. Δt is the time for this phase of the motion, $t_2 - t_1 = 6.0$ s.

From t_2 to t_3 the gazelle moves at a constant speed, so we can use the uniform motion equation, Equation 2.4, to find its final position:

The gazelle begins this phase of the motion at $x_{2G} = 240$ m. Δt for this phase of the motion is $t_3 - t_2 = 9.0$ s.

$$x_{3G} = x_{2G} + (v_x)_{2G}\ \Delta t = 240 \text{ m} + (27 \text{ m/s})(9.0 \text{ s}) = 480 \text{ m}$$

x_{3C} is 450 m; x_{3G} is 480 m. The gazelle is 30 m ahead of the cheetah when the cheetah has to break off the chase, so the gazelle escapes.

ASSESS Does our solution make sense? Let's look at the final result. The numbers in the problem statement are realistic, so we expect our results to mirror real life. The speed for the gazelle is close to that of the cheetah, which seems reasonable for two animals known for their speed. And the result is the most common occurrence—the chase is close, but the gazelle gets away.

FIGURE 2.40 Visual overview for the cheetah and for the gazelle.

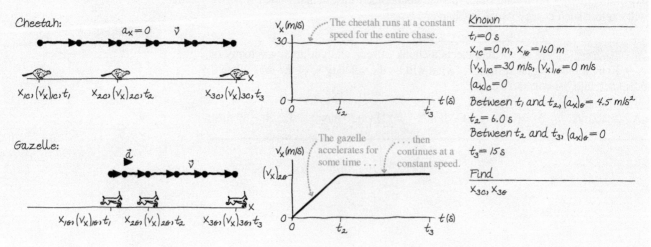

SUMMARY

Goal: To describe and analyze linear motion.

GENERAL STRATEGIES

Problem-Solving Strategy

Our general problem-solving strategy has three parts:

PREPARE Set up the problem:

- Draw a picture.
- Collect necessary information.
- Do preliminary calculations.

SOLVE Do the necessary mathematics or reasoning.

ASSESS Check your answer to see if it is complete in all details and makes physical sense.

Visual Overview

A visual overview consists of several pieces that completely specify a problem. This may include any or all of the elements below:

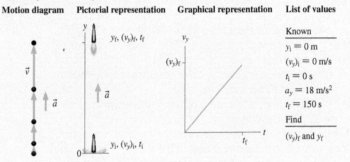

Motion diagram   Pictorial representation   Graphical representation   List of values

Known
$y_i = 0$ m
$(v_y)_i = 0$ m/s
$t_i = 0$ s
$a_y = 18$ m/s^2
$t_f = 150$ s

Find
$(v_y)_f$ and y_f

IMPORTANT CONCEPTS

Velocity is the rate of change of position:

$$v_x = \frac{\Delta x}{\Delta t}$$

Acceleration is the rate of change of velocity:

$$a_x = \frac{\Delta v_x}{\Delta t}$$

The units of acceleration are m/s^2.

An object is speeding up if v_x and a_x have the same sign, slowing down if they have opposite signs.

A **position-versus-time graph** plots position on the vertical axis against time on the horizontal axis.

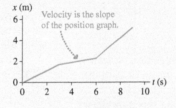

Velocity is the slope of the position graph.

A **velocity-versus-time graph** plots velocity on the vertical axis against time on the horizontal axis.

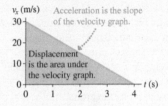

Acceleration is the slope of the velocity graph.

Displacement is the area under the velocity graph.

APPLICATIONS

Uniform motion

An object in uniform motion has a constant velocity. Its velocity graph is a horizontal line; its position graph is linear.

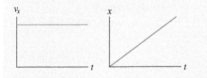

Kinematic equation for uniform motion:

$$x_f = x_i + v_x \, \Delta t$$

Uniform motion is a special case of constant-acceleration motion, with $a_x = 0$.

Motion with constant acceleration

An object with constant acceleration has a constantly changing velocity. Its velocity graph is linear; its position graph is a parabola.

Kinematic equations for motion with constant acceleration:

$$(v_x)_f = (v_x)_i + a_x \, \Delta t$$

$$x_f = x_i + (v_x)_i \, \Delta t + \tfrac{1}{2} a_x (\Delta t)^2$$

$$(v_x)_f^2 = (v_x)_i^2 + 2 a_x \, \Delta x$$

Free fall

Free fall is a special case of constant-acceleration motion. The acceleration has magnitude $g = 9.80$ m/s^2 and is always directed vertically downward whether an object is moving up or down.

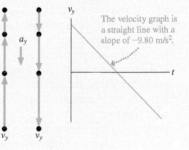

The velocity graph is a straight line with a slope of -9.80 m/s^2.

QUESTIONS

Conceptual Questions

1. A person gets in an elevator on the ground floor and rides it to the top floor of a building. Sketch a velocity-versus-time graph for this motion.

2. a. Give an example of a vertical motion with a positive velocity and a negative acceleration.
   b. Give an example of a vertical motion with a negative velocity and a negative acceleration.

3. BIO Figure Q2.3 shows growth rings in the trunk of a pine tree. You can clearly see the wide and the narrow rings that correspond to years of fast and slow growth. You can think of these rings as a motion diagram for the growth of the tree. If we define an axis as shown, with x measured out from the center of the tree, use the appearance of the rings to sketch a velocity-versus-time graph for the radial growth of the tree.

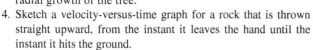

FIGURE Q2.3

4. Sketch a velocity-versus-time graph for a rock that is thrown straight upward, from the instant it leaves the hand until the instant it hits the ground.

5. You are driving down the road at a constant speed. Another car going a bit faster catches up with you and passes you. Draw a position graph for both vehicles on the same set of axes, and note the point on the graph where the other vehicle passes you.

6. A car is traveling north. Can its acceleration vector ever point south? Explain.

7. BIO Certain animals are capable of running at great speeds; other animals are capable of tremendous accelerations. Speculate on which would be more beneficial to a predator—large maximum speed or large acceleration.

8. A ball is thrown straight up into the air. At each of the following instants, is the ball's acceleration a_y equal to g, $-g$, 0, $< g$, or $> g$?
   a. Just after leaving your hand?
   b. At the very top (maximum height)?
   c. Just before hitting the ground?

9. A rock is *thrown* (not dropped) straight down from a bridge into the river below.
   a. Immediately after being released, is the magnitude of the rock's acceleration greater than g, less than g, or equal to g? Explain.
   b. Immediately before hitting the water, is the magnitude of the the rock's acceleration greater than g, less than g, or equal to g? Explain.

10. Figure Q2.10 shows an object's position-versus-time graph. The letters A to E correspond to various segments of the motion in which the graph has constant slope.
    a. Write a realistic motion short story for an object that would have this position graph.
    b. In which segment(s) is the object at rest?
    c. In which segment(s) is the object moving to the right?
    d. Is the speed of the object during segment C greater than, equal to, or less than its speed during segment E? Explain.

FIGURE Q2.10

11. Figure Q2.11 shows the position graph for an object moving along the horizontal axis.
    a. Write a realistic motion short story for an object that would have this position graph.
    b. Draw the corresponding velocity graph.

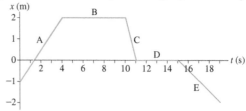

FIGURE Q2.11

12. Figure Q2.12 shows the position-versus-time graphs for two objects, A and B, that are moving along the same axis.
    a. At the instant $t = 1$ s, is the speed of A greater than, less than, or equal to the speed of B? Explain.
    b. Do objects A and B ever have the *same* speed? If so, at what time or times? Explain.

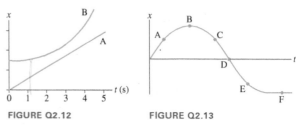

FIGURE Q2.12   FIGURE Q2.13

13. Figure Q2.13 shows a position-versus-time graph. At which lettered point or points is the object
    a. Moving the fastest?   b. Moving to the left?
    c. Speeding up?   d. Slowing down?
    e. Turning around?

14. Figure Q2.14 is the velocity-versus-time graph for an object moving along the *x*-axis.
    a. During which segment(s) is the velocity constant?
    b. During which segment(s) is the object speeding up?
    c. During which segment(s) is the object slowing down?
    d. During which segment(s) is the object standing still?
    e. During which segment(s) is the object moving to the right?

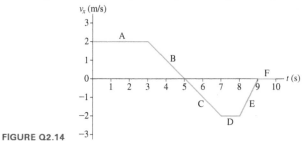

FIGURE Q2.14

Multiple-Choice Questions

15. | Figure Q2.15 shows the position graph of a car traveling on a straight road. At which labeled instant is the speed of the car greatest?

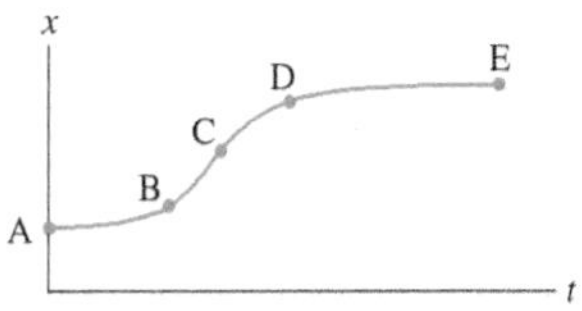

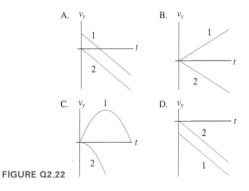

FIGURE Q2.15

16. | Figure Q2.16 shows the position graph of a car traveling on a straight road. The velocity at instant 1 is _____ and the velocity at instant 2 is _____.
    A. positive, negative
    B. positive, positive
    C. negative, negative
    D. negative, zero
    E. positive, zero

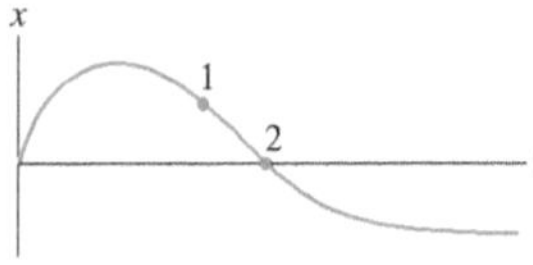

FIGURE Q2.16

17. | Figure Q2.17 shows an object's position-versus-time graph. What is the velocity of the object at $t = 6$ s?
    A. 0.67 m/s   B. 0.83 m/s
    C. 3.3 m/s   D. 4.2 m/s
    E. 25 m/s

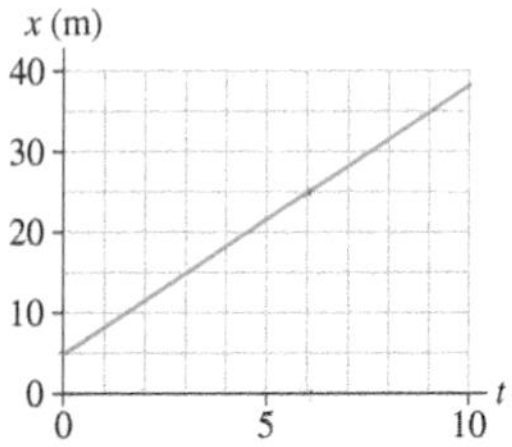

FIGURE Q2.17

18. | The following options describe the motion of four cars A–D. Which car has the largest acceleration?
    A. Goes from 0 m/s to 10 m/s in 5.0 s
    B. Goes from 0 m/s to 5.0 m/s in 2.0 s
    C. Goes from 0 m/s to 20 m/s in 7.0 s
    D. Goes from 0 m/s to 3.0 m/s in 1.0 s

19. | A car is traveling at $v_x = 20$ m/s. The driver applies the brakes, and the car slows with $a_x = -4.0$ m/s². What is the stopping distance?
    A. 5.0 m   B. 25 m   C. 40 m   D. 50 m

20. ‖ Velocity-versus-time graphs for three drag racers are shown in Figure Q2.20. At $t = 5.0$ s, which car has traveled the farthest?
    A. Andy   B. Betty   C. Carl
    D. All have traveled the same distance

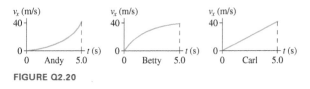

FIGURE Q2.20

21. | Which of the three drag racers in Question 20 had the greatest acceleration at $t = 0$ s?
    A. Andy   B. Betty   C. Carl
    D. All had the same acceleration

22. ‖ Chris is holding two softballs while standing on a balcony. She throws ball 1 straight up in the air and, at the same instant, releases her grip on ball 2, letting it drop over the side of the building. Which velocity graph in Figure Q2.22 best represents the motion of the two balls?

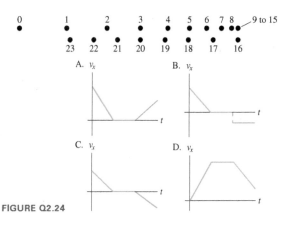

FIGURE Q2.22

23. ‖ Suppose a plane accelerates from rest for 30 s, achieving a takeoff speed of 80 m/s after traveling a distance of 1200 m down the runway. A smaller plane with the same acceleration has a takeoff speed of 40 m/s. Starting from rest, after what distance will this smaller plane reach its takeoff speed?
    A. 300 m   B. 600 m   C. 900 m   D. 1200 m

24. ‖ Figure Q2.24 shows a motion diagram with the clock reading (in seconds) shown at each position. From $t = 9$ s to $t = 15$ s the object is at the same position. After that, it returns along the same track. The positions of the dots for $t \geq 16$ s are offset for clarity. Which graph best represents the object's *velocity?*

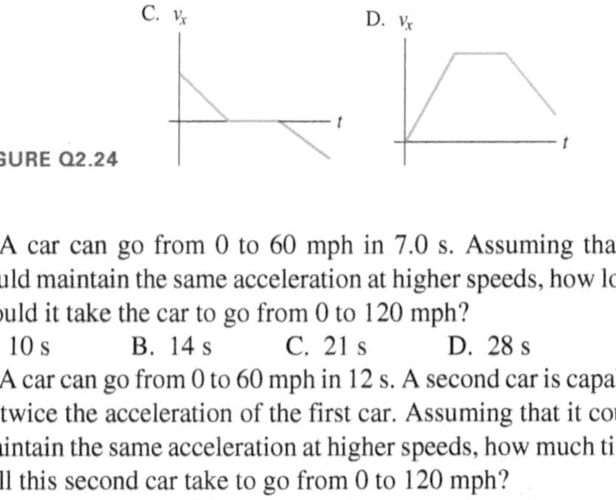

FIGURE Q2.24

25. ‖ A car can go from 0 to 60 mph in 7.0 s. Assuming that it could maintain the same acceleration at higher speeds, how long would it take the car to go from 0 to 120 mph?
    A. 10 s   B. 14 s   C. 21 s   D. 28 s

26. ‖ A car can go from 0 to 60 mph in 12 s. A second car is capable of twice the acceleration of the first car. Assuming that it could maintain the same acceleration at higher speeds, how much time will this second car take to go from 0 to 120 mph?
    A. 12 s   B. 9.0 s   C. 6.0 s   D. 3.0 s

PROBLEMS

Section 2.1 Describing Motion

1. ▥ Figure P2.1 shows a motion diagram of a car traveling down a street. The camera took one frame every second. A distance scale is provided.
   a. Measure the x-value of the car at each dot. Place your data in a table, similar to Table 2.1, showing each position and the instant of time at which it occurred.
   b. Make a graph of x versus t, using the data in your table. Because you have data only at certain instants of time, your graph should consist of dots that are not connected together.

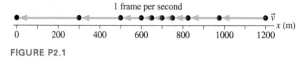

FIGURE P2.1

2. ▌ For each motion diagram in Figure P2.2, determine the sign (positive or negative) of the position and the velocity.

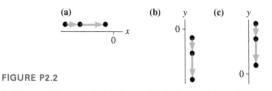

FIGURE P2.2

3. ▥ The position graph of Figure P2.3 shows a dog slowly sneaking up on a squirrel, then putting on a burst of speed.
   a. For how many seconds does the dog move at the slower speed?
   b. Draw the dog's velocity-versus-time graph. Include a numerical scale on both axes.

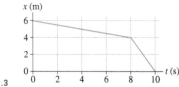

FIGURE P2.3

4. ▥ A rural mail carrier is driving slowly, putting mail in mailboxes near the road. He overshoots one mailbox, stops, shifts into reverse, and then backs up until he is at the right spot. The velocity graph of Figure P2.4 represents his motion.
   a. Draw the mail carrier's position-versus-time graph. Assume that $x = 0$ m at $t = 0$ s.
   b. What is the position of the mailbox?

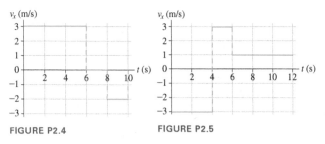

FIGURE P2.4   **FIGURE P2.5**

5. ▥ For the velocity-versus-time graph of Figure P2.5:
   a. Draw the corresponding position-versus-time graph. Assume that $x = 0$ m at $t = 0$ s.
   b. What is the object's position at $t = 12$ s?
   c. Describe a moving object that could have these graphs.

6. ▥ A bicyclist has the position-versus-time graph shown in Figure P2.6. What is the bicyclist's velocity at $t = 10$ s, at $t = 25$ s, and at $t = 35$ s?

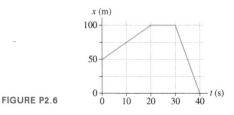

FIGURE P2.6

Section 2.2 Uniform Motion

7. ▌ In major league baseball, the pitcher's mound is 60 feet from the batter. If a pitcher throws a 95 mph fastball, how much time elapses from when the ball leaves the pitcher's hand until the ball reaches the batter?

8. ▌ In college softball, the distance from the pitcher's mound to the batter is 43 feet. If the ball leaves the bat at 100 mph, how much time elapses between the hit and the ball reaching the pitcher?

9. ▥ Alan leaves Los Angeles at 8:00 AM to drive to San Francisco, 400 mi away. He travels at a steady 50 mph. Beth leaves Los Angeles at 9:00 AM and drives a steady 60 mph.
   a. Who gets to San Francisco first?
   b. How long does the first to arrive have to wait for the second?

10. ▥ Richard is driving home to visit his parents. 125 mi of the trip are on the interstate highway where the speed limit is 65 mph. Normally Richard drives at the speed limit, but today he is running late and decides to take his chances by driving at 70 mph. How many minutes does he save?

11. ▥ In a 5.00 km race, one runner runs at a steady 12.0 km/h and another runs at 14.5 km/h. How long does the faster runner have to wait at the finish line to see the slower runner cross?

12. ▥ In an 8.00 km race, one runner runs at a steady 11.0 km/h and another runs at 14.0 km/h. How far from the finish line is the slower runner when the faster runner finishes the race?

13. ▥ A car moves with constant velocity along a straight road. Its position is $x_1 = 0$ m at $t_1 = 0$ s and is $x_2 = 30$ m at $t_2 = 3.0$ s. Answer the following by considering ratios, without computing the car's velocity.
   a. What is the car's position at $t = 1.5$ s?
   b. What will be its position at $t = 9.0$ s?

14. ▥ While running a marathon, a long-distance runner uses a stopwatch to time herself over a distance of 100 m. She finds that she runs this distance in 18 s. Answer the following by considering ratios, without computing her velocity.
   a. If she maintains her speed, how much time will it take her to run the next 400 m?
   b. How long will it take her to run a mile at this speed?

Section 2.3 Instantaneous Velocity

15. ▌ Figure P2.15 shows the position graph of a particle.
   a. Draw the particle's velocity graph for the interval $0 \text{ s} \leq t \leq 4 \text{ s}$.
   b. Does this particle have a turning point or points? If so, at what time or times?

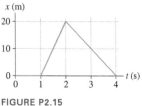

FIGURE P2.15

16. ‖ A somewhat idealized graph of the speed of the blood in
BIO the ascending aorta during one beat of the heart appears as in
Figure P2.16.
   a. Approximately how far, in cm, does the blood move during
   one beat?
   b. Assume similar data for the motion of the blood in your
   aorta, and make a rough estimate of the distance from your
   heart to your brain. Estimate how many beats of the heart it
   takes for blood to travel from your heart to your brain.

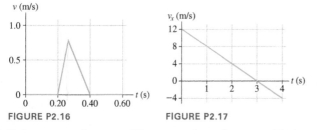

FIGURE P2.16   FIGURE P2.17

17. ‖‖ A car starts from $x_i = 10$ m at $t_i = 0$ s and moves with the
velocity graph shown in Figure P2.17.
   a. What is the car's position at $t = 2$ s, 3 s, and 4 s?
   b. Does this car ever change direction? If so, at what time?
18. ‖ Figure P2.18 shows a graph of actual position-versus-time
data for a particular type of drag racer known as a "funny car."
   a. Estimate the car's velocity at 2.0 s.
   b. Estimate the car's velocity at 4.0 s.

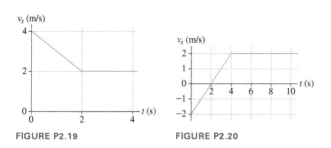

FIGURE P2.18

Section 2.4 Acceleration

19. ‖ Figure P2.19 shows the velocity graph of a bicycle. Draw the
bicycle's acceleration graph for the interval $0 \text{ s} \le t \le 4$ s. Give
both axes an appropriate numerical scale.

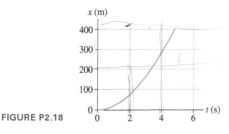

FIGURE P2.19   FIGURE P2.20

20. ‖‖ We set the origin of a coordinate system so that the position
of a train is $x = 0$ m at $t = 0$ s. Figure P2.20 shows the train's
velocity graph.
   a. Draw position and acceleration graphs for the train.
   b. Find the acceleration of the train at $t = 3.0$ s.
21. ∣ For each motion diagram shown earlier in Figure P2.2, determine the sign (positive or negative) of the acceleration.
22. ‖ Figure P2.16 showed data for the speed of blood in the aorta.
BIO Determine the magnitude of the acceleration for both phases,
speeding up and slowing down.

23. ‖ Figure P2.23 is a somewhat simplified velocity graph for
Olympic sprinter Carl Lewis starting a 100 m dash. Estimate his
acceleration during each of the intervals A, B, and C.

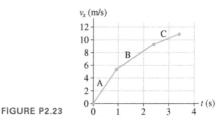

FIGURE P2.23

24. ∣ Small frogs that are good jumpers are capable of remarkable
BIO accelerations. One species reaches a takeoff speed of 3.7 m/s in
60 ms. What is the frog's acceleration during the jump?
25. ∣ A Thomson's gazelle can reach a speed of 13 m/s in 3.0 s. A
BIO lion can reach a speed of 9.5 m/s in 1.0 s. A trout can reach a speed
of 2.8 m/s in 0.12 s. Which animal has the largest acceleration?
26. ‖‖ When striking, the pike, a
BIO predatory fish, can accelerate
from rest to a speed of 4.0 m/s
in 0.11 s.
   a. What is the acceleration of
   the pike during this strike?
   b. How far does the pike move
   during this strike?

Section 2.5 Motion with Constant Acceleration

27. ‖ a. What constant acceleration, in SI units, must a car have to
   go from zero to 60 mph in 10 s?
   b. What fraction of g is this?
   c. How far has the car traveled when it reaches 60 mph?
   Give your answer both in SI units and in feet.
28. ‖ When jumping, a flea rapidly extends its legs, reaching a
BIO takeoff speed of 1.0 m/s over a distance of 0.50 mm.
   a. What is the flea's acceleration as it extends its legs?
   b. How long does it take the flea to leave the ground after it
   begins pushing off?
29. ‖ A car traveling at speed v takes distance d to stop after the
brakes are applied. What is the stopping distance if the car is
initially traveling at speed $2v$? Assume that the acceleration due
to the braking is the same in both cases.
30. ‖ Light-rail passenger trains that provide transportation within
and between cities speed up and slow down with a nearly constant (and quite modest) acceleration. A train travels through
a congested part of town at 5.0 m/s. Once free of this area, it
speeds up to 12 m/s in 8.0 s. At the edge of town, the driver
again accelerates, with the same acceleration, for another 16 s
to reach a higher cruising speed. What is the final speed?
31. ‖‖ A cross-country skier is skiing along at a zippy 8.0 m/s. She
stops pushing and simply glides along, slowing to a reduced
speed of 6.0 m/s after gliding for 5.0 m. What is the magnitude
of her acceleration as she slows?
32. ‖ A small propeller airplane can comfortably achieve a high
enough speed to take off on a runway that is 1/4 mile long. A
large, fully loaded passenger jet has about the same acceleration
from rest, but it needs to achieve twice the speed to take off.
What is the minimum runway length that will serve? **Hint:** You
can solve this problem using ratios without having any additional information.

33. ‖ Formula One racers speed up much more quickly than normal passenger vehicles, and they also can stop in a much shorter distance. A Formula One racer traveling at 90 m/s can stop in a distance of 110 m. What is the magnitude of the car's acceleration as it slows during braking?

34. ‖ Figure P2.34 shows a velocity-versus-time graph for a particle moving along the x-axis. At $t = 0$ s, assume that $x = 0$ m.
    a. What are the particle's position, velocity, and acceleration at $t = 1.0$ s?
    b. What are the particle's position, velocity, and acceleration at $t = 3.0$ s?

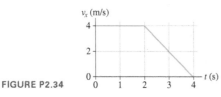

FIGURE P2.34

Section 2.6 Solving One-Dimensional Motion Problems

35. ‖ A driver has a reaction time of 0.50 s, and the maximum deceleration of her car is 6.0 m/s². She is driving at 20 m/s when suddenly she sees an obstacle in the road 50 m in front of her. Can she stop the car in time to avoid a collision?

36. ‖ Chameleons catch insects with their tongues, which they can BIO rapidly extend to great lengths. In a typical strike, the chameleon's tongue accelerates at a remarkable 250 m/s² for 20 ms, then travels at constant speed for another 30 ms. During this total time of 50 ms, 1/20 of a second, how far does the tongue reach?

37. ‖‖ You're driving down the highway late one night at 20 m/s when a deer steps onto the road 35 m in front of you. Your reaction time before stepping on the brakes is 0.50 s, and the maximum deceleration of your car is 10 m/s².
    a. How much distance is between you and the deer when you come to a stop?
    b. What is the maximum speed you could have and still not hit the deer?

38. ‖‖ A light-rail train going from one station to the next on a straight section of track accelerates from rest at 1.1 m/s² for 20 s. It then proceeds at constant speed for 1100 m before slowing down at 2.2 m/s² until it stops at the station.
    a. What is the distance between the stations?
    b. How much time does it take the train to go between the stations?

39. ‖ A car is traveling at a steady 80 km/h in a 50 km/h zone. A police motorcycle takes off at the instant the car passes it, accelerating at a steady 8.0 m/s².
    a. How much time elapses before the motorcycle is moving as fast as the car?
    b. How far is the motorcycle from the car when it reaches this speed?

40. ‖ When a jet lands on an aircraft carrier, a hook on the tail of the plane grabs a wire that quickly brings the plane to a halt before it overshoots the deck. In a typical landing, a jet touching down at 240 km/h is stopped in a distance of 95 m.
    a. What is the magnitude of the jet's acceleration as it is brought to rest?
    b. How much time does the landing take?

41. ‖‖ A simple model for a person running the 100 m dash is to assume the sprinter runs with constant acceleration until reaching top speed, then maintains that speed through the finish line. If a sprinter reaches his top speed of 11.2 m/s in 2.14 s, what will be his total time?

Section 2.7 Free Fall

42. ‖‖ Ball bearings can be made by letting spherical drops of molten metal fall inside a tall tower—called a *shot tower*—and solidify as they fall.
    a. If a bearing needs 4.0 s to solidify enough for impact, how high must the tower be?
    b. What is the bearing's impact velocity?

43. ‖ Here's an interesting chal-
    BIO lenge you can give to a friend. Hold a $1 (or larger!) bill by an upper corner. Have a friend prepare to pinch a lower corner, putting her fingers near but not touching the bill. Tell her to try to catch the bill when you drop it by simply closing her fingers. This seems like it should be easy, but it's not. After she sees that you have released the bill, it will take her about 0.25 s to react and close her fingers—which is  not fast enough to catch the bill. How much time does it take for the bill to fall beyond her grasp? The length of a bill is 16 cm.

44. ‖‖ In the preceding problem we saw that a person's reaction
    BIO time is generally not quick enough to allow the person to catch a $1 bill dropped between the fingers. The 16 cm length of the bill passes through a student's fingers before she can grab it if she has a typical 0.25 s reaction time. How long would a bill need to be for her to have a good chance of catching it?

45. ‖ A gannet is a seabird that fishes by diving from a great height.
    BIO If a gannet hits the water at 32 m/s (which they do), what height did it dive from? Assume that the gannet was motionless before starting its dive.

46. ‖ A student at the top of a building of height h throws ball A straight upward with speed v_0 and throws ball B straight downward with the same initial speed.
    a. Compare the balls' accelerations, both direction and magnitude, immediately after they leave her hand. Is one acceleration larger than the other? Or are the magnitudes equal?
    b. Compare the final speeds of the balls as they reach the ground. Is one larger than the other? Or are they equal?

47. ‖ Excellent human jumpers can leap straight up to a height of
    BIO 110 cm off the ground. To reach this height, with what speed would a person need to leave the ground?

48. ‖ A football is kicked straight up into the air; it hits the ground 5.2 s later.
    a. What was the greatest height reached by the ball? Assume it is kicked from ground level.
    b. With what speed did it leave the kicker's foot?

49. ‖‖‖ In an action movie, the villain is rescued from the ocean by grabbing onto the ladder hanging from a helicopter. He is so intent on gripping the ladder that he lets go of his briefcase of counterfeit money when he is 130 m above the water. If the briefcase hits the water 6.0 s later, what was the speed at which the helicopter was ascending?

50. ‖‖ Spud Webb was, at 5 ft 8 in, one of the shortest basketball
    BIO players to play in the NBA. But he had an amazing vertical leap; he could jump to a height of 1.1 m off the ground, so he could easily dunk a basketball. For such a leap, what was his "hang time"—the time spent in the air after leaving the ground and before touching down again?

51. ⦀ A rock climber stands on top of a 50-m-high cliff overhanging a pool of water. He throws two stones vertically downward 1.0 s apart and observes that they cause a single splash. The initial speed of the first stone was 2.0 m/s.
    a. How long after the release of the first stone does the second stone hit the water?
    b. What was the initial speed of the second stone?
    c. What is the speed of each stone as it hits the water?

General Problems

52. ⦀ Actual velocity data for a
BIO lion pursuing prey are shown in Figure P2.52. Estimate:
    a. The initial acceleration of the lion.
    b. The acceleration of the lion at 2 s and at 4 s.
    c. The distance traveled by the lion between 0 s and 8 s.

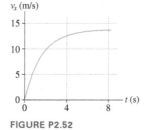

FIGURE P2.52

53. ⦀ A truck driver has a shipment of apples to deliver to a destination 440 miles away. The trip usually takes him 8 hours. Today he finds himself daydreaming and realizes 120 miles into his trip that he is running 15 minutes later than his usual pace at this point. At what speed must he drive for the remainder of the trip to complete the trip in the usual amount of time?

54. ⦀ When you sneeze, the air in your lungs accelerates from rest
BIO to approximately 150 km/h in about 0.50 seconds.
    a. What is the acceleration of the air in m/s^2?
    b. What is this acceleration, in units of g?

55. ⦀ Figure P2.55 shows the motion diagram, made at two frames of film per second, of a ball rolling along a track. The track has a 3.0-m-long sticky section.

FIGURE P2.55

    a. Use the scale to determine the positions of the center of the ball. Place your data in a table, similar to Table 2.1, showing each position and the instant of time at which it occurred.
    b. Make a graph of x versus t for the ball. Because you have data only at certain instants of time, your graph should consist of dots that are not connected together.
    c. What is the *change* in the ball's position from $t = 0$ s to $t = 1.0$ s?
    d. What is the *change* in the ball's position from $t = 2.0$ s to $t = 4.0$ s?
    e. What is the ball's velocity before reaching the sticky section?
    f. What is the ball's velocity after passing the sticky section?
    g. Determine the ball's acceleration on the sticky section of the track.

56. ⦀ Julie drives 100 mi to Grandmother's house. On the way to Grandmother's, Julie drives half the *distance* at 40 mph and half the distance at 60 mph. On her return trip, she drives half the *time* at 40 mph and half the time at 60 mph.
    a. How long does it take Julie to complete the trip to Grandmother's house?
    b. How long does the return trip take?

57. ⦀ The takeoff speed for an Airbus A320 jetliner is 80 m/s. Velocity data measured during takeoff are as shown in the table.

t (s)	v_x (m/s)
0	0
10	23
20	46
30	69

    a. What is the jetliner's acceleration during takeoff, in m/s^2 and in g's?
    b. At what time do the wheels leave the ground?
    c. For safety reasons, in case of an aborted takeoff, the length of the runway must be three times the takeoff distance. What is the minimum length runway this aircraft can use?

58. ⦀ Does a real automobile have constant acceleration? Measured data for a Porsche 944 Turbo at maximum acceleration are as shown in the table.

t (s)	v_x (mph)
0	0
2	28
4	46
6	60
8	70
10	78

    a. Convert the velocities to m/s, then make a graph of velocity versus time. Based on your graph, is the acceleration constant? Explain.
    b. Draw a smooth curve through the points on your graph, then use your graph to *estimate* the car's acceleration at 2.0 s and 8.0 s. Give your answer in SI units. **Hint:** Remember that acceleration is the slope of the velocity graph.

59. ⦀ People hoping to travel to other worlds are faced with huge challenges. One of the biggest is the time required for a journey. The nearest star is 4.1×10^{16} m away. Suppose you had a spacecraft that could accelerate at $1.0g$ for half a year, then continue at a constant speed. (This is far beyond what can be achieved with any known technology.) How long would it take you to reach the nearest star to earth?

60. ⦀ You are driving to the grocery store at 20 m/s. You are 110 m from an intersection when the traffic light turns red. Assume that your reaction time is 0.70 s and that your car brakes with constant acceleration.
    a. How far are you from the intersection when you begin to apply the brakes?
    b. What acceleration will bring you to rest right at the intersection?
    c. How long does it take you to stop?

61. ⦀ When you blink your eye, the upper lid goes from rest with
BIO your eye open to completely covering your eye in a time of 0.024 s.
    a. Estimate the distance that the top lid of your eye moves during a blink.
    b. What is the acceleration of your eyelid? Assume it to be constant.
    c. What is your upper eyelid's final speed as it hits the bottom eyelid?

62. ⦀ A bush baby, an African
BIO primate, is capable of a remarkable vertical leap. The bush baby goes into a crouch and extends its legs, pushing upward for a distance of 0.16 m. After this upward acceleration, the bush baby leaves the ground and travels upward for 2.3 m. What is the acceleration during the pushing-off phase? Give your answer in m/s^2 and in g's.

63. |||| When jumping, a flea reaches a takeoff speed of 1.0 m/s over
BIO a distance of 0.50 mm.
    a. What is the flea's acceleration during the jump phase?
    b. How long does the acceleration phase last?
    c. If the flea jumps straight up, how high will it go? (Ignore air resistance for this problem; in reality, air resistance plays a large role, and the flea will not reach this height.)

64. ||| Certain insects can achieve
BIO seemingly impossible accelerations while jumping. The
click beetle accelerates at an
astonishing 400g over a distance of 0.60 cm as it rapidly
bends its thorax, making the
"click" that gives it its name.

    a. Assuming the beetle jumps straight up, at what speed does it leave the ground?
    b. How much time is required for the beetle to reach this speed?
    c. Ignoring air resistance, how high would it go?

65. |||| A student standing on the ground throws a ball straight up. The ball leaves the student's hand with a speed of 15 m/s when the hand is 2.0 m above the ground. How long is the ball in the air before it hits the ground? (The student moves her hand out of the way.)

66. |||| A rock is tossed straight up with a speed of 20 m/s. When it returns, it falls into a hole 10 m deep.
    a. What is the rock's velocity as it hits the bottom of the hole?
    b. How long is the rock in the air, from the instant it is released until it hits the bottom of the hole?

67. |||| A 200 kg weather rocket is loaded with 100 kg of fuel and fired straight up. It accelerates upward at 30.0 m/s^2 for 30.0 s, then runs out of fuel. Ignore any air resistance effects.
    a. What is the rocket's maximum altitude?
    b. How long is the rocket in the air?
    c. Draw a velocity-versus-time graph for the rocket from liftoff until it hits the ground.

68. |||| A hotel elevator ascends 200 m with a maximum speed of 5.0 m/s. Its acceleration and deceleration both have a magnitude of 1.0 m/s^2.
    a. How far does the elevator move while accelerating to full speed from rest?
    b. How long does it take to make the complete trip from bottom to top?

69. |||| A car starts from rest at a stop sign. It accelerates at 2.0 m/s^2 for 6.0 seconds, coasts for 2.0 s, and then slows down at a rate of 1.5 m/s^2 for the next stop sign. How far apart are the stop signs?

70. |||| A toy train is pushed forward and released at $x_i = 2.0$ m with a speed of 2.0 m/s. It rolls at a steady speed for 2.0 s, then one wheel begins to stick. The train comes to a stop 6.0 m from the point at which it was released. What is the train's acceleration after its wheel begins to stick?

71. || Heather and Jerry are standing on a bridge 50 m above a river. Heather throws a rock straight down with a speed of 20 m/s. Jerry, at exactly the same instant of time, throws a rock straight up with the same speed. Ignore air resistance.
    a. How much time elapses between the first splash and the second splash?
    b. Which rock has the faster speed as it hits the water?

72. || A Thomson's gazelle can run at very high speeds, but its
BIO acceleration is relatively modest. A reasonable model for the sprint of a gazelle assumes an acceleration of 4.2 m/s^2 for 6.5 s, after which the gazelle continues at a steady speed.
    a. What is the gazelle's top speed?
    b. A human would win a very short race with a gazelle. The best time for a 30 m sprint for a human runner is 3.6 s. How much time would the gazelle take for a 30 m race?
    c. A gazelle would win a longer race. The best time for a 200 m sprint for a human runner is 19.3 s. How much time would the gazelle take for a 200 m race?

73. ||| We've seen that a man's higher initial acceleration means that
BIO a man can outrun a horse over a very short race. A simple—but plausible—model for a sprint by a man and a horse uses the following assumptions: The man accelerates at 6.0 m/s^2 for 1.8 s and then runs at a constant speed. A horse accelerates at a more modest 5.0 m/s^2 but continues accelerating for 4.8 s and then continues at a constant speed. A man and a horse are competing in a 200 m race. The man is given a 100 m head start, so he begins 100 m from the finish line. How much time does the man take to complete the race? How much time does the horse take? Who wins the race?

74. || A pole-vaulter is nearly motionless as he clears the bar, set 4.2 m above the ground. He then falls onto a thick pad. The top of the pad is 80 cm above the ground, and it compresses by 50 cm as he comes to rest. What is his acceleration as he comes to rest on the pad?

75. || A Porsche challenges a Honda to a 400 m race. Because the Porsche's acceleration of 3.5 m/s^2 is larger than the Honda's 3.0 m/s^2, the Honda gets a 100-m head start—it is only 300 m from the finish line. Assume, somewhat unrealistically, that both cars can maintain these accelerations the entire distance. Who wins, and by how much time?

76. |||| The minimum stopping distance for a car traveling at a speed of 30 m/s is 60 m, including the distance traveled during the driver's reaction time of 0.50 s.
    a. Draw a position-versus-time graph for the motion of the car. Assume the car is at $x_i = 0$ m when the driver first sees the emergency situation ahead that calls for a rapid halt.
    b. What is the minimum stopping distance for the same car traveling at a speed of 40 m/s?

77. |||| A rocket is launched straight up with constant acceleration. Four seconds after liftoff, a bolt falls off the side of the rocket. The bolt hits the ground 6.0 s later. What was the rocket's acceleration?

MCAT-Style Passage Problems

Free Fall on Different Worlds

Objects in free fall on the earth have acceleration $a_y = -9.8$ m/s^2. On the moon, free-fall acceleration is approximately 1/6 of the acceleration on earth. This changes the scale of problems involving free fall. For instance, suppose you jump straight upward, leaving the ground with velocity v_i and then steadily slowing until reaching zero velocity at your highest point. Because your initial velocity is determined mostly by the strength of your leg muscles, we can assume your initial velocity would be the same on the moon. But considering the final equation in Synthesis 2.1 we can see that, with a smaller free-fall acceleration, your maximum height would be greater. The following questions ask you to think about how certain athletic feats might be performed in this reduced-gravity environment.

78. | If an astronaut can jump straight up to a height of 0.50 m on earth, how high could he jump on the moon?
    A. 1.2 m   B. 3.0 m   C. 3.6 m   D. 18 m

79. | On the earth, an astronaut can safely jump to the ground from a height of 1.0 m; her velocity when reaching the ground is slow enough to not cause injury. From what height could the astronaut safely jump to the ground on the moon?
    A. 2.4 m   B. 6.0 m   C. 7.2 m   D. 36 m

80. | On the earth, an astronaut throws a ball straight upward; it stays in the air for a total time of 3.0 s before reaching the ground again. If a ball were to be thrown upward with the same initial speed on the moon, how much time would pass before it hit the ground?
    A. 7.3 s   B. 18 s   C. 44 s   D. 108 s

STOP TO THINK ANSWERS

Chapter Preview Stop to Think: B. The bicycle is moving to the left, so the velocity vectors must point to the left. The speed is increasing, so successive velocity vectors must get longer.

Stop to Think 2.1: D. The motion consists of two constant-velocity phases, and the second one has a higher velocity. The correct graph has two straight-line segments, with the second one having a steeper slope.

Stop to Think 2.2: B. The displacement is the area under a velocity-versus-time curve. In all four cases, the graph is a straight line, so the area under the curve is a rectangle. The area is the product of the length and the height, so the largest displacement belongs to the graph with the largest product of the length (the time interval, in s) and the height (the velocity, in m/s).

Stop to Think 2.3: C. Consider the slope of the position-versus-time graph. It starts out positive and constant, then decreases to zero. Thus the velocity graph must start with a constant positive value, then decrease to zero.

Stop to Think 2.4: B. The elevator is moving down, so $v_y < 0$. It is slowing down, so the magnitude of the velocity is decreasing. As time goes on, the velocity graph should get closer to the origin. This means that the acceleration is positive, and the slope of the graph is positive.

Stop to Think 2.5: C. Acceleration is the slope of the velocity-versus-time graph. The largest magnitude of the slope is at point C.

Stop to Think 2.6: E. An acceleration of 1.2 m/s^2 corresponds to an increase of 1.2 m/s every second. At this rate, the cruising speed of 6.0 m/s will be reached after 5.0 s.

Stop to Think 2.7: D. The final velocity will have the same *magnitude* as the initial velocity, but the velocity is negative because the rock will be moving downward.

3 Vectors and Motion in Two Dimensions

Once the leopard jumps, its trajectory is fixed by the initial speed and angle of the jump. How can we work out where the leopard will land?

LOOKING AHEAD »

Goal: To learn more about vectors and to use vectors as a tool to analyze motion in two dimensions.

Vectors and Components

The dark green vector is the ball's initial velocity. The light green **component vectors** show initial horizontal and vertical velocity.

You'll learn how to find components of vectors and how to use these components to solve problems.

Projectile Motion

A leaping fish's parabolic arc is an example of **projectile motion.** The details are the same for a fish or a basketball.

You'll see how to solve projectile motion problems, determining how long an object is in the air and how far it travels.

Circular Motion

The riders move in a circle at a constant speed, but they have an acceleration because the direction is constantly changing.

You'll learn how to determine the magnitude and the direction of the acceleration for an object in circular motion.

LOOKING BACK «

Free Fall

You learned in Section 2.7 that an object tossed straight up is in free fall. The acceleration is the same whether the object is going up or coming back down.

For an object in projectile motion, the vertical component of the motion is also free fall. you'll use your knowledge of free fall to solve projectile motion problems.

$\vec{a}$ $\vec{a}$

STOP TO THINK

A player kicks a football straight up into the air. The ball takes 2.0 s to reach its highest point. Approximately how fast was the ball moving when it left the player's foot?

A. 5 m/s
B. 10 m/s
C. 15 m/s
D. 20 m/s

3.1 Using Vectors

In Chapter 2, we solved problems in which an object moved in a straight-line path. In this chapter, objects move in curving paths—motion in two dimensions. Because the direction of motion will be so important, we need to develop an appropriate mathematical language to describe it—the language of vectors.

We introduced the concept of a vector in ◀ SECTION 1.5. In the next few sections we will develop techniques for working with vectors as a tool for studying motion in two dimensions. A vector is a quantity with both a size (magnitude) and a direction. FIGURE 3.1 shows how to represent a particle's velocity as a vector $\vec{v}$. The particle's speed at this point is 5 m/s *and* it is moving in the direction indicated by the arrow. The magnitude of a vector is represented by the letter without an arrow. In this case, the particle's speed—the magnitude of the velocity vector $\vec{v}$—is $v = 5$ m/s. The magnitude of a vector, a *scalar* quantity, cannot be a negative number.

NOTE ▶ Although the vector arrow is drawn across the page, from its tail to its tip, this arrow does *not* indicate that the vector "stretches" across this distance. Instead, the arrow tells us the value of the vector quantity only at the one point where the tail of the vector is placed. ◀

We saw in Chapter 1 that the displacement of an object is a vector drawn from its initial position to its position at some later time. Because displacement is an easy concept to think about, we will use it to introduce some of the properties of vectors. However, **all the properties we will discuss in this chapter (addition, subtraction, multiplication, components) apply to all types of vectors, not just to displacement.**

Suppose that Sam, our old friend from Chapter 1, starts from his front door, walks across the street, and ends up 200 ft to the northeast of where he started. Sam's displacement, which we will label $\vec{d}_S$, is shown in FIGURE 3.2a. The displacement vector is a straight-line connection from his initial to his final position, not necessarily his actual path. The dashed line indicates a possible route Sam might have taken, but his displacement is the vector $\vec{d}_S$.

To describe a vector we must specify both its magnitude and its direction. We can write Sam's displacement as

$$\vec{d}_S = (200 \text{ ft, northeast})$$

where the first number specifies the magnitude and the second item gives the direction. The magnitude of Sam's displacement is $d_S = 200$ ft, the distance between his initial and final points.

Sam's next-door neighbor Becky also walks 200 ft to the northeast, starting from her own front door. Becky's displacement $\vec{d}_B = (200 \text{ ft, northeast})$ has the same magnitude and direction as Sam's displacement $\vec{d}_S$. Because vectors are defined by their magnitude and direction, **two vectors are equal if they have the same magnitude and direction.** This is true regardless of the individual starting points of the vectors. Thus the two displacements in FIGURE 3.2b are equal to each other, and we can write $\vec{d}_B = \vec{d}_S$.

Vector Addition

As we saw in Chapter 1, we can combine successive displacements by vector addition. Let's review and extend this concept. FIGURE 3.3 shows the displacement of a hiker who starts at point P and ends at point S. She first hikes 4 miles to the east, then 3 miles to the north. The first leg of the hike is described by the displacement vector $\vec{A} = (4 \text{ mi, east})$. The second leg of the hike has displacement $\vec{B} = (3 \text{ mi, north})$. By definition, a vector from her initial position P to her final position S is also a displacement. This is vector $\vec{C}$ on the figure. $\vec{C}$ is the *net displacement* because it describes the net result of the hiker's having first displacement $\vec{A}$, then displacement $\vec{B}$.

The word "net" implies addition. The net displacement $\vec{C}$ is an initial displacement $\vec{A}$ *plus* a second displacement $\vec{B}$, or

$$\vec{C} = \vec{A} + \vec{B} \tag{3.1}$$

FIGURE 3.1 The velocity vector $\vec{v}$ has both a magnitude and a direction.

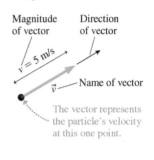

FIGURE 3.2 Displacement vectors.

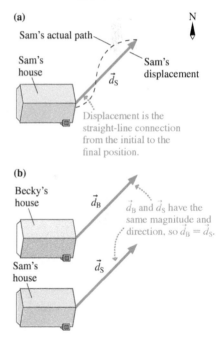

FIGURE 3.3 The net displacement $\vec{C}$ resulting from two displacements $\vec{A}$ and $\vec{B}$.

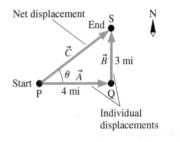

The sum of two vectors is called the **resultant vector**. Vector addition is commutative: $\vec{A} + \vec{B} = \vec{B} + \vec{A}$. You can add vectors in any order you wish.

Look back at ◀ TACTICS BOX 1.4 to review the three-step procedure for adding two vectors. This tip-to-tail method for adding vectors, which is used to find $\vec{C} = \vec{A} + \vec{B}$ in Figure 3.3, is called *graphical addition*.

When two vectors are to be added, it is often convenient to draw them with their tails together, as shown in FIGURE 3.4a. To evaluate $\vec{D} + \vec{E}$, you could move vector $\vec{E}$ over to where its tail is on the tip of $\vec{D}$, then use the tip-to-tail rule of graphical addition. This gives vector $\vec{F} = \vec{D} + \vec{E}$ in FIGURE 3.4b. Alternatively, FIGURE 3.4c shows that the vector sum $\vec{D} + \vec{E}$ can be found as the diagonal of the parallelogram defined by $\vec{D}$ and $\vec{E}$. This method is called the *parallelogram rule* of vector addition.

FIGURE 3.4 Two vectors can be added using the tip-to-tail rule or the parallelogram rule.

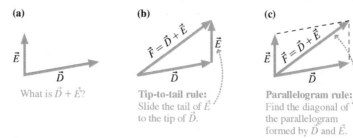

Multiplication by a Scalar

FIGURE 3.5 Multiplication of a vector by a positive scalar.

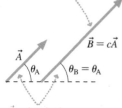

The length of $\vec{B}$ is "stretched" by the factor c; that is, $B = cA$.

$\vec{B}$ points in the same direction as $\vec{A}$.

The hiker in Figure 3.3 started with displacement $\vec{A}_1 = (4 \text{ mi, east})$. Suppose a second hiker walks twice as far to the east. The second hiker's displacement will then certainly be $\vec{A}_2 = (8 \text{ mi, east})$. The words "twice as" indicate a multiplication, so we can say

$$\vec{A}_2 = 2\vec{A}_1$$

Multiplying a vector by a positive scalar gives another vector of *different magnitude* but pointing in the *same direction*.

Let the vector $\vec{A}$ be specified as a magnitude A and a direction θ_A; that is, $\vec{A} = (A, \theta_A)$. Now let $\vec{B} = c\vec{A}$, where c is a positive scalar constant. Then

$$\vec{B} = c\vec{A} \text{ means that } (B, \theta_B) = (cA, \theta_A) \tag{3.2}$$

The vector is stretched or compressed by the factor c (i.e., vector $\vec{B}$ has magnitude $B = cA$), but $\vec{B}$ points in the same direction as $\vec{A}$. This is illustrated in FIGURE 3.5.

Suppose we multiply $\vec{A}$ by zero. Using Equation 3.2, we get

$$0 \cdot \vec{A} = \vec{0} = (0 \text{ m, direction undefined}) \tag{3.3}$$

FIGURE 3.6 Vector $-\vec{A}$.

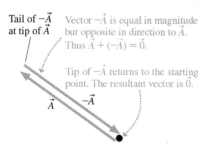

Tail of $-\vec{A}$ at tip of $\vec{A}$

Vector $-\vec{A}$ is equal in magnitude but opposite in direction to $\vec{A}$. Thus $\vec{A} + (-\vec{A}) = \vec{0}$.

Tip of $-\vec{A}$ returns to the starting point. The resultant vector is $\vec{0}$.

$\vec{A}$ $-\vec{A}$

The product is a vector having zero length or magnitude. This vector is known as the **zero vector,** denoted $\vec{0}$. The direction of the zero vector is irrelevant; you cannot describe the direction of an arrow of zero length!

What happens if we multiply a vector by a negative number? Equation 3.2 does not apply if $c < 0$ because vector $\vec{B}$ cannot have a negative magnitude. Consider the vector $-\vec{A}$, which is equivalent to multiplying $\vec{A}$ by -1. Because

$$\vec{A} + (-\vec{A}) = \vec{0} \tag{3.4}$$

the vector $-\vec{A}$ must be such that, when it is added to $\vec{A}$, the resultant is the zero vector $\vec{0}$. In other words, the *tip* of $-\vec{A}$ must return to the *tail* of $\vec{A}$, as shown in FIGURE 3.6. This will be true only if $-\vec{A}$ is equal in magnitude to $\vec{A}$ but opposite in direction. Thus we can conclude that

$$-\vec{A} = (A, \text{ direction opposite } \vec{A}) \tag{3.5}$$

FIGURE 3.7 Vectors $\vec{A}$, $2\vec{A}$, and $-3\vec{A}$.

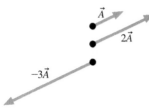

Multiplying a vector by -1 reverses its direction without changing its length.

As an example, FIGURE 3.7 shows vectors $\vec{A}$, $2\vec{A}$, and $-3\vec{A}$. Multiplication by 2 doubles the length of the vector but does not change its direction. Multiplication by -3 stretches the length by a factor of 3 *and* reverses the direction.

Vector Subtraction

How might we *subtract* vector $\vec{B}$ from vector $\vec{A}$ to form the vector $\vec{A} - \vec{B}$? With numbers, subtraction is the same as the addition of a negative number. That is, $5 - 3$ is the same as $5 + (-3)$. Similarly, $\vec{A} - \vec{B} = \vec{A} + (-\vec{B})$. We can use the rules for vector addition and the fact that $-\vec{B}$ is a vector opposite in direction to $\vec{B}$ to form rules for vector subtraction.

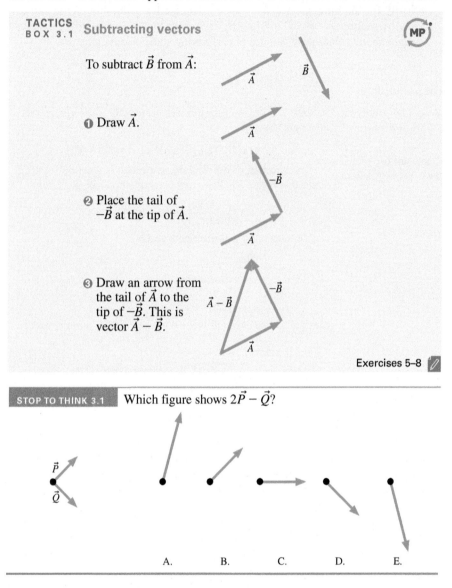

TACTICS
BOX 3.1 Subtracting vectors

To subtract $\vec{B}$ from $\vec{A}$:

❶ Draw $\vec{A}$.

❷ Place the tail of $-\vec{B}$ at the tip of $\vec{A}$.

❸ Draw an arrow from the tail of $\vec{A}$ to the tip of $-\vec{B}$. This is vector $\vec{A} - \vec{B}$.

Exercises 5–8

STOP TO THINK 3.1   Which figure shows $2\vec{P} - \vec{Q}$?

A.   B.   C.   D.   E.

3.2 Using Vectors on Motion Diagrams

In Chapter 2, we defined velocity for one-dimensional motion as an object's displacement—the change in position—divided by the time interval in which the change occurs:

$$v_x = \frac{\Delta x}{\Delta t} = \frac{x_f - x_i}{\Delta t}$$

In two dimensions, an object's displacement is a vector. Suppose an object undergoes displacement $\vec{d}$ during the time interval Δt. Let's define an object's velocity *vector* to be

$$\vec{v} = \frac{\vec{d}}{\Delta t} = \left(\frac{d}{\Delta t}, \text{same direction as } \vec{d}\right) \qquad (3.6)$$

Definition of velocity in two or more dimensions

Notice that we've multiplied a vector by a scalar: The velocity vector is simply the displacement vector multiplied by the scalar $1/\Delta t$. Consequently **the velocity vector points in the direction of the displacement.** As a result, we can use the dot-to-dot vectors on a motion diagram to visualize the velocity.

NOTE ▶ Strictly speaking, the velocity defined in Equation 3.6 is the *average* velocity for the time interval Δt. This is adequate for using motion diagrams to visualize motion. As we did in Chapter 2, when we make Δt very small, we get an *instantaneous* velocity we can use in performing some calculations. ◀

EXAMPLE 3.1   **Finding the velocity of an airplane**

A small plane is 100 km due east of Denver. After 1 hour of flying at a constant speed in the same direction, it is 200 km due north of Denver. What is the plane's velocity?

PREPARE The initial and final positions of the plane are shown in FIGURE 3.8; the displacement $\vec{d}$ is the vector that points from the initial to the final position.

FIGURE 3.8 Displacement vector for an airplane.

End

200 km $\vec{d}$

θ

Denver   Start
100 km

SOLVE The length of the displacement vector is the hypotenuse of a right triangle:

$$d = \sqrt{(100 \text{ km})^2 + (200 \text{ km})^2} = 224 \text{ km}$$

The direction of the displacement vector is described by the angle θ in Figure 3.8. From trigonometry, this angle is

$$\theta = \tan^{-1}\left(\frac{200 \text{ km}}{100 \text{ km}}\right) = \tan^{-1}(2.00) = 63.4°$$

Thus the plane's displacement vector is

$$\vec{d} = (224 \text{ km}, 63.4° \text{ north of west})$$

Because the plane undergoes this displacement during 1 hour, its velocity is

$$\vec{v} = \left(\frac{d}{\Delta t}, \text{ same direction as } \vec{d}\right) = \left(\frac{224 \text{ km}}{1 \text{ h}}, 63.4° \text{ north of west}\right)$$

$$= (224 \text{ km/h}, 63.4° \text{ north of west})$$

ASSESS The plane's *speed* is the magnitude of the velocity, $v = 224$ km/h. This is approximately 140 mph, which is a reasonable speed for a small plane.

We defined an object's acceleration in one dimension as $a_x = \Delta v_x / \Delta t$. In two dimensions, we need to use a vector to describe acceleration. The vector definition of acceleration is a straightforward extension of the one-dimensional version:

$$\vec{a} = \frac{\vec{v}_f - \vec{v}_i}{t_f - t_i} = \frac{\Delta \vec{v}}{\Delta t} \tag{3.7}$$

Definition of acceleration in two or more dimensions

There is an acceleration whenever there is a *change* in velocity. Because velocity is a vector, it can change in either or both of two possible ways:

1. The magnitude can change, indicating a change in speed.
2. The direction of motion can change.

In Chapter 2 we saw how to compute an acceleration vector for the first case, in which an object speeds up or slows down while moving in a straight line. In this chapter we will examine the second case, in which an object changes its direction of motion.

Suppose an object has an initial velocity $\vec{v}_i$ at time t_i and later, at time t_f, has velocity $\vec{v}_f$. The fact that the velocity *changes* tells us the object undergoes an acceleration during the time interval $\Delta t = t_f - t_i$. We see from Equation 3.7 that the acceleration points in the same direction as the vector $\Delta\vec{v}$. This vector is the change in the velocity $\Delta\vec{v} = \vec{v}_f - \vec{v}_i$, so to know which way the acceleration vector points, we have to perform the vector subtraction $\vec{v}_f - \vec{v}_i$. Tactics Box 3.1 showed how to perform vector subtraction. Tactics Box 3.2 shows how to use vector subtraction to find the acceleration vector.

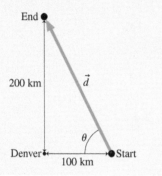

Lunging versus veering BIO   The top photo shows a barracuda, a type of fish that catches prey with a rapid linear acceleration, a quick change in speed. The barracuda's body shape is optimized for such a straight-line strike. The butterfly fish in the bottom photo has a very different appearance. It can't rapidly change its speed, but its body shape lets it quickly change its direction.

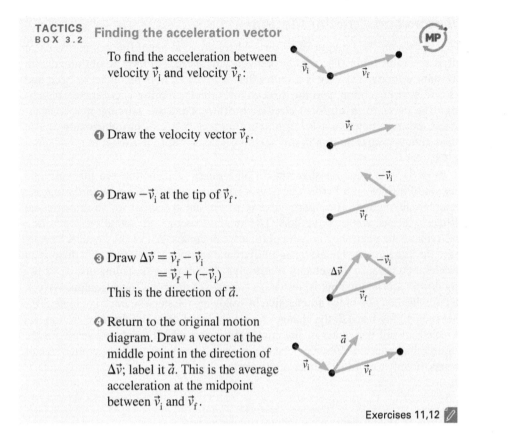

TACTICS
BOX 3.2 Finding the acceleration vector

To find the acceleration between velocity $\vec{v}_i$ and velocity $\vec{v}_f$:

❶ Draw the velocity vector $\vec{v}_f$.

❷ Draw $-\vec{v}_i$ at the tip of $\vec{v}_f$.

❸ Draw $\Delta \vec{v} = \vec{v}_f - \vec{v}_i$
$\qquad = \vec{v}_f + (-\vec{v}_i)$
This is the direction of $\vec{a}$.

❹ Return to the original motion diagram. Draw a vector at the middle point in the direction of $\Delta \vec{v}$; label it $\vec{a}$. This is the average acceleration at the midpoint between $\vec{v}_i$ and $\vec{v}_f$.

Exercises 11, 12 🖉

Now that we know how to determine acceleration vectors, we can make a complete motion diagram with dots showing the position of the object, average velocity vectors found by connecting the dots with arrows, and acceleration vectors found using Tactics Box 3.2. Note that there is *one* acceleration vector linking each *two* velocity vectors, and $\vec{a}$ is drawn at the dot between the two velocity vectors it links.

EXAMPLE 3.2 **Drawing the acceleration for a Mars descent**

A spacecraft slows as it safely descends to the surface of Mars. Draw a complete motion diagram for the last few seconds of the descent.

PREPARE FIGURE 3.9 shows two versions of a motion diagram: a professionally drawn version like you generally find in this text and a simpler version similar to what you might draw for a homework assignment. As the spacecraft slows in its descent, the dots get closer together and the velocity vectors get shorter.

SOLVE The inset in Figure 3.9 shows how Tactics Box 3.2 is used to determine the acceleration at one point. All the other acceleration vectors will be similar, because for each pair of velocity vectors the earlier one is longer than the later one.

ASSESS As the spacecraft slows, the acceleration vectors and velocity vectors point in opposite directions, consistent with what we learned about the sign of the acceleration in Chapter 2.

FIGURE 3.9 Motion diagram for a descending spacecraft.

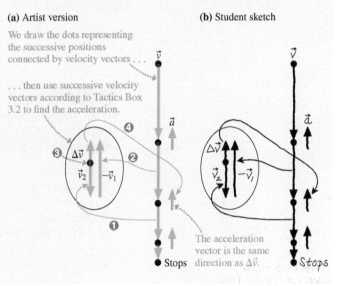

(a) Artist version

(b) Student sketch

We draw the dots representing the successive positions connected by velocity vectors . . .

. . . then use successive velocity vectors according to Tactics Box 3.2 to find the acceleration.

The acceleration vector is the same direction as $\Delta \vec{v}$.

The London Eye Ferris wheel.

Vectors and Circular Motion

The 32 cars on the London Eye Ferris wheel move at about 0.5 m/s in a vertical circle of radius 65 m. The cars have a constant *speed*, but they do not move with constant *velocity*. Velocity is a vector that depends on both an object's speed and its direction of motion, and the direction of circular motion is constantly changing. The cars are in **uniform circular motion**: They are moving at a constant speed, but in a continuously changing direction. We will introduce some basic ideas about circular motion in this section and then return to treat it in more detail in Chapter 6.

FIGURE 3.10 is a motion diagram for uniform circular motion—in this case, the motion of a rider on a Ferris wheel—showing 10 points of the motion during one complete revolution. The riders move at a constant speed, so we've drawn equal distances between successive dots. The velocity vectors are found by connecting each dot to the next; they are straight lines, not curves. All of the velocity vectors have the same length, but each has a different direction. This tells us that there is an acceleration because the velocity is changing. This is not a "speeding up" or "slowing down" acceleration but is, instead, a "change of direction" acceleration. We can deduce the direction of the acceleration by following the steps outlined in Figure 3.10. Between points 1 and 2, the change in velocity—and therefore the acceleration—is directed toward the center of the circle. No matter which points we select on the motion diagram, the velocities change in such a way that the acceleration vector points directly toward the center of the circle.

FIGURE 3.10 Motion diagram for uniform circular motion.

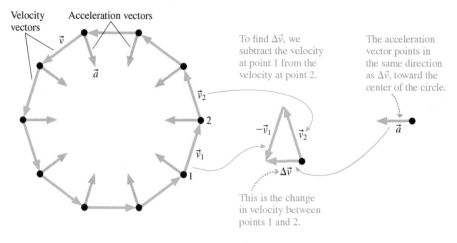

STOP TO THINK 3.2　The figure shows a car rounding a corner. Using what you've learned in Figure 3.10, what is the direction of the car's acceleration at the instant shown in the figure?

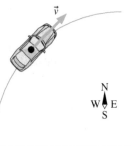

A. Northeast
B. Southeast
C. Southwest
D. Northwest

3.3 Coordinate Systems and Vector Components

In the past two sections, we have seen how to add and subtract vectors graphically, using these operations to deduce important details of motion. But the graphical combination of vectors is not an especially good way to find quantitative results. In this section we will introduce a *coordinate description* of vectors that will be the basis for doing vector calculations.

A coordinate system for an archaeological excavation.

Coordinate Systems

As we saw in Chapter 1, the world does not come with a coordinate system attached to it. A coordinate system is an artificially imposed grid that you place on a problem in order to make quantitative measurements. The right choice of coordinate system will make a problem easier to solve. We will generally use **Cartesian coordinates,** the familiar rectangular grid with perpendicular axes, as illustrated in FIGURE 3.11.

Coordinate axes have a positive end and a negative end, separated by zero at the origin where the two axes cross. When you draw a coordinate system, it is important to label the axes. This is done by placing *x* and *y* labels at the *positive* ends of the axes, as in Figure 3.11.

FIGURE 3.11 A Cartesian coordinate system.

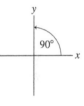

Component Vectors

FIGURE 3.12 shows a vector $\vec{A}$ and an *xy*-coordinate system that we've chosen. Once the directions of the axes are known, we can define two new vectors *parallel to the axes* that we call the **component vectors** of $\vec{A}$. Vector $\vec{A}_x$, called the *x-component vector,* is the projection of $\vec{A}$ along the *x*-axis. Vector $\vec{A}_y$, the *y-component vector,* is the projection of $\vec{A}$ along the *y*-axis.

You can see, using the parallelogram rule, that $\vec{A}$ is the vector sum of the two component vectors:

$$\vec{A} = \vec{A}_x + \vec{A}_y \tag{3.8}$$

In essence, we have "broken" vector $\vec{A}$ into two perpendicular vectors that are parallel to the coordinate axes. We say that we have **decomposed** or **resolved** vector $\vec{A}$ into its component vectors.

> NOTE ▶ It is not necessary for the tail of $\vec{A}$ to be at the origin. All we need to know is the *orientation* of the coordinate system so that we can draw $\vec{A}_x$ and $\vec{A}_y$ parallel to the axes. ◀

FIGURE 3.12 Component vectors $\vec{A}_x$ and $\vec{A}_y$ are drawn parallel to the coordinate axes such that $\vec{A} = \vec{A}_x + \vec{A}_y$.

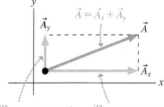

The *y*-component vector is parallel to the *y*-axis.

The *x*-component vector is parallel to the *x*-axis.

Components

Our convention is to give the one-dimensional kinematic variable v_x a positive sign if the velocity vector $\vec{v}$ points toward the positive end of the *x*-axis and a negative sign if $\vec{v}$ points in the negative *x*-direction. The basis of this rule is that v_x is the *x-component* of $\vec{v}$. We need to extend this idea to vectors in general.

Suppose we have a vector $\vec{A}$ that has been decomposed into component vectors $\vec{A}_x$ and $\vec{A}_y$ parallel to the coordinate axes. We can describe each component vector with a single number (a scalar) called the **component.** The *x-component* and *y-component* of vector $\vec{A}$, denoted A_x and A_y, are determined as follows:

TACTICS BOX 3.3 Determining the components of a vector   (MP)

❶ The absolute value $|A_x|$ of the *x*-component A_x is the magnitude of the component vector $\vec{A}_x$.
❷ The *sign* of A_x is positive if $\vec{A}_x$ points in the positive *x*-direction, negative if $\vec{A}_x$ points in the negative *x*-direction.
❸ The *y*-component A_y is determined similarly.

Exercises 16–18

FIGURE 3.13 Determining the components of a vector.

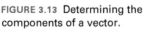

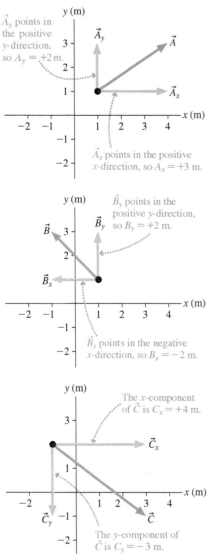

In other words, the component A_x tells us two things: how big $\vec{A}_x$ is and which end of the axis $\vec{A}_x$ points toward. FIGURE 3.13 shows three examples of determining the components of a vector.

NOTE ▶ $\vec{A}_x$ and $\vec{A}_y$ are *component vectors*—they have a magnitude and a direction. A_x and A_y are simply *components*. The components A_x and A_y are scalars—just numbers (with units) that can be positive or negative. ◀

Much of physics is expressed in the language of vectors. We will frequently need to decompose a vector into its components or to "reassemble" a vector from its components, moving back and forth between the graphical and the component representations of a vector.

Let's start with the problem of decomposing a vector into its x- and y-components. FIGURE 3.14a shows a vector $\vec{A}$ at an angle θ above horizontal. It is *essential* to use a picture or diagram such as this to define the angle you are using to describe a vector's direction. $\vec{A}$ points to the right and up, so Tactics Box 3.3 tells us that the components A_x and A_y are both positive.

FIGURE 3.14 Breaking a vector into components.

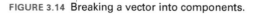

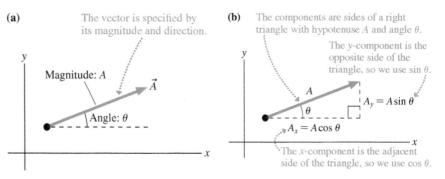

We can find the components using trigonometry, as illustrated in FIGURE 3.14b. For this case, we find that

$$A_x = A \cos \theta$$
$$A_y = A \sin \theta \tag{3.9}$$

where A is the magnitude, or length, of $\vec{A}$. These equations convert the length and angle description of vector $\vec{A}$ into the vector's components, but they are correct *only* if θ is measured from horizontal.

Alternatively, if we are given the components of a vector, we can determine the length and angle of the vector from the x- and y-components, as shown in FIGURE 3.15. Because A in Figure 3.15 is the hypotenuse of a right triangle, its length is given by the Pythagorean theorem:

$$A = \sqrt{A_x^2 + A_y^2} \tag{3.10}$$

FIGURE 3.15 Specifying a vector from its components.

Similarly, the tangent of angle θ is the ratio of the opposite side to the adjacent side, so

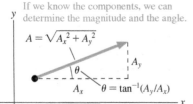

$$\theta = \tan^{-1}\left(\frac{A_y}{A_x}\right) \tag{3.11}$$

Equations 3.10 and 3.11 can be thought of as the "inverse" of Equations 3.9.

How do things change if the vector isn't pointing to the right and up—that is, if one of the components is negative? FIGURE 3.16 shows vector $\vec{C}$ pointing to the right and down. In this case, the component vector $\vec{C}_y$ is pointing *down*, in the negative *y*-direction, so the *y*-component C_y is a *negative* number. The angle ϕ is drawn measured from the *y*-axis, so the components of $\vec{C}$ are

$$C_x = C \sin \phi$$
$$C_y = -C \cos \phi \tag{3.12}$$

The roles of sine and cosine are reversed from those in Equations 3.9 because the angle ϕ is measured with respect to vertical, not horizontal.

NOTE ▶ Whether the *x*- and *y*-components use the sine or cosine depends on how you define the vector's angle. As noted above, you *must* draw a diagram to define the angle that you use, and you must be sure to refer to the diagram when computing components. Don't use Equations 3.9 or 3.12 as general rules—they aren't! They appear as they do because of how we defined the angles. ◀

Next, let's look at the "inverse" problem for this case: determining the length and direction of the vector given the components. The signs of the components don't matter for determining the length; the Pythagorean theorem always works to find the length or magnitude of a vector because the squares eliminate any concerns over the signs. The length of the vector in Figure 3.16 is simply

$$C = \sqrt{C_x^2 + C_y^2} \tag{3.13}$$

When we determine the direction of the vector from its components, we must consider the signs of the components. Finding the angle of vector $\vec{C}$ in Figure 3.16 requires the length of C_y *without* the minus sign, so vector $\vec{C}$ has direction

$$\phi = \tan^{-1}\left(\frac{C_x}{|C_y|}\right) \tag{3.14}$$

Notice that the roles of *x* and *y* differ from those in Equation 3.11.

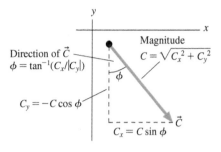

FIGURE 3.16 Relationships for a vector with a negative component.

EXAMPLE 3.3 **Finding the components of an acceleration vector**

Find the *x*- and *y*-components of the acceleration vector $\vec{a}$ shown in FIGURE 3.17.

FIGURE 3.17 Acceleration vector $\vec{a}$ of Example 3.4.

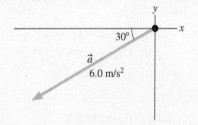

FIGURE 3.18 The components of the acceleration vector.

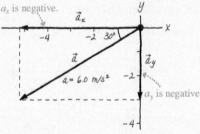

PREPARE It's important to *draw* the vectors. Making a sketch is crucial to setting up this problem. FIGURE 3.18 shows the original vector $\vec{a}$ decomposed into component vectors parallel to the axes.

SOLVE The acceleration vector $\vec{a} = (6.0 \text{ m/s}^2, 30°$ below the negative *x*-axis) points to the left (negative *x*-direction) and down (negative *y*-direction), so the components a_x and a_y are both negative:

$$a_x = -a\cos 30° = -(6.0 \text{ m/s}^2)\cos 30° = -5.2 \text{ m/s}^2$$
$$a_y = -a\sin 30° = -(6.0 \text{ m/s}^2)\sin 30° = -3.0 \text{ m/s}^2$$

ASSESS The magnitude of the *y*-component is less than that of the *x*-component, as seems to be the case in Figure 3.18, a good check on our work. The units of a_x and a_y are the same as the units of vector $\vec{a}$. Notice that we had to insert the minus signs manually by observing that the vector points down and to the left.

STOP TO THINK 3.3 | What are the x- and y-components C_x and C_y of vector $\vec{C}$?

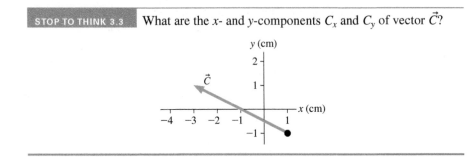

Working with Components

FIGURE 3.19 Using components to add vectors.

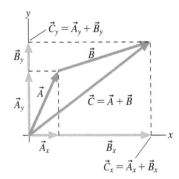

We've seen how to add vectors graphically, but there's an easier way: using components. To illustrate, let's look at the vector sum $\vec{C} = \vec{A} + \vec{B}$ for the vectors shown in FIGURE 3.19. You can see that the component vectors of $\vec{C}$ are the sums of the component vectors of $\vec{A}$ and $\vec{B}$. The same is true of the components: $C_x = A_x + B_x$ and $C_y = A_y + B_y$.

In general, if $\vec{D} = \vec{A} + \vec{B} + \vec{C} + \cdots$, then the x- and y-components of the resultant vector $\vec{D}$ are

$$D_x = A_x + B_x + C_x + \cdots$$
$$D_y = A_y + B_y + C_y + \cdots \qquad (3.15)$$

This method of vector addition is called *algebraic addition*.

EXAMPLE 3.4 | **Using algebraic addition to find a bird's displacement**

A bird flies 100 m due east from a tree, then 200 m northwest (that is, 45° north of west). What is the bird's net displacement?

PREPARE FIGURE 3.20a shows the displacement vectors $\vec{A} = (100 \text{ m, east})$ and $\vec{B} = (200 \text{ m, northwest})$ and also the net displacement $\vec{C}$. We draw vectors tip-to-tail if we are going to add them graphically, but it's usually easier to draw them all from the origin if we are going to use algebraic addition. FIGURE 3.20b redraws the vectors with their tails together.

FIGURE 3.20 Finding the net displacement.

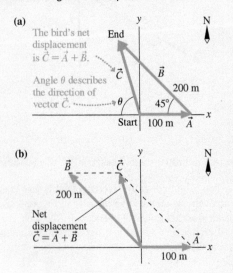

SOLVE To add the vectors algebraically we must know their components. From the figure these are seen to be

$$A_x = 100 \text{ m}$$
$$A_y = 0 \text{ m}$$

$$B_x = -(200 \text{ m})\cos 45° = -141 \text{ m}$$
$$B_y = (200 \text{ m})\sin 45° = 141 \text{ m}$$

We learned *from the figure* that $\vec{B}$ has a negative x-component. Adding $\vec{A}$ and $\vec{B}$ by components gives

$$C_x = A_x + B_x = 100 \text{ m} - 141 \text{ m} = -41 \text{ m}$$
$$C_y = A_y + B_y = 0 \text{ m} + 141 \text{ m} = 141 \text{ m}$$

The magnitude of the net displacement $\vec{C}$ is

$$C = \sqrt{C_x^2 + C_y^2} = \sqrt{(-41 \text{ m})^2 + (141 \text{ m})^2} = 147 \text{ m}$$

The angle θ, as defined in Figure 3.20, is

$$\theta = \tan^{-1}\left(\frac{C_y}{|C_x|}\right) = \tan^{-1}\left(\frac{141 \text{ m}}{41 \text{ m}}\right) = 74°$$

Thus the bird's net displacement is $\vec{C} = (147 \text{ m}, 74° \text{ north of west})$.

ASSESS The final values of C_x and C_y match what we would expect from the sketch in Figure 3.20. The geometric addition was a valuable check on the answer we found by algebraic addition.

Vector subtraction and the multiplication of a vector by a scalar are also easily performed using components. To find $\vec{D} = \vec{P} - \vec{Q}$ we would compute

$$D_x = P_x - Q_x$$
$$D_y = P_y - Q_y$$

(3.16)

Similarly, $\vec{T} = c\vec{S}$ has components

$$T_x = cS_x$$
$$T_y = cS_y$$

(3.17)

The next few chapters will make frequent use of *vector equations*. For example, you will learn that the equation to calculate the net force on a car skidding to a stop is

$$\vec{F} = \vec{n} + \vec{w} + \vec{f}$$

(3.18)

Equation 3.18 is really just a shorthand way of writing the two simultaneous equations:

$$F_x = n_x + w_x + f_x$$
$$F_y = n_y + w_y + f_y$$

(3.19)

In other words, a vector equation is interpreted as meaning: Equate the *x*-components on both sides of the equals sign, then equate the *y*-components. Vector notation allows us to write these two equations in a more compact form.

Tilted Axes

Although we are used to having the *x*-axis horizontal, there is no requirement that it has to be that way. In Chapter 1, we saw that for motion on a slope, it is often most convenient to put the *x*-axis along the slope. When we add the *y*-axis, this gives us a tilted coordinate system such as that shown in FIGURE 3.21.

Finding components with tilted axes is no harder than what we have done so far. Vector $\vec{C}$ in Figure 3.21 can be decomposed into component vectors $\vec{C}_x$ and $\vec{C}_y$, with $C_x = C\cos\theta$ and $C_y = C\sin\theta$.

FIGURE 3.21 A coordinate system with tilted axes.

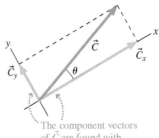

The component vectors of $\vec{C}$ are found with respect to the tilted axes.

STOP TO THINK 3.4 Angle ϕ that specifies the direction of $\vec{C}$ is computed as

A. $\tan^{-1}(C_x/C_y)$.
B. $\tan^{-1}(C_x/|C_y|)$.
C. $\tan^{-1}(|C_x|/|C_y|)$.
D. $\tan^{-1}(C_y/C_x)$.
E. $\tan^{-1}(C_y/|C_x|)$.
F. $\tan^{-1}(|C_y|/|C_x|)$.

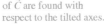

3.4 Motion on a Ramp

In this section, we will examine the problem of motion on a ramp or incline. There are three reasons to look at this problem. First, it will provide good practice at using vectors to analyze motion. Second, it is a simple problem for which we can find an exact solution. Third, this seemingly abstract problem has real and important applications.

We begin with a constant-velocity example to give us some practice with vectors and components before moving on to the more general case of accelerated motion.

Class Video

EXAMPLE 3.5   Finding the height gained on a slope

A car drives up a steep 10° slope at a constant speed of 15 m/s. After 10 s, how much height has the car gained?

PREPARE FIGURE 3.22 is a visual overview, with x- and y-axes defined. The velocity vector $\vec{v}$ points up the slope. We are interested in the vertical motion of the car, so we decompose $\vec{v}$ into component vectors $\vec{v}_x$ and $\vec{v}_y$ as shown.

FIGURE 3.22   Visual overview of a car moving up a slope.

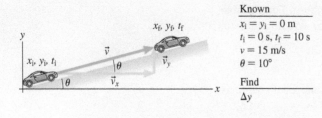

Known
$x_i = y_i = 0$ m
$t_i = 0$ s, $t_f = 10$ s
$v = 15$ m/s
$\theta = 10°$

Find
Δy

SOLVE The velocity component we need is v_y, which describes the vertical motion of the car. Using the rules for finding components outlined above, we find

$$v_y = v \sin\theta = (15 \text{ m/s}) \sin 10° = 2.6 \text{ m/s}$$

Because the velocity is constant, the car's vertical displacement (i.e., the height gained) during 10 s is

$$\Delta y = v_y \, \Delta t = (2.6 \text{ m/s})(10 \text{ s}) = 26 \text{ m}$$

ASSESS The car is traveling at a pretty good clip—15 m/s is a bit faster than 30 mph—up a steep slope, so it should climb a respectable height in 10 s. 26 m, or about 80 ft, seems reasonable.

Accelerated Motion on a Ramp

FIGURE 3.23   Acceleration on an inclined plane.

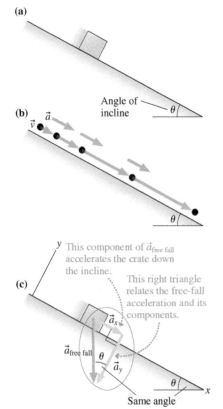

(a)

Angle of incline θ

(b) $\vec{a}$ $\vec{v}$ θ

(c) y This component of $\vec{a}_{\text{free fall}}$ accelerates the crate down the incline.

This right triangle relates the free-fall acceleration and its components.

$\vec{a}_x$
$\vec{a}_{\text{free fall}}$ θ $\vec{a}_y$
θ x

Same angle

FIGURE 3.23a shows a crate sliding down a frictionless (i.e., smooth) ramp tilted at angle θ. The crate accelerates due to the action of gravity, but it is *constrained* to accelerate parallel to the surface. What is the acceleration?

A motion diagram for the crate is drawn in FIGURE 3.23b. There is an acceleration because the velocity is changing, with both the acceleration and velocity vectors parallel to the ramp. We can take advantage of the properties of vectors to find the crate's acceleration. To do so, FIGURE 3.23c sets up a coordinate system with the x-axis along the ramp and the y-axis perpendicular. All motion will be along the x-axis.

If the incline suddenly vanished, the object would have a free-fall acceleration $\vec{a}_{\text{free fall}}$ straight down. As Figure 3.23c shows, this acceleration vector can be decomposed into two component vectors: a vector $\vec{a}_x$ that is *parallel* to the incline and a vector $\vec{a}_y$ that is *perpendicular* to the incline. The vector addition rules studied earlier in this chapter tell us that $\vec{a}_{\text{free fall}} = \vec{a}_x + \vec{a}_y$.

The motion diagram shows that the object's actual acceleration $\vec{a}_x$ is parallel to the incline. The surface of the incline somehow "blocks" the other component of the acceleration $\vec{a}_y$, through a process we will examine in Chapter 5, but $\vec{a}_x$ is unhindered. It is this component of $\vec{a}_{\text{free fall}}$, parallel to the incline, that accelerates the object.

We can use trigonometry to work out the magnitude of this acceleration. Figure 3.23c shows that the three vectors $\vec{a}_{\text{free fall}}$, $\vec{a}_y$, and $\vec{a}_x$ form a right triangle with angle θ as shown. This angle is the same as the angle of the incline. By definition, the magnitude of $\vec{a}_{\text{free fall}}$ is g. This vector is the hypotenuse of the right triangle. The vector we are interested in, $\vec{a}_x$, is opposite angle θ. Thus the value of the acceleration along a frictionless slope is

$$a_x = \pm g \sin\theta \qquad (3.20)$$

NOTE ▶ The correct sign depends on the direction in which the ramp is tilted. The acceleration in Figure 3.23 is $+g \sin\theta$, but upcoming examples will show situations in which the acceleration is $-g \sin\theta$. ◀

Let's look at Equation 3.20 to verify that it makes sense. A good way to do this is to consider some **limiting cases** in which the angle is at one end of its range. In these cases, the physics is clear and we can check our result. Let's look at two such possibilities:

1. Suppose the plane is perfectly horizontal, with $\theta = 0°$. If you place an object on a horizontal surface, you expect it to stay at rest with no acceleration. Equation 3.20 gives $a_x = 0$ when $\theta = 0°$, in agreement with our expectations.

2. Now suppose you tilt the plane until it becomes vertical, with $\theta = 90°$. You know what happens—the object will be in free fall, parallel to the vertical surface. Equation 3.20 gives $a_x = g$ when $\theta = 90°$, again in agreement with our expectations.

Video Tutor
Demo

NOTE ▶ Checking your answer by looking at such limiting cases is a very good way to see if your answer makes sense. We will often do this in the Assess step of a solution. ◀

▶**Extreme physics** A speed skier, on wide skis with little friction, wearing an aerodynamic helmet and crouched low to minimize air resistance, moves in a straight line down a steep slope—pretty much like an object sliding down a frictionless ramp. There is a maximum speed that a skier could possibly achieve at the end of the slope.

EXAMPLE 3.6   **Maximum possible speed for a skier**

The Willamette Pass ski area in Oregon was the site of the 1993 U.S. National Speed Skiing Competition. The skiers started from rest and then accelerated down a stretch of the mountain with a reasonably constant slope, aiming for the highest possible speed at the end of this run. During this acceleration phase, the skiers traveled 360 m while dropping a vertical distance of 170 m. What is the fastest speed a skier could achieve at the end of this run?

PREPARE We begin with the visual overview in **FIGURE 3.24**. The motion diagram shows the acceleration of the skier and the pictorial representation gives an overview of the problem including the dimensions of the slope. As before, we put the x-axis along the slope.

FIGURE 3.24 Visual overview of a skier accelerating down a slope.

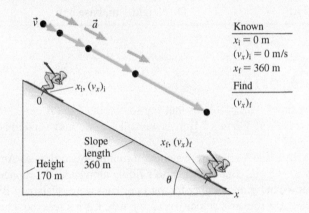

Known
$x_i = 0$ m
$(v_x)_i = 0$ m/s
$x_f = 360$ m

Find
$(v_x)_f$

SOLVE The fastest possible run would be one without any friction or air resistance, meaning the acceleration down the slope is given by Equation 3.20. The acceleration is in the positive x-direction, so we use the positive sign. What is the angle in Equation 3.20? Figure 3.24 shows that the 360-m-long slope is the hypotenuse of a triangle of height 170 m, so we use trigonometry to find

$$\sin\theta = \frac{170 \text{ m}}{360 \text{ m}}$$

which gives $\theta = \sin^{-1}(170/360) = 28°$. Equation 3.20 then gives

$$a_x = +g\sin\theta = (9.8 \text{ m/s}^2)(\sin 28°) = 4.6 \text{ m/s}^2$$

For linear motion with constant acceleration, we can use the third of the kinematic equations in Synthesis 2.1: $(v_x)_f^2 = (v_x)_i^2 + 2a_x\,\Delta x$. The initial velocity $(v_x)_i$ is zero; thus

This is the distance along the slope, the length of the run.

$$(v_x)_f = \sqrt{2a_x\Delta x} = \sqrt{2(4.6 \text{ m/s}^2)(360 \text{ m})} = 58 \text{ m/s}$$

This is the fastest that any skier could hope to be moving at the end of the run. Any friction or air resistance would decrease this speed.

ASSESS The final speed we calculated is 58 m/s, which is about 130 mph, reasonable because we expect a high speed for this sport. In the competition noted, the actual winning speed was 111 mph, not much slower than the result we calculated. Obviously, efforts to minimize friction and air resistance are working!

Skis on snow have very little friction, but there are other ways to reduce the friction between surfaces. For instance, a roller coaster car rolls along a track on low-friction wheels. No drive force is applied to the cars after they are released at the top of the first hill: the speed changes due to gravity alone. The cars speed up as they go down hills and slow down as they climb.

EXAMPLE 3.7   Speed of a roller coaster

A classic wooden coaster has cars that go down a big first hill, gaining speed. The cars then ascend a second hill with a slope of 30°. If the cars are going 25 m/s at the bottom and it takes them 2.0 s to climb this hill, how fast are they going at the top?

PREPARE We start with the visual overview in FIGURE 3.25, which includes a motion diagram, a pictorial representation, and a list of values. We've done this with a sketch such as you might

FIGURE 3.25 The coaster's speed decreases as it goes up the hill.

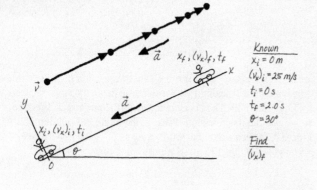

draw for your homework. Notice how the motion diagram of Figure 3.25 differs from that of Example 3.6: The velocity decreases as the car moves up the hill, so the acceleration vector is opposite the direction of the velocity vector. The motion is along the x-axis, as before, but the acceleration vector points in the negative x-direction, so the component a_x is negative. In the motion diagram, notice that we drew only a single acceleration vector—a reasonable shortcut because we know that the acceleration is constant. One vector can represent the acceleration for the entire motion.

SOLVE To determine the final speed, we need to know the acceleration. We will assume that there is no friction or air resistance, so the magnitude of the roller coaster's acceleration is given by Equation 3.20 using the minus sign, as noted:

$$a_x = -g\sin\theta = -(9.8 \text{ m/s}^2)\sin 30° = -4.9 \text{ m/s}^2$$

The speed at the top of the hill can then be computed using our kinematic equation for velocity:

$$(v_x)_f = (v_x)_i + a_x\,\Delta t = 25 \text{ m/s} + (-4.9 \text{ m/s}^2)(2.0 \text{ s}) = 15 \text{ m/s}$$

ASSESS The speed is less at the top of the hill than at the bottom, as it should be, but the coaster is still moving at a pretty good clip at the top—almost 35 mph. This seems reasonable—A fast ride is a fun ride.

STOP TO THINK 3.5   A block of ice slides down a ramp. For which height and base length is the acceleration the greatest?

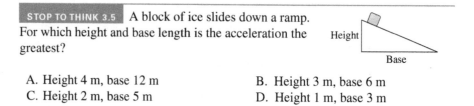

A. Height 4 m, base 12 m   B. Height 3 m, base 6 m
C. Height 2 m, base 5 m   D. Height 1 m, base 3 m

3.5 Relative Motion

You've now dealt many times with problems that say something like "A car travels at 30 m/s" or "A plane travels at 300 m/s." But, as we will see, we may need to be a bit more specific.

In FIGURE 3.26, Amy, Bill, and Carlos are watching a runner. According to Amy, the runner's velocity is $v_x = 5$ m/s. But to Bill, who's riding alongside, the runner is lifting his legs up and down but going neither forward nor backward relative to Bill. As far as Bill is concerned, the runner's velocity is $v_x = 0$ m/s. Carlos sees the runner receding in his rearview mirror, in the *negative x*-direction, getting 10 m farther away from him every second. According to Carlos, the runner's velocity is $v_x = -10$ m/s. Which is the runner's *true* velocity?

Velocity is not a concept that can be true or false. The runner's velocity *relative to Amy* is 5 m/s; that is, his velocity is 5 m/s in a coordinate system attached to Amy and in which Amy is at rest. The runner's velocity relative to Bill is 0 m/s, and the velocity relative to Carlos is −10 m/s. These are all valid descriptions of the runner's motion.

FIGURE 3.26 Amy, Bill, and Carlos each measure the velocity of the runner. The velocities are shown relative to Amy.

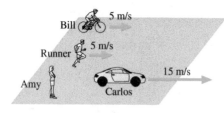

Relative Velocity

Suppose we know that the runner's velocity relative to Amy is 5 m/s; we will call this velocity $(v_x)_{RA}$. The second subscript "RA" means "**R**unner relative to **A**my." We also know that the velocity of Carlos relative to Amy is 15 m/s; we write this as $(v_x)_{CA} = 15$ m/s. It is equally valid to compute Amy's velocity relative to Carlos. From Carlos's point of view, Amy is moving to the left at 15 m/s. We write Amy's velocity relative to Carlos as $(v_x)_{AC} = -15$ m/s. Note that $(v_x)_{AC} = -(v_x)_{CA}$.

Given the runner's velocity relative to Amy and Amy's velocity relative to Carlos, we can compute the runner's velocity relative to Carlos by combining the two velocities we know. The subscripts as we have defined them are our guide for this combination:

$$(v_x)_{RC} = (v_x)_{RA} + (v_x)_{AC} \qquad (3.21)$$

The "A" appears on the right of the first expression and on the left of the second; when we combine these velocities, we "cancel" the A to get $(v_x)_{RC}$.

Generally, you can add two relative velocities in this manner, by "canceling" subscripts as in Equation 3.21. In Chapter 27, when we learn about relativity, we will have a more rigorous scheme for computing relative velocities, but this technique will serve our purposes at present.

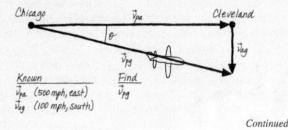

Throwing for the gold An athlete throwing the javelin does so while running. It's harder to throw the javelin on the run, but there's a very good reason to do so. The distance of the throw will be determined by the velocity of the javelin with respect to the ground—which is the sum of the velocity of the throw plus the velocity of the athlete. A faster run means a longer throw.

EXAMPLE 3.8   **Speed of a seabird**

Researchers doing satellite tracking of albatrosses in the Southern Ocean observed a bird maintaining sustained flight speeds of 35 m/s—nearly 80 mph! This seems surprisingly fast until you realize that this particular bird was flying with the wind, which was moving at 23 m/s. What was the bird's airspeed—its speed relative to the air? This is a truer measure of its flight speed.

PREPARE FIGURE 3.27 shows the wind and the albatross moving to the right, so all velocities will be positive. We've shown the

velocity $(v_x)_{bw}$ of the bird with respect to the water, which is the measured flight speed, and the velocity $(v_x)_{aw}$ of the air with respect to the water, which is the known wind speed. We want to find the bird's airspeed—the speed of the bird with respect to the air.

SOLVE We've noted three different velocities that are important in the problem: $(v_x)_{bw}$, $(v_x)_{aw}$, and $(v_x)_{ba}$. We can combine these in the usual way:

$$(v_x)_{bw} = (v_x)_{ba} + (v_x)_{aw}$$

Then, to solve for $(v_x)_{ba}$, we can rearrange the terms:

$$(v_x)_{ba} = (v_x)_{bw} - (v_x)_{aw} = 35 \text{ m/s} - 23 \text{ m/s} = 12 \text{ m/s}$$

ASSESS 12 m/s—about 25 mph—is a reasonable airspeed for a bird. And it's slower than the observed flight speed, which makes sense because the bird is flying with the wind.

FIGURE 3.27 Relative velocities for the albatross and the wind for Example 3.8.

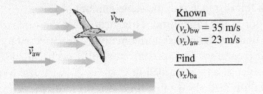

Known
$(v_x)_{bw} = 35$ m/s
$(v_x)_{aw} = 23$ m/s

Find
$(v_x)_{ba}$

This technique for finding relative velocities also works for two-dimensional situations, another good exercise in working with vectors.

EXAMPLE 3.9   **Finding the ground speed of an airplane**

Cleveland is approximately 300 miles east of Chicago. A plane leaves Chicago flying due east at 500 mph. The pilot forgot to check the weather and doesn't know that the wind is blowing to the south at 100 mph. What is the plane's velocity relative to the ground?

PREPARE FIGURE 3.28 is a visual overview of the situation. We are given the speed of the plane relative to the air ($\vec{v}_{pa}$) and the speed of the air relative to the ground ($\vec{v}_{ag}$); the speed of the plane relative to the ground will be the vector sum of these velocities:

$$\vec{v}_{pg} = \vec{v}_{pa} + \vec{v}_{ag}$$

This vector sum is shown in Figure 3.28.

FIGURE 3.28 The wind causes a plane flying due east in the air to move to the southeast relative to the ground.

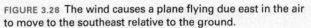

Known	Find
$\vec{v}_{pa}$ (500 mph, east)	$\vec{v}_{pg}$
$\vec{v}_{ag}$ (100 mph, south)	

Continued

SOLVE The plane's speed relative to the ground is the hypotenuse of the right triangle in Figure 3.28; thus:

$$v_{pg} = \sqrt{v_{pa}^2 + v_{ag}^2} = \sqrt{(500 \text{ mph})^2 + (100 \text{ mph})^2} = 510 \text{ mph}$$

The plane's direction can be specified by the angle θ measured from due east:

$$\theta = \tan^{-1}\left(\frac{100 \text{ mph}}{500 \text{ mph}}\right) = \tan^{-1}(0.20) = 11°$$

The velocity of the plane relative to the ground is thus

$$\vec{v}_{pg} = (510 \text{ mph}, 11° \text{ south of east})$$

ASSESS The good news is that the wind is making the plane move a bit faster relative to the ground. The bad news is that the wind is making the plane move in the wrong direction!

STOP TO THINK 3.6 The water in a river flows downstream at 3.0 m/s. A boat is motoring upstream against the flow at 5.0 m/s relative to the water. What is the boat's speed relative to the riverbank?

A. 8.0 m/s   B. 5.0 m/s   C. 3.0 m/s   D. 2.0 m/s

3.6 Motion in Two Dimensions: Projectile Motion

Balls flying through the air, long jumpers, and cars doing stunt jumps are all examples of the two-dimensional motion that we call **projectile motion.** Projectile motion is an extension to two dimensions of the free-fall motion we studied in ◀ SECTION 2.7. **A projectile is an object that moves in two dimensions under the influence of gravity and nothing else.** Although real objects are also influenced by air resistance, the effect of air resistance is small for reasonably dense objects moving at modest speeds, so we can ignore it for the cases we consider in this chapter. As long as we can neglect air resistance, any projectile will follow the same type of path. Because the form of the motion will always be the same, the strategies we develop to solve one projectile problem can be applied to others as well.

FIGURE 3.29 is a strobe photograph of two balls, one shot horizontally and the other released from rest at the same instant. The *vertical* motions of the two balls are identical, and they hit the floor simultaneously. Neither ball has any initial motion in the vertical direction, and both fall distance *h* in the same amount of time.

The two balls hit the floor at the same time. This means that the *vertical* motion of the yellow ball, the one that was launched horizontally, is not affected by the fact that the ball is moving horizontally. The vertical motion of each ball is free fall, which is the same for all objects, as we learned in Chapter 2. A careful look at the *horizontal* motion of the yellow ball shows that it is uniform motion; the horizontal motion continues as if the ball were not falling.

So, for an object in projectile motion, the initial horizontal velocity has *no* influence over the vertical motion, and vice versa. We'd find something similar if we looked in detail at any projectile motion, so we can state a general rule: **The horizontal and vertical components of an object undergoing projectile motion are independent of each other.**

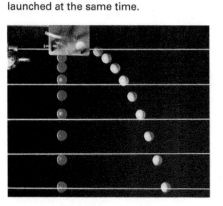

FIGURE 3.29 The motions of two balls launched at the same time.

CONCEPTUAL EXAMPLE 3.10 Time and distance for balls rolled off the table

Two balls are rolling toward the edge of a table, with ball 1 rolling twice as fast as ball 2. Both balls leave the edge table at the same time. Which ball hits the ground first? Which ball goes farther?

REASON The vertical motion of both balls is the same—free fall—and they both fall from the same height. Both balls are in the air for the same time interval and hit the ground at the exact same time. During this time interval, the two balls continue moving horizontally at the speed with which they left the table. Ball 2 has twice the horizontal speed of ball 1; it will therefore go twice as far.

ASSESS This result makes sense. The vertical and horizontal motions are independent of each other, so we can analyze them separately. If you drop two objects from the same height, they hit the ground at the same time, so both balls should land at the same time. And as they drop, the one that was moving faster horizontally will continue to do so—thus it will go farther.

We've shown the independence of the horizontal and vertical motions for the case of a ball launched horizontally, but this result applies even if a projectile is launched at an angle. FIGURE 3.30 shows the motion diagram for a ball tossed into the air at an angle. The acceleration vector points in the same direction as the change in velocity, which we can compute using the techniques of Tactics Box 3.2. The acceleration vector points straight down. A careful analysis would show that it has magnitude 9.80 m/s². This is just what we'd expect: The acceleration of a projectile is the same as the acceleration of an object falling straight down—namely, the free-fall acceleration.

As the projectile moves, the free-fall acceleration will change the *vertical* component of the velocity, but there will be no change to the *horizontal* component of the velocity. **The vertical component of acceleration a_y for all projectile motion is just the familiar $-g$ of free fall, while the horizontal component a_x is zero.**

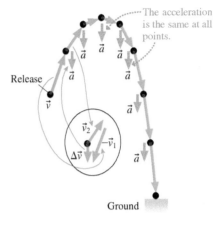

FIGURE 3.30 The motion of a tossed ball.

Analyzing Projectile Motion

Suppose you toss a basketball down the court, as shown in FIGURE 3.31. To study the resulting projectile motion, we've established a coordinate system with the x-axis horizontal and the y-axis vertical. Because we call the start of a projectile's motion the launch, the angle of the initial velocity above the horizontal (i.e., above the x-axis) is the **launch angle**. As you learned in Section 3.3, the initial velocity vector can be expressed in terms of the x- and y-components, which are shown in the figure.

> NOTE ▶ The components $(v_x)_i$ and $(v_y)_i$ are not always positive. A projectile launched at an angle *below* the horizontal (such as a ball thrown downward from the roof of a building) has *negative* values for θ and $(v_y)_i$. However, the *speed* v_i is always positive. ◀

Once the basketball leaves your hand, its subsequent motion is determined by the initial components of the velocity and the acceleration. To see how this plays out in practice, let's look at a specific case with definite numbers. FIGURE 3.32 shows a projectile launched at a speed of 22.0 m/s at an angle of 63° from the horizontal. In Figure 3.32a, the initial velocity vector is broken into its horizontal and vertical components. In Figure 3.32b, the velocity vector and its component vectors are shown every 1.0 s. Because there is no horizontal acceleration ($a_x = 0$) the value of v_x never changes. In contrast, v_y decreases by 9.8 m/s every second. (This is what it *means* to accelerate at $a_y = -9.8$ m/s² $= (-9.8$ m/s) per second.) There is nothing *pushing* the projectile along the curve. As the ball moves, the downward acceleration changes the velocity vector as shown, causing it to tip downward as the motion proceeds. At the end of the arc, when the ball is at the same height as it started, v_y is -19.6 m/s, the negative of its initial value. **The ball finishes its motion moving downward at the same speed as it started moving upward,** just as we saw in the case of one-dimensional free fall in Chapter 2.

You can see from Figure 3.32 that **projectile motion is made up of two independent motions: uniform motion at constant velocity in the horizontal direction and free-fall motion in the vertical direction.** The two motions are independent, but because they occur together they must be analyzed together, as we'll see.

In Chapter 2, we saw kinematic equations for constant-velocity and constant-acceleration motion. We can adapt these equations to this current case. The horizontal motion is constant-velocity motion at $(v_x)_i$, and the vertical motion is constant-acceleration motion with initial velocity $(v_y)_i$ and an acceleration of $a_y = -g$.

Let's summarize and synthesize everything we've learned to this point about projectile motion, including the equations from Chapter 2 for the horizontal and vertical motion. We'll use this information to solve some problems in the next section.

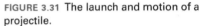

FIGURE 3.31 The launch and motion of a projectile.

$\vec{v}_i$ is the initial velocity.

$(v_y)_i = v_i \sin\theta$ ... Initial velocity

θ - Launch angle

$(v_x)_i = v_i \cos\theta$

We compute components of the initial velocity as shown.

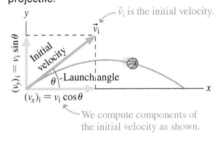

FIGURE 3.32 The velocity and acceleration vectors of a projectile.

(a) $\vec{v}_i = (22.0$ m/s, 63° from horizontal)

The initial velocity is decomposed into horizontal and vertical components.

$(v_y)_i = 19.6$ m/s

63°

$(v_x)_i = 10.0$ m/s

(b)

The vertical component of velocity decreases by 9.8 m/s every second.

The horizontal component of velocity is constant throughout the motion.

$a_y = -9.8$ m/s per s

Velocity vectors are shown every 1 s. Values are in m/s.

When the particle returns to its initial height, v_y is opposite its initial value.

SYNTHESIS 3.1 Projectile motion

The horizontal and vertical components of projectile motion are independent, but must be analyzed together.

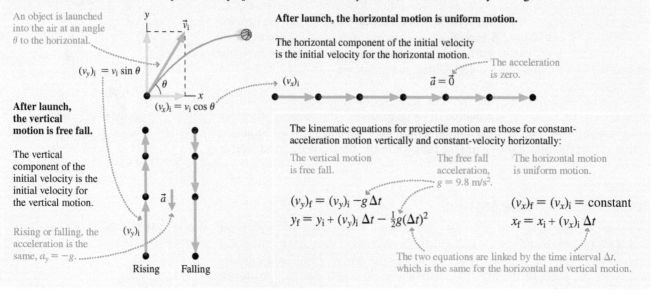

An object is launched into the air at an angle θ to the horizontal.

$(v_y)_i = v_i \sin\theta$

$(v_x)_i = v_i \cos\theta$

After launch, the vertical motion is free fall.

The vertical component of the initial velocity is the initial velocity for the vertical motion.

Rising or falling, the acceleration is the same, $a_y = -g$.

$(v_y)_i$

Rising   Falling

After launch, the horizontal motion is uniform motion.

The horizontal component of the initial velocity is the initial velocity for the horizontal motion.

$(v_x)_i$

The acceleration is zero.

$\vec{a} = \vec{0}$

The kinematic equations for projectile motion are those for constant-acceleration motion vertically and constant-velocity horizontally:

The vertical motion is free fall.

The free fall acceleration, $g = 9.8 \text{ m/s}^2$.

The horizontal motion is uniform motion.

$$(v_y)_f = (v_y)_i - g\,\Delta t$$
$$y_f = y_i + (v_y)_i\,\Delta t - \tfrac{1}{2}g(\Delta t)^2$$

$$(v_x)_f = (v_x)_i = \text{constant}$$
$$x_f = x_i + (v_x)_i\,\Delta t$$

The two equations are linked by the time interval Δt, which is the same for the horizontal and vertical motion.

Video Tutor Demo

Video Tutor Demo

Video Tutor Demo

Video Tutor Demo

STOP TO THINK 3.7 A 100 g ball rolls off a table and lands 2 m from the base of the table. A 200 g ball rolls off the same table with the same speed. How far does it land from the base of the table?

A. <1 m   B. 1 m
C. Between 1 m and 2 m   D. 2 m
E. Between 2 m and 4 m   F. 4 m

3.7 Projectile Motion: Solving Problems

Now that we have a good idea of how projectile motion works, we can use that knowledge to solve some true two-dimensional motion problems.

EXAMPLE 3.11   Dock jumping

In the sport of dock jumping, dogs run at full speed off the end of a dock that sits a few feet above a pool of water. The winning dog is the one that lands farthest from the end of the dock. If a dog runs at 8.5 m/s (a pretty typical speed for this event) straight off the end of a dock that is 0.61 m (2 ft, a standard height) above the water, how far will the dog go before splashing down?

PREPARE We start with a visual overview of the situation in FIGURE 3.33. We have chosen to put the origin of the coordinate system at the base of the dock. The dog runs horizontally off the end of the dock, so the initial components of the velocity are $(v_x)_i = 8.5$ m/s and $(v_y)_i = 0$ m/s. We can treat this as a projectile motion problem, so we can use the details and equations presented in Synthesis 3.1 above.

We know that the horizontal and vertical motions are independent. The fact that the dog is falling toward the water doesn't affect its horizontal motion. When the dog leaves the end of

FIGURE 3.33 Visual overview for Example 3.11.

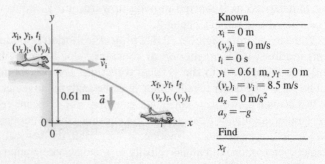

Known
$x_i = 0$ m
$(v_y)_i = 0$ m/s
$t_i = 0$ s
$y_i = 0.61$ m, $y_f = 0$ m
$(v_x)_i = v_i = 8.5$ m/s
$a_x = 0$ m/s^2
$a_y = -g$

Find
x_f

the dock, it will continue to move horizontally at 8.5 m/s. The vertical motion is free fall. The jump ends when the dog hits the water—that is, when it has dropped by 0.61 m. We are ultimately interested in how far the dog goes, but to make this determination we'll need to find the time interval Δt that the dog is in the air.

SOLVE We'll start by solving for the time interval Δt, the time the dog is in the air. This time is determined by the vertical motion, which is free fall with an initial velocity $(v_y)_i = 0$ m/s. We use the vertical-position equation from Synthesis 3.1 to find the time interval:

$$y_f = y_i + (v_y)_i \, \Delta t - \frac{1}{2} g(\Delta t)^2$$

$$0 \text{ m} = 0.61 \text{ m} + (0 \text{ m/s})\Delta t - \frac{1}{2}(9.8 \text{ m/s}^2)(\Delta t)^2$$

Rearranging terms to solve for Δt, we find that

$$\Delta t = 0.35 \text{ s}$$

This is how long it takes the dog's vertical motion to reach the water. During this time interval, the dog's horizontal motion is uniform motion at the initial velocity. We can use the horizontal-position equation with the initial speed and $\Delta t = 0.35$ s to find how far the dog travels. This is the distance we are looking for:

$$x_f = x_i + (v_x)_i \, \Delta t$$

$$= 0 \text{ m} + (8.5 \text{ m/s})(0.35 \text{ s}) = 3.0 \text{ m}$$

The dog hits the water 3.0 m from the edge of the dock.

ASSESS 3.0 m is about 10 feet. This seems like a reasonable distance for a dog running at a very fast clip off the end of a 2-foot-high dock. Indeed, this is a typical distance for dogs in such competitions.

In Example 3.11, the dog ran horizontally off the end of the dock. Much greater distances are possible if the dog goes off the end of the dock at an angle above the horizontal, which gives much more time in the air. A launch at an angle involves an initial vertical velocity, but this is still an example of projectile motion, and the general problem-solving strategy is the same.

PROBLEM-SOLVING STRATEGY 3.1 **Projectile motion problems** (MP)

We can solve projectile motion problems by considering the horizontal and vertical motions as separate but related problems.

PREPARE There are a number of steps that you should go through in setting up the solution to a projectile motion problem:

- Make simplifying assumptions. Whether the projectile is a car or a basketball, the motion will be the same.
- Draw a visual overview including a pictorial representation showing the beginning and ending points of the motion.
- Establish a coordinate system with the x-axis horizontal and the y-axis vertical. In this case, you know that the horizontal acceleration will be zero and the vertical acceleration will be free fall: $a_x = 0$ and $a_y = -g$.
- Define symbols and write down a list of known values. Identify what the problem is trying to find.

SOLVE There are two sets of kinematic equations for projectile motion, one for the horizontal component and one for the vertical:

Horizontal	Vertical
$x_f = x_i + (v_x)_i \, \Delta t$	$y_f = y_i + (v_y)_i \, \Delta t - \frac{1}{2}g(\Delta t)^2$
$(v_x)_f = (v_x)_i = \text{constant}$	$(v_y)_f = (v_y)_i - g \, \Delta t$

Δt is the same for the horizontal and vertical components of the motion. Find Δt by solving for the vertical or the horizontal component of the motion; then use that value to complete the solution for the other component.

ASSESS Check that your result has the correct units, is reasonable, and answers the question.

EXAMPLE 3.12 **Checking the feasibility of a Hollywood stunt**

The main characters in the movie *Speed* are on a bus that has been booby-trapped to explode if its speed drops below 50 mph. But there is a problem ahead: A 50 ft section of a freeway overpass is missing. They decide to jump the bus over the gap. The road leading up to the break has an angle of about 5°. A view of the speedometer just before the jump shows that the bus is traveling at 67 mph. The movie bus makes the jump and survives. Is this realistic, or movie fiction?

PREPARE We begin by converting speed and distance to SI units. The initial speed is $v_i = 30$ m/s and the size of the gap is $L = 15$ m. Next, following the problem-solving strategy, we make a sketch, the visual overview shown in FIGURE 3.34, and a list of values.

FIGURE 3.34 Visual overview of the bus jumping the gap.

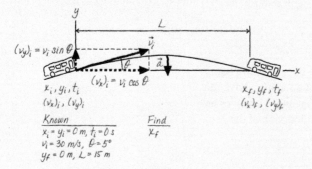

In choosing our axes, we've placed the origin at the point where the bus starts its jump. The initial velocity vector is tilted 5° above horizontal, so the components of the initial velocity are

$$(v_x)_i = v_i \cos \theta = (30 \text{ m/s})(\cos 5°) = 30 \text{ m/s}$$
$$(v_y)_i = v_i \sin \theta = (30 \text{ m/s})(\sin 5°) = 2.6 \text{ m/s}$$

How do we specify the "end" of the problem? By setting $y_f = 0$ m, we'll solve for the horizontal distance x_f at which the bus returns to its initial height. If x_f exceeds 50 ft, the bus successfully clears the gap. We have optimistically drawn our diagram as if the bus makes the jump, but . . .

SOLVE Problem-Solving Strategy 3.1 suggests using one component of the motion to solve for Δt. We will begin with the vertical motion. The kinematic equation for the vertical position is

$$y_f = y_i + (v_y)_i \, \Delta t - \tfrac{1}{2}g(\Delta t)^2$$

We know that $y_f = y_i = 0$ m. If we factor out Δt, the position equation becomes

$$0 = \Delta t \left((v_y)_i - \tfrac{1}{2}g \, \Delta t \right)$$

One solution to this equation is $\Delta t = 0$ s. This is a legitimate solution, but it corresponds to the instant when $y = 0$ at the beginning of the trajectory. We want the second solution, for $y = 0$ at the end of the trajectory, which is when

$$0 = (v_y)_i - \tfrac{1}{2}g \, \Delta t = (2.6 \text{ m/s}) - \tfrac{1}{2}(9.8 \text{ m/s}^2) \, \Delta t$$

which gives

$$\Delta t = \frac{2 \times (2.6 \text{ m/s})}{9.8 \text{ m/s}^2} = 0.53 \text{ s}$$

During the 0.53 s that the bus is moving vertically it is also moving horizontally. The final horizontal position of the bus is $x_f = x_i + (v_x)_i \, \Delta t$, or

$$x_f = 0 \text{ m} + (30 \text{ m/s})(0.53 \text{ s}) = 16 \text{ m}$$

This is how far the bus has traveled horizontally when it returns to its original height. 16 m is a bit more than the width of the gap, so a bus coming off a 5° ramp at the noted speed would make it—just barely!

ASSESS We can do a quick check on our math by noting that the bus takes off and lands at the same height. This means, as we saw in Figure 3.32b, that the y-velocity at the landing should be the negative of its initial value. We can use the velocity equation for the vertical component of the motion to compute the final value and see that the final velocity value is as we predict:

$$(v_y)_f = (v_y)_i - g \, \Delta t$$
$$= (2.6 \text{ m/s}) - (9.8 \text{ m/s}^2)(0.53 \text{ s}) = -2.6 \text{ m/s}$$

During the filming of the movie, the filmmakers really did jump a bus over a gap in an overpass! The actual jump was a bit more complicated than our example because a real bus, being an extended object rather than a particle, will start rotating as the front end comes off the ramp. The actual stunt jump used an extra ramp to give a boost to the front end of the bus. Nonetheless, our example shows that the filmmakers did their homework and devised a situation in which the physics was correct.

FIGURE 3.35 Trajectories of a projectile for different launch angles, assuming air resistance can be neglected.

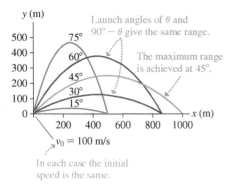

The Range of a Projectile

When the quarterback throws a football down the field, how far will it go? What will be the **range** for this particular projectile motion—the horizontal distance traveled?

Example 3.12 was a range problem: For a given speed and a given angle, we wanted to know how far the bus would go. The speed and the angle are the two variables that determine the range. A higher speed means a greater range, of course. But how does angle figure in?

FIGURE 3.35 shows the trajectory that a projectile launched at 100 m/s will follow for different launch angles. At very small or very large angles, the range is quite small. If you throw a ball at a 75° angle, it will do a great deal of up-and-down motion, but it won't achieve much horizontal travel. If you throw a ball at a 15° angle, the ball won't be in the air long enough to go very far. These cases both have the same range, as Figure 3.35 shows.

If the angle is too small or too large, the range is shorter than it could be. The "just right" case that gives the maximum range when landing at the same elevation as the launch is a launch angle of 45°, as Figure 3.35 shows.

For real-life projectiles, such as golf balls and baseballs, the optimal angle may be less than 45° because of air resistance. Up to this point we've ignored air resistance, but for small objects traveling at high speeds, air resistance is critical. Aerodynamic forces come into play, causing the projectile's trajectory to deviate from a parabola. The maximum range for a golf ball comes at an angle much less than 45°, as you no doubt know if you have ever played golf.

Video Tutor
Demo

▶ **A long long jump** A 45° angle gives the greatest range for a projectile, so why do long jumpers take off at a much shallower angle? Two of the assumptions that lead to the 45° optimal angle don't apply here. The athlete changes the position of his legs in the air—he doesn't land at the same height as that from which he took off. Also, athletes can't keep the same launch speed for different angles—they can jump faster at smaller angles. The gain from the faster speed outweighs the effect of the smaller angle.

STOP TO THINK 3.8 A baseball player is taking batting practice. He hits two successive pitches at different angles but at exactly the same speed. Ball 1 and ball 2 follow the paths shown. Which ball is in the air for a longer time? Assume that you can ignore air resistance for this problem.

A. Ball 1   B. Ball 2
C. Both balls are in the air for the same amount of time.

3.8 Motion in Two Dimensions: Circular Motion

It may seem strange to think that an object moving with constant speed can be accelerating, but that's exactly what an object in uniform circular motion is doing. It is accelerating because its velocity is changing as its direction of motion changes. What is the acceleration in this case? We saw in Section 3.2 that for circular motion at a constant speed, **the acceleration vector $\vec{a}$ points toward the center of the circle.** This is an idea that is worth reviewing. As you can see in **FIGURE 3.36**, the velocity is always tangent to the circle, so $\vec{v}$ and $\vec{a}$ are perpendicular to each other at all points on the circle.

An acceleration that always points directly toward the center of a circle is called a **centripetal acceleration**. The word "centripetal" comes from a Greek root meaning "center seeking."

NOTE ▶ Centripetal acceleration is not a new type of acceleration. All we are doing is *naming* an acceleration that corresponds to a particular type of motion. The magnitude of the centripetal acceleration is constant because each successive $\Delta\vec{v}$ in the motion diagram has the same length. ◀

To complete our description of circular motion, we need to find a quantitative relationship between the magnitude of the acceleration a and the speed v. Let's return to the case of the Ferris wheel. During a time Δt in which a car on the Ferris wheel moves around the circle from point 1 to point 2, the car moves through an angle θ and

FIGURE 3.36 The velocity and acceleration vectors for circular motion.

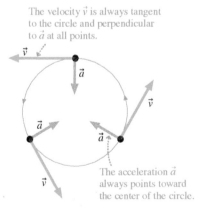

The velocity $\vec{v}$ is always tangent to the circle and perpendicular to $\vec{a}$ at all points.

The acceleration $\vec{a}$ always points toward the center of the circle.

undergoes a displacement $\vec{d}$, as shown in FIGURE 3.37a. We've chosen a relatively large angle θ for our drawing so that angular relationships can be clearly seen, but for a small angle the displacement is essentially identical to the actual distance traveled, and we'll make this approximation.

FIGURE 3.37 Changing position and velocity for an object in circular motion.

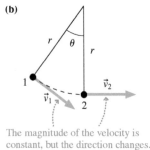

(a)

As the car moves from point 1 to point 2, the displacement is $\vec{d}$.

(b)

The magnitude of the velocity is constant, but the direction changes.

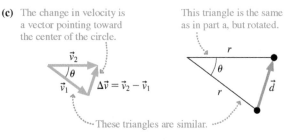

(c) The change in velocity is a vector pointing toward the center of the circle.

$\Delta\vec{v} = \vec{v}_2 - \vec{v}_1$

This triangle is the same as in part a, but rotated.

These triangles are similar.

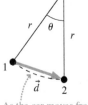

Video Tutor Demo   Class Video

FIGURE 3.37b shows how the velocity changes as the car moves, and FIGURE 3.37c shows the vector calculation of the change in velocity. The triangle we use to make this calculation is geometrically *similar* to the one that shows the displacement, as Figure 3.37c shows. This is a key piece of information: You'll remember from geometry that similar triangles have equal ratios of their sides, so we can write

$$\frac{\Delta v}{v} = \frac{d}{r} \qquad (3.22)$$

where Δv is the magnitude of the velocity-change vector $\Delta\vec{v}$. We've used the unsubscripted speed v for the length of a side of the first triangle because it is the same for velocities $\vec{v}_1$ and $\vec{v}_2$.

Now we're ready to compute the acceleration. The displacement is just the speed v times the time interval Δt, so we can write

$$d = v\Delta t$$

We can substitute this for d in Equation 3.22 to obtain

$$\frac{\Delta v}{v} = \frac{v\Delta t}{r}$$

which we can rearrange like so:

$$\frac{\Delta v}{\Delta t} = \frac{v^2}{r}$$

We recognize the left-hand side of the equation as the acceleration, so this becomes

$$a = \frac{v^2}{r}$$

Combining this magnitude with the direction we noted above, we can write the centripetal acceleration as

$$\vec{a} = \left(\frac{v^2}{r}, \text{ toward center of circle}\right) \qquad (3.23)$$

Centripetal acceleration of object moving in a circle of radius r at speed v

QUADRATIC

CONCEPTUAL EXAMPLE 3.13 **Acceleration on a swing**

A child is riding a playground swing. The swing rotates in a segment of a circle around a central point where the rope or chain for the swing is attached. The speed isn't changing at the lowest point of the motion, but the direction is—this is circular motion, with an acceleration directed upward, as shown in FIGURE 3.38. More acceleration will mean a more exciting ride.

FIGURE 3.38 A child at the lowest point of motion on a swing.

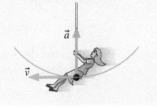

What change could the child make to increase the acceleration she experiences?

REASON The acceleration the child experiences is the "changing direction" acceleration of circular motion, given by Equation 3.23. The acceleration depends on the speed and the radius of the circle. The radius of the circle is determined by the length of the chain or rope, so the only easy way to change the acceleration is to change the speed, which she could do by swinging higher. Because the acceleration is proportional to the square of the speed, doubling the speed means a fourfold increase in the acceleration.

ASSESS If you have ever ridden a swing, you know that the acceleration you experience is greater the faster you go—so our answer makes sense.

EXAMPLE 3.14 **Acceleration in the turn**

World-class female short-track speed skaters can cover the 500 m of a race in 45 s. The most challenging elements of the race are the turns, which are very tight, with a radius of approximately 11 m. Estimate the magnitude of the skater's centripetal acceleration in a turn.

PREPARE The centripetal acceleration depends on two quantities: the radius of the turn (given as approximately 11 m) and the

speed. The speed varies during the race, but we can make a good estimate of the speed by using the total distance and time:

$$v \simeq \frac{500 \text{ m}}{45 \text{ s}} = 11 \text{ m/s}$$

SOLVE We can use these values to estimate the magnitude of the acceleration:

$$a = \frac{v^2}{r} \simeq \frac{(11 \text{m/s})^2}{11 \text{ m}} = 11 \text{ m/s}^2$$

ASSESS This is a large acceleration—a bit more than *g*—but the photo shows the skaters leaning quite hard into the turn, so such a large acceleration seems quite reasonable.

What Comes Next: Forces

Kinematics, the mathematical description of motion, is a good place to start our study of physics because motion is very visible and very familiar. But what actually *causes* motion? We noted that the skaters in the above example were "leaning quite hard into the turn"; this is a statement about the forces acting on them, the forces that cause the acceleration of the turn. In the next chapter, we'll explore the nature of forces and the connection to motion, giving us the tools to treat a much broader range of problems.

STOP TO THINK 3.9 Which of the following particles has the greatest centripetal acceleration?

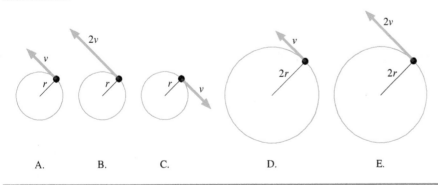

A.   B.   C.   D.   E.

World-record jumpers BIO

Frogs, with their long, strong legs, are excellent jumpers. And thanks to the good folks of Calaveras County, California, who have a jumping frog contest every year in honor of a Mark Twain story, we have very good data as to just  how far a determined frog can jump. The current record holder is Rosie the Ribeter, a bullfrog who made a leap of 6.5 m from a standing start. This compares favorably with the world record for a human, which is a mere 3.7 m.

Typical data for a serious leap by a bullfrog look like this: The frog goes into a crouch, then rapidly extends its legs by 15 cm as it pushes off, leaving the ground at an angle of 30° to the horizontal. It's in the air for 0.68 s before landing at the same height from which it took off. Given this leap, what is the acceleration while the frog is pushing off? How far does the frog jump?

PREPARE The problem really has two parts: the leap through the air and the acceleration required to produce this leap. We'll need to analyze the leap—the projectile motion—first, which will give us the frog's launch speed and the distance of the jump. Once we know the velocity with which the frog leaves the ground, we can calculate its acceleration while pushing off the ground. Let's start with a visual overview of the two parts, as shown in FIGURE 3.39. Notice that the second part of the problem uses a different x-axis, tilted as we did earlier for motion on a ramp.

SOLVE The "flying through the air" part shown in Figure 3.39a is projectile motion. The frog lifts off at a 30° angle with a speed v_i. The x- and y-components of the initial velocity are

$$(v_x)_i = v_i \cos 30°$$

$$(v_y)_i = v_i \sin 30°$$

The vertical motion can be analyzed as we did in Example 3.12. The kinematic equation is

$$y_f = y_i + (v_y)_i \Delta t + \tfrac{1}{2}a_y(\Delta t)^2$$

We know that $y_f = y_i = 0$, so this reduces to

$$(v_y)_i = -\tfrac{1}{2}a_y \Delta t = -\tfrac{1}{2}(-9.8 \text{ m/s}^2)(0.68 \text{ s}) = 3.3 \text{ m/s}$$

We know the y-component of the velocity and the angle, so we can find the magnitude of the velocity and the x-component:

$$v_i = \frac{(v_y)_i}{\sin 30°} = \frac{3.3 \text{ m/s}}{\sin 30°} = 6.6 \text{ m/s}$$

$$(v_x)_i = v_i \cos 30° = (6.6 \text{ m/s}) \cos 30° = 5.7 \text{ m/s}$$

The horizontal motion is uniform motion, so the frog's horizontal position when it returns to the ground is

$$x_f = x_i + (v_x)_i \Delta t = 0 + (5.7 \text{ m/s})(0.68 \text{ s}) = 3.9 \text{ m}$$

This is the length of the jump.

Now that we know how fast the frog is going when it leaves the ground, we can calculate the acceleration necessary to produce this jump—the "pushing off the ground" part shown in Figure 3.39b. We've drawn the x-axis along the direction of motion, as we did for problems of motion on a ramp. We know the displacement Δx of the jump but not the time, so we can use the third equation in Synthesis 2.1:

$$(v_x)_f^2 = (v_x)_i^2 + 2a_x \Delta x$$

The initial velocity is zero, the final velocity is $(v_x)_f = 6.6 \text{ m/s}$, and the displacement is the 15 cm (or 0.15 m) stretch of the legs during the jump. Thus the frog's acceleration while pushing off is

$$a_x = \frac{(v_x)_f^2}{2 \Delta x} = \frac{(6.6 \text{ m/s})^2}{2(0.15 \text{ m})} = 150 \text{ m/s}^2$$

ASSESS A 3.9 m jump is more than a human can achieve, but it's less than the record for a frog, so the final result for the distance seems reasonable. Such a long jump must require a large acceleration during the pushing-off phase, which is what we found.

FIGURE 3.39 A visual overview for the leap of a frog.

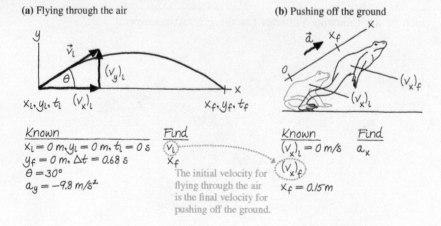

(a) Flying through the air

Known
$x_i = 0$ m, $y_i = 0$ m, $t_i = 0$ s
$y_f = 0$ m, $\Delta t = 0.68$ s
$\theta = 30°$
$a_y = -9.8$ m/s²

Find
v_i
x_f

The initial velocity for flying through the air is the final velocity for pushing off the ground.

(b) Pushing off the ground

Known
$(v_x)_i = 0$ m/s
$(v_x)_f$
$x_f = 0.15$ m

Find
a_x

SUMMARY

Goal: To learn more about vectors and to use vectors as a tool to analyze motion in two dimensions.

GENERAL PRINCIPLES

Projectile Motion

A projectile is an object that moves through the air under the influence of gravity and nothing else.

The path of the motion is a parabola.

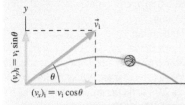

The motion consists of two pieces:

1. Vertical motion with free-fall acceleration, $a_y = -g$

2. Horizontal motion with constant velocity

Kinematic equations:

$$x_f = x_i + (v_x)_i \, \Delta t$$

$$(v_x)_f = (v_x)_i = \text{constant}$$

$$y_f = y_i + (v_y)_i \, \Delta t - \tfrac{1}{2} g (\Delta t)^2$$

$$(v_y)_f = (v_y)_i - g \, \Delta t$$

Circular Motion

An object moving in a circle at a constant speed has a velocity that is constantly changing direction, and so experiences an acceleration:

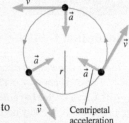

- The velocity is tangent to the circular path.

- The acceleration points toward the center of the circle and has magnitude

$$a = \frac{v^2}{r}$$

IMPORTANT CONCEPTS

Vectors and Components

A vector can be decomposed into x- and y-**components**.

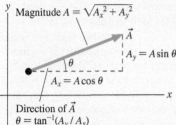

The magnitude and direction of a vector can be expressed in terms of its components.

The sign of the components depends on the direction of the vector:

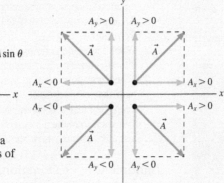

The Acceleration Vector

We define the acceleration vector as

$$\vec{a} = \frac{\vec{v}_f - \vec{v}_i}{t_f - t_i} = \frac{\Delta \vec{v}}{\Delta t}$$

We find the acceleration vector on a motion diagram as follows:

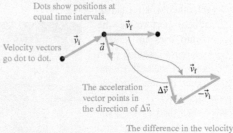

The difference in the velocity vectors is found by adding the negative of $\vec{v}_i$ to $\vec{v}_f$.

APPLICATIONS

Relative motion

Velocities can be expressed relative to an observer. We can add relative velocities to convert to another observer's point of view.

c = car, r = runner, g = ground

The speed of the car with respect to the runner is

$$(v_x)_{cr} = (v_x)_{cg} + (v_x)_{gr}$$

Motion on a ramp

An object sliding down a ramp will accelerate parallel to the ramp:

$$a_x = \pm g \sin\theta$$

The correct sign depends on the direction in which the ramp is tilted.

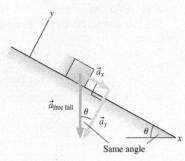

Problem difficulty is labeled as | (straightforward) to ||||| (challenging). Problems labeled INT integrate significant material from earlier chapters; BIO are of biological or medical interest.

For assigned homework and other learning materials, go to MasteringPhysics®

Scan this QR code to launch a Video Tutor Solution that will help you solve problems for this chapter.

QUESTIONS

Conceptual Questions

1. a. Can a vector have nonzero magnitude if a component is zero? If no, why not? If yes, give an example.
   b. Can a vector have zero magnitude and a nonzero component? If no, why not? If yes, give an example.
2. Is it possible to add a scalar to a vector? If so, demonstrate. If not, explain why not.
3. Suppose two vectors have unequal magnitudes. Can their sum be $\vec{0}$? Explain.
4. Suppose $\vec{C} = \vec{A} + \vec{B}$.
   a. Under what circumstances does $C = A + B$?
   b. Could $C = A - B$? If so, how? If not, why not?
5. For a projectile, which of the following quantities are constant during the flight: x, y, v_x, v_y, v, a_x, a_y? Which of the quantities are zero throughout the flight?
6. A baseball player throws a ball at a 40° angle to the ground. The ball lands on the ground some distance away.
   a. Is there any point on the trajectory where $\vec{v}$ and $\vec{a}$ are parallel to each other? If so, where?
   b. Is there any point where $\vec{v}$ and $\vec{a}$ are perpendicular to each other? If so, where?
7. An athlete performing the long jump tries to achieve the maximum distance from the point of takeoff to the first point of touching the ground. After the jump, rather than land upright, she extends her legs forward as in the photo. How does this affect the time in the air? How does this give the jumper a longer range?

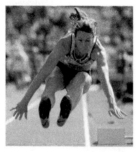

8. A person trying to throw a ball as far as possible will run forward during the throw. Explain why this increases the distance of the throw.
9. If you kick a football, at what angle to the ground should you kick the ball for the maximum range—that is, the greatest distance down the field? At what angle to the ground should you kick the ball for the maximum "hang time"—that is, the maximum time in the air?
10. A passenger on a jet airplane claims to be able to walk at a speed in excess of 500 mph. Can this be true? Explain.
11. If you go to a ski area, you'll likely find that the beginner's slope has the smallest angle. Use the concept of acceleration on a ramp to explain why this is so.
12. In an amusement-park ride, cars rolling along at high speed suddenly head up a long, straight ramp. They roll up the ramp, reverse direction at the highest point, then roll backward back down the ramp. In each of the following segments of the motion, which way does the acceleration vector point?
    a. As the cars roll up the ramp.
    b. At the highest point on the ramp.
    c. As the cars roll back down the ramp.
13. There are competitions in which pilots fly small planes low over the ground and drop weights, trying to hit a target. A pilot flying low and slow drops a weight; it takes 2.0 s to hit the ground, during which it travels a horizontal distance of 100 m. Now the pilot does a run at the same height but twice the speed. How much time does it take the weight to hit the ground? How far does it travel before it lands?
14. A cyclist goes around a level, circular track at constant speed. Do you agree or disagree with the following statement: "Because the cyclist's speed is constant, her acceleration is zero." Explain.
15. You are cycling around a circular track at a constant speed. Does the magnitude of your acceleration change? The direction?
16. An airplane has been directed to fly in a clockwise circle, as seen from above, at constant speed until another plane has landed. When the plane is going north, is it accelerating? If so, in what direction does the acceleration vector point? If not, why not?
17. When you go around a corner in your car, your car follows a path that is a segment of a circle. To turn safely, you should keep your car's acceleration below some safe upper limit. If you want to make a "tighter" turn—that is, turn in a circle with a smaller radius—how should you adjust your speed? Explain.

Multiple-Choice Questions

18. || Which combination of the vectors shown in Figure Q3.18 has the largest magnitude?
    A. $\vec{A} + \vec{B} + \vec{C}$
    B. $\vec{B} + \vec{A} - \vec{C}$
    C. $\vec{A} - \vec{B} + \vec{C}$
    D. $\vec{B} - \vec{A} - \vec{C}$

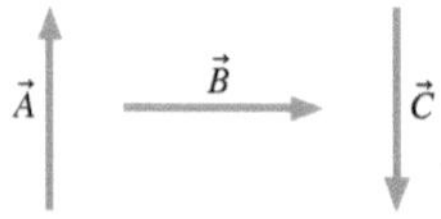

FIGURE Q3.18

19. || Two vectors appear as in Figure Q3.19. Which combination points directly to the left?
    A. $\vec{P} + \vec{Q}$
    B. $\vec{P} - \vec{Q}$
    C. $\vec{Q} - \vec{P}$
    D. $-\vec{Q} - \vec{P}$

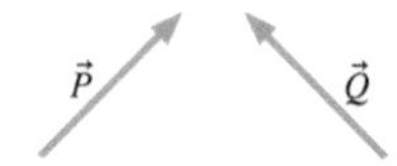

FIGURE Q3.19

20. | The gas pedal in a car is sometimes referred to as "the accelerator." Which other controls on the vehicle can be used to produce acceleration?
    A. The brakes.
    B. The steering wheel.
    C. The gear shift.
    D. All of the above.

21. | A car travels at constant speed along the curved path shown from above in Figure Q3.21. Five possible vectors are also shown in the figure; the letter E represents the zero vector. Which vector best represents
    a. The car's velocity at position 1?
    b. The car's acceleration at position 1?
    c. The car's velocity at position 2?
    d. The car's acceleration at position 2?
    e. The car's velocity at position 3?
    f. The car's acceleration at position 3?

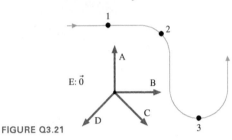

FIGURE Q3.21

22. | A ball is fired from a cannon at point 1 and follows the trajectory shown in Figure Q3.22. Air resistance may be neglected. Five possible vectors are also shown in the figure; the letter E represents the zero vector. Which vector best represents
    a. The ball's velocity at position 2?
    b. The ball's acceleration at position 2?
    c. The ball's velocity at position 3?
    d. The ball's acceleration at position 3?

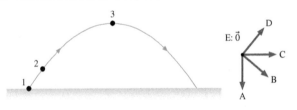

FIGURE Q3.22

23. | A ball thrown at an initial angle of 37.0° and initial velocity of 23.0 m/s reaches a maximum height h, as shown in Figure Q3.23. With what initial speed must a ball be thrown *straight up* to reach the same maximum height h?
    A. 13.8 m/s   B. 17.3 m/s
    C. 18.4 m/s   D. 23.0 m/s

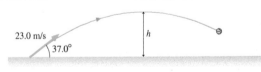

FIGURE Q3.23

24. | A cannon elevated at 40° is fired at a wall 300 m away on level ground, as shown in Figure Q3.24. The initial speed of the cannonball is 89 m/s.
    a. How long does it take for the ball to hit the wall?
        A. 1.3 s   B. 3.3 s   C. 4.4 s
        D. 6.8 s   E. 7.2 s
    b. At what height h does the ball hit the wall?
        A. 39 m   B. 47 m   C. 74 m
        D. 160 m   E. 210 m

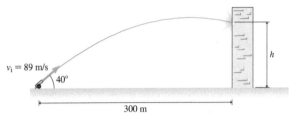

FIGURE Q3.24

25. | A car drives horizontally off a 73-m-high cliff at a speed of 27 m/s. Ignore air resistance.
    a. How long will it take the car to hit the ground?
        A. 2.0 s   B. 3.2 s   C. 3.9 s
        D. 4.9 s   E. 5.0 s
    b. Approximately how far from the base of the cliff will the car hit?
        A. 75 m   B. 90 m   C. 100 m
        D. 170 m   E. 280 m

26. | A football is kicked at an angle of 30° with a speed of 20 m/s. To the nearest second, how long will the ball stay in the air?
    A. 1 s   B. 2 s
    C. 3 s   D. 4 s

27. | A football is kicked at an angle of 30° with a speed of 20 m/s. To the nearest 5 m, how far will the ball travel?
    A. 15 m   B. 25 m
    C. 35 m   D. 45 m

28. | Riders on a Ferris wheel move in a circle with a speed of 4.0 m/s. As they go around, they experience a centripetal acceleration of 2.0 m/s². What is the diameter of this particular Ferris wheel?
    A. 4.0 m   B. 6.0 m   C. 8.0 m
    D. 16 m   E. 24 m

29. | Formula One race cars are capable of remarkable accelerations when speeding up, slowing down, and turning corners. At one track, cars round a corner that is a segment of a circle of radius 95 m at a speed of 68 m/s. What is the approximate magnitude of the centripetal acceleration, in units of g?
    A. 1g B. 2g C. 3g
    D. 4g E. 5g

PROBLEMS

Section 3.1 Using Vectors

1. ‖ Trace the vectors in Figure P3.1 onto your paper. Then use graphical methods to draw the vectors (a) $\vec{A} + \vec{B}$ and (b) $\vec{A} - \vec{B}$.

FIGURE P3.1

2. ‖ Trace the vectors in Figure P3.2 onto your paper. Then use graphical methods to draw the vectors (a) $\vec{A} + \vec{B}$ and (b) $\vec{A} - \vec{B}$.

FIGURE P3.2

Section 3.2 Using Vectors on Motion Diagrams

3. | A car goes around a corner in a circular arc at constant speed. Draw a motion diagram including positions, velocity vectors, and acceleration vectors.

4. ⦀ Figure 3.10 showed the motion diagram for a rider on a Ferris wheel that was turning at a constant speed. The inset to the figure showed how to find the acceleration vector at the rightmost point. Use a similar analysis to find the rider's acceleration vector at the leftmost position of the motion diagram, then at one of the highest positions and at one of the lowest positions. Use a ruler so that your analysis is accurate.

Section 3.3 Coordinate Systems and Vector Components

5. ‖ A position vector with magnitude 10 m points to the right and up. Its x-component is 6.0 m. What is the value of its y-component?

6. ⦀ A velocity vector 40° above the positive x-axis has a y-component of 10 m/s. What is the value of its x-component?

7. ‖ Jack and Jill ran up the hill at 3.0 m/s. The horizontal component of Jill's velocity vector was 2.5 m/s.
   a. What was the angle of the hill?
   b. What was the vertical component of Jill's velocity?

8. | Josh is climbing up a steep 34° slope, moving at a steady 0.75 m/s along the ground. How many meters of elevation does he gain in one minute of this climb?

9. ‖ A cannon tilted upward at 30° fires a cannonball with a speed of 100 m/s. At that instant, what is the component of the cannonball's velocity parallel to the ground?

10. ‖ a. What are the x- and y-components of vector $\vec{E}$ of Figure P3.10 in terms of the angle θ and the magnitude E?
    b. For the same vector, what are the x- and y-components in terms of the angle ϕ and the magnitude E?

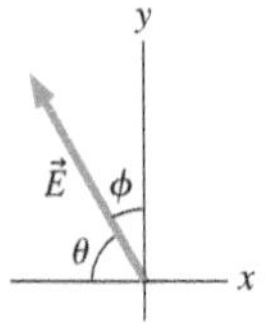

FIGURE P3.10

11. | Draw each of the following vectors, then find its x- and y-components.
    a. $\vec{d} = (100\ m, 45°\ \text{below} +x\text{-axis})$
    b. $\vec{v} = (300\ m/s, 20°\ \text{above} +x\text{-axis})$
    c. $\vec{a} = (5.0\ m/s^2, -y\text{-direction})$

12. ‖ Draw each of the following vectors, then find its x- and y-components.
    a. $\vec{d} = (2.0\ km, 30°\ \text{left of} +y\text{-axis})$
    b. $\vec{v} = (5.0\ cm/s, -x\text{-direction})$
    c. $\vec{a} = (10\ m/s^2, 40°\ \text{left of} -y\text{-axis})$

13. | Each of the following vectors is given in terms of its x- and y-components. Draw the vector, label an angle that specifies the vector's direction, then find the vector's magnitude and direction.
    a. $v_x = 20\ m/s$, $v_y = 40\ m/s$
    b. $a_x = 2.0\ m/s^2$, $a_y = -6.0\ m/s^2$

14. | Each of the following vectors is given in terms of its x- and y-components. Draw the vector, label an angle that specifies the vector's direction, then find the vector's magnitude and direction.
    a. $v_x = 10\ m/s$, $v_y = 30\ m/s$
    b. $a_x = 20\ m/s^2$, $a_y = 10\ m/s^2$

15. ‖ A wildlife researcher is tracking a flock of geese. The geese fly 4.0 km due west, then turn toward the north by 40° and fly another 4.0 km. How far west are they of their initial position? What is the magnitude of their displacement?

Section 3.4 Motion on a Ramp

16. ⦀ You begin sliding down a 15° ski slope. Ignoring friction and air resistance, how fast will you be moving after 10 s?

17. ⦀ A car traveling at 30 m/s runs out of gas while traveling up a 5.0° slope. How far will it coast before starting to roll back down?

18. ‖ In the Soapbox Derby, young participants build non-motorized cars with very low-friction wheels. Cars race by rolling down a hill. The track at Akron's Derby Downs, where the national championship is held, begins with a 55-ft-long section tilted 13° below horizontal.

   a. What is the maximum possible acceleration of a car moving down this stretch of track?
   b. If a car starts from rest and undergoes this acceleration for the full 55 ft, what is its final speed in m/s?

19. ⦀ A piano has been pushed to the top of the ramp at the back of a moving van. The workers think it is safe, but as they walk away, it begins to roll down the ramp. If the back of the truck is 1.0 m above the ground and the ramp is inclined at 20°, how much time do the workers have to get to the piano before it reaches the bottom of the ramp?

20. ‖ A car turns into a driveway that slopes upward at a 9.0° angle. The car is moving at 6.5 m/s. If the driver lets the car coast, how far along the slope will the car roll before being instantaneously at rest and then starting to roll back?

Section 3.5 Relative Motion

21. | Anita is running to the right at 5 m/s, as shown in Figure P3.21. Balls 1 and 2 are thrown toward her at 10 m/s by friends standing on the ground. According to Anita, what is the speed of each ball?

FIGURE P3.21

22. ‖ An airplane cruises at 880 km/h relative to the air. It is flying from Denver, Colorado, due west to Reno, Nevada, a distance of 1200 km, and will then return. There is a steady 90 km/h wind blowing to the east. What is the difference in flight time between the two legs of the trip?

23. | Anita is running to the right at 5 m/s, as shown in Figure P3.23. Balls 1 and 2 are thrown toward her by friends standing on the ground. According to Anita, both balls are approaching her at 10 m/s. According to her friends, with what speeds were the balls thrown?

FIGURE P3.23

24. ‖ Raindrops are falling straight down at 12 m/s when suddenly the wind starts blowing horizontally at a brisk 4.0 m/s. From your point of view, the rain is now coming down at an angle. What is the angle?

25. ‖‖ A boat takes 3.0 h to travel 30 km down a river, then 5.0 h to return. How fast is the river flowing?

26. ‖ Two children who are bored while waiting for their flight at the airport decide to race from one end of the 20-m-long moving sidewalk to the other and back. Phillippe runs on the sidewalk at 2.0 m/s (relative to the sidewalk). Renee runs on the floor at 2.0 m/s. The sidewalk moves at 1.5 m/s relative to the floor. Both make the turn instantly with no loss of speed.
    a. Who wins the race?
    b. By how much time does the winner win?

Section 3.6 Motion in Two Dimensions: Projectile Motion

Section 3.7 Projectile Motion: Solving Problems

27. ‖‖ A ball is thrown horizontally from a 20-m-high building with a speed of 5.0 m/s.
    a. Make a sketch of the ball's trajectory.
    b. Draw a graph of v_x, the horizontal velocity, as a function of time. Include units on both axes.
    c. Draw a graph of v_y, the vertical velocity, as a function of time. Include units on both axes.
    d. How far from the base of the building does the ball hit the ground?

28. ‖ A ball with a horizontal speed of 1.25 m/s rolls off a bench 1.00 m above the floor.
    a. How long will it take the ball to hit the floor?
    b. How far from a point on the floor directly below the edge of the bench will the ball land?

29. ‖ A pipe discharges storm water into a creek. Water flows horizontally out of the pipe at 1.5 m/s, and the end of the pipe is 2.5 m above the creek. How far out from the end of the pipe is the point where the stream of water meets the creek?

30. ‖ On a day when the water is flowing relatively gently, water in the Niagara River is moving horizontally at 4.5 m/s before shooting over Niagara Falls. After moving over the edge, the water drops 53 m to the water below. If we ignore air resistance, how much time does it take for the water to go from the top of the falls to the bottom? How far does the water move horizontally during this time?

31. ‖ Two spheres are both launched horizontally from a 1.0-m-high table. Sphere A is launched with an initial speed of 5.0 m/s. Sphere B is launched with an initial speed of 2.5 m/s.
    a. What are the times for each sphere to hit the floor?
    b. What are the distances that each travels from the edge of the table?

32. ‖‖ A rifle is aimed horizontally at a target 50 m away. The bullet hits the target 2.0 cm below the aim point.
    a. What was the bullet's flight time?
    b. What was the bullet's speed as it left the barrel?

33. ‖‖ A gray kangaroo can bound across a flat stretch of ground
    BIO with each jump carrying it 10 m from the takeoff point. If the kangaroo leaves the ground at a 20° angle, what are its (a) take-off speed and (b) horizontal speed?

34. ‖‖ On the Apollo 14 mission to the moon, astronaut Alan Shepard hit a golf ball with a golf club improvised from a tool. The free-fall acceleration on the moon is 1/6 of its value on earth. Suppose he hit the ball with a speed of 25 m/s at an angle 30° above the horizontal.
    a. How long was the ball in flight?
    b. How far did it travel?
    c. Ignoring air resistance, how much farther would it travel on the moon than on earth?

35. ‖ A sprinkler mounted on the ground sends out a jet of water at a 30° angle to the horizontal. The water leaves the nozzle at a speed of 12 m/s. How far does the water travel before it hits the ground?

36. ‖ A good quarterback can throw a football at 27 m/s (about 60 mph). If we assume that the ball is caught at the same height from which it is thrown, and if we ignore air resistance, what is the maximum range in meters (which is approximately the same as the range in yards) of a pass at this speed? How long is the ball in the air?

Section 3.8 Motion in Two Dimensions: Circular Motion

37. ‖ Racing greyhounds are capable of rounding corners at very
    BIO high speeds. A typical greyhound track has turns that are 45-m-diameter semicircles. A greyhound can run around these turns at a constant speed of 15 m/s. What is its acceleration in m/s² and in units of g?

38. ‖ To withstand "g-forces" of up to 10 g's, caused by suddenly
    BIO pulling out of a steep dive, fighter jet pilots train on a "human centrifuge." 10 g's is an acceleration of 98 m/s². If the length of the centrifuge arm is 12 m, at what speed is the rider moving when she experiences 10 g's?

39. ‖ The Scion iQ is a compact car that is capable of very tight turns—it can spin around in a circle 8.0 m in diameter. If a driver goes around such a circle at 5 m/s (a bit faster than 10 mph), what is the magnitude of his acceleration?

40. ‖ In a roundabout (or traffic circle), cars go around a 25-m-diameter circle. If a car's tires will skid when the car experiences a centripetal acceleration greater than 0.60g, what is the maximum speed of the car in this roundabout?

41. ‖ A particle rotates in a circle with centripetal acceleration $a = 8.0$ m/s². What is a if
    a. The radius is doubled without changing the particle's speed?
    b. The speed is doubled without changing the circle's radius?

42. ‖ Entrance and exit ramps for freeways are often circular stretches of road. As you go around one at a constant speed, you will experience a constant acceleration. Suppose you drive through an entrance ramp at a modest  speed and your acceleration is 3.0 m/s². What will be the acceleration if you double your speed?

43. ‖ A peregrine falcon in a tight, circular turn can attain a cen-
    BIO tripetal acceleration 1.5 times the free-fall acceleration. If the falcon is flying at 20 m/s, what is the radius of the turn?

General Problems

44. | Suppose $\vec{C} = \vec{A} + \vec{B}$ where vector $\vec{A}$ has components $A_x = 5$, $A_y = 2$ and vector $\vec{B}$ has components $B_x = -3, B_y = -5$.
    a. What are the x- and y-components of vector $\vec{C}$?
    b. Draw a coordinate system and on it show vectors $\vec{A}$, $\vec{B}$, and $\vec{C}$.
    c. What are the magnitude and direction of vector $\vec{C}$?

45. | Suppose $\vec{D} = \vec{A} - \vec{B}$ where vector $\vec{A}$ has components $A_x = 5$, $A_y = 2$ and vector $\vec{B}$ has components $B_x = -3, B_y = -5$.
    a. What are the x- and y-components of vector $\vec{D}$?
    b. Draw a coordinate system and on it show vectors $\vec{A}$, $\vec{B}$, and $\vec{D}$.
    c. What are the magnitude and direction of vector $\vec{D}$?

46. || Suppose $\vec{E} = 2\vec{A} + 3\vec{B}$ where vector $\vec{A}$ has components $A_x = 5, A_y = 2$ and vector $\vec{B}$ has components $B_x = -3, B_y = -5$.
    a. What are the x- and y-components of vector $\vec{E}$?
    b. Draw a coordinate system and on it show vectors $\vec{A}$, $\vec{B}$, and $\vec{E}$.
    c. What are the magnitude and direction of vector $\vec{E}$?

47. || For the three vectors shown in Figure P3.47, the vector sum $\vec{D} = \vec{A} + \vec{B} + \vec{C}$ has components $D_x = 2$ and $D_y = 0$.
    a. What are the x- and y-components of vector $\vec{B}$?
    b. Write $\vec{B}$ as a magnitude and a direction.

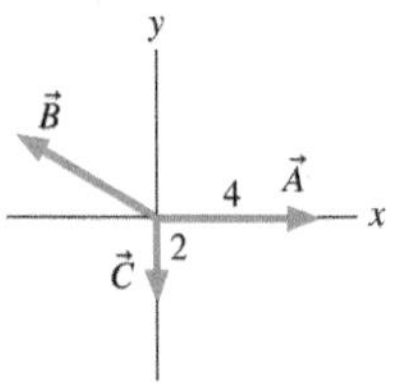

FIGURE P3.47

48. || Let $\vec{A} = (3.0 \text{ m}, 20° \text{ south of east})$, $\vec{B} = (2.0 \text{ m, north})$, and $\vec{C} = (5.0 \text{ m}, 70° \text{ south of west})$.
    a. Draw and label $\vec{A}$, $\vec{B}$, and $\vec{C}$ with their tails at the origin. Use a coordinate system with the x-axis to the east.
    b. Write the x- and y-components of vectors $\vec{A}$, $\vec{B}$, and $\vec{C}$.
    c. Find the magnitude and the direction of $\vec{D} = \vec{A} + \vec{B} + \vec{C}$.

49. || A typical set of stairs is angled at 38°. You climb a set of stairs at a speed of 3.5 m/s.
    a. How much height will you gain in 2.0 s?
    b. How much horizontal distance will you cover in 2.0 s?

50. ||| A pilot in a small plane encounters shifting winds. He flies 26.0 km northeast, then 45.0 km due north. From this point, he flies an additional distance in an unknown direction, only to find himself at a small airstrip that his map shows to be 70.0 km directly north of his starting point. What were the length and direction of the third leg of his trip?

51. || A small plane, 100 km due south of the equator, is flying at 150 km/h with respect to the ground along a straight line 30° to the west of north. In how many minutes will the plane cross the equator?

52. || The bacterium *Escherichia coli* (or *E. coli*) is a single-celled
BIO organism that lives in the gut of healthy humans and animals. When grown in a uniform medium rich in salts and amino acids, these bacteria swim along zig-zag paths at a constant speed of 20 μm/s. Figure P3.52 shows the trajectory of an *E. coli* as it moves from point A to point E. Each segment of the motion can be identified by two letters, such as segment BC.
    a. For each of the four segments in the bacterium's trajectory, calculate the x- and y-components of its displacement and of its velocity.
    b. Calculate both the total distance traveled and the magnitude of the net displacement for the entire motion.
    c. What are the magnitude and the direction of the bacterium's average velocity for the entire trip?

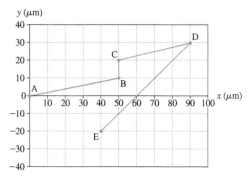

FIGURE P3.52

53. |||| A skier gliding across the snow at 3.0 m/s suddenly starts down a 10° incline, reaching a speed of 15 m/s at the bottom. Friction between the snow and her freshly waxed skies is negligible.
    a. What is the length of the incline?
    b. How long does it take her to reach the bottom?

54. ||| A block slides along the frictionless track shown in Figure P3.54 with an initial speed of 5.0 m/s. Assume it turns all the corners smoothly, with no loss of speed. What is the block's speed as it goes over the top?

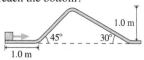

FIGURE P3.54

55. ||| When the moving sidewalk at the airport is broken, as it often seems to be, it takes you 50 s to walk from your gate to the baggage claim. When it is working and you stand on the moving sidewalk the entire way, without walking, it takes 75 s to travel the same distance. How long will it take you to travel from the gate to baggage claim if you walk while riding on the moving sidewalk?

56. |||| Ships A and B leave port together. For the next two hours, ship A travels at 20 mph in a direction 30° west of north while ship B travels 20° east of north at 25 mph.
    a. What is the distance between the two ships two hours after they depart?
    b. What is the speed of ship A as seen by ship B?

57. ||| A flock of ducks is trying to migrate south for the winter, but they keep being blown off course by a wind blowing from the west at 12 m/s. A wise elder duck finally realizes that the solution is to fly at an angle to the wind. If the ducks can fly at 16 m/s relative to the air, in what direction should they head in order to move directly south?

58. ||| A kayaker needs to paddle north across a 100-m-wide harbor. The tide is going out, creating a tidal current flowing east at 2.0 m/s. The kayaker can paddle with a speed of 3.0 m/s.
    a. In which direction should he paddle in order to travel straight across the harbor?
    b. How long will it take him to cross?

59. |||| A plane has an airspeed of 200 mph. The pilot wishes to reach a destination 600 mi due east, but a wind is blowing at 50 mph in the direction 30° north of east.
    a. In what direction must the pilot head the plane in order to reach her destination?
    b. How long will the trip take?

60. ||| The Gulf Stream off the east coast of the United States can flow at a rapid 3.6 m/s to the north. A ship in this current has a cruising speed of 10 m/s. The captain would like to reach land at a point due west from the current position.
    a. In what direction with respect to the water should the ship sail?
    b. At this heading, what is the ship's speed with respect to land?

61. ‖ A physics student on Planet Exidor throws a ball, and it follows the parabolic trajectory shown in Figure P3.61. The ball's position is shown at 1.0 s intervals until $t = 3.0$ s. At $t = 1.0$ s, the ball's velocity has components $v_x = 2.0$ m/s, $v_y = 2.0$ m/s.

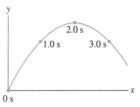

FIGURE P3.61

    a. Determine the x- and y-components of the ball's velocity at $t = 0.0$ s, 2.0 s, and 3.0 s.
    b. What is the value of g on Planet Exidor?
    c. What was the ball's launch angle?

62. ‖ A ball thrown horizontally at 25 m/s travels a horizontal distance of 50 m before hitting the ground. From what height was the ball thrown?

63. ‖ In 1780, in what is now referred to as "Brady's Leap," Captain Sam Brady of the U.S. Continental Army escaped certain death from his enemies by running horizontally off the edge of the cliff above Ohio's Cuyahoga River, which is confined at that spot to a gorge. He landed safely on the far side of the river. It was reported that he leapt 22 ft across while falling 20 ft. Tall tale, or possible?

FIGURE P3.63

    a. What is the minimum speed with which he'd need to run off the edge of the cliff to make it safely to the far side of the river?
    b. The world-record time for the 100 m dash is approximately 10 s. Given this, is it reasonable to expect Brady to be able to run fast enough to achieve Brady's leap?

64. ‖ The longest recorded pass in an NFL game traveled 83 yards in the air from the quarterback to the receiver. Assuming that the pass was thrown at the optimal 45° angle, what was the speed at which the ball left the quarterback's hand?

65. ‖ A spring-loaded gun, fired vertically, shoots a marble 6.0 m straight up in the air. What is the marble's range if it is fired horizontally from 1.5 m above the ground?

66. ‖ Small-plane pilots regularly compete in "message drop" competitions, dropping heavy weights (for which air resistance can be ignored) from their low-flying planes and scoring points for having the weights land close to a target. A plane 60 m above the ground is flying directly toward a target at 45 m/s.
    a. At what distance from the target should the pilot drop the weight?
    b. The pilot looks down at the weight after she drops it. Where is the plane located at the instant the weight hits the ground—not yet over the target, directly over the target, or past the target?

67. ‖ In a shot-put event, an athlete throws the shot with an initial speed of 12.0 m/s at a 40.0° angle from the horizontal. The shot leaves her hand at a height of 1.80 m above the ground.
    a. How far does the shot travel?
    b. Repeat the calculation of part a for angles 42.5°, 45.0°, and 47.5°. Put all your results, including 40.0°, in a table. At what angle of release does she throw the farthest?

68. ‖ Trained dolphins are capable of a vertical leap of 7.0 m straight up from the surface of the water—an impressive feat. Suppose you could train a dolphin to launch itself out of the water at this same speed but at an angle. What maximum horizontal range could the dolphin achieve?

69. ‖ A tennis player hits a ball 2.0 m above the ground. The ball leaves his racquet with a speed of 20 m/s at an angle 5.0° above the horizontal. The horizontal distance to the net is 7.0 m, and the net is 1.0 m high. Does the ball clear the net? If so, by how much? If not, by how much does it miss?

70. ‖ The shot put is a track-and-field event in which athletes throw a heavy ball—the shot—as far as possible. The best athletes can throw the shot as far as 23 m. Athletes who use the "glide" technique push the shot outward in a reasonably straight line, accelerating it over a distance of about 2.0 m. What acceleration do they provide to the shot as they push on it? Assume that the shot is launched at an angle of 37°, a reasonable value for an excellent throw. You can assume that the shot lands at the same height from which it is thrown; this simplifies the calculation considerably, and makes only a small difference in the final result.

71. ‖ Water at the top of Horseshoe Falls (part of Niagara Falls) is moving horizontally at 9.0 m/s as it goes off the edge and plunges 53 m to the pool below. If you ignore air resistance, at what angle is the falling water moving as it enters the pool?

72. ‖ A supply plane needs to drop a package of food to scientists working on a glacier in Greenland. The plane flies 100 m above the glacier at a speed of 150 m/s. How far short of the target should it drop the package?

73. ‖ A child slides down a frictionless 3.0-m-long playground slide tilted upward at an angle of 40°. At the end of the slide, there is an additional section that curves so that the child is launched off the end of the slide horizontally.
    a. How fast is the child moving at the bottom of the slide?
    b. If the end of the slide is 0.40 m above the ground, how far from the end does she land?

74. ‖ A sports car is advertised as capable of "reaching 60 mph in 5 seconds flat, cornering at 0.85g, and stopping from 70 mph in only 168 feet." In which of those three situations is the magnitude of the car's acceleration the largest? In which is it the smallest?

75. ‖ A Ford Mustang can accelerate from 0 to 60 mph in a time of 5.6 s. A Mini Cooper isn't capable of such a rapid start, but it can turn in a very small circle 34 ft in diameter. How fast would you need to drive the Mini Cooper in this tight circle to match the magnitude of the Mustang's acceleration?

76. ‖ The "Screaming Swing" is a carnival ride that is—not surprisingly—a giant swing. It's actually two swings moving in opposite directions. At the bottom of its arc, a rider in one swing is moving at 30 m/s with respect to the ground in a 50-m-diameter circle. The rider in the other swing is moving in a similar circle at the same speed, but in the exact opposite direction.
    a. What is the acceleration, in m/s² and in units of g, that riders experience?
    b. At the bottom of the ride, as they pass each other, how fast do the riders move with respect to each other?

77. ‖ On an otherwise straight stretch of road near Moffat, Colorado, the road suddenly turns. This bend in the road is a segment of a circle with radius 110 m. Drivers are cautioned to slow down to 40 mph as they navigate the curve.
    a. If you heed the sign and slow to 40 mph, what will be your acceleration going around the curve at this constant speed? Give your answer in m/s² and in units of g.
    b. At what speed would your acceleration be double that at the recommended speed?

MCAT-Style Passage Problems

Riding the Water Slide

A rider on a water slide goes through three different kinds of motion, as illustrated in Figure P3.78. Use the data and details from the figure to answer the following questions.

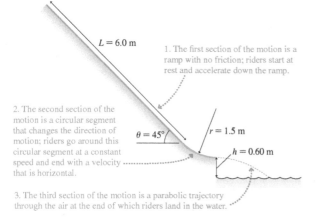

2. The second section of the motion is a circular segment that changes the direction of motion; riders go around this circular segment at a constant speed and end with a velocity that is horizontal.

1. The first section of the motion is a ramp with no friction; riders start at rest and accelerate down the ramp.

$L = 6.0$ m

$\theta = 45°$

$r = 1.5$ m

$h = 0.60$ m

3. The third section of the motion is a parabolic trajectory through the air at the end of which riders land in the water.

FIGURE P3.78

78. | At the end of the first section of the motion, riders are moving at what approximate speed?
    A. 3 m/s   B. 6 m/s
    C. 9 m/s   D. 12 m/s
79. | Suppose the acceleration during the second section of the motion is too large to be comfortable for riders. What change could be made to decrease the acceleration during this section?
    A. Reduce the radius of the circular segment.
    B. Increase the radius of the circular segment.
    C. Increase the angle of the ramp.
    D. Increase the length of the ramp.
80. | What is the vertical component of the velocity of a rider as he or she hits the water?
    A. 2.4 m/s   B. 3.4 m/s
    C. 5.2 m/s   D. 9.1 m/s
81. | Suppose the designers of the water slide want to adjust the height h above the water so that riders land twice as far away from the bottom of the slide. What would be the necessary height above the water?
    A. 1.2 m   B. 1.8 m
    C. 2.4 m   D. 3.0 m
82. | During which section of the motion is the magnitude of the acceleration experienced by a rider the greatest?
    A. The first.   B. The second.
    C. The third.   D. It is the same in all sections.

STOP TO THINK ANSWERS

Chapter Preview Stop to Think: D. The ball took 2.0 s to reach the highest point, at which time it is (momentarily) at rest. As it rises, its vertical velocity is changing at $a_y = -g = -9.8$ m/s² ≈ -10 m/s². The change in speed is $\Delta v_y \approx (-10$ m/s²$)(2.0$ s$) = -20$ m/s, so the initial speed is approximately 20 m/s.

Stop to Think 3.1: A. The graphical construction of $2\vec{P} - \vec{Q}$ is shown at right.

Stop to Think 3.2: B. The acceleration is directed toward the center of the circle. At the instant shown, the acceleration is down and to the right—to the southeast.

Stop to Think 3.3: From the axes on the graph, we can see that the x- and y-components are −4 cm and +2 cm, respectively.

Stop to Think 3.4: C. Vector $\vec{C}$ points to the left and down, so both C_x and C_y are negative. C_x is in the numerator because it is the side opposite ϕ.

Stop to Think 3.5: B. The angle of the slope is greatest in this case, leading to the greatest acceleration.

Stop to Think 3.6: D. Let's define upstream as positive and downstream as negative. Then, the speed of the boat with respect to the water is $v_{bw} = +5$ m/s, and the speed of the water with respect to the riverbank is $v_{wr} = -3$ m/s. Thus the speed of the boat with respect to the riverbank is

$$v_{br} = v_{bw} + v_{wr}$$
$$= 5 \text{ m/s} - 3 \text{ m/s} = 2 \text{ m/s}$$

Stop to Think 3.7: D. Mass does not appear in the kinematic equations, so the mass has no effect. The balls follow the same path.

Stop to Think 3.8: A. The time in the air is determined by the vertical component of the velocity. Ball 1 has a higher vertical velocity, so it will be in the air for a longer time.

Stop to Think 3.9: B. The magnitude of the acceleration is v^2/r. Acceleration is largest for the combination of highest speed and smallest radius.

4 Forces and Newton's Laws of Motion

We don't normally think of turtles as speedy, but the snake-necked turtle is an ambush predator, catching prey by surprise with a very rapid acceleration of its head. How is it able to achieve this feat?

LOOKING AHEAD ▶

Goal: To establish a connection between force and motion.

Forces

A force is a push or a pull. It is an interaction between two objects, the **agent** (the woman) and the **object** (the car).

In this chapter, you'll learn how to identify different forces, and you'll learn their properties.

Forces and Motion

Acceleration is caused by forces. A forward acceleration of the sled requires a forward force.

A larger acceleration requires a larger force. You'll learn this connection between force and motion, part of Newton's second law.

Reaction Forces

The hammer exerts a downward force on the nail. Surprisingly, the nail exerts an equal force on the hammer, directed upward.

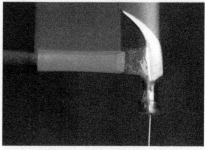

You'll learn how to identify and reason with **action/reaction pairs** of forces.

LOOKING BACK ◀

Acceleration

You learned in Chapters 2 and 3 that acceleration is a vector pointing in the direction of the change in velocity.

If the velocity is changing, there is an acceleration. And so, as you'll learn in this chapter, there must be a net force.

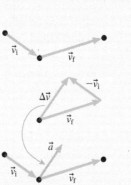

STOP TO THINK

A swan is landing on an icy lake, sliding across the ice and gradually coming to a stop. As the swan slides, the direction of the acceleration is

A. To the left.
B. To the right.
C. Upward.
D. Downward.

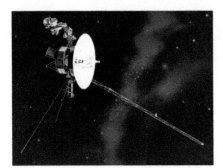

Interstellar coasting A nearly perfect example of Newton's first law is the pair of Voyager space probes launched in 1977. Both spacecraft long ago ran out of fuel and are now coasting through the frictionless vacuum of space. Although not entirely free of influence from the sun's gravity, they are now so far from the sun and other stars that gravitational influences are very nearly zero, and the probes will continue their motion for billions of years.

4.1 Motion and Forces

The snake-necked turtle in the photo at the beginning of the chapter can accelerate its head forward at 40 m/s² to capture prey. In Chapters 1 through 3, we've learned how to *describe* this and other types of motion with pictures, graphs, and equations, and you know enough about the scale of things to see that this acceleration—which is about 4*g*—is quite impressive. But, until now, we've said nothing to *explain* the motion, to say how the turtle is able to achieve this feat. In this chapter, we turn our attention to the *cause* of motion—**forces.** This topic is called **dynamics,** which joins with kinematics to form **mechanics,** the general science of motion. We'll begin our study of dynamics qualitatively in this chapter and then add quantitative detail over the next four chapters.

What Causes Motion?

Let's start with a basic question: Do you need to keep pushing on something—to keep applying a force—to keep it going? Your daily experience might suggest that the answer is Yes. If you slide your textbook across the desk and then stop pushing, the book will quickly come to rest. Other objects will continue to move for a longer time: When a hockey puck is sliding across the ice, it keeps going for a long time, but it, too, comes to rest at some point. Now let's take a closer look to see whether this idea holds up to scrutiny.

FIGURE 4.1 shows a series of motion experiments. Tyler slides down a hill on his sled and then out onto a horizontal patch of smooth snow, as shown in Figure 4.1a. Even if the snow is smooth, the friction between the sled and the snow will soon cause the sled to come to rest. What if Tyler slides down the hill onto some very slick ice, as in Figure 4.1b? The friction is much less, so the sled could slide for quite a distance before stopping. Now, *imagine* the situation in Figure 4.1c, where the sled slides on idealized *frictionless* ice. In this case, the sled, once started in its motion, would continue in motion forever, moving in a straight line with no loss of speed.

FIGURE 4.1 A sled sliding on increasingly smooth surfaces.

(a) Smooth snow
On smooth snow, the sled soon comes to rest.

(b) Slick ice
On slick ice, the sled slides farther.

(c) Frictionless surface
If friction could be reduced to zero, the sled would *never* stop.

In the absence of friction, **if the sled is moving, it will stay in motion.** It's also true that if the sled were sitting still, it wouldn't start moving on its own; if the sled is at rest, it will stay at rest. Careful experiments done over the past few centuries, notably by Galileo and then by Isaac Newton, verify that this is in fact the way the world works. What we have concluded for the sled is actually a general rule that applies to other similar situations. We call the generalization Newton's first law of motion:

> **Newton's first law** An object has no forces acting on it. If it is at rest, it will remain at rest. If it is moving, it will continue to move in a straight line at a constant speed.

As an important application of Newton's first law, consider the crash test of FIGURE 4.2. As the car contacts the wall, the wall exerts a force on the car and the car

FIGURE 4.2 Newton's first law tells us: Wear your seatbelts!

At the instant of impact, the car and driver are moving at the same speed.

The car slows as it hits, but the driver continues at the same speed . . .

. . . until he hits the now-stationary dashboard. Ouch!

begins to slow. But the wall is a force on the *car,* not on the dummy. In accordance with Newton's first law, the unbelted dummy continues to move straight ahead at his original speed. Sooner or later, a force will act to bring the dummy to rest. The only questions are when and how large the force will be. In the case shown, the dummy comes to rest in a short, violent collision with the dashboard of the stopped car. Seatbelts and air bags slow a dummy or a rider at a much lower rate and provide a much gentler stop.

Forces

Newton's first law tells us that an object in motion subject to no forces will continue to move in a straight line forever. But this law does not explain in any detail exactly what a force *is.* The concept of force is best introduced by looking at examples of some common forces and considering the basic properties shared by all forces. Let's begin by examining the properties that all forces have in common, as presented in the table below.

What is a force?

A force is a push or a pull.

Our commonsense idea of a **force** is that it is a *push* or a *pull.* We will refine this idea as we go along, but it is an adequate starting point. Notice our careful choice of words: We refer to "*a* force" rather than simply "force." We want to think of a force as a very specific *action,* so that we can talk about a single force or perhaps about two or three individual forces that we can clearly distinguish—hence the concrete idea of "a force" acting on an object.

A force acts on an object.

Implicit in our concept of force is that **a force acts on an object.** In other words, pushes and pulls are applied *to* something—an object. From the object's perspective, it has a force *exerted* on it. Forces do not exist in isolation from the object that experiences them.

A force requires an agent.

Every force has an **agent,** something that acts or pushes or pulls; that is, a force has a specific, identifiable *cause.* As you throw a ball, it is your hand, while in contact with the ball, that is the agent or the cause of the force exerted on the ball. *If* a force is being exerted on an object, you must be able to identify a specific cause (i.e., the agent) of that force. Conversely, a force is not exerted on an object *unless* you can identify a specific cause or agent. Note that an agent can be an inert object such as a tabletop or a wall. Such agents are the cause of many common forces.

A force is a vector.

If you push an object, you can push either gently or very hard. Similarly, you can push either left or right, up or down. To quantify a push, we need to specify both a magnitude *and* a direction. It should thus come as no surprise that a force is a vector quantity. The general symbol for a force is the vector symbol $\vec{F}$. The size or strength of such a force is its magnitude F.

A force can be either a contact force . . .

There are two basic classes of forces, depending on whether the agent touches the object or not. **Contact forces** are forces that act on an object by touching it at a point of contact. The bat must touch the ball to hit it. A string must be tied to an object to pull it. The majority of forces that we will examine are contact forces.

. . . or a long-range force.

Long-range forces are forces that act on an object without physical contact. Magnetism is an example of a long-range force. You have undoubtedly held a magnet over a paper clip and seen the paper clip leap up to the magnet. A coffee cup released from your hand is pulled to the earth by the long-range force of gravity.

There's one more important aspect of forces. If you push against a door (the object) to close it, the door pushes back against your hand (the agent). If a tow rope pulls on a car (the object), the car pulls back on the rope (the agent). In general, if an agent exerts a force on an object, the object exerts a force on the agent. We really need to think of a force as an *interaction* between two objects. Although the interaction perspective is a more exact way to view forces, it adds complications that we would like to avoid for now. Our approach will be to start by focusing on how a single object responds to forces exerted on it. Later in this chapter, we'll return to the larger issue of how two or more objects interact with each other.

Force Vectors

We can use a simple diagram to visualize how forces are exerted on objects. Because we are using the particle model, in which objects are treated as points, the process of drawing a force vector is straightforward:

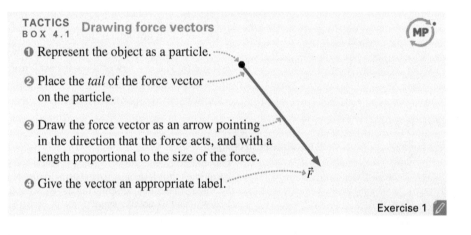

TACTICS
BOX 4.1   **Drawing force vectors**

❶ Represent the object as a particle.

❷ Place the *tail* of the force vector on the particle.

❸ Draw the force vector as an arrow pointing in the direction that the force acts, and with a length proportional to the size of the force.

❹ Give the vector an appropriate label. $\vec{F}$

Exercise 1

Step 2 may seem contrary to what a "push" should do (it may look as if the force arrow is *pulling* the object rather than *pushing* it), but recall that moving a vector does not change it as long as the length and angle do not change. The vector $\vec{F}$ is the same regardless of whether the tail or the tip is placed on the particle. Our reason for using the tail will become clear when we consider how to combine several forces.

FIGURE 4.3 shows three examples of force vectors. One is a pull, one a push, and one a long-range force, but in all three the *tail* of the force vector is placed on the particle that represents the object.

FIGURE 4.3 Three force vectors.

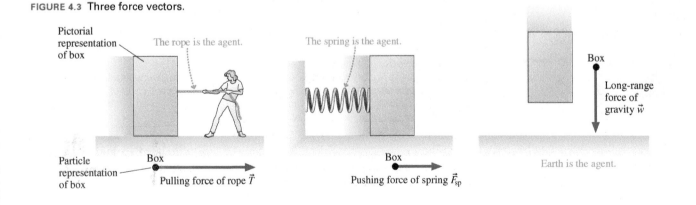

Pictorial representation of box

The rope is the agent.

The spring is the agent.

Box

Long-range force of gravity $\vec{w}$

Earth is the agent.

Particle representation of box

Box
Pulling force of rope $\vec{T}$

Box
Pushing force of spring $\vec{F}_{sp}$

Combining Forces

Force is a vector quantity. We saw in ◀ SECTION 3.1 how to combine displacement vectors, and we noted that the techniques of vector addition we introduced would work equally well for other vectors. FIGURE 4.4a shows a top view of a box being pulled by two ropes, each exerting a force on the box. How will the box respond? Experiments show that when several forces $\vec{F}_1, \vec{F}_2, \vec{F}_3, \ldots$ are exerted on an object, they combine to form a **net force** that is the *vector* sum of all the forces.

$$\vec{F}_{net} = \vec{F}_1 + \vec{F}_2 + \vec{F}_3 + \cdots \tag{4.1}$$

That is, the single force $\vec{F}_{net}$ causes the exact same motion of the object as the combination of original forces $\vec{F}_1, \vec{F}_2, \vec{F}_3, \ldots$. Mathematically, this summation is called a *superposition* of forces. The net force is sometimes called the *resultant force*. FIGURE 4.4b shows the net force on the box.

> NOTE ▶ It is important to realize that the net force $\vec{F}_{net}$ is not a new force acting *in addition* to the original forces $\vec{F}_1, \vec{F}_2, \vec{F}_3, \ldots$. Instead, we should think of the original forces being *replaced* by $\vec{F}_{net}$. ◀

FIGURE 4.4 Two forces applied to a box.

(a)

(b)

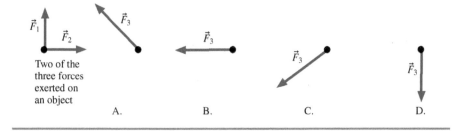

STOP TO THINK 4.1 Two of the three forces exerted on an object are shown. The net force points directly to the left. Which is the missing third force?

4.2 A Short Catalog of Forces

There are many forces we will deal with over and over. This section will introduce you to some of them and to the symbols we use to represent them.

Weight

A falling rock is pulled toward the earth by the long-range force of gravity. Gravity is what keeps you in your chair, keeps the planets in their orbits around the sun, and shapes the large-scale structure of the universe. We'll have a thorough look at gravity in Chapter 6. For now we'll concentrate on objects on or near the surface of the earth (or other planet).

The gravitational pull of the earth on an object on or near the surface of the earth is called **weight**. The symbol for weight is $\vec{w}$. Weight is the only long-range force we will encounter in the next few chapters. The agent for the weight force is the *entire earth* pulling on an object. The weight force is in some ways the simplest force we'll study. As FIGURE 4.5 shows, **an object's weight vector always points vertically downward,** no matter how the object is moving.

> NOTE ▶ We often refer to "the weight" of an object. This is an informal expression for w, the magnitude of the weight force exerted on the object. Note that **weight is not the same thing as mass.** We will briefly examine mass later in the chapter and explore the connection between weight and mass in Chapter 5. ◀

FIGURE 4.5 Weight always points vertically downward.

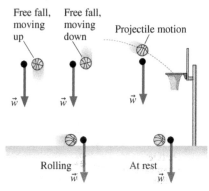

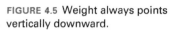

Springs come in many forms. When deflected, they push or pull with a spring force.

Spring Force

Springs exert one of the most basic contact forces. A spring can either push (when compressed) or pull (when stretched). FIGURE 4.6 shows the **spring force**. In both cases, pushing and pulling, the tail of the force vector is placed on the particle in the force diagram. There is no special symbol for a spring force, so we simply use a subscript label: $\vec{F}_{sp}$.

FIGURE 4.6 The spring force is parallel to the spring.

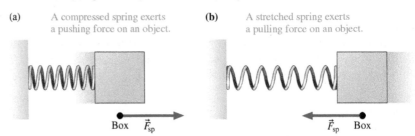

Although you may think of a spring as a metal coil that can be stretched or compressed, this is only one type of spring. Hold a ruler, or any other thin piece of wood or metal, by the ends and bend it slightly. It flexes. When you let go, it "springs" back to its original shape. This is just as much a spring as is a metal coil.

Tension Force

FIGURE 4.7 Tension is parallel to the rope.

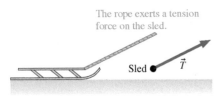

When a string or rope or wire pulls on an object, it exerts a contact force that we call the **tension force,** represented by $\vec{T}$. **The direction of the tension force is always in the direction of the string or rope,** as you can see in FIGURE 4.7. When we speak of "the tension" in a string, this is an informal expression for T, the size or magnitude of the tension force. Note that the tension force can only *pull* in the direction of the string; if you try to *push* with a string, it will go slack and be unable to exert a force.

We can think about the tension force using a microscopic picture. If you were to use a very powerful microscope to look inside a rope, you would "see" that it is made of *atoms* joined together by *molecular bonds*. Molecular bonds are not rigid connections between the atoms. They are more accurately thought of as tiny *springs* holding the atoms together, as in FIGURE 4.8. Pulling on the ends of a string or rope stretches the molecular springs ever so slightly. The tension within a rope and the tension force experienced by an object at the end of the rope are really the net spring force exerted by billions and billions of microscopic springs.

FIGURE 4.8 An atomic model of tension.

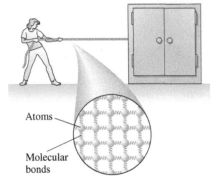

This atomic-level view of tension introduces a new idea: a microscopic **atomic model** for understanding the behavior and properties of **macroscopic** (i.e., containing many atoms) objects. We will frequently use atomic models to obtain a deeper understanding of our observations.

The atomic model of tension also helps to explain one of the basic properties of ropes and strings. When you pull on a rope tied to a heavy box, the rope in turn exerts a tension force on the box. If you pull harder, the tension force on the box becomes greater. How does the box "know" that you are pulling harder on the other end of the rope? According to our atomic model, when you pull harder on the rope, its microscopic springs stretch a bit more, increasing the spring force they exert on each other—and on the box they're attached to.

Normal Force

If you sit on a bed, the springs in the mattress compress and, as a consequence of the compression, exert an upward force on you. Stiffer springs would show less compression but would still exert an upward force. The compression of extremely stiff springs might be measurable only by sensitive instruments. Nonetheless, the springs would compress ever so slightly and exert an upward spring force on you.

FIGURE 4.9 shows a book resting on top of a sturdy table. The table may not visibly flex or sag, but–just as you do to the bed–the book compresses the molecular springs in the table. The compression is very small, but it is not zero. As a consequence, the compressed molecular springs *push upward* on the book. We say that "the table" exerts the upward force, but it is important to understand that the pushing is *really* done by molecular springs. Similarly, an object resting on the ground compresses the molecular springs holding the ground together and, as a consequence, the ground pushes up on the object.

We can extend this idea. Suppose you place your hand on a wall and lean against it, as shown in FIGURE 4.10. Does the wall exert a force on your hand? As you lean, you compress the molecular springs in the wall and, as a consequence, they push outward *against* your hand. So the answer is Yes, the wall does exert a force on you. It's not hard to see this if you examine your hand as you lean: You can see that your hand is slightly deformed, and becomes more so the harder you lean. This deformation is direct evidence of the force that the wall exerts on your hand. Consider also what would happen if the wall suddenly vanished. Without the wall there to push against you, you would topple forward.

The force the table surface exerts is vertical, while the force the wall exerts is horizontal. In all cases, the force exerted on an object that is pressing against a surface is in a direction *perpendicular* to the surface. Mathematicians refer to a line that is perpendicular to a surface as being *normal* to the surface. In keeping with this terminology, we define the **normal force** as the force exerted by a surface (the agent) against an object that is pressing against the surface. The symbol for the normal force is $\vec{n}$.

We're not using the word "normal" to imply that the force is an "ordinary" force or to distinguish it from an "abnormal force." A surface exerts a force *perpendicular* (i.e., normal) to itself as the molecular springs press *outward*. FIGURE 4.11 shows an object on an inclined surface, a common situation. Notice how the normal force $\vec{n}$ is perpendicular to the surface.

The normal force is a very real force arising from the very real compression of molecular bonds. It is in essence just a spring force, but one exerted by a vast number of microscopic springs acting at once. The normal force is responsible for the "solidness" of solids. It is what prevents you from passing right through the chair you are sitting in and what causes the pain and the lump if you bang your head into a door. Your head can then tell you that the force exerted on it by the door was very real!

Friction

You've certainly observed that a rolling or sliding object, if not pushed or propelled, slows down and eventually stops. You've probably discovered that you can slide better across a sheet of ice than across asphalt. And you also know that most objects stay in place on a table without sliding off even if the table is tilted a bit. The force responsible for these sorts of behavior is **friction**. The symbol for friction is $\vec{f}$.

Friction, like the normal force, is exerted by a surface. Unlike the normal force, however, **the frictional force is always** *parallel* **to the surface,** not perpendicular to it. (In many cases, a surface will exert *both* a normal and a frictional force.) On a microscopic level, friction arises as atoms from the object and atoms on the surface run into each other. The rougher the surface is, the more these atoms are forced into close proximity and, as a result, the larger the friction force. We will develop a simple model of friction in the next chapter that will be sufficient for our needs. For now, it is useful to distinguish between two kinds of friction:

■ *Kinetic friction*, denoted $\vec{f}_k$, acts as an object slides across a surface. Kinetic friction is a force that always "opposes the motion," meaning that the friction force $\vec{f}_k$ on a sliding object points in the direction opposite the direction of the object's motion.

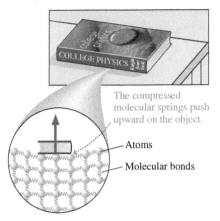

FIGURE 4.9 An atomic model of the force exerted by a table.

The compressed molecular springs push upward on the object.

Atoms

Molecular bonds

FIGURE 4.10 The wall pushes outward against your hand.

The compressed molecular springs in the wall press outward against her hand.

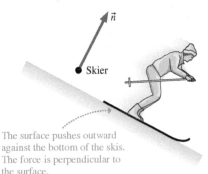

FIGURE 4.11 The normal force is perpendicular to the surface.

$\vec{n}$

Skier

The surface pushes outward against the bottom of the skis. The force is perpendicular to the surface.

■ *Static friction*, denoted $\vec{f}_s$, is the force that keeps an object "stuck" on a surface and prevents its motion relative to the surface. Finding the direction of $\vec{f}_s$ is a little trickier than finding the direction of $\vec{f}_k$. Static friction points opposite the direction in which the object *would* move if there were no friction; that is, it points in the direction necessary to *prevent* motion.

FIGURE 4.12 shows examples of kinetic and static friction.

FIGURE 4.12 **Kinetic and static friction are parallel to the surface.**

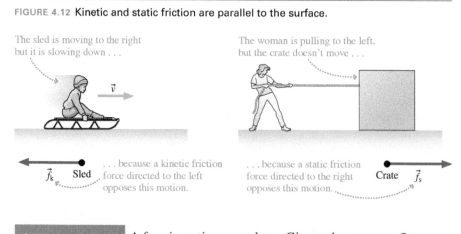

The sled is moving to the right but it is slowing down . . .

The woman is pulling to the left, but the crate doesn't move . . .

$\vec{f}_k$ Sled   . . . because a kinetic friction force directed to the left opposes this motion.

. . . because a static friction force directed to the right opposes this motion.   Crate $\vec{f}_s$

STOP TO THINK 4.2   A frog is resting on a slope. Given what you learned in Figure 4.12, what can you say about the friction force acting on the frog?

A. There is no friction force.
B. There is a kinetic friction force directed up the slope.
C. There is a static friction force directed up the slope.
D. There is a kinetic friction force directed down the slope.
E. There is a static friction force directed down the slope.

FIGURE 4.13 **Air resistance is an example of drag.**

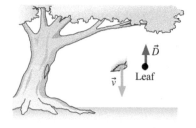

Air resistance is a significant force on falling leaves. It points opposite the direction of motion.

$\vec{D}$

$\vec{v}$ Leaf

FIGURE 4.14 **The thrust force on a rocket is opposite the direction of the expelled gases.**

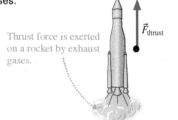

Thrust force is exerted on a rocket by exhaust gases.

$\vec{F}_{\text{thrust}}$

Drag

Friction at a surface is one example of a *resistive force*, a force that opposes or resists motion. Resistive forces are also experienced by objects moving through *fluids*—gases (like air) and liquids (like water). This kind of resistive force—the force of a fluid on a moving object—is called **drag** and is symbolized as $\vec{D}$. Like kinetic friction, **drag points opposite the direction of motion.** FIGURE 4.13 shows an example of drag.

Drag can be a large force for objects moving at high speeds or in dense fluids. Hold your arm out the window as you ride in a car and feel how hard the air pushes against your arm. Note also how the air resistance against your arm increases rapidly as the car's speed increases. Drop a lightweight bead into a beaker of water and watch how slowly it settles to the bottom. The drag force of the water on the bead is significant.

On the other hand, for objects that are heavy and compact, moving in air, and with a speed that is not too great, the drag force of air resistance is fairly small. To keep things as simple as possible, **you can neglect air resistance in all problems unless a problem explicitly asks you to include it.** The error introduced into calculations by this approximation is generally pretty small.

Thrust

A jet airplane obviously has a force that propels it forward; likewise for the rocket in FIGURE 4.14. This force, called **thrust**, occurs when a jet or rocket engine expels gas molecules at high speed. Thrust is a contact force, with the exhaust gas being the agent that pushes on the engine. The process by which thrust is generated is rather subtle and requires an appreciation of Newton's third law, introduced later in this chapter. For now, we need only consider that **thrust is a force opposite the direction in which the exhaust gas is expelled.** There's no special symbol for thrust, so we will call it $\vec{F}_{thrust}$.

Electric and Magnetic Forces

Electricity and magnetism, like gravity, exert long-range forces. The forces of electricity and magnetism act on charged particles. We will study electric and magnetic forces in detail in Part VI of this book. These forces—and the forces inside the nucleus, which we will also see later in the text—won't be important for the dynamics problems we consider in the next several chapters.

STOP TO THINK 4.3 A boy is using a rope to pull a sled to the right. What are the directions of the tension force and the friction force on the sled, respectively?

A. Right, right
C. Left, right
B. Right, left
D. Left, left

It's not just rocket science BIO Rockets are propelled by thrust, but many animals are as well. Scallops are shellfish with no feet and no fins, but they can escape from predators or move to new territory by using a form of jet propulsion. A scallop forcibly ejects water from the rear of its shell, resulting in a thrust force that moves it forward.

4.3 Identifying Forces

A typical physics problem describes an object that is being pushed and pulled in various directions. Some forces are given explicitly, while others are only implied. In order to proceed, it is necessary to determine all the forces that act on the object. It is also necessary to avoid including forces that do not really exist. Now that you have learned the properties of forces and seen a catalog of typical forces, we can develop a step-by-step method for identifying each force in a problem. A list of the most common forces we'll come across in the next few chapters is given in Table 4.1.

Class Video

NOTE ▶ Occasionally, you'll see labels for forces that aren't included in Table 4.1. For instance, if you push a book across a table, we might simply refer to the force you apply as $\vec{F}_{hand}$. When two objects are in contact, the force between them is a mix of friction forces and normal forces—forces that are listed in Table 4.1. But, if we are concerned with only the magnitude and direction of the force and not the exact nature of the force, we may use a more general notation. ◀

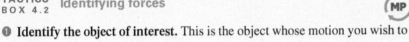

TACTICS BOX 4.2 **Identifying forces** (MP)

❶ **Identify the object of interest.** This is the object whose motion you wish to study.

❷ **Draw a picture of the situation.** Show the object of interest and all other objects—such as ropes, springs, and surfaces—that touch it.

❸ **Draw a closed curve around the object.** Only the object of interest is inside the curve; everything else is outside.

❹ **Locate every point on the boundary of this curve where other objects touch the object of interest.** These are the points where *contact forces* are exerted on the object.

❺ **Name and label each contact force acting on the object.** There is at least one force at each point of contact; there may be more than one. When necessary, use subscripts to distinguish forces of the same type.

❻ **Name and label each long-range force acting on the object.** For now, the only long-range force is weight.

Exercises 4–8

TABLE 4.1 Common forces and their notation

Force	Notation
General force	$\vec{F}$
Weight	$\vec{w}$
Spring force	$\vec{F}_{sp}$
Tension	$\vec{T}$
Normal force	$\vec{n}$
Static friction	$\vec{f}_s$
Kinetic friction	$\vec{f}_k$
Drag	$\vec{D}$
Thrust	$\vec{F}_{thrust}$

CONCEPTUAL EXAMPLE 4.1 **Identifying forces on a bungee jumper**

A bungee jumper has leapt off a bridge and is nearing the bottom of her fall. What forces are being exerted on the bungee jumper?

REASON   FIGURE 4.15 Forces on a bungee jumper.

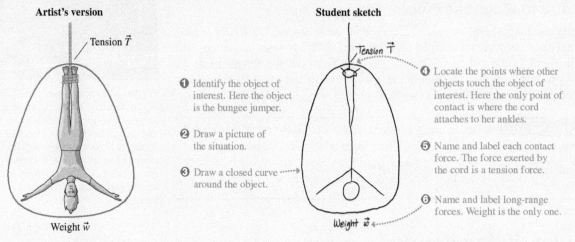

Artist's version

Tension $\vec{T}$

Weight $\vec{w}$

Student sketch

❶ Identify the object of interest. Here the object is the bungee jumper.

❷ Draw a picture of the situation.

❸ Draw a closed curve around the object.

❹ Locate the points where other objects touch the object of interest. Here the only point of contact is where the cord attaches to her ankles.

❺ Name and label each contact force. The force exerted by the cord is a tension force.

❻ Name and label long-range forces. Weight is the only one.

CONCEPTUAL EXAMPLE 4.2 **Identifying forces on a skier**

A skier is being towed up a snow-covered hill by a tow rope. What forces are being exerted on the skier?

REASON   FIGURE 4.16 Forces on a skier.

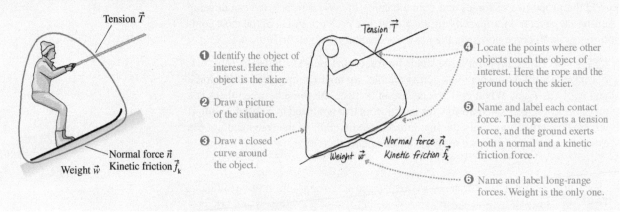

Tension $\vec{T}$

Normal force $\vec{n}$
Kinetic friction $\vec{f}_k$
Weight $\vec{w}$

❶ Identify the object of interest. Here the object is the skier.

❷ Draw a picture of the situation.

❸ Draw a closed curve around the object.

❹ Locate the points where other objects touch the object of interest. Here the rope and the ground touch the skier.

❺ Name and label each contact force. The rope exerts a tension force, and the ground exerts both a normal and a kinetic friction force.

❻ Name and label long-range forces. Weight is the only one.

NOTE ▶ You might have expected two friction forces and two normal forces in Example 4.2, one on each ski. Keep in mind, however, that we're working within the particle model, which represents the skier by a single point. A particle has only one contact with the ground, so there is a single normal force and a single friction force. The particle model is valid if we want to analyze the motion of the skier as a whole, but we would have to go beyond the particle model to find out what happens to each ski. ◀

CONCEPTUAL EXAMPLE 4.3 **Identifying forces on a rocket**

A rocket is flying upward through the air, high above the ground. Air resistance is not negligible. What forces are being exerted on the rocket?

REASON

FIGURE 4.17 Forces on a rocket.

Drag $\vec{D}$

Weight $\vec{w}$

Thrust $\vec{F}_{thrust}$

You've just kicked a rock, and it is now sliding across the ground about 2 meters in front of you. Which of these are forces acting on the rock? List all that apply.

A. Gravity, acting downward
B. The normal force, acting upward
C. The force of the kick, acting in the direction of motion
D. Friction, acting opposite the direction of motion
E. Air resistance, acting opposite the direction of motion

4.4 What Do Forces Do?

The fundamental question is: How does an object move when a force is exerted on it? The only way to answer this question is to do experiments. To do experiments, however, we need a way to reproduce the same force again and again, and we need a standard object so that our experiments are repeatable.

FIGURE 4.18 shows how you can use your fingers to stretch a rubber band to a certain length–say, 10 centimeters–that you can measure with a ruler. We'll call this the *standard length*. You know that a stretched rubber band exerts a force because your fingers *feel* the pull. Furthermore, this is a reproducible force. The rubber band exerts the same force every time you stretch it to the standard length. We'll call the magnitude of this force the *standard force F*. Not surprisingly, two identical rubber bands, each stretched to the standard length, exert twice the force of one rubber band; three rubber bands exert three times the force; and so on.

We'll also need several identical standard objects to which the force will be applied. As we learned in Chapter 1, the SI unit of mass is the kilogram (kg). For our standard objects, we will make ourselves several identical objects, each with a mass of 1 kg.

Now we're ready to start a virtual experiment. First, place one of the 1 kg blocks on a frictionless surface. (In a real experiment, we can nearly eliminate friction by floating the block on a cushion of air.) Second, attach a rubber band to the block and stretch the band to the standard length. Then the block experiences the same force F as your finger did. As the block starts to move, in order to keep the pulling force constant you must *move your hand* in just the right way to keep the length of the rubber band—and thus the force—*constant*. FIGURE 4.19 shows the experiment being carried out. Once the motion is complete, you can use motion diagrams and kinematics to analyze the block's motion.

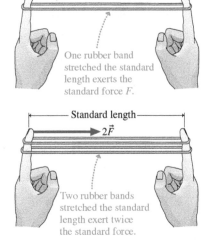

FIGURE 4.18 A reproducible force.

Standard length

One rubber band stretched the standard length exerts the standard force F.

Standard length

Two rubber bands stretched the standard length exert twice the standard force.

FIGURE 4.19 Measuring the motion of a 1 kg block that is pulled with a constant force.

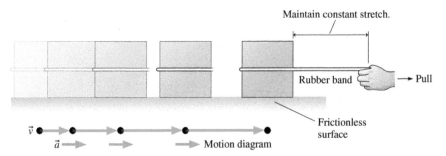

Maintain constant stretch.

Rubber band ——> Pull

Frictionless surface

$\vec{v}$
$\vec{a}$ —— Motion diagram

The motion diagram in Figure 4.19 shows that the velocity vectors are getting longer, so the velocity is increasing: The block is *accelerating*. Furthermore, a close inspection of the motion diagram shows that the acceleration vectors are all the same length. This is the first important finding of this experiment: **An object pulled with a constant force moves with a constant acceleration.** This finding could not have been anticipated in advance. It's conceivable that the object would speed up for a

FIGURE 4.20 Graph of acceleration versus force.

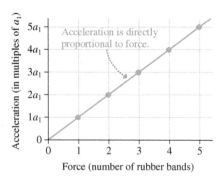

FIGURE 4.21 Graph of acceleration versus number of blocks.

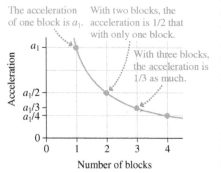

while and then move with a steady speed. Or that it would continue to speed up, but that the *rate* of increase, the acceleration, would steadily decline. But these descriptions do not match what happens. Instead, the object continues *with a constant acceleration* for as long as you pull it with a constant force. We'll call this constant acceleration of *one* block pulled by *one* band a_1.

What happens if you increase the force by using several rubber bands? To find out, use two rubber bands. Stretch both to the standard length to double the force to $2F$, then measure the acceleration. Measure the acceleration due to three rubber bands, then four, and so on. FIGURE 4.20 is a graph of the results. Force is the independent variable, the one you can control, so we've placed force on the horizontal axis to make an acceleration-versus-force graph. The graph reveals our second important finding: **Acceleration is directly proportional to force.**

The final question for our virtual experiment is: How does the acceleration of an object depend on the mass of the object? To find out, we'll glue two of our 1 kg blocks together, so that we have a block with twice as much matter as a 1 kg block—that is, a 2 kg block. Now apply the same force—a single rubber band—as you applied to the single 1 kg block. FIGURE 4.21 shows that the acceleration is *one-half* as great as that of the single block. If we glue three blocks together, making a 3 kg object, we find that the acceleration is only *one-third* of the 1 kg block's acceleration. In general, we find that the acceleration is proportional to the *inverse* of the mass of the object. So our third important result is: **Acceleration is *inversely proportional* to an object's mass.**

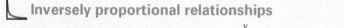

⌐ Inversely proportional relationships   (MP)

Two quantities are said to be **inversely proportional** to each other if one quantity is proportional to the *inverse* of the other. Mathematically, this means that

$$y = \frac{A}{x}$$

y is inversely proportional to x

Here, A is a proportionality constant. This relationship is sometimes written as $y \propto 1/x$.

[Graph showing $y = \dfrac{A}{x}$ with annotations: "If x is halved, y doubles." "Original point x, y" "If x is doubled, y is halved."]

SCALING ■ If you double x, you halve y.
■ If you triple x, y is reduced by a factor of 3.
■ If you halve x, y doubles.
■ If you reduce x by a factor of 3, y becomes 3 times as large.

RATIOS For any two values of x—say, x_1 and x_2—we have

$$y_1 = \frac{A}{x_1} \quad \text{and} \quad y_2 = \frac{A}{x_2}$$

Dividing the y_1 equation by the y_2 equation, we find

$$\frac{y_1}{y_2} = \frac{A/x_1}{A/x_2} = \frac{A}{x_1}\frac{x_2}{A} = \frac{x_2}{x_1}$$

That is, the ratio of y-values is the inverse of the ratio of the corresponding values of x.

LIMITS ■ As x gets very large, y approaches zero.
■ As x approaches zero, y gets very large.

Exercise 10 ✎

You're familiar with this idea: It's much harder to get your car rolling by pushing it than to get your bicycle rolling, and it's harder to stop a heavily loaded grocery cart than to stop a skateboard. This tendency to resist a change in velocity (i.e., to resist speeding up or slowing down) is called **inertia.** Thus we can say that more massive objects have more inertia.

EXAMPLE 4.4 **Finding the mass of an unknown block**

When a rubber band is stretched to pull on a 1.0 kg block with a constant force, the acceleration of the block is measured to be 3.0 m/s². When a block with an unknown mass is pulled with the same rubber band, using the same force, its acceleration is 5.0 m/s². What is the mass of the unknown block?

PREPARE Each block's acceleration is inversely proportional to its mass.

SOLVE We can use the result of the Inversely Proportional Relationships box to write

$$\frac{3.0 \text{ m/s}^2}{5.0 \text{ m/s}^2} = \frac{m}{1.0 \text{ kg}}$$

or

$$m = \frac{3.0 \text{ m/s}^2}{5.0 \text{ m/s}^2} \times (1.0 \text{ kg}) = 0.60 \text{ kg}$$

ASSESS With the same force applied, the unknown block had a *larger* acceleration than the 1.0 kg block. It makes sense, then, that its mass—its resistance to acceleration—is *less* than 1.0 kg.

Feel the difference Because of its high sugar content, a can of regular soda has a mass about 4% greater than that of a can of diet soda. If you try to judge which can is more massive by simply holding one in each hand, this small difference is almost impossible to detect. If you *move* the cans up and down, however, the difference becomes subtly but noticeably apparent: People evidently are more sensitive to how the mass of each can resists acceleration than they are to the cans' weights alone.

Class Video

STOP TO THINK 4.5 Two rubber bands stretched to the standard length cause an object to accelerate at 2 m/s². Suppose another object with twice the mass is pulled by four rubber bands stretched to the standard length. What is the acceleration of this second object?

A. 1 m/s²   B. 2 m/s²   C. 4 m/s²   D. 8 m/s²   E. 16 m/s²

4.5 Newton's Second Law

We can now summarize the results of our experiments. We've seen that **a force causes an object to accelerate. The acceleration** a **is directly proportional to the force** F **and inversely proportional to the mass** m. We can express both these relationships in equation form as

$$a = \frac{F}{m} \tag{4.2}$$

Class Video

Video Tutor Demo

Note that if we double the size of the force F, the acceleration a will double, as we found experimentally. And if we triple the mass m, the acceleration will be only one-third as great, again agreeing with experiment.

Equation 4.2 tells us the magnitude of an object's acceleration in terms of its mass and the force applied. But our experiments also had another important finding: The *direction* of the acceleration was the same as the direction of the force. We can express this fact by writing Equation 4.2 in *vector* form as

$$\vec{a} = \frac{\vec{F}}{m} \tag{4.3}$$

Video Tutor
Demo

Finally, our experiment was limited to looking at an object's response to a *single* applied force acting in a single direction. Realistically, an object is likely to be subjected to several distinct forces $\vec{F}_1, \vec{F}_2, \vec{F}_3, \ldots$ that may point in different directions. What happens then? Experiments show that the acceleration of the object is determined by the *net force* acting on it. Recall from Figure 4.4 and Equation 4.1 that the net force is the *vector sum* of all forces acting on the object. So if several forces are acting, we use the *net* force in Equation 4.4.

Newton was the first to recognize these connections between force and motion. This relationship is known today as Newton's second law.

Video Tutor
Demo

Newton's second law   An object of mass m subjected to forces $\vec{F}_1, \vec{F}_2, \vec{F}_3, \ldots$ will undergo an acceleration $\vec{a}$ given by

$$\vec{a} = \frac{\vec{F}_{\text{net}}}{m} \qquad (4.4)$$

where the net force $\vec{F}_{\text{net}} = \vec{F}_1 + \vec{F}_2 + \vec{F}_3 + \cdots$ is the vector sum of all forces acting on the object. **The acceleration vector $\vec{a}$ points in the same direction as the net force vector $\vec{F}_{\text{net}}$.**

While some relationships are found to apply only in special circumstances, others seem to have universal applicability. Those equations that appear to apply at all times and under all conditions have come to be called "laws of nature." Newton's second law is a law of nature; you will meet others as we go through this book.

We can rewrite Newton's second law in the form

$$\vec{F}_{\text{net}} = m\vec{a} \qquad (4.5)$$

which is how you'll see it presented in many textbooks and how, in practice, we'll often use the second law. Equations 4.4 and 4.5 are mathematically equivalent, but Equation 4.4 better describes the central idea of Newtonian mechanics: A force applied to an object causes the object to accelerate and the acceleration is in the direction of the net force.

NOTE ▶ When several forces act on an object, be careful not to think that the strongest force "overcomes" the others to determine the motion on its own. It is $\vec{F}_{\text{net}}$, the sum of *all* the forces, that determines the acceleration $\vec{a}$. ◀

Size matters? Race car driver Danica Patrick was the subject of controversial comments by drivers who thought her relatively small mass of 45 kg gave her an unfair advantage. Because every driver's car must have the same mass, Patrick's overall racing mass was lower than any other driver's, so her car could be expected to have a slightly greater acceleration.

CONCEPTUAL EXAMPLE 4.5   **Acceleration of a wind-blown basketball**

A basketball is released from rest in a stiff breeze directed to the right. In what direction does the ball accelerate?

REASON Wind is just air in motion. If the air is moving to the *right* with respect to the ball, then the ball is moving to the *left* with respect to the air. There will be a drag force opposite the velocity of the ball relative to the air, to the right. So, as FIGURE 4.22a shows, two forces are acting on the ball: its weight $\vec{w}$ directed downward and the drag force $\vec{D}$ directed to the right. Newton's second law tells us that the direction of the acceleration is the same as the direction of the net force $\vec{F}_{\text{net}}$. In FIGURE 4.22b we find $\vec{F}_{\text{net}}$ by graphical vector addition of $\vec{w}$ and $\vec{D}$. We see that $\vec{F}_{\text{net}}$ and therefore $\vec{a}$ point downward and to the right.

FIGURE 4.22   A basketball falling in a strong breeze.

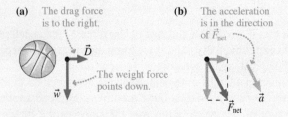

(a) The drag force is to the right. $\vec{D}$ The weight force points down. $\vec{w}$

(b) The acceleration is in the direction of $\vec{F}_{\text{net}}$ $\vec{F}_{\text{net}}$ $\vec{a}$

ASSESS This makes sense on the basis of your experience. Weight pulls the ball down, and the wind pushes the ball to the right. The net result is an acceleration down and to the right.

Units of Force

Because $\vec{F}_{net} = m\vec{a}$, the unit of force must be mass units multiplied by acceleration units. We've previously specified the SI unit of mass as the kilogram. We can now define the basic unit of force as "the force that causes a 1 kg mass to accelerate at 1 m/s²." From Newton's second law, this force is

$$1 \text{ basic unit of force} = (1 \text{ kg}) \times (1 \text{ m/s}^2) = 1 \, \frac{\text{kg·m}}{\text{s}^2}$$

This basic unit of force is called a *newton:* One **newton** is the force that causes a 1 kg mass to accelerate at 1 m/s². The abbreviation for newton is N. Mathematically, $1 \text{ N} = 1 \text{ kg·m/s}^2$. Table 4.2 lists some typical forces.

The newton is a *secondary unit,* meaning that it is defined in terms of the *primary units* of kilograms, meters, and seconds.

The unit of force in the English system is the *pound* (abbreviated lb). Although the definition of the pound has varied, it is now defined in terms of the newton:

$$1 \text{ pound} = 1 \text{ lb} = 4.45 \text{ N}$$

You very likely associate pounds with kilograms rather than with newtons. Everyday language often confuses the ideas of mass and weight, but we're going to need to make a clear distinction between them. We'll have more to say about this in the next chapter.

TABLE 4.2 Approximate magnitude of some typical forces

Force	Approximate magnitude (newtons)
Weight of a U.S. nickel	0.05
Weight of ¼ cup of sugar	0.5
Weight of a 1 pound object	5
Weight of a typical house cat	50
Weight of a 110 pound person	500
Propulsion force of a car	5000
Thrust force of a small jet engine	50,000
Pulling force of a locomotive	500,000

EXAMPLE 4.6 **Racing down the runway**

A Boeing 737—a small, short-range jet with a mass of 51,000 kg—sits at rest. The pilot turns the pair of jet engines to full throttle, and the thrust accelerates the plane down the runway. After traveling 940 m, the plane reaches its takeoff speed of 70 m/s and leaves the ground. What is the thrust of each engine?

PREPARE If we assume that the plane undergoes a constant acceleration (a reasonable assumption), we can use kinematics to find the magnitude of that acceleration. Then we can use Newton's second law to find the force—the thrust—that produced this acceleration. FIGURE 4.23 is a visual overview of the airplane's motion.

FIGURE 4.23 Visual overview of the accelerating airplane.

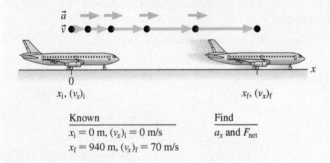

Known	Find
$x_i = 0$ m, $(v_x)_i = 0$ m/s	a_x and F_{net}
$x_f = 940$ m, $(v_x)_f = 70$ m/s	

SOLVE We don't know how much time it took the plane to reach its takeoff speed, but we do know that it traveled a distance of 940 m. We can solve for the acceleration by using the third constant-acceleration equation in Synthesis 2.1:

$$(v_x)_f^2 = (v_x)_i^2 + 2a_x \Delta x$$

The displacement is $\Delta x = x_f - x_i = 940$ m, and the initial velocity is 0. We can rearrange the equation to solve for the acceleration:

$$a_x = \frac{(v_x)_f^2}{2 \Delta x} = \frac{(70 \text{ m/s})^2}{2(940 \text{ m})} = 2.61 \text{ m/s}^2$$

We've kept an extra significant figure because this isn't our final result—we are asked to solve for the thrust. We complete the solution by using Newton's second law:

$$F = ma_x = (51,000 \text{ kg})(2.61 \text{ m/s}^2) = 133,000 \text{ N}$$

The thrust of each engine is half of this total force:

$$\text{Thrust of one engine} = 67,000 \text{ N} = 67 \text{ kN}$$

ASSESS An acceleration of about ¼g seems reasonable for an airplane: It's zippy, but it's not a thrill ride. And the final value we find for the thrust of each engine is close to the value given in Table 4.2. This gives us confidence that our final result makes good physical sense.

STOP TO THINK 4.6 Three forces act on an object. In which direction does the object accelerate?

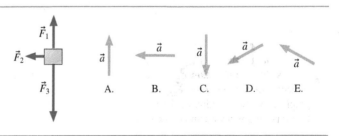

4.6 Free-Body Diagrams

When we solve a dynamics problem, it is useful to assemble all of the information about the forces that act on an object into a single diagram called a **free-body diagram**. A free-body diagram represents the object as a particle and shows *all* of the forces that act on the object. Learning how to draw a correct free-body diagram is a very important skill, one that in the next chapter will become a critical part of our strategy for solving motion problems.

TACTICS BOX 4.3   Drawing a free-body diagram

❶ **Identify all forces acting on the object.** This step was described in Tactics Box 4.2.
❷ **Draw a coordinate system.** Use the axes defined in your pictorial representation (Tactics Box 2.2). If those axes are tilted, for motion along an incline, then the axes of the free-body diagram should be similarly tilted.
❸ **Represent the object as a dot at the origin of the coordinate axes.** This is the particle model.
❹ **Draw vectors representing each of the identified forces.** This was described in Tactics Box 4.1. Be sure to label each force vector.
❺ **Draw and label the** *net force* **vector** $\vec{F}_{net}$. Draw this vector beside the diagram, not on the particle. Then check that $\vec{F}_{net}$ points in the same direction as the acceleration vector $\vec{a}$ on your motion diagram. Or, if appropriate, write $\vec{F}_{net} = \vec{0}$.

Exercises 17–22 ✏

EXAMPLE 4.7   **Forces on an elevator**

An elevator, suspended by a cable, speeds up as it moves upward from the ground floor. Draw a free-body diagram of the elevator.

PREPARE The elevator is moving upward, and its speed is increasing. This means that the acceleration is directed upward—that's enough to say about acceleration for the purposes of this problem. Next, we continue with the forces. **FIGURE 4.24** illustrates the steps listed in Tactics Box 4.3. We know that the acceleration is directed upward, so $\vec{F}_{net}$ must be directed upward as well.

FIGURE 4.24 Free-body diagram of an elevator accelerating upward.

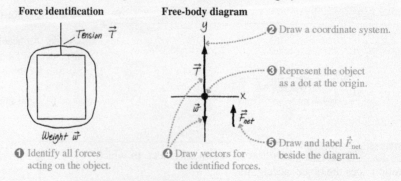

ASSESS Let's take a look at our picture and see if it makes sense. The coordinate axes, with a vertical *y*-axis, are the ones we would use in a pictorial representation of the motion, so we've chosen the correct axes. $\vec{F}_{net}$ is directed upward. For this to be true, the magnitude of $\vec{T}$ must be greater than the magnitude of $\vec{w}$, which is just what we've drawn.

EXAMPLE 4.8 Forces on a carbon dioxide racer

A high-school class has built lightweight model cars that are propelled by the thrust from carbon dioxide canisters. The students will race the cars, so they've made every effort to minimize friction and drag. These forces are so small compared to the thrust force that we can ignore them. A car starts from rest, the cylinder that propels it is punctured, and the car accelerates along the track. Draw a visual overview—motion diagram, force identification diagram, and free-body diagram—for the car.

PREPARE We can treat the car as a particle. The visual overview consists of a motion diagram to determine $\vec{a}$, a force identification picture, and a free-body diagram. The statement of the situation tells us that friction and drag are negligible. We can draw these three pictures using Problem-Solving Strategy 1.1 for the motion diagram, Tactics Box 4.2 to identify the forces, and Tactics Box 4.3 to draw the free-body diagrams. These pictures are shown in **FIGURE 4.25**.

ASSESS The motion diagram tells us that the acceleration is in the positive x-direction. According to the rules of vector addition, this can be true only if the upward-pointing $\vec{n}$ and the downward-pointing $\vec{w}$ are equal in magnitude and thus cancel each other. The vectors have been drawn accordingly, and this leaves the net force vector pointing toward the right, in agreement with $\vec{a}$ from the motion diagram.

FIGURE 4.25 Visual overview for a carbon dioxide racer.

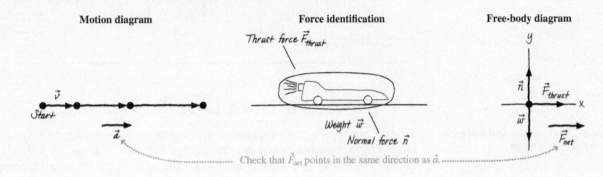

EXAMPLE 4.9 Forces on a towed skier

A tow rope pulls a skier up a snow-covered hill at a constant speed. Draw a full visual overview of the skier.

PREPARE This is Example 4.2 again with the additional information that the skier is moving at a constant speed. If we were doing a kinematics problem, the pictorial representation would use a tilted coordinate system with the x-axis parallel to the slope, so we use these same tilted coordinate axes for the free-body diagram. The motion diagram, force identification diagram, and free-body diagram are shown in **FIGURE 4.26**.

ASSESS We have shown $\vec{T}$ pulling parallel to the slope and $\vec{f}_k$, which opposes the direction of motion, pointing down the slope. The normal force $\vec{n}$ is perpendicular to the surface and thus along the y-axis. Finally, and this is important, the weight $\vec{w}$ is *vertically* downward, *not* along the negative y-axis.

The skier moves in a straight line with constant speed, so $\vec{a} = \vec{0}$. Newton's second law then tells us that $\vec{F}_{net} = m\vec{a} = \vec{0}$. Thus we have drawn the vectors such that the forces add to zero. We'll learn more about how to do this in Chapter 5.

FIGURE 4.26 Visual overview for a skier being towed at a constant speed.

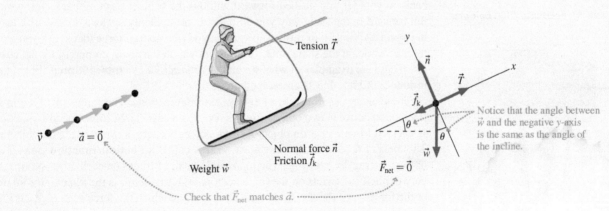

Free-body diagrams will be our major tool for the next several chapters. Careful practice with the workbook exercises and homework in this chapter will pay immediate benefits in the next chapter. Indeed, it is not too much to assert that a problem is more than half solved when you correctly complete the free-body diagram.

STOP TO THINK 4.7 An elevator suspended by a cable is moving upward and slowing to a stop. Which free-body diagram is correct?

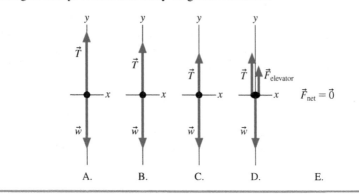

A.   B.   C.   D.   E.

4.7 Newton's Third Law

Thus far, we've focused on the motion of a single particle responding to well-defined forces exerted by other objects or to long-range forces. A skier sliding downhill, for instance, is subject to frictional and normal forces from the slope and the pull of gravity on his body. Once we have identified these forces, we can use Newton's second law to calculate the acceleration, and hence the overall motion, of the skier.

But motion in the real world often involves two or more objects *interacting* with each other. Consider the hammer and nail in FIGURE 4.27. As the hammer hits the nail, the nail pushes back on the hammer. A bat and a ball, your foot and a soccer ball, and the earth–moon system are other examples of interacting objects. We need to consider how the forces on these interacting objects are related to each other.

Interacting Objects

Think about the hammer and nail in Figure 4.27 again. The hammer certainly exerts a force on the nail as it drives the nail forward. At the same time, the nail exerts a force on the hammer. If you are not sure that it does, imagine hitting the nail with a glass hammer. It's the force of the nail on the hammer that would cause the glass to shatter.

Indeed, any time that object A pushes or pulls on object B, object B pushes or pulls back on object A. As you push on a filing cabinet to move it, the cabinet pushes back on you. (If you pushed forward without the cabinet pushing back, you would fall forward in the same way you do if someone suddenly opens a door you're leaning against.) Your chair pushes upward on you (the normal force that keeps you from falling) while, at the same time, you push down on the chair, compressing the cushion. These are examples of what we call an *interaction*. An **interaction** is the mutual influence of two objects on each other.

These examples illustrate a key aspect of interactions: The forces involved in an interaction between two objects always occur as a *pair*. To be more specific, if object A exerts a force $\vec{F}_{A \text{ on } B}$ on object B, then object B exerts a force $\vec{F}_{B \text{ on } A}$ on object A. This pair of forces, shown in FIGURE 4.28, is called an **action/reaction pair**. Two objects interact by exerting an action/reaction pair of forces on each other. Notice the very explicit subscripts on the force vectors. The first letter is the *agent*—the source of the force—and the second letter is the *object* on which the force acts. $\vec{F}_{A \text{ on } B}$ is thus the force exerted *by* A *on* B.

FIGURE 4.27 The hammer and nail each exert a force on the other.

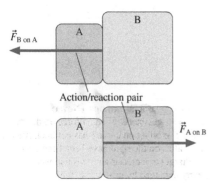

The hammer exerts a force on the nail . . .

. . . but the *nail* also exerts a force on the *hammer*.

FIGURE 4.28 An action/reaction pair of forces.

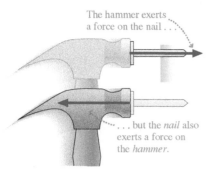

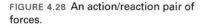

NOTE ▶ The name "action/reaction pair" is somewhat misleading. The forces occur simultaneously, and we cannot say which is the "action" and which the "reaction." Neither is there any implication about cause and effect: The action does not cause the reaction. **An action/reaction pair of forces exists as a pair, or not at all.** For action/reaction pairs, the labels are the key: Force $\vec{F}_{A \text{ on } B}$ is paired with force $\vec{F}_{B \text{ on } A}$. ◀

Reasoning with Newton's Third Law

Two objects always interact via an action/reaction pair of forces. Newton was the first to recognize how the two members of an action/reaction pair of forces are related to each other. Today we know this as Newton's third law:

> **Newton's third law** Every force occurs as one member of an action/reaction pair of forces.
> - The two members of an action/reaction pair act on two *different* objects.
> - The two members of an action/reaction pair point in *opposite* directions and are *equal in magnitude*.

Newton's third law is often stated: "For every action there is an equal but opposite reaction." While this is a catchy phrase, it lacks the preciseness of our preferred version. In particular, it fails to capture an essential feature of the two members of an action/reaction pair—that each acts on a *different* object. This is shown in **FIGURE 4.29**, where a hammer hitting a nail exerts a force $\vec{F}_{\text{hammer on nail}}$ on the nail, and by the third law, the nail must exert a force $\vec{F}_{\text{nail on hammer}}$ to complete the action/reaction pair.

Figure 4.29 also illustrates that these two forces point in *opposite directions*. This feature of the third law is also in accord with our experience. If the hammer hits the nail with a force directed to the right, the force of the nail on the hammer is directed to the left. If the force of my chair on me pushes up, the force of me on the chair pushes down.

Finally, Figure 4.29 shows that, according to Newton's third law, the two members of an action/reaction pair have *equal* magnitudes, so that $F_{\text{hammer on nail}} = F_{\text{nail on hammer}}$. This is something new, and it is by no means obvious.

Video Tutor Demo   Video Tutor Demo

FIGURE 4.29 Newton's third law.

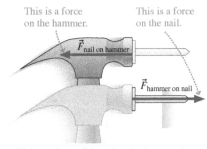

Each force in an action/reaction pair acts on a *different* object.

This is a force on the hammer.   This is a force on the nail.

$\vec{F}_{\text{nail on hammer}}$

$\vec{F}_{\text{hammer on nail}}$

The members of the pair point in *opposite directions*, but are of *equal magnitude*.

CONCEPTUAL EXAMPLE 4.10   **Forces in a collision**

A 10,000 kg truck has a head-on collision with a 1000 kg compact car. During the collision, is the force of the truck on the car greater than, less than, or equal to the force of the car on the truck?

REASON Newton's third law tells us that the magnitude of the force of the car on the truck must be equal to that of the truck on the car! How can this be, when the car is so small compared to the truck? The source of puzzlement in problems like this is that Newton's third law equates the sizes of the *forces* acting on the two objects, not their *accelerations*. The acceleration of each object depends not only on the force applied to it but also, according to Newton's second law, on its mass. The car and the truck do in fact feel forces of equal strength from the other, but the car, with its smaller mass, undergoes a much greater acceleration than the more massive truck.

ASSESS This is a type of question for which your intuition may need to be refined. When you think about questions of this sort, be sure to separate the *effects* (the accelerations) from the *causes* (the forces themselves). Because two interacting objects can have very different masses, their accelerations can be very different. Don't let this dissuade you from realizing that the interaction forces are of the same strength.

Revenge of the target We normally think of the damage that the force of a bullet inflicts on its target. But according to Newton's third law, the target exerts an equal force on the bullet. The bullet on the left has not been fired. The bullets on the right have been fired into a test target at increasing speeds and were clearly damaged by this interaction.

FIGURE 4.30 Examples of propulsion.

(a)

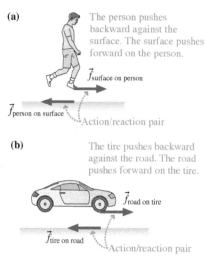

The person pushes backward against the surface. The surface pushes forward on the person.

$\vec{f}_{\text{surface on person}}$

$\vec{f}_{\text{person on surface}}$

Action/reaction pair

(b)

The tire pushes backward against the road. The road pushes forward on the tire.

$\vec{f}_{\text{road on tire}}$

$\vec{f}_{\text{tire on road}}$

Action/reaction pair

FIGURE 4.31 Rocket propulsion.

The rocket pushes the hot gases backward. The gases push the rocket forward.

$\vec{F}_{\text{gases on rocket}}$

Action/reaction pair

$\vec{F}_{\text{rocket on gases}}$

Runners and Rockets

A runner starts from rest and then begins to move down the track. Because he's accelerating, there must be a force on him in the forward direction. The *energy* to put his body into motion comes from inside his body (we'll consider this type of problem in detail in Chapter 11). But where does the *force* come from?

If you tried to walk across a frictionless floor, your foot would slip and slide *backward*. In order for you to walk, the floor needs to have friction so that your foot *sticks* to the floor as you straighten your leg, moving your body forward. The friction that prevents slipping is *static* friction. Static friction, you will recall, acts in the direction that prevents slipping, so the static friction force $\vec{f}_{\text{S on P}}$ (for **Surface on Person**) has to point in the *forward* direction to prevent your foot from slipping backward. As shown in FIGURE 4.30a, it is this forward-directed static friction force that propels you forward! The force of your foot on the floor, $\vec{f}_{\text{P on S}}$, is the other half of the action/reaction pair, and it points in the opposite direction as you push backward against the floor. So, when the runner starts down the track—or when you start walking across the floor—it is the static friction force between the ground and the runner that provides the acceleration. This may seem surprising, but imagine that the race was held on an icy pond. The runners would have much more trouble getting started.

NOTE ▶ A counterintuitive notion for many students is that it is *static* friction that pushes you forward. You are moving, so how can this be static friction? It's true that your body is in motion, but your feet are not sliding, so this is not kinetic friction, but static. ◀

Similarly, the car in FIGURE 4.30b uses static friction to propel itself. The car uses its motor to turn the tires, causing the tires to push backward against the road ($\vec{f}_{\text{tire on road}}$). The road surface responds by pushing the car forward ($\vec{f}_{\text{road on tire}}$). Again, the forces involved are *static* friction forces. The tire is rolling, but the bottom of the tire, where it contacts the road, is instantaneously at rest. If it weren't, you would leave one giant skid mark as you drove and would burn off the tread within a few miles.

For a system that has an internal source of energy, a force that drives the system is a force of propulsion. Rocket motors provide propulsion as well, but there is a difference from the earlier cases: A rocket doesn't need to push against the ground or even the atmosphere. That's why rocket propulsion works in the vacuum of space. Instead, the rocket engine pushes hot, expanding gases out of the back of the rocket, as shown in FIGURE 4.31. In response, the exhaust gases push the rocket forward with the force we've called *thrust*.

The rocket pushes hot gases out the back, and this results in a forward force on the rocket. But let's consider the opposite case—a pull *in* rather than a push *out*. The chapter opened with a photo of a snake-necked turtle. The muscles in the turtle's neck aren't strong enough to snap its head forward with the observed acceleration. Instead, the turtle uses a different approach. The turtle opens its mouth and forcefully pulls water into its throat. The turtle's head and the water form an action/reaction pair: The water is pulled backward, and this results in a forward force on the turtle's head. This is just the reverse of what happens in a rocket, and it is a surprisingly effective technique, enabling the turtle to strike more rapidly than many predatory fish.

Now we've assembled all the pieces we need in order to start solving problems in dynamics. We have seen what forces are and how to identify them, and we've learned how forces cause objects to accelerate according to Newton's second law. We've also found how Newton's third law governs the interaction forces between two objects. Our goal in the next several chapters is to apply Newton's laws to a variety of problems involving straight-line and circular motion.

STOP TO THINK 4.8   A small car is pushing a larger truck that has a dead battery.
The mass of the truck is greater than the
mass of the car. The car and the truck are
moving to the right and are speeding up.
Which of the following statements is true?

A. The car exerts a force on the truck, but the truck doesn't exert a force on the car.
B. The car exerts a larger force on the truck than the truck exerts on the car.
C. The car exerts the same amount of force on the truck as the truck exerts on the car.
D. The truck exerts a larger force on the car than the car exerts on the truck.
E. The truck exerts a force on the car, but the car doesn't exert a force on the truck.

INTEGRATED EXAMPLE 4.11   **Pulling an excursion train**

An engine slows as it pulls two cars of
an excursion train up a mountain.
Draw a visual overview (motion diagram, force identification diagram,
and free-body diagram) for the car just
behind the engine. Ignore friction.

PREPARE Because the train is slowing
down, the motion diagram consists of a series of particle positions that become closer together at successive times; the corresponding velocity vectors become shorter and shorter. To identify
the forces acting on the car we use the steps of Tactics Box 4.2.
Finally, we can draw a free-body diagram using Tactics Box 4.3.

SOLVE Finding the forces acting on car 1 can be tricky. The engine exerts a forward force $\vec{F}_{\text{engine on 1}}$ on car 1 where the engine
touches the front of car 1. At its back, car 1 touches car 2, so car 2

must also exert a force on car 1. The direction of this force can be
understood from Newton's third law. Car 1 exerts an uphill force
on car 2 in order to pull it up the mountain. Thus, by Newton's
third law, car 2 must exert an oppositely directed *downhill* force
on car 1. This is the force we label $\vec{F}_{\text{2 on 1}}$. The three diagrams that
make up the full visual overview are shown in FIGURE 4.32.

ASSESS Correctly preparing the three diagrams illustrated in this
example is critical for solving problems using Newton's laws.
The motion diagram allows you to determine the direction of
the acceleration and hence of $\vec{F}_{\text{net}}$. Using the force identification
diagram, you will correctly identify all the forces acting on the
object and, just as important, not add any extraneous forces. And
by properly drawing these force vectors in a free-body diagram,
you'll be ready for the quantitative application of Newton's laws
that is the focus of Chapter 5.

FIGURE 4.32 Visual overview for a slowing train car being pulled up a mountain.

Motion diagram	Force identification (Numbered steps from Tactics Box 4.2)	Free-body diagram (Numbered steps from Tactics Box 4.3)

Force identification
① The object of interest is car 1.
② Draw a picture.
③ Draw a closed curve around the object.
④ Locate the points where the object
   touches other objects.
⑤ Name and label each contact force.
⑥ Weight is the only long-range force.

Free-body diagram
① Identify all forces (already done).
② Draw a coordinate system. Because the motion
   here is along an incline, we tilt our *x*-axis to match.
③ Represent the object as a dot at the origin.
④ Draw vectors representing each identified force.
⑤ Draw the net force vector. Check that
   it points in the same direction as $\vec{a}$.

Because the train is slowing
down, its acceleration vector
points in the direction opposite
its motion.

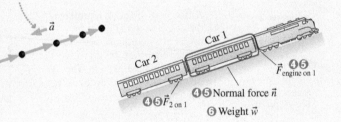

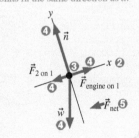

SUMMARY

Goal: To establish a connection between force and motion.

GENERAL PRINCIPLES

Newton's First Law

Consider an object with no force acting on it. If it is at rest, it will remain at rest. If it is in motion, then it will continue to move in a straight line at a constant speed.

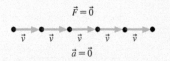

$$\vec{F} = \vec{0}$$
$$\vec{a} = \vec{0}$$

The first law tells us that an object that experiences no force will experience no acceleration.

Newton's Second Law

An object with mass m will undergo acceleration

$$\vec{a} = \frac{\vec{F}_{\text{net}}}{m}$$

where the net force $\vec{F}_{\text{net}} = \vec{F}_1 + \vec{F}_2 + \vec{F}_3 + \cdots$ is the vector sum of all the individual forces acting on the object.

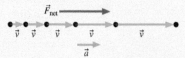

$$\vec{F}_{\text{net}}$$
$$\vec{a}$$

The second law tells us that a net force causes an object to accelerate. This is the connection between force and motion. The acceleration points in the direction of $\vec{F}_{\text{net}}$.

Newton's Third Law

Every force occurs as one member of an **action/reaction** pair of forces. The two members of an action/reaction pair:

- act on two *different* objects.

- point in opposite directions and are equal in magnitude:

$$\vec{F}_{\text{A on B}} = -\vec{F}_{\text{B on A}}$$

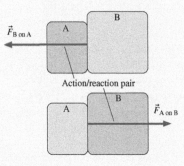

Action/reaction pair

IMPORTANT CONCEPTS

Force is a push or pull on an object.

- Force is a vector, with a magnitude and a direction.

- A force requires an agent.

- A force is either a contact force or a long-range force.

The SI unit of force is the **newton** (N). A 1 N force will cause a 1 kg mass to accelerate at 1 m/s².

Net force is the vector sum of all the forces acting on an object.

$$\vec{F}_1 \quad \vec{F}_2$$
$$\vec{F}_3$$
$$\vec{F}_{\text{net}} = \vec{F}_1 + \vec{F}_2 + \vec{F}_3$$

Mass is the property of an object that determines its resistance to acceleration.

If the same force is applied to objects A and B, then the ratio of their accelerations is related to the ratio of their masses as

$$\frac{a_{\text{A}}}{a_{\text{B}}} = \frac{m_{\text{B}}}{m_{\text{A}}}$$

The mass of objects can be determined in terms of their accelerations.

APPLICATIONS

Identifying Forces

Forces are identified by locating the points where other objects touch the object of interest. These are points where contact forces are exerted. In addition, objects feel a long-range weight force.

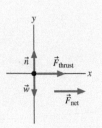

Tension $\vec{T}$

Normal force $\vec{n}$
Kinetic friction $\vec{f}_k$
Weight $\vec{w}$

Free-Body Diagrams

A free-body diagram represents the object as a particle at the origin of a coordinate system. Force vectors are drawn with their tails on the particle. The net force vector is drawn beside the diagram.

y
$\vec{n}$ $\vec{F}_{\text{thrust}}$
x
$\vec{w}$
$\vec{F}_{\text{net}}$

Problem difficulty is labeled as | (straightforward) to |||| (challenging). Problems labeled INT integrate significant material from earlier chapters; BIO are of biological or medical interest.

For assigned homework and other learning materials, go to

MasteringPhysics®

Scan this QR code to launch a Video Tutor Solution that will help you solve problems for this chapter.

QUESTIONS

Conceptual Questions

1. If an object is not moving, does that mean that there are no forces acting on it? Explain.

2. An object moves in a straight line at a constant speed. Is it true that there must be no forces of any kind acting on this object? Explain.

3. If you know all of the forces acting on a moving object, can you tell in which direction the object is moving? If the answer is Yes, explain how. If the answer is No, give an example.

4. Three arrows are shot horizontally. They have left the bow and are traveling parallel to the ground as shown in Figure Q4.4. Air resistance is negligible. Rank in order, from largest to smallest, the magnitudes of the *horizontal* forces F_1, F_2, and F_3 acting on the arrows. Some may be equal. State your reasoning.

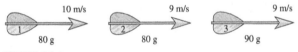

FIGURE Q4.4

5. A carpenter wishes to tighten the heavy head of his hammer onto its light handle. Which method shown in Figure Q4.5 will better tighten the head? Explain.

FIGURE Q4.5

6. Internal injuries in vehicular accidents may be due to what is called the "third collision." The first collision is the vehicle hitting the external object. The second collision is the person hitting something on the inside of the car, such as the dashboard or windshield. This may cause external lacerations. The third collision, possibly the most damaging to the body, is when organs, such as the heart or brain, hit the ribcage, skull, or other confines of the body, bruising the tissues on the leading edge and tearing the organ from its supporting structures on the trailing edge.
   a. Why is there a third collision? In other words, why are the organs still moving after the second collision?
   b. If the vehicle was traveling at 60 mph before the first collision, would the organs be traveling faster than, equal to, or slower than 60 mph just before the third collision?

7. Here's a great everyday use of the physics described in this chapter. If you are trying to get ketchup out of the bottle, the best way to do it is to turn the bottle upside down and give the bottle a sharp *upward* smack, forcing the bottle rapidly upward. Think about what subsequently happens to the ketchup, which is initially at rest, and use Newton's first law to explain why this technique is so successful.

8. a. Give an example of the motion of an object in which the frictional force on the object is directed opposite to the motion.
   b. Give an example of the motion of an object in which the frictional force on the object is in the same direction as the motion.

9. Suppose you are an astronaut in deep space, far from any source of gravity. You have two objects that look identical, but one has a large mass and the other a small mass. How can you tell the difference between the two?

10. Jonathan accelerates away from a stop sign. His eight-year-old daughter sits in the passenger seat. On whom does the back of the seat exert a greater force?

11. Normally, jet engines push air out the back of the engine, resulting in forward thrust, but commercial aircraft often have thrust reversers that can change the direction of the ejected air, sending it forward. How does this affect the direction of thrust? When might these thrust reversers be useful in practice?

12. If you are standing still, the upward normal force on you from the floor is equal in magnitude to the weight force that acts on you. But it's possible to move so that the normal force is greater than your weight. Explain how this could be done.

13. Josh and Taylor, standing face-to-face on frictionless ice, push off each other, causing each to slide backward. Josh is much bigger than Taylor. After the push, which of the two is moving faster?

14. A person sits on a sloped hillside. Is it ever possible to have the static friction force on this person point down the hill? Explain.

15. Walking without slipping requires a static friction force between your feet (or footwear) and the floor. As described in this chapter, the force on your foot as you push off the floor is forward while the force exerted by your foot on the floor is backward. But what about your *other* foot, the one moved during a stride? What is the direction of the force on that foot as it comes into contact with the floor? Explain.

16. Figure 4.30b showed a situation in which the force of the road on the car's tire points forward. In other situations, the force points backward. Give an example of such a situation.

17. Alyssa pushes to the right on a filing cabinet; the friction force from the floor pushes on it to the left. Because the cabinet doesn't move, these forces have the same magnitude. Do they form an action/reaction pair? Explain.

18. A very smart three-year-old child is given a wagon for her birthday. She refuses to use it. "After all," she says, "Newton's third law says that no matter how hard I pull, the wagon will exert an equal but opposite force on me. So I will never be able to get it to move forward." What would you say to her in reply?

19. The tire on this drag racer is severely twisted: The force of the road on the tire is quite large (most likely several times the weight of the car) and is directed forward as shown. Is the car speeding up or slowing down? Explain.

$\vec{F}_{\text{road on tire}}$

20. Suppose that, while in a squatting position, you stand on your hands, and then you pull up on your feet with a great deal of force. You are applying a large force to the bottoms of your feet, but no matter how strong you are, you will never be able to lift yourself off the ground. Use your understanding of force and motion to explain why this is not possible.

Multiple-Choice Questions

21. | A block has acceleration a when pulled by a string. If two identical blocks are glued together and pulled with twice the original force, their acceleration will be
    A. $(1/4)a$ B. $(1/2)a$ C. a D. $2a$ E. $4a$

22. | A 5.0 kg block has an acceleration of 0.20 m/s^2 when a force is exerted on it. A second block has an acceleration of 0.10 m/s^2 when subject to the same force. What is the mass of the second block?
    A. 10 kg   B. 5.0 kg   C. 2.5 kg   D. 7.5 kg

23. | Tennis balls experience a large drag force. A tennis ball is hit so that it goes straight up and then comes back down. The direction of the drag force is
    A. Always up.   B. Up and then down.
    C. Always down.   D. Down and then up.

24. ‖ A group of students is making model cars that will be propelled by model rocket engines. These engines provide a nearly constant thrust force. The cars are light—most of the weight comes from the rocket engine—and friction and drag are very small. As the engine fires, it uses fuel, so it is much lighter at the end of the run than at the start. A student ignites the engine in a car, and the car accelerates. As the fuel burns and the car continues to speed up, the magnitude of the acceleration will
    A. Increase.   B. Stay the same.   C. Decrease.

25. | A person gives a box a shove so that it slides up a ramp, then reverses its motion and slides down. The direction of the force of friction is
    A. Always down the ramp.
    B. Up the ramp and then down the ramp.
    C. Always down the ramp.
    D. Down the ramp and then up the ramp.

26. | A person is pushing horizontally on a box with a constant force, causing it to slide across the floor with a constant speed. If the person suddenly stops pushing on the box, the box will
    A. Immediately come to a stop.
    B. Continue moving at a constant speed for a while, then gradually slow down to a stop.
    C. Immediately change to a slower but constant speed.
    D. Immediately begin slowing down and eventually stop.

27. | As shown in the chapter, scallops use jet propulsion to move
BIO from one place to another. Their shells make them denser than
INT water, so they normally rest on the ocean floor. If a scallop wishes to remain stationary, hovering a fixed distance above the ocean floor, it must eject water _____ so that the thrust force on the scallop is _____.
    A. upward, upward   B. upward, downward
    C. downward, upward   D. downward, downward

28. ‖ Dave pushes his four-year-old son Thomas across the snow on a sled. As Dave pushes, Thomas speeds up. Which statement is true?
    A. The force of Dave on Thomas is larger than the force of Thomas on Dave.
    B. The force of Thomas on Dave is larger than the force of Dave on Thomas.
    C. Both forces have the same magnitude.
    D. It depends on how hard Dave pushes on Thomas.

29. | Figure Q4.29 shows block A sitting on top of block B. A constant force $\vec{F}$ is exerted on block B, causing block B to accelerate to the right. Block A rides on block B without slipping. Which statement is true?
    A. Block B exerts a friction force on block A, directed to the left.
    B. Block B exerts a friction force on block A, directed to the right.
    C. Block B does not exert a friction force on block A.

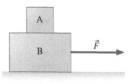

FIGURE Q4.29

PROBLEMS

Section 4.1 Motion and Forces

1. | Whiplash injuries during an automobile accident are caused
BIO by the inertia of the head. If someone is wearing a seatbelt, her body will tend to move with the car seat. However, her head is free to move until the neck restrains it, causing damage to the neck. Brain damage can also occur.

   Figure P4.1 shows two sequences of head and neck motion for a passenger in an auto accident. One corresponds to a head-on collision, the other to a rear-end collision. Which is which? Explain.

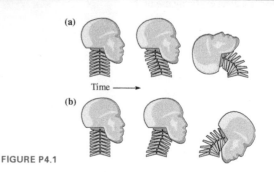

(a)

Time ⟶

(b)

FIGURE P4.1

2. | An automobile has a head-on
BIO collision. A passenger in the car
experiences a compression injury
to the brain. Is this injury most
likely to be in the front or rear
portion of the brain? Explain.

3. | In a head-on collision, an infant
is much safer in a child safety seat
when the seat is installed facing
the rear of the car. Explain.

Problems 4 through 6 show two forces acting on an object at rest.
Redraw the diagram, then add a third force that will allow the object
to remain at rest. Label the new force $\vec{F}_3$.

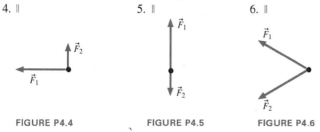

4. ‖   5. ‖   6. ‖

FIGURE P4.4   FIGURE P4.5   FIGURE P4.6

Section 4.2 A Short Catalog of Forces

Section 4.3 Identifying Forces

7. ‖ A mountain climber is hanging from a vertical rope, far above
the ground and far from the rock face. Identify the forces on the
mountain climber.

8. ‖ You look up from your textbook and observe a spider,
BIO motionless above you, suspended from a strand of spider silk
attached to the ceiling. You distract yourself by identifying the
forces acting on the spider. What are they?

9. ‖‖ A baseball player is sliding into second base. Identify the
forces on the baseball player.

10. ‖‖ A jet plane is speeding down the runway during takeoff. Air
resistance is not negligible. Identify the forces on the jet.

11. | A skier is sliding down a 15° slope. Friction is not negligible.
Identify the forces on the skier.

12. ‖ A falcon is hovering above the ground, then suddenly pulls in
BIO its wings and begins to fall toward the ground. Air resistance is
not negligible. Identify the forces on the falcon.

Section 4.4 What Do Forces Do?

13. ‖‖ Figure P4.13 shows an acceleration-versus-force graph for three
objects pulled by rubber bands. The mass of object 2 is 0.20 kg.
What are the masses of objects 1 and 3? Explain your reasoning.

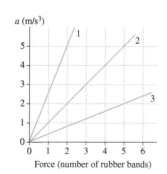

FIGURE P4.13   Force (number of rubber bands)

14. | A constant force applied to object A causes it to accelerate at
5 m/s². The same force applied to object B causes an acceleration
of 3 m/s². Applied to object C, it causes an acceleration of 8 m/s².
  a. Which object has the largest mass?
  b. Which object has the smallest mass?
  c. What is the ratio of mass A to mass B (m_A/m_B)?

15. ‖ A compact car has a maximum acceleration of 4.0 m/s² when
it carries only the driver and has a total mass of 1200 kg. What is
its maximum acceleration after picking up four passengers and
their luggage, adding an additional 400 kg of mass?

16. | A constant force is applied to an object, causing the object to
accelerate at 10 m/s². What will the acceleration be if
  a. The force is halved?
  b. The object's mass is halved?
  c. The force and the object's mass are both halved?
  d. The force is halved and the object's mass is doubled?

17. | A constant force is applied to an object, causing the object to
accelerate at 8.0 m/s². What will the acceleration be if
  a. The force is doubled?
  b. The object's mass is doubled?
  c. The force and the object's mass are both doubled?
  d. The force is doubled and the object's mass is halved?

18. ‖‖ A man pulling an empty wagon causes it to accelerate at
1.4 m/s². What will the acceleration be if he pulls with the same
force when the wagon contains a child whose mass is three
times that of the wagon?

19. | A car has a maximum acceleration of 5.0 m/s². What will the
maximum acceleration be if the car is towing another car of the
same mass?

Section 4.5 Newton's Second Law

20. ‖ Scallops eject water from their shells to provide a thrust
BIO force. The graph shows a smoothed graph of actual data for the
INT initial motion of a 25 g scallop speeding up to escape a predator.
What is the magnitude of the net force needed to achieve this
motion? How does this force compare to the 0.25 N weight of
the scallop?

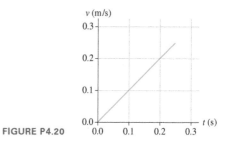

FIGURE P4.20

21. | Figure P4.21 shows an object's acceleration-versus-force
graph. What is the object's mass?

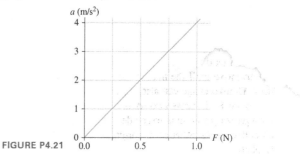

FIGURE P4.21

22. ‖ In t-ball, young players use a bat to hit a stationary ball off a
ⁿ stand. The 140 g ball has about the same mass as a baseball, but
it is larger and softer. In one hit, the ball leaves the bat at 12 m/s
after being in contact with the bat for 2.0 ms. Assume constant
acceleration during the hit.
   a. What is the acceleration of the ball?
   b. What is the net force on the ball during the hit?
23. ‖ Two children fight over a 200 g stuffed bear. The 25 kg boy
pulls to the right with a 15 N force and the 20 kg girl pulls to the
left with a 17 N force. Ignore all other forces on the bear (such
as its weight).
   a. At this instant, can you say what the velocity of the bear is?
If so, what are the magnitude and direction of the velocity?
   b. At this instant, can you say what the acceleration of the bear
is? If so, what are the magnitude and direction of the acceleration?
24. ‖ A 1500 kg car is traveling along a straight road at 20 m/s.
ⁿ Two seconds later its speed is 21 m/s. What is the magnitude of
the net force acting on the car during this time?
25. ‖ The motion of a very massive object can be minimally
ⁿ affected by what would seem to be a substantial force. Consider an oil supertanker with mass 3.0×10^8 kg. Suppose you
strapped two jet engines (with thrust as given in Table 4.2) onto
the sides of the tanker. Ignoring the drag of the water (which, in
reality, is not a very good approximation), how long will it take
the tanker, starting from rest, to reach a typical cruising speed
of 6.0 m/s?
26. ‖ Very small forces can have tremendous effects on the motion
ⁿ of very small objects. This is particularly apparent at the scale
of the atom. An electron, mass 9.1×10^{-31} kg, experiences a
force of 1.6×10^{-17} N in a typical electric field at the earth's
surface. From rest, how much time would it take for the electron
to reach a speed of 3.0×10^6 m/s, 1% of the speed of light?

Section 4.6 Free-Body Diagrams

Problems 27 through 29 show a free-body diagram. For each problem, (a) redraw the free-body diagram and (b) write a short description of a real object for which this is the correct free-body diagram.
Use the situations described in Conceptual Examples 4.1, 4.2, and
4.3 as models of what a description should be like.

27. |

28. |

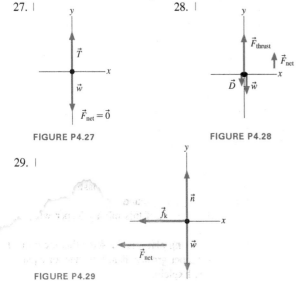

FIGURE P4.27   FIGURE P4.28

29. |

FIGURE P4.29

Problems 30 through 38 describe a situation. For each problem,
identify all the forces acting on the object and draw a free-body
diagram of the object.
30. ‖ Your car is sitting in the parking lot.
31. ‖ Your car is accelerating from a stop.
32. ‖ Your car is skidding to a stop from a high speed.
33. ‖ Your physics textbook is sliding across the table.
34. ‖ An ascending elevator, hanging from a
cable, is coming to a stop.
35. | You are driving on the highway, and you
come to a steep downhill section. As you roll
down the hill, you take your foot off the gas
pedal. You can ignore friction, but you can't
ignore air resistance.
36. ‖ You hold a picture motionless against a
wall by pressing on it, as shown in Figure
P4.36.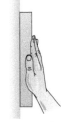
                                              FIGURE P4.36
37. | A box is being dragged across the floor at
a constant speed by a rope pulling horizontally on it. Friction is
not negligible.
38. | A skydiver has his parachute open and is floating downward
through the air at a constant speed.

Section 4.7 Newton's Third Law

39. ‖ Three ice skaters, numbered 1, 2, and 3, stand in a line, each
with her hands on the shoulders of the skater in front. Skater 3,
at the rear, pushes on skater 2. Identify all the action/reaction
pairs of forces between the three skaters. Draw a free-body diagram for skater 2, in the middle. Assume the ice is frictionless.
40. | A girl stands on a sofa. Identify all the action/reaction pairs of
forces between the girl and the sofa.
41. ‖ A car is skidding to a stop on a level stretch of road. Identify
all the action/reaction pairs of forces between the car and the
road surface. Then draw a free-body diagram for the car.
42. ‖ Squid use jet propulsion for rapid escapes. A squid pulls
ᴮᴵᴼ water into its body and then rapidly ejects the water backward
to propel itself forward. A 1.5 kg squid (not including water
mass) can accelerate at 20 m/s² by ejecting 0.15 kg of water.
   a. What is the magnitude of the thrust force on the squid?
   b. What is the magnitude of the force on the water being
ejected?
   c. What acceleration is experienced by the water?

General Problems

43. | Redraw the motion dia-
ⁿ gram shown in Figure
P4.43, then draw a vector
beside it to show the direction of the net force acting
on the object. Explain your
reasoning.
44. | Redraw the motion dia-
ⁿ gram shown in Figure
P4.44, then draw a vector
beside it to show the direction of the net force acting
on the object. Explain your
reasoning.

FIGURE P4.43   FIGURE P4.44

45. | Redraw the motion diagram shown in Figure P4.45, then
IN draw a vector beside it to show the direction of the net force
    acting on the object. Explain your reasoning.

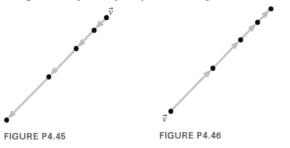

FIGURE P4.45   **FIGURE P4.46**

46. | Redraw the motion diagram shown in Figure P4.46, then
IN draw a vector beside it to show the direction of the net force
    acting on the object. Explain your reasoning.

47. ‖‖‖‖ A student draws the flawed free-body diagram shown in
    Figure P4.47 to represent the forces acting on a car traveling
    at constant speed on a level road. Identify the errors in the dia-
    gram, then draw a correct free-body diagram for this situation.

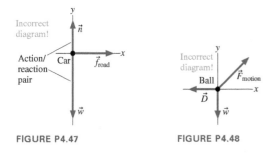

FIGURE P4.47   **FIGURE P4.48**

48. ‖‖‖ A student draws the flawed free-body diagram shown in Figure
    P4.48 to represent the forces acting on a golf ball that is traveling
    upward and to the right a very short time after being hit off the tee.
    Air resistance is assumed to be relevant. Identify the errors in the
    diagram, then draw a correct free-body diagram for this situation.

Problems 49 through 61 describe a situation. For each problem,
draw a motion diagram, a force identification diagram, and a free-
body diagram.

49. ‖ An elevator, suspended by a single cable, has just left the tenth
    floor and is speeding up as it descends toward the ground floor.
50. ‖‖‖ A rocket is being launched straight up. Air resistance is not
    negligible.
51. ‖‖‖ A jet plane is speeding down the runway during takeoff. Air
    resistance is not negligible.
52. ‖ You've slammed on the brakes and your car is skidding to a
    stop while going down a 20° hill.
53. ‖ A sprinter has just started a race and is speeding up as she
    runs down the track.
54. ‖ A basketball player is getting ready to jump, pushing off the
    ground and accelerating upward.
55. ‖ A bale of hay sits on the bed of a trailer. The trailer is starting
    to accelerate forward, and the bale is slipping toward the back
    of the trailer.
56. ‖ A Styrofoam ball has just been shot straight up. Air resistance
    is not negligible.
57. ‖‖‖ A spring-loaded gun shoots a plastic ball. The trigger has just
    been pulled and the ball is starting to move down the barrel. The
    barrel is horizontal.

58. ‖ A person on a bridge throws a rock straight down toward the
    water. The rock has just been released.
59. ‖‖‖ A gymnast has just landed on a trampoline. She's still mov-
    ing downward as the trampoline stretches.
60. ‖‖‖ A heavy box is in the back of a truck. The truck is accelerat-
    ing to the right. Apply your analysis to the box.
61. ‖‖‖ A bag of groceries is on the back seat of your car as you stop
    for a stop light. The bag does not slide. Apply your analysis to
    the bag.
62. ‖ A car has a mass of 1500 kg. If the driver applies the brakes
IN while on a gravel road, the maximum friction force that the tires
    can provide without skidding is about 7000 N. If the car is mov-
    ing at 20 m/s, what is the shortest distance in which the car can
    stop safely?
63. ‖ A rubber ball bounces. We'd like to understand *how* the ball
    bounces.
    a. A rubber ball has been dropped and is bouncing off the
       floor. Draw a motion diagram of the ball during the brief
       time interval that it is in contact with the floor. Show 4 or 5
       frames as the ball compresses, then another 4 or 5 frames as
       it expands. What is the direction of $\vec{a}$ during each of these
       parts of the motion?
    b. Draw a picture of the ball in contact with the floor and iden-
       tify all forces acting on the ball.
    c. Draw a free-body diagram of the ball during its contact with
       the ground. Is there a net force acting on the ball? If so, in
       which direction?
    d. During contact, is the force of the ground on the ball larger
       than, smaller than, or equal to the weight of the ball? Use
       your answers to parts a–c to explain your reasoning.
64. ‖ If a car stops suddenly, you feel "thrown forward." We'd like
    to understand what happens to the passengers as a car stops.
    Imagine yourself sitting on a *very* slippery bench inside a car.
    This bench has no friction, no seat back, and there's nothing for
    you to hold on to.
    a. Draw a picture and identify all of the forces acting on you as
       the car travels in a straight line at a perfectly steady speed on
       level ground.
    b. Draw your free-body diagram. Is there a net force on you? If
       so, in which direction?
    c. Repeat parts a and b with the car slowing down.
    d. Describe what happens to you as the car slows down.
    e. Use Newton's laws to explain why you seem to be "thrown
       forward" as the car stops. Is there really a force pushing you
       forward?
65. ‖‖‖ The fastest pitched baseball was clocked at 46 m/s. If the
BIO pitcher exerted his force (assumed to be horizontal and constant)
    over a distance of 1.0 m, and a baseball has a mass of 145 g,
    a. Draw a free-body diagram of the ball during the pitch.
    b. What force did the pitcher exert on the ball during this
       record-setting pitch?
    c. Estimate the force in part b as a fraction of the pitcher's
       weight.
66. | The froghopper, champion leaper of the insect world, can
BIO jump straight up at 4.0 m/s. The jump itself lasts a mere 1.0 ms
    before the insect is clear of the ground.
    a. Draw a free-body diagram of this mighty leaper while the
       jump is taking place.
    b. While the jump is taking place, is the force that the ground
       exerts on the froghopper greater than, less than, or equal to
       the insect's weight? Explain.

67. ‖ A beach ball is thrown straight up, and some time later it lands on the sand. Is the magnitude of the net force on the ball greatest when it is going up or when it is on the way down? Or is it the same in both cases? Explain. Air resistance should not be neglected for a large, light object.

MCAT-Style Passage Problems

A Simple Solution for a Stuck Car

If your car is stuck in the mud and you don't have a winch to pull it out, you can use a piece of rope and a tree to do the trick. First, you tie one end of the rope to your car and the other to a tree, then pull as hard as you can on the middle of the rope, as shown in Figure P4.68a. This technique applies a force to the car much larger than the force that you can apply directly. To see why the car experiences such a large force, look at the forces acting on the center point of the rope, as shown in Figure P4.68b. The sum of the forces is zero, thus the tension is much greater than the force you apply. It is this tension force that acts on the car and, with luck, pulls it free.

68. | The sum of the three forces acting on the center point of the rope is assumed to be zero because
    A. This point has a very small mass.
    B. Tension forces in a rope always cancel.
    C. This point is not accelerating.
    D. The angle of deflection is very small.

(a)

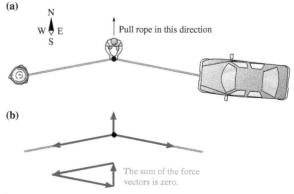

(b)

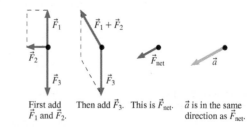

FIGURE P4.68

69. | When you are pulling on the rope as shown, what is the approximate direction of the tension force on the tree?
    A. North   B. South   C. East   D. West
70. | Assume that you are pulling on the rope but the car is not moving. What is the approximate direction of the force of the mud on the car?
    A. North   B. South   C. East   D. West
71. | Suppose your efforts work, and the car begins to move forward out of the mud. As it does so, the force of the car on the rope is
    A. Zero.
    B. Less than the force of the rope on the car.
    C. Equal to the force of the rope on the car.
    D. Greater than the force of the rope on the car.

STOP TO THINK ANSWERS

Chapter Preview Stop to Think: B. The swan's velocity is to the left. Its speed is decreasing, so the acceleration is opposite the velocity, or to the right.

Stop to Think 4.1: C.

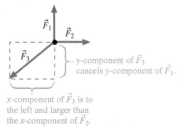

Stop to Think 4.2: C. The frog isn't moving, so a static friction force is keeping it at rest. If there was no friction, the weight force would cause the frog to slide down the slope. The static friction force opposes this, so it must be directed up the slope.

Stop to Think 4.3: B. The tension force is the force pulling the sled to the right. The friction force is opposing the motion, so it is directed to the left.

Stop to Think 4.4: A, B, and D. Friction and the normal force are the only contact forces. Nothing is touching the rock to provide a "force of the kick." We've agreed to ignore air resistance unless a problem specifically calls for it.

Stop to Think 4.5: B. Acceleration is proportional to force, so doubling the number of rubber bands doubles the acceleration of the original object from 2 m/s^2 to 4 m/s^2. But acceleration is also inversely proportional to mass. Doubling the mass cuts the acceleration in half, back to 2 m/s^2.

Stop to Think 4.6: D

Stop to Think 4.7: C. The acceleration vector points downward as the elevator slows. $\vec{F}_{net}$ points in the same direction as $\vec{a}$, so $\vec{F}_{net}$ also points downward. This will be true if the tension is less than the weight: $T < w$.

Stop to Think 4.8: C. Newton's third law says that the force of A on B is *equal* and opposite to the force of B on A. This is always true. The mass of the objects isn't relevant, nor is the fact that the car and truck are accelerating.

5 Applying Newton's Laws

This frog has leaped from a tall tree, but it is floating gently to the ground and will land safely. Why does holding its feet and toes this way help slow its fall?

LOOKING AHEAD ▶

Goal: To use Newton's laws to solve equilibrium and dynamics problems.

Working with Forces

In this chapter you'll learn expressions for the different forces we've seen, and you'll learn how to use them to solve problems.

You'll learn how a balance between weight and drag forces leads to a maximum speed for a skydiver.

Equilibrium Problems

The boy is pushing as hard as he can, but the sofa isn't going anywhere. It's in **equilibrium**—the sum of the forces on it is zero.

You'll learn to solve equilibrium problems by using the fact that there is no net force.

Dynamics Problems

Newton's laws allow us to relate the forces acting on an object to its motion, and so to solve a wide range of **dynamics** problems.

This skier is picking up speed. You'll see how her acceleration is determined by the forces acting on her.

LOOKING BACK ◀

Free-Body Diagrams

In Section 4.6 you learned to draw a free-body diagram showing the magnitudes and directions of the forces acting on an object.

In this chapter, you'll use free-body diagrams as an essential problem-solving tool for single objects and interacting objects.

An elevator is suspended from a cable. It is moving upward at a steady speed. Which is the correct free-body diagram for this situation?

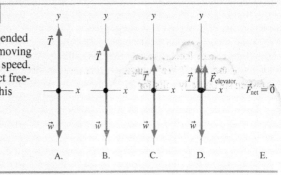

This human tower is in equilibrium because the net force on each man is zero.

5.1 Equilibrium

In this chapter, we will use Newton's laws to solve force and motion problems. We'll treat objects that are at rest or that move in a straight line. We start with objects that are at rest, which is a surprisingly interesting class of problem.

We say that an object at rest is in **static equilibrium.** If an object is at rest, $\vec{a} = \vec{0}$. But $\vec{a} = \vec{0}$ for objects that are moving in a straight line at a constant speed as well. Such an object is in **dynamic equilibrium.**

In either type of equilibrium, Newton's second law, $\vec{F} = m\vec{a}$, tells us that there is no net force acting on the object. In both types, $\vec{F}_{net} = \vec{0}$. Recall that $\vec{F}_{net}$ is the vector sum

$$\vec{F}_{net} = \vec{F}_1 + \vec{F}_2 + \vec{F}_3 + \cdots$$

where $\vec{F}_1$, $\vec{F}_2$, and so on are the individual forces, such as tension or friction, acting on the object. We found in ◀ SECTION 3.3 that vector sums can be evaluated in terms of the x- and y-components of the vectors; that is, the x-component of the net force is $(F_{net})_x = F_{1x} + F_{2x} + F_{3x} + \cdots$. If we restrict ourselves to problems where all the forces are in the xy-plane, then the equilibrium requirement $\vec{F}_{net} = \vec{0}$ is a shorthand way of writing two simultaneous equations:

$$(F_{net})_x = F_{1x} + F_{2x} + F_{3x} + \cdots = 0$$
$$(F_{net})_y = F_{1y} + F_{2y} + F_{3y} + \cdots = 0$$

Recall from your math classes that the Greek letter Σ (sigma) stands for "the sum of." It will be convenient to abbreviate the sum of the x-components of all forces as

$$F_{1x} + F_{2x} + F_{3x} + \cdots = \sum F_x$$

With this notation, Newton's second law for an object in equilibrium, with $\vec{a} = \vec{0}$, can be written as the two equations

$$\sum F_x = ma_x = 0 \quad \text{and} \quad \sum F_y = ma_y = 0 \tag{5.1}$$

In equilibrium, the sums of the x- and y-components of the force are zero

These equations are the basis for a strategy for solving equilibrium problems.

PROBLEM-SOLVING STRATEGY 5.1   Equilibrium problems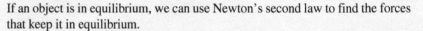

If an object is in equilibrium, we can use Newton's second law to find the forces that keep it in equilibrium.

PREPARE   First check that the object is in equilibrium: Does $\vec{a} = \vec{0}$?

- An object at rest is in static equilibrium.
- An object moving at a constant velocity is in dynamic equilibrium.

Then identify all forces acting on the object and show them on a free-body diagram. Determine which forces you know and which you need to solve for.

SOLVE   An object in equilibrium must satisfy Newton's second law for the case where $\vec{a} = \vec{0}$. In component form, the requirement is

$$\sum F_x = ma_x = 0 \quad \text{and} \quad \sum F_y = ma_y = 0$$

You can find the force components that go into these sums directly from your free-body diagram. From these two equations, solve for the unknown forces in the problem.

ASSESS   Check that your result has the correct units, is reasonable, and answers the question.

Static Equilibrium

EXAMPLE 5.1 | **Forces supporting an orangutan**

An orangutan weighing 500 N hangs from a vertical rope. What is the tension in the rope?

PREPARE The orangutan is at rest, so it is in static equilibrium. The net force on it must then be zero. **FIGURE 5.1** first identifies the forces acting on the orangutan: the upward force of the tension in the rope and the downward, long-range force of gravity. These forces are then shown on a free-body diagram, where it's noted that equilibrium requires $\vec{F}_{net} = \vec{0}$.

FIGURE 5.1 The forces on an orangutan.

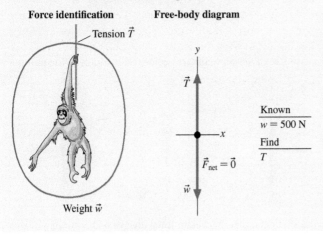

Force identification   Free-body diagram

Tension $\vec{T}$

Weight $\vec{w}$

Known
$w = 500 \text{ N}$

Find
T

SOLVE Neither force has an x-component, so we need to examine only the y-components of the forces. In this case, the y-component of Newton's second law is

$$\sum F_y = T_y + w_y = ma_y = 0$$

You might have been tempted to write $T_y - w_y$ because the weight force points down. But remember that T_y and w_y are *components* of vectors and can thus be positive (for a vector such as $\vec{T}$ that points up) or negative (for a vector such as $\vec{w}$ that points down). The fact that $\vec{w}$ points down is taken into account when we *evaluate* the components—that is, when we write them in terms of the *magnitudes* T and w of the vectors $\vec{T}$ and $\vec{w}$.

Because the tension vector $\vec{T}$ points straight up, in the positive y-direction, its y-component is $T_y = T$. Because the weight vector $\vec{w}$ points straight down, in the negative y-direction, its y-component is $w_y = -w$. This is where the signs enter. With these components, Newton's second law becomes

$$T - w = 0$$

This equation is easily solved for the tension in the rope:

$$T = w = 500 \text{ N}$$

ASSESS It's not surprising that the tension in the rope equals the weight of the orangutan. That gives us confidence in our solution.

EXAMPLE 5.2 | **Readying a wrecking ball**

A wrecking ball weighing 2500 N hangs from a cable. Prior to swinging, it is pulled back to a 20° angle by a second, horizontal cable. What is the tension in the horizontal cable?

PREPARE Because the ball is not moving, it hangs in static equilibrium, with $\vec{a} = \vec{0}$, until it is released. In **FIGURE 5.2**, we start by identifying all the forces acting on the ball: a tension force from each cable and the ball's weight. We've used different symbols $\vec{T}_1$ and $\vec{T}_2$ for the two different tension forces. We then construct a free-body diagram for these three forces, noting that $\vec{F}_{net} = m\vec{a} = \vec{0}$. We're looking for the magnitude T_1 of the tension force $\vec{T}_1$ in the horizontal cable.

FIGURE 5.2 Visual overview of a wrecking ball just before release.

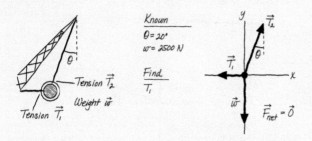

Known
$\theta = 20°$
$w = 2500 \text{ N}$

Find
T_1

Tension $\vec{T}_2$
Weight $\vec{w}$
Tension $\vec{T}_1$

SOLVE The requirement of equilibrium is $\vec{F}_{net} = m\vec{a} = \vec{0}$. In component form, we have the two equations:

$$\sum F_x = T_{1x} + T_{2x} + w_x = ma_x = 0$$
$$\sum F_y = T_{1y} + T_{2y} + w_y = ma_y = 0$$

As always, we *add* the force components together. Now we're ready to write the components of each force vector in terms of the magnitudes and directions of those vectors. We learned how to do this in Section 3.3. With practice you'll learn to read the components directly off the free-body diagram, but to begin it's worthwhile to organize the components into a table.

Force	Name of x-component	Value of x-component	Name of y-component	Value of y-component
$\vec{T}_1$	T_{1x}	$-T_1$	T_{1y}	0
$\vec{T}_2$	T_{2x}	$T_2 \sin\theta$	T_{2y}	$T_2 \cos\theta$
$\vec{w}$	w_x	0	w_y	$-w$

We see from the free-body diagram that $\vec{T}_1$ points along the negative x-axis, so $T_{1x} = -T_1$ and $T_{1y} = 0$. We need to be careful with our trigonometry as we find the components of $\vec{T}_2$. Remembering that the side adjacent to the angle is related to the cosine,

Continued

we see that the vertical (y) component of $\vec{T}_2$ is $T_2 \cos\theta$. Similarly, the horizontal (x) component is $T_2 \sin\theta$. The weight vector points straight down, so its y-component is $-w$. Notice that negative signs enter as we evaluate the components of the vectors, *not* when we write Newton's second law. This is a critical aspect of solving force and motion problems. With these components, Newton's second law now becomes

$$-T_1 + T_2 \sin\theta + 0 = 0 \quad \text{and} \quad 0 + T_2 \cos\theta - w = 0$$

We can rewrite these equations as

$$T_2 \sin\theta = T_1 \quad \text{and} \quad T_2 \cos\theta = w$$

These are two simultaneous equations with two unknowns: T_1 and T_2. To eliminate T_2 from the two equations, we solve the second equation for T_2, giving $T_2 = w/\cos\theta$. Then we insert this expression for T_2 into the first equation to get

$$T_1 = \frac{w}{\cos\theta}\sin\theta = \frac{\sin\theta}{\cos\theta}w = w\tan\theta = (2500\ \text{N})\tan 20° = 910\ \text{N}$$

where we made use of the fact that $\tan\theta = \sin\theta/\cos\theta$.

ASSESS It seems reasonable that to pull the ball back to this modest angle, a force substantially less than the ball's weight will be required.

CONCEPTUAL EXAMPLE 5.3 **Forces in static equilibrium**

A rod is free to slide on a frictionless sheet of ice. One end of the rod is lifted by a string. If the rod is at rest, which diagram in FIGURE 5.3 shows the correct angle of the string?

FIGURE 5.3 Which is the correct angle of the string?

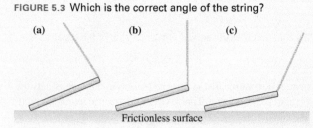

Frictionless surface

REASON Let's start by identifying the forces that act on the rod. In addition to the weight force, the string exerts a tension force and the ice exerts an upward normal force. What can we say about these forces? If the rod is to hang motionless, it must be in static equilibrium with $\sum F_x = ma_x = 0$ and $\sum F_y = ma_y = 0$. FIGURE 5.4 shows free-body diagrams for the three string orientations. Remember that tension always acts along the direction of

FIGURE 5.4 Free-body diagrams for three angles of the string.

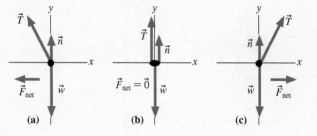

the string and that the weight force always points straight down. The ice pushes up with a normal force perpendicular to the surface, but frictionless ice cannot exert any horizontal force. If the string is angled, we see that its horizontal component exerts a net force on the rod. Only in case b, where the tension and the string are vertical, can the net force be zero.

ASSESS If friction were present, the rod could in fact hang as in cases a and c. But without friction, the rods in these cases would slide until they came to rest as in case b.

Dynamic Equilibrium

EXAMPLE 5.4 **Tension in towing a car**

A car with a mass of 1500 kg is being towed at a steady speed by a rope held at a 20° angle from the horizontal. A friction force of 320 N opposes the car's motion. What is the tension in the rope?

PREPARE The car is moving in a straight line at a constant speed ($\vec{a} = \vec{0}$) so it is in dynamic equilibrium and must have

$\vec{F}_{\text{net}} = m\vec{a} = \vec{0}$. FIGURE 5.5 shows three contact forces acting on the car—the tension force $\vec{T}$, friction $\vec{f}$, and the normal force $\vec{n}$— and the long-range force of gravity $\vec{w}$. These four forces are shown on the free-body diagram.

FIGURE 5.5 Visual overview of a car being towed.

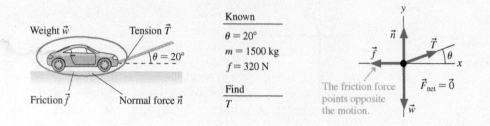

SOLVE This is still an equilibrium problem, even though the car is moving, so our problem-solving procedure is unchanged. With four forces, the requirement of equilibrium is

$$\sum F_x = n_x + T_x + f_x + w_x = ma_x = 0$$

$$\sum F_y = n_y + T_y + f_y + w_y = ma_y = 0$$

We can again determine the horizontal and vertical components of the forces by "reading" the free-body diagram. The results are shown in the table.

Force	Name of x-component	Value of x-component	Name of y-component	Value of y-component
$\vec{n}$	n_x	0	n_y	n
$\vec{T}$	T_x	$T\cos\theta$	T_y	$T\sin\theta$
$\vec{f}$	f_x	$-f$	f_y	0
$\vec{w}$	w_x	0	w_y	$-w$

With these components, Newton's second law becomes

$$T\cos\theta - f = 0$$

$$n + T\sin\theta - w = 0$$

The first equation can be used to solve for the tension in the rope:

$$T = \frac{f}{\cos\theta} = \frac{320\text{ N}}{\cos 20°} = 340\text{ N}$$

to two significant figures. It turned out that we did not need the y-component equation in this problem. We would need it if we wanted to find the normal force $\vec{n}$.

ASSESS Had we pulled the car with a horizontal rope, the tension would need to exactly balance the friction force of 320 N. Because we are pulling at an angle, however, part of the tension in the rope pulls *up* on the car instead of in the forward direction. Thus we need a little more tension in the rope when it's at an angle, so our result seems reasonable.

STOP TO THINK 5.1 A ball of weight 200 N is suspended from two cables, one horizontal and one at a 60° angle, as shown. Which of the following must be true of the tension in the angled cable?

A. $T > 200$ N   B. $T = 200$ N   C. $T < 200$ N

5.2 Dynamics and Newton's Second Law

Newton's second law is the essential link between force and motion. The essence of Newtonian mechanics can be expressed in two steps:

- The forces acting on an object determine its acceleration $\vec{a} = \vec{F}_{net}/m$.
- The object's motion can be found by using $\vec{a}$ in the equations of kinematics.

We want to develop a strategy to solve a variety of problems in mechanics, but first we need to write the second law in terms of its components. To do so, let's first rewrite Newton's second law in the form

$$\vec{F}_{net} = \vec{F}_1 + \vec{F}_2 + \vec{F}_3 + \cdots = m\vec{a}$$

where $\vec{F}_1, \vec{F}_2, \vec{F}_3$, and so on are the forces acting on an object. To write the second law in component form merely requires that we use the x- and y-components of the acceleration. Thus Newton's second law, $\vec{F}_{net} = m\vec{a}$, is

Class Video

$$\sum F_x = ma_x \quad \text{and} \quad \sum F_y = ma_y \qquad (5.2)$$

Newton's second law in component form

The first equation says that **the component of the acceleration in the x-direction is determined by the sum of the x-components of the forces acting on the object.** A similar statement applies to the y-direction.

PROBLEM-SOLVING
STRATEGY 5.2   Dynamics problems

There are two basic ways to solve problems in mechanics: You either start from known forces to find acceleration (and from that positions and velocities using kinematics) or start from kinematics to find acceleration (and from that solve for unknown forces). In both cases, the strategy for the solution is the same.

PREPARE   Sketch a visual overview consisting of:

- A list of values that identifies known quantities and what the problem is trying to find.
- A force identification diagram to help you identify all the forces acting on the object.
- A free-body diagram that shows all the forces acting on the object.

If you'll need to use kinematics to find velocities or positions, you'll also need to sketch:

- A motion diagram to determine the direction of the acceleration.
- A pictorial representation that establishes a coordinate system, shows important points in the motion, and defines symbols.

It's OK to go back and forth between these steps as you visualize the situation.

SOLVE   Write Newton's second law in component form as

$$\sum F_x = ma_x \quad \text{and} \quad \sum F_y = ma_y$$

You can find the components of the forces directly from your free-body diagram. Depending on the problem, either:

- Solve for the acceleration, then use kinematics to find velocities and positions.
- Use kinematics to determine the acceleration, then solve for unknown forces.

ASSESS   Check that your result has the correct units, is reasonable, and answers the question.

Exercise 24 ✐

EXAMPLE 5.5   **Putting a golf ball**

A golfer putts a 46 g ball with a speed of 3.0 m/s. Friction exerts a 0.020 N retarding force on the ball, slowing it down. Will her putt reach the hole, 10 m away?

PREPARE FIGURE 5.6 is a visual overview of the problem. We've collected the known information, drawn a sketch, and identified what we want to find. The motion diagram shows that the ball is

slowing down as it rolls to the right, so the acceleration vector points to the left. Next, we identify the forces acting on the ball and show them on a free-body diagram. Note that the net force points to the left, as it must because the acceleration points to the left.

FIGURE 5.6 Visual overview of a golf putt.

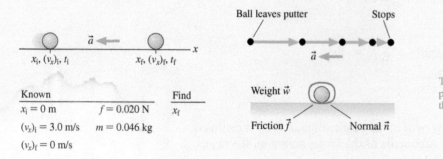
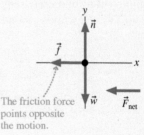

Known
$x_i = 0$ m $f = 0.020$ N
$(v_x)_i = 3.0$ m/s $m = 0.046$ kg
$(v_x)_f = 0$ m/s

Find
x_f

SOLVE Newton's second law in component form is

$$\sum F_x = n_x + f_x + w_x = 0 - f + 0 = ma_x$$

$$\sum F_y = n_y + f_y + w_y = n + 0 - w = ma_y = 0$$

We've written the equations as sums, as we did with equilibrium problems, then "read" the values of the force components from the free-body diagram. The components are simple enough in this problem that we don't really need to show them in a table. It is particularly important to notice that we set $a_y = 0$ in the second equation. This is because the ball does not move in the y-direction, so it can't have any acceleration in the y-direction. This will be an important step in many problems.

The first equation is $-f = ma_x$, from which we find

$$a_x = -\frac{f}{m} = \frac{-(0.020 \text{ N})}{0.046 \text{ kg}} = -0.435 \text{ m/s}^2$$

To avoid rounding errors we keep an extra digit in this intermediate step in the calculation. The negative sign shows that the acceleration is directed to the left, as expected.

Now that we know the acceleration, we can use kinematics to find how far the ball will roll before stopping. We don't have any information about the time it takes for the ball to stop, so we'll use the kinematic equation $(v_x)_f^2 = (v_x)_i^2 + 2a_x(x_f - x_i)$. This gives

$$x_f = x_i + \frac{(v_x)_f^2 - (v_x)_i^2}{2a_x} = 0 \text{ m} + \frac{(0 \text{ m/s})^2 - (3.0 \text{ m/s})^2}{2(-0.435 \text{ m/s}^2)} = 10.3 \text{ m}$$

If her aim is true, the ball will just make it into the hole.

ASSESS It seems reasonable that a ball putted on grass with an initial speed of 3 m/s—about jogging speed—would travel roughly 10 m.

EXAMPLE 5.6 **Towing a car with acceleration**

A car with a mass of 1500 kg is being towed by a rope held at a 20° angle to the horizontal. A friction force of 320 N opposes the car's motion. What is the tension in the rope if the car goes from rest to 12 m/s in 10 s?

PREPARE You should recognize that this problem is almost identical to Example 5.4. The difference is that the car is now accelerating, so it is no longer in equilibrium. This means, as shown in **FIGURE 5.7**, that the net force is not zero. We've already identified all the forces in Example 5.4.

SOLVE Newton's second law in component form is

$$\sum F_x = n_x + T_x + f_x + w_x = ma_x$$

$$\sum F_y = n_y + T_y + f_y + w_y = ma_y = 0$$

We've again used the fact that $a_y = 0$ for motion that is purely along the x-axis. The components of the forces were worked out in Example 5.4. With that information, Newton's second law in component form is

$$T\cos\theta - f = ma_x$$

$$n + T\sin\theta - w = 0$$

Because the car speeds up from rest to 12 m/s in 10 s, we can use kinematics to find the acceleration:

$$a_x = \frac{\Delta v_x}{\Delta t} = \frac{(v_x)_f - (v_x)_i}{t_f - t_i} = \frac{(12 \text{ m/s}) - (0 \text{ m/s})}{(10 \text{ s}) - (0 \text{ s})} = 1.2 \text{ m/s}^2$$

We can now use the first Newton's-law equation above to solve for the tension. We have

$$T = \frac{ma_x + f}{\cos\theta} = \frac{(1500 \text{ kg})(1.2 \text{ m/s}^2) + 320 \text{ N}}{\cos 20°} = 2300 \text{ N}$$

ASSESS The tension is substantially greater than the 340 N found in Example 5.4. It takes much more force to accelerate the car than to keep it rolling at a constant speed.

FIGURE 5.7 Visual overview of a car being towed.

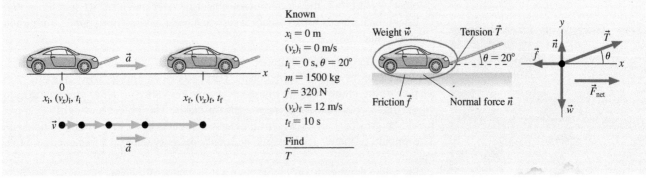

These first examples have shown all the details of our problem-solving strategy. Our purpose has been to demonstrate how the strategy is put into practice. Future examples will be briefer, but the basic *procedure* will remain the same.

A Martian lander is approaching the surface. It is slowing its descent by firing its rocket motor. Which is the correct free-body diagram for the lander?

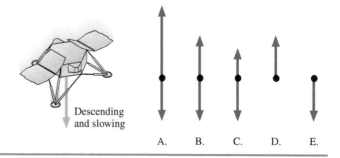

Descending
and slowing

A.   B.   C.   D.   E.

5.3 Mass and Weight

When the doctor asks what you weigh, what does she really mean? We do not make much distinction in our ordinary use of language between the terms "weight" and "mass," but in physics their distinction is of critical importance.

Mass, you'll recall from Chapter 4, is a quantity that describes an object's inertia, its tendency to resist being accelerated. Loosely speaking, it also describes the amount of matter in an object. Mass, measured in kilograms, is an intrinsic property of an object; it has the same value wherever the object may be and whatever forces might be acting on it.

Weight, on the other hand, is a *force*. Specifically, it is the gravitational force exerted on an object by a planet. Weight is a vector, not a scalar, and the vector's direction is always straight down. Weight is measured in newtons.

Mass and weight are not the same thing, but they are related. FIGURE 5.8 shows the free-body diagram of an object in free fall. The *only* force acting on this object is its weight $\vec{w}$, the downward pull of gravity. The object is in free fall, so, as we saw in ◄ SECTION 2.7, the acceleration is vertical, with $a_y = -g$, where g is the free-fall acceleration, 9.80 m/s². Newton's second law for this object is thus

$$\sum F_y = -w = -mg$$

which tells us that

$$w = mg \qquad (5.3)$$

The magnitude of the weight force, which we call simply "the weight," is directly proportional to the mass, with g as the constant of proportionality.

NOTE ► Although we derived the relationship between mass and weight for an object in free fall, the weight of an object is *independent* of its state of motion. Equation 5.3 holds for an object at rest on a table, sliding horizontally, or moving in any other way. ◄

Because an object's weight depends on g, and the value of g varies from planet to planet, weight is not a fixed, constant property of an object. The value of g at the surface of the moon is about one-sixth its earthly value, so an object on the moon would have only one-sixth its weight on earth. The object's weight on Jupiter would be greater than its weight on earth. Its mass, however, would be the same. The amount of matter has not changed, only the gravitational force exerted on that matter.

So, when the doctor asks what you weigh, she really wants to know your *mass*. That's the amount of matter in your body. You can't really "lose weight" by going to the moon, even though you would weigh less there!

We need to make a clarification here. When you give your weight, you most likely give it in pounds, which is the unit of force in the English system. (We noted in Chapter 4 that the pound is defined as 1 lb = 4.45 N.) You might then "convert"

FIGURE 5.8 The free-body diagram of an object in free fall.

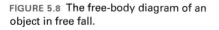

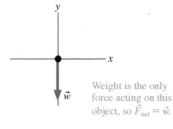

Weight is the only force acting on this object, so $\vec{F}_{net} = \vec{w}$.

On the moon, astronaut John Young jumps 2 feet straight up, despite his spacesuit that weighs 370 pounds on earth. On the moon, where $g = 1.6$ m/s², he and his suit together weighed only 90 pounds.

this to kilograms. But a kilogram is a unit of mass, not a unit of force. An object that weighs 1 pound, meaning $w = mg = 4.45$ N, has a mass of

$$m = \frac{w}{g} = \frac{4.45 \text{ N}}{9.80 \text{ m/s}^2} = 0.454 \text{ kg}$$

This calculation is different from converting, for instance, feet to meters. Both feet and meters are units of length, and it's always true that 1 m = 3.28 ft. When you "convert" from pounds to kilograms, you are determining the mass that has a certain weight—two fundamentally different quantities—and this calculation depends on the value of g. But we are usually working on the earth, where we assume that $g = 9.80$ m/s². In this case, a given mass always corresponds to the same weight, and we can use the relationships listed in Table 5.1.

TABLE 5.1 Mass, weight, force

Conversion between force units:

1 pound = 4.45 N
1 N = 0.225 pound

Correspondence between mass and weight, assuming $g = 9.80$ m/s²:

1 kg ↔ 2.20 lb
1 lb ↔ 0.454 kg = 454 g

EXAMPLE 5.7 **Typical masses and weights**

What are the weight, in N, and the mass, in kg, of a 90 pound gymnast, a 150 pound professor, and a 240 pound football player?

PREPARE We can use the conversions and correspondences in Table 5.1.

SOLVE We will use the correspondence between mass and weight just as we use the conversion factor between different forces:

$$w_{\text{gymnast}} = 90 \text{ lb} \times \frac{4.45 \text{ N}}{1 \text{ lb}} = 400 \text{ N} \qquad m_{\text{gymnast}} = 90 \text{ lb} \times \frac{0.454 \text{ kg}}{1 \text{ lb}} = 41 \text{ kg}$$

$$w_{\text{prof}} = 150 \text{ lb} \times \frac{4.45 \text{ N}}{1 \text{ lb}} = 670 \text{ N} \qquad m_{\text{prof}} = 150 \text{ lb} \times \frac{0.454 \text{ kg}}{1 \text{ lb}} = 68 \text{ kg}$$

$$w_{\text{player}} = 240 \text{ lb} \times \frac{4.45 \text{ N}}{1 \text{ lb}} = 1070 \text{ N} \qquad m_{\text{player}} = 240 \text{ lb} \times \frac{0.454 \text{ kg}}{1 \text{ lb}} = 110 \text{ kg}$$

ASSESS We can use the information in this problem to assess the results of future problems. If you get an answer of 1000 N, you now know that this is approximately the weight of a football player, which can help with your assessment.

Apparent Weight

The weight of an object is the force of gravity on that object. You may never have thought about it, but gravity is not a force that you can feel or sense directly. Your *sensation* of weight—how heavy you *feel*—is due to *contact forces* supporting you. As you read this, your sensation of weight is due to the normal force exerted on you by the chair in which you are sitting. The chair's surface touches you and activates nerve endings in your skin. You sense the magnitude of this force, and this is your sensation of weight. When you stand, you feel the contact force of the floor pushing against your feet. If you are hanging from a rope, you feel the friction force between the rope and your hands.

Let's define your **apparent weight** w_{app} in terms of the force you feel:

$$w_{\text{app}} = \text{magnitude of supporting contact forces} \qquad (5.4)$$

Definition of apparent weight

If you are in equilibrium, your weight and apparent weight are generally the same. But if you undergo an acceleration, this is not necessarily the case. For instance, you feel "heavy" when an elevator you are riding in suddenly accelerates upward, and you feel lighter than normal as the upward-moving elevator brakes to a halt. Your true weight $w = mg$ has not changed during these events, but your *sensation* of your weight has.

Physics students can't jump If you ride in an elevator and try to jump up into the air just as the elevator starts to rise, you'll feel like you can hardly get off the ground because your apparent weight is greater than your actual weight. On a fast elevator with a large acceleration, it's like trying to jump with an extra 30 or 40 pounds!

FIGURE 5.9 A man in an accelerating elevator.

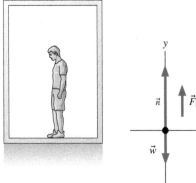

The man feels heavier than normal while accelerating upward.

Let's look at the details for this case. Imagine a man standing in an elevator as it accelerates upward. As FIGURE 5.9 shows, the only forces acting on the man are the upward normal force of the floor and the downward weight force. Because the man has an acceleration $\vec{a}$, according to Newton's second law there must be a net force acting on the man in the direction of $\vec{a}$.

Looking at the free-body diagram in Figure 5.9, we see that the y-component of Newton's second law is

$$\sum F_y = n_y + w_y = n - w = ma_y = ma \qquad (5.5)$$

where m is the man's mass. Solving Equation 5.5 for n gives

$$n = w + ma \qquad (5.6)$$

The normal force is the contact force supporting the man, so, given the definition of Equation 5.4, we can rewrite Equation 5.6 as

$$w_{app} = w + ma$$

Thus $w_{app} > w$ and the man feels heavier than normal. If the elevator is accelerating downward, the acceleration vector $\vec{a}$ points downward and $a_y = -a$. If we repeated the solution, we'd find that $w_{app} = w - ma < w$; the man would feel lighter.

The apparent weight isn't just a sensation, though. You can measure it with a scale. When you stand on a bathroom scale, the scale reading is the upward force of the scale on you. If you aren't accelerating (usually a pretty good assumption!), the upward force of the scale on you equals your weight, so the scale reading is your true weight. But if you are accelerating, this correspondence may not hold. In the example above, if the man were standing on a bathroom scale in the elevator, the contact force supporting him would be the upward force of the scale. This force is equal to the scale reading, so if the elevator is accelerating upward, the scale shows an increased weight: The scale reading is equal to w_{app}.

Apparent weight can be measured by a scale, so it's no surprise that apparent weight has real, physical implications. Astronauts are nearly crushed by their apparent weight during a rocket launch when a is much greater than g. Much of the thrill of amusement park rides, such as roller coasters, comes from rapid changes in your apparent weight.

EXAMPLE 5.8 Apparent weight in an elevator

Anjay's mass is 70 kg. He is standing on a scale in an elevator that is moving at 5.0 m/s. As the elevator stops, the scale reads 750 N. Before it stopped, was the elevator moving up or down? How long did the elevator take to come to rest?

PREPARE The scale reading as the elevator comes to rest, 750 N, is Anjay's *apparent* weight. Anjay's *actual* weight is

$$w = mg = (70 \text{ kg})(9.80 \text{ m/s}^2) = 686 \text{ N}$$

This is an intermediate step in the calculation, so we are keeping an extra significant figure. Anjay's apparent weight, which is the upward force of the scale on him, is greater than his actual weight, so there is a net upward force on Anjay. His acceleration must be upward as well. This is exactly the situation of Figure 5.9, so we can use this figure as the free-body diagram for this problem. We can find the net force on Anjay, and then we can use this net force to determine his acceleration. Once we know the acceleration, we can use kinematics to determine the time it takes for the elevator to stop.

SOLVE We can read components of vectors from the figure. The vertical component of Newton's second law for Anjay's motion is

$$\sum F_y = n - w = ma_y$$

n is the normal force, which is the scale force on Anjay, 750 N. w is his weight, 686 N. We can thus solve for a_y:

$$a_y = \frac{n - w}{m} = \frac{750 \text{ N} - 686 \text{ N}}{70 \text{ kg}} = +0.91 \text{ m/s}^2$$

The acceleration is positive and so is directed upward, exactly as we assumed—a good check on our work. The elevator is slowing down, but the acceleration is directed upward. This means that the elevator was moving *downward*, with a negative velocity, before it stopped.

To find the stopping time, we can use the kinematic equation

$$(v_y)_f = (v_y)_i + a_y \Delta t$$

The elevator is initially moving downward, so $(v_y)_i = -5.0$ m/s, and it then comes to a halt, so $(v_y)_f = 0$. We know the acceleration, so the time interval is

$$\Delta t = \frac{(v_y)_f - (v_y)_i}{a_y} = \frac{0 - (-5.0 \text{ m/s})}{0.91 \text{ m/s}^2} = 5.5 \text{ s}$$

ASSESS Think back to your experiences riding elevators. If the elevator is moving downward and then comes to rest, you "feel heavy." This gives us confidence that our analysis of the motion is correct. And 5.0 m/s is a pretty fast elevator: At this speed, the elevator will be passing more than one floor per second. If you've been in a fast elevator in a tall building, you know that 5.5 s is reasonable for the time it takes for the elevator to slow to a stop.

Weightlessness

Let's return to the elevator example above. Suppose Anjay's apparent weight was 0 N—meaning that the scale read *zero*. How could this happen? Recall that Anjay's apparent weight is the contact force supporting him, so we are saying that the upward force from the scale is zero. If we use $n = 0$ in the equation for the acceleration, we find $a_y = -9.8$ m/s^2. This is a case we've seen before—it is free fall! This means that **a person in free fall has zero apparent weight.**

Think about this carefully. Suppose, as the elevator falls, the man inside releases a ball from his hand. In the absence of air resistance, both the man and the ball would fall at the same rate. From the man's perspective, the ball would appear to "float" beside him. Similarly, the scale would float beneath him and not press against his feet. He is what we call *weightless*.

"Weightless" does *not* mean "no weight." An object that is **weightless** has no *apparent* weight. The distinction is significant. The man's weight is still *mg* because gravity is still pulling down on him, but he has no *sensation* of weight as he free falls. The term "weightless" is a very poor one because it implies that objects have no weight. As we see, that is not the case.

You've seen films of astronauts and various objects floating inside the International Space Station as it orbits the earth. If an astronaut tries to stand on a scale, it does not exert any force against her feet and reads zero. She is said to be weightless. But if the criterion to be weightless is to be in free fall, and if astronauts orbiting the earth are weightless, does this mean that they are in free fall? This is a very interesting question to which we shall return in Chapter 6.

A weightless experience As we learned in Chapter 3, objects undergoing projectile motion are in free fall. This specially adapted plane flies in the same parabolic trajectory as would a projectile with no air resistance. Objects inside, such as these passengers, thus move along a free-fall trajectory. They feel weightless, and then float with respect to the plane's interior, until the plane resumes normal flight, up to 30 seconds later.

STOP TO THINK 5.3 You're bouncing up and down on a trampoline. After you have left the trampoline and are moving upward, your apparent weight is

A. More than your true weight.
B. Less than your true weight.
C. Equal to your true weight.
D. Zero.

5.4 Normal Forces

In Chapter 4 we saw that an object at rest on a table is subject to an upward force due to the table. This force is called the *normal force* because it is always directed normal, or perpendicular, to the surface of contact. As we saw, the normal force has its origin in the atomic "springs" that make up the surface. The harder the object bears down on the surface, the more these springs are compressed and the harder they push back. Thus the normal force *adjusts* itself so that the object stays on the surface without penetrating it. This fact is key in solving for the normal force.

EXAMPLE 5.9   **Normal force on a pressed book**

A 1.2 kg book lies on a table. The book is pressed down from above with a force of 15 N. What is the normal force acting on the book from the table below?

PREPARE The book is not moving and is thus in static equilibrium. We need to identify the forces acting on the book and prepare a free-body diagram showing these forces. These steps are illustrated in **FIGURE 5.10**.

FIGURE 5.10 Finding the normal force on a book pressed from above.

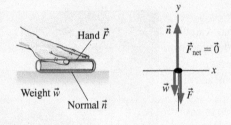

SOLVE Because the book is in static equilibrium, the net force on it must be zero. The only forces acting are in the y-direction, so Newton's second law is

$$\sum F_y = n_y + w_y + F_y = n - w - F = ma_y = 0$$

We learned in the last section that the weight force is $w = mg$. The weight of the book is thus

$$w = mg = (1.2 \text{ kg})(9.8 \text{ m/s}^2) = 12 \text{ N}$$

With this information, we see that the normal force exerted by the table is

$$n = F + w = 15 \text{ N} + 12 \text{ N} = 27 \text{ N}$$

ASSESS The magnitude of the normal force is *larger* than the weight of the book. From the table's perspective, the extra force from the hand pushes the book further into the atomic springs of the table. These springs then push back harder, giving a normal force that is greater than the weight of the book.

A common situation is an object on a ramp or incline. If friction is ignored, there are only two forces acting on the object: gravity and the normal force. However, we need to carefully work out the components of these two forces in order to solve dynamics problems. **FIGURE 5.11a** shows how. Be sure you avoid the two common errors shown in **FIGURE 5.11b**.

FIGURE 5.11 The forces on an object on an incline.

(a) Analyzing forces on an incline

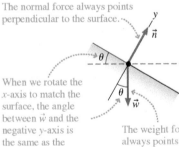

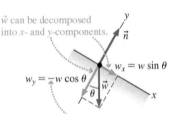

(b) Two common mistakes to avoid

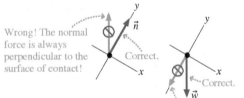

STOP TO THINK 5.4   A mountain biker is climbing a steep 20° slope at a constant speed. The cyclist and bike have a combined weight of 800 N. Referring to Figure 5.11 for guidance, what can you say about the magnitude of the normal force of the ground on the bike?

A. $n > 800$ N
B. $n = 800$ N
C. $n < 800$ N

EXAMPLE 5.10 Acceleration of a downhill skier

A skier slides down a steep 27° slope. On a slope this steep, friction is much smaller than the other forces at work and can be ignored. What is the skier's acceleration?

PREPARE FIGURE 5.12 is a visual overview. We choose a coordinate system tilted so that the *x*-axis points down the slope. This greatly simplifies the analysis because with this choice $a_y = 0$ (the skier does not move in the *y*-direction at all). The free-body diagram is based on the information in Figure 5.11.

FIGURE 5.12 Visual overview of a downhill skier.

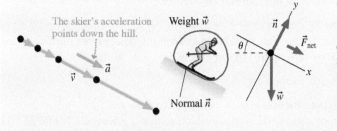

SOLVE We can now use Newton's second law in component form to find the skier's acceleration:

$$\sum F_x = w_x + n_x = ma_x$$
$$\sum F_y = w_y + n_y = ma_y$$

Because $\vec{n}$ points directly in the positive *y*-direction, $n_y = n$ and $n_x = 0$. Figure 5.11a showed the important fact that the angle between $\vec{w}$ and the negative *y*-axis is the *same* as the slope angle θ. With this information, the components of $\vec{w}$ are $w_x = w \sin \theta = mg \sin \theta$ and $w_y = -w \cos \theta = -mg \cos \theta$, where we used the fact that $w = mg$. With these components in hand, Newton's second law becomes

$$\sum F_x = w_x + n_x = mg \sin \theta = ma_x$$
$$\sum F_y = w_y + n_y = -mg \cos \theta + n = ma_y = 0$$

In the second equation we used the fact that $a_y = 0$. The *m* cancels in the first of these equations, leaving us with

$$a_x = g \sin \theta$$

This is the expression for acceleration on a frictionless surface that we presented, without proof, in Chapter 3. Now we've justified our earlier assertion. We can use this to calculate the skier's acceleration:

$$a_x = g \sin \theta = (9.8 \text{ m/s}^2) \sin 27° = 4.4 \text{ m/s}^2$$

ASSESS Our result shows that when $\theta = 0$, so that the slope is horizontal, the skier's acceleration is zero, as it should be. Further, when $\theta = 90°$ (a vertical slope), his acceleration is *g*, which makes sense because he's in free fall when $\theta = 90°$. Notice that the mass canceled out, so we didn't need to know the skier's mass. We first saw the formula for the acceleration in ◀ SECTION 3.4, but now we see the physical reasons behind it.

5.5 Friction

In everyday life, friction is everywhere. Friction is absolutely essential for many things we do. Without friction you could not walk, drive, or even sit down (you would slide right off the chair!). It is sometimes useful to think about idealized frictionless situations, but it is equally necessary to understand a real world where friction is present. Although friction is a complicated force, many aspects of friction can be described with a simple model.

Class Video

FIGURE 5.13 Static friction keeps an object from slipping.

Static Friction

Chapter 4 defined static friction $\vec{f_s}$ as the force that a surface exerts on an object to keep it from slipping across that surface. Consider the woman pushing on the box in FIGURE 5.13a. Because the box is not moving with respect to the floor, the woman's push to the right must be balanced by a static friction force $\vec{f_s}$ pointing to the left. This is the general rule for finding the *direction* of $\vec{f_s}$: Decide which way the object *would* move if there were no friction. The static friction force $\vec{f_s}$ then points in the opposite direction, to prevent motion relative to the surface.

Determining the *magnitude* of $\vec{f_s}$ is a bit trickier. Because the box is at rest, it's in static equilibrium. From the free-body diagram of FIGURE 5.13b, this means that the static friction force must exactly balance the pushing force, so that $f_s = F_{push}$. As shown in FIGURES 5.14a and 5.14b on the next page, the harder the woman pushes, the harder the friction force from the floor pushes back. If she reduces her pushing force, the friction force will automatically be reduced to match. Static friction acts in *response* to an applied force.

(a) Force identification

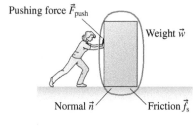

(b) Free-body diagram

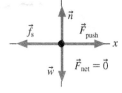

FIGURE 5.14 Static friction acts in *response* to an applied force.

(a) Pushing gently: friction pushes back gently.

$\vec{f_s}$ balances $\vec{F}_{push}$ and the box does not move.

(b) Pushing harder: friction pushes back harder.

$\vec{f_s}$ grows as $\vec{F}_{push}$ increases, but the two still cancel and the box remains at rest.

(c) Pushing harder still: $\vec{f_s}$ is now pushing back as hard as it can.

Now the magnitude of f_s has reached its maximum value $f_{s\,max}$. If $\vec{F}_{push}$ gets any bigger, the forces will *not* cancel and the box will start to accelerate.

But there's clearly a limit to how big $\vec{f_s}$ can get. If the woman pushes hard enough, the box will slip and start to move across the floor. In other words, the static friction force has a *maximum* possible magnitude $f_{s\,max}$, as illustrated in **FIGURE 5.14c**. Experiments with friction show that $f_{s\,max}$ is proportional to the magnitude of the normal force between the surface and the object; that is,

$$f_{s\,max} = \mu_s n \qquad (5.7)$$

where μ_s is called the **coefficient of static friction**. The coefficient is a number that depends on the materials from which the object and the surface are made. The higher the coefficient of static friction, the greater the "stickiness" between the object and the surface, and the harder it is to make the object slip. Table 5.2 lists some approximate values of coefficients of friction.

NOTE ▶ Equation 5.7 does not say $f_s = \mu_s n$. The value of f_s depends on the force or forces that static friction has to balance to keep the object from moving. It can have any value from zero up to, but not exceeding, $\mu_s n$. ◀

So our rules for static friction are:

- The direction of static friction is such as to oppose motion.
- The magnitude f_s of static friction adjusts itself so that the net force is zero and the object doesn't move.
- The magnitude of static friction cannot exceed the maximum value $f_{s\,max}$ given by Equation 5.7. If the friction force needed to keep the object stationary is greater than $f_{s\,max}$, the object slips and starts to move.

Kinetic Friction

Once the box starts to slide, as in **FIGURE 5.15**, the static friction force is replaced by a kinetic (or sliding) friction force $\vec{f_k}$. Kinetic friction is in some ways simpler than static friction: The direction of $\vec{f_k}$ is always opposite the direction in which an object slides across the surface, and experiments show that kinetic friction, unlike static friction, has a nearly *constant* magnitude, given by

$$f_k = \mu_k n \qquad (5.8)$$

where μ_k is called the **coefficient of kinetic friction**. Equation 5.8 also shows that kinetic friction, like static friction, is proportional to the magnitude of the normal force n. Notice that **the magnitude of the kinetic friction force does not depend on how fast the object is sliding.**

FIGURE 5.15 The kinetic friction force is *opposite* the direction of motion.

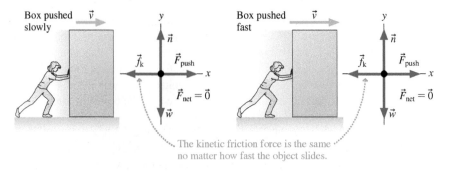

The kinetic friction force is the same no matter how fast the object slides.

TABLE 5.2 Coefficients of friction

Materials	Static μ_s	Kinetic μ_k	Rolling μ_r
Rubber on concrete	1.00	0.80	0.02
Steel on steel (dry)	0.80	0.60	0.002
Steel on steel (lubricated)	0.10	0.05	
Wood on wood	0.50	0.20	
Wood on snow	0.12	0.06	
Ice on ice	0.10	0.03	

Table 5.2 includes approximate values of μ_k. You can see that $\mu_k < \mu_s$, which explains why it is easier to keep a box moving than it is to start it moving.

Rolling Friction

If you slam on the brakes hard enough, your car tires slide against the road surface and leave skid marks. This is kinetic friction because the tire and the road are *sliding* against each other. A wheel *rolling* on a surface also experiences friction, but not kinetic friction: The portion of the wheel that contacts the surface is stationary with respect to the surface, not sliding. The photo in FIGURE 5.16 was taken with a stationary camera. Note that the part of the wheel touching the ground is not blurred, indicating that this part of the wheel is not moving with respect to the ground.

The interaction between a rolling wheel and the road involves adhesion between and deformation of surfaces and can be quite complicated, but in many cases we can treat it like another type of friction force that opposes the motion, one defined by a **coefficient of rolling friction** μ_r:

$$f_r = \mu_r n \qquad (5.9)$$

Rolling friction acts very much like kinetic friction, but values of μ_r (see Table 5.2) are much lower than values of μ_k. It's easier to roll something on wheels than to slide it!

FIGURE 5.16 The bottom of the wheel is stationary.

STOP TO THINK 5.5   Rank in order, from largest to smallest, the size of the friction forces $\vec{f}_A$ to $\vec{f}_E$ in the five different situations (one or more friction forces could be zero). The box and the floor are made of the same materials in all situations.

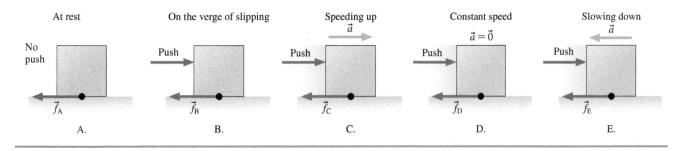

At rest	On the verge of slipping	Speeding up	Constant speed	Slowing down

A.   B.   C.   D.   E.

Working with Friction Forces

These ideas can be summarized in a *model* of friction:

Static: $\vec{f}_s =$ (magnitude $\leq f_{s\,max} = \mu_s n$,
      direction as necessary to prevent motion)

Kinetic: $\vec{f}_k = (\mu_k n$, direction opposite the motion) (5.10)

Rolling: $\vec{f}_r = (\mu_r n$, direction opposite the motion)

PROPORTIONAL   p. 34

Here "motion" means "motion relative to the surface." The maximum value of static friction $f_{s\,max} = \mu_s n$ occurs when the object slips and begins to move.

NOTE ▸ Equations 5.10 are a "model" of friction, not a "law" of friction. These equations provide a reasonably accurate, but not perfect, description of how friction forces act. They are a simplification of reality that works reasonably well, which is what we mean by a "model." They are not a "law of nature" on a level with Newton's laws. ◂

To skid or not to skid If you brake as hard as you can without skidding, the force that stops your car is the static friction force between your tires and the road. This force is bigger than the kinetic friction force, so if you skid, not only can you lose control of your car but you also end up taking a longer distance to stop. Antilock braking systems brake the car as hard as possible without skidding, stopping the car in the shortest possible distance while also retaining control.

**TACTICS
BOX 5.1**   **Working with friction forces**   (MP)

❶ If the object is *not moving* relative to the surface it's in contact with, then the friction force is **static friction**. Draw a free-body diagram of the object. The *direction* of the friction force is such as to oppose sliding of the object relative to the surface. Then use Problem-Solving Strategy 5.1 to solve for f_s. If f_s is greater than $f_{s\,max} = \mu_s n$, then static friction cannot hold the object in place. The assumption that the object is at rest is not valid, and you need to redo the problem using kinetic friction.

❷ If the object is *sliding* relative to the surface, then **kinetic friction** is acting. From Newton's second law, find the normal force n. Equation 5.10 then gives the magnitude and direction of the friction force.

❸ If the object is *rolling* along the surface, then **rolling friction** is acting. From Newton's second law, find the normal force n. Equation 5.10 then gives the magnitude and direction of the friction force.

Exercises 20, 21

EXAMPLE 5.11   Finding the force to slide a sofa

Carol wants to move her 32 kg sofa to a different room in the house. She places "sofa sliders," slippery disks with $\mu_k = 0.080$, on the carpet, under the feet of the sofa. She then pushes the sofa at a steady 0.40 m/s across the floor. How much force does she apply to the sofa?

PREPARE Let's assume the sofa slides to the right. In this case, a kinetic friction force $\vec{f}_k$, opposes the motion by pointing to the left. In **FIGURE 5.17** we identify the forces acting on the sofa and construct a free-body diagram.

FIGURE 5.17 Forces on a sofa being pushed across a floor.

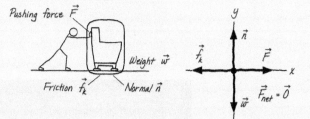

SOLVE The sofa is moving at a constant speed, so it is in dynamic equilibrium with $\vec{F}_{net} = \vec{0}$. This means that the x- and y-components of the net force must be zero:

$$\sum F_x = n_x + w_x + F_x + (f_k)_x = 0 + 0 + F - f_k = 0$$
$$\sum F_y = n_y + w_y + F_y + (f_k)_y = n - w + 0 + 0 = 0$$

In the first equation, the x-component of $\vec{f}_k$ is equal to $-f_k$ because $\vec{f}_k$ is directed to the left. Similarly, $w_y = -w$ because the weight force points down.

From the first equation, we see that Carol's pushing force is $F = f_k$. To evaluate this, we need f_k. Here we can use our model for kinetic friction:

$$f_k = \mu_k n$$

Let's look at the vertical motion first. The second equation ultimately reduces to

$$n - w = 0$$

The weight force $w = mg$, so we can write

$$n = mg$$

This is a common result we'll see again. The force that Carol pushes with is equal to the friction force, and this depends on the normal force and the coefficient of kinetic friction, $\mu_k = 0.080$:

$$F = f_k = \mu_k n = \mu_k mg$$
$$= (0.080)(32\,\text{kg})(9.80\,\text{m/s}^2) = 25\,\text{N}$$

ASSESS The speed with which Carol pushes the sofa does not enter into the answer. This makes sense because the kinetic friction force doesn't depend on speed. The final result of 25 N is a rather small force—only about $5\frac{1}{2}$ pounds—but we expect this because Carol has used slippery disks to move the sofa.

CONCEPTUAL EXAMPLE 5.12   To push or pull a lawn roller?

A lawn roller is a heavy cylinder used to flatten a bumpy lawn, as shown in **FIGURE 5.18**. Is it easier to push or pull such a roller? Which is more effective for flattening the lawn: pushing or pulling? Assume that the pushing or pulling force is directed along the handle of the roller.

FIGURE 5.18 Pushing and pulling a lawn roller.

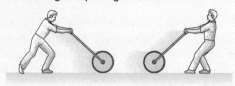

REASON FIGURE 5.19 shows free-body diagrams for the two cases. We assume that the roller is pushed at a constant speed so that it is in dynamic equilibrium with $\vec{F}_{net} = \vec{0}$. Because the roller does not move in the y-direction, the y-component of the net force must be zero. According to our model, the magnitude f_r of rolling friction is proportional to the magnitude n of the normal force. If we *push* on the roller, our pushing force $\vec{F}$ will have a downward y-component. To compensate for this, the normal force must increase and, because $f_r = \mu_r n$, the rolling friction will increase as well. This makes the roller harder to move. If we *pull* on the roller, the now upward y-component of $\vec{F}$ will lead to a *reduced* value of n and hence of f_r. Thus the roller is easier to pull than to push.

However, the purpose of the roller is to flatten the soil. If the normal force $\vec{n}$ of the ground on the roller is greater, then by Newton's third law the force of the roller on the ground will be greater as well. So for smoothing your lawn, it's better to push.

FIGURE 5.19 Free-body diagrams for the lawn roller.

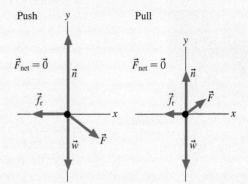

ASSESS You've probably experienced this effect while using an upright vacuum cleaner. The vacuum is harder to push on the forward stroke than when drawing it back.

EXAMPLE 5.13 **How to dump a file cabinet**

A 50.0 kg steel file cabinet is in the back of a dump truck. The truck's bed, also made of steel, is slowly tilted. What is the magnitude of the static friction force on the cabinet when the bed is tilted 20°? At what angle will the file cabinet begin to slide?

PREPARE We'll use our model of static friction. The file cabinet will slip when the static friction force reaches its maximum possible value $f_{s\,max}$. FIGURE 5.20 shows the visual overview when the truck bed is tilted at angle θ. We can make the analysis easier if we tilt the coordinate system to match the bed of the truck. To prevent the file cabinet from slipping, the static friction force must point *up* the slope.

SOLVE Before it slips, the file cabinet is in static equilibrium. Newton's second law gives

$$\sum F_x = n_x + w_x + (f_s)_x = 0$$

$$\sum F_y = n_y + w_y + (f_s)_y = 0$$

From the free-body diagram we see that f_s has only a negative x-component and that n has only a positive y-component. We also have $w_x = w\sin\theta$ and $w_y = -w\cos\theta$. Thus the second law becomes

$$\sum F_x = w\sin\theta - f_s = mg\sin\theta - f_s = 0$$

$$\sum F_y = n - w\cos\theta = n - mg\cos\theta = 0$$

The x-component equation allows us to determine the magnitude of the static friction force when $\theta = 20°$:

$$f_s = mg\sin\theta = (50.0\ \text{kg})(9.80\ \text{m/s}^2)\sin 20° = 168\ \text{N}$$

This value does not require that we know μ_s. The coefficient of static friction enters only when we want to find the angle at which the file cabinet slips. Slipping occurs when the static friction force reaches its maximum value:

$$f_s = f_{s\,max} = \mu_s n$$

From the y-component of Newton's second law we see that $n = mg\cos\theta$. Consequently,

$$f_{s\,max} = \mu_s mg\cos\theta$$

The x-component of the second law gave

$$f_s = mg\sin\theta$$

Setting $f_s = f_{s\,max}$ then gives

$$mg\sin\theta = \mu_s mg\cos\theta$$

The mg in both terms cancels, and we find

$$\frac{\sin\theta}{\cos\theta} = \tan\theta = \mu_s$$

$$\theta = \tan^{-1}\mu_s = \tan^{-1}(0.80) = 39°$$

ASSESS Steel doesn't slide all that well on unlubricated steel, so a fairly large angle is not surprising. The answer seems reasonable. It is worth noting that $n = mg\cos\theta$ in this example. A common error is to use simply $n = mg$. Be sure to evaluate the normal force within the context of each particular problem.

FIGURE 5.20 Visual overview of a file cabinet in a tilted dump truck.

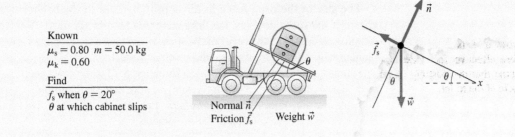

Known	
$\mu_s = 0.80$	$m = 50.0$ kg
$\mu_k = 0.60$	

Find

f_s when $\theta = 20°$
θ at which cabinet slips

Normal $\vec{n}$
Friction $\vec{f}_s$ Weight $\vec{w}$

FIGURE 5.21 A microscopic view of friction.

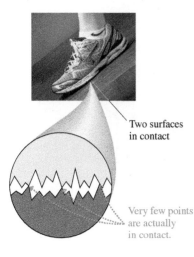

Two surfaces in contact

Very few points are actually in contact.

Causes of Friction

It is worth a brief pause to look at the *causes* of friction. All surfaces, even those quite smooth to the touch, are very rough on a microscopic scale. When two objects are placed in contact, they do not make a smooth fit. Instead, as FIGURE 5.21 shows, the high points on one surface become jammed against the high points on the other surface, while the low points are not in contact at all. Only a very small fraction (typically 10^{-4}) of the surface area is in actual contact. The amount of contact depends on how hard the surfaces are pushed together, which is why friction forces are proportional to n.

For an object to slip, you must push it hard enough to overcome the forces exerted at these contact points. Once the two surfaces are sliding against each other, their high points undergo constant collisions, deformations, and even brief bonding that lead to the resistive force of kinetic friction.

5.6 Drag

The air exerts a drag force on objects as they move through it. You experience drag forces every day as you jog, bicycle, ski, or drive your car. The drag force $\vec{D}$:

- Is opposite in direction to the velocity $\vec{v}$.
- Increases in magnitude as the object's speed increases.

At relatively low speeds, the drag force in air is small and can usually be ignored, but drag plays an important role as speeds increase. Fortunately, we can use a fairly simple *model* of drag if the following three conditions are met:

- The object's size (diameter) is between a few millimeters and a few meters.
- The object's speed is less than a few hundred meters per second.
- The object is moving through the air near the earth's surface.

These conditions are usually satisfied for balls, people, cars, and many other objects in our everyday experience. Under these conditions, the drag force can be written as:

$$\vec{D} = \left(\tfrac{1}{2} C_D \rho A v^2, \text{ direction opposite the motion} \right) \qquad (5.11)$$

Here, ρ is the density of air ($\rho = 1.2$ kg/m^3 at sea level), A is the cross-section area of the object (in m^2), and the **drag coefficient** C_D depends on the details of the object's shape. However, the value of C_D for everyday moving objects is roughly 1/2, so a good approximation to the drag force is

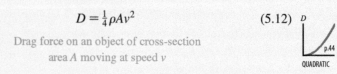

$$D = \tfrac{1}{4} \rho A v^2 \qquad (5.12)$$

Drag force on an object of cross-section area A moving at speed v

D | p.44 | v
QUADRATIC

This is the expression for the magnitude of the drag force that we'll use in this chapter.

The size of the drag force in air is proportional to the *square* of the object's speed: If the speed doubles, the drag increases by a factor of 4. This model of drag fails for objects that are very small (such as dust particles) or very fast (such as jet planes) or that move in other media (such as water).

FIGURE 5.22 shows that the area A in Equation 5.12 is the cross section of the object as it "faces into the wind." It's interesting to note that the magnitude of the drag force depends on the object's *size and shape* but not on its *mass*. This has important consequences for the motion of falling objects.

FIGURE 5.22 How to calculate the cross-section area A.

FIGURE 5.22 How to calculate the cross-section area A.

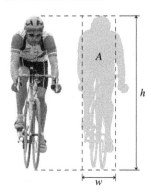

A is the cross-section area of the cyclist as seen from the front. This area is approximated by the rectangle shown, with area $A = h \times w$.

The cross-section area of a sphere is a circle. For this soccer ball, $A = \pi r^2$.

FIGURE 5.23 A falling object eventually reaches terminal speed.

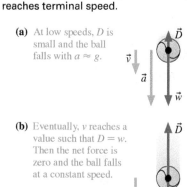

(a) At low speeds, D is small and the ball falls with $a \approx g$.

(b) Eventually, v reaches a value such that $D = w$. Then the net force is zero and the ball falls at a constant speed.

Terminal Speed

Just after an object is released from rest, its speed is low and the drag force is small (as shown in FIGURE 5.23a). Because the net force is nearly equal to the weight, the object will fall with an acceleration only a little less than g. As it falls farther, its speed and hence the drag force increase. Now the net force is smaller, so the acceleration is smaller. Eventually the speed will increase to a point such that the magnitude of the drag force *equals* the weight (as shown in FIGURE 5.23b). The net force and hence the acceleration at this speed are then *zero*, and the object falls with a *constant* speed. The speed at which the exact balance between the upward drag force and the downward weight force causes an object to fall without acceleration is called the **terminal speed**. **Once an object has reached terminal speed, it will continue falling at that speed until it hits the ground.**

EXAMPLE 5.14 **Terminal speeds of a skydiver and a mouse**

A skydiver and his pet mouse jump from a plane. Estimate their terminal speeds, assuming that they both fall in a prone position with limbs extended.

PREPARE There is no net force on a man or a mouse that has reached terminal speed. This is the situation shown in Figure 5.23b, where the drag force D and the weight w are equal in magnitude. Equating expressions for these two forces, we find that

$$\frac{1}{4}\rho A v^2 = mg$$

To solve this for the terminal speed v for both the man and the mouse, we need to estimate the mass m and cross-section area A of each. FIGURE 5.24 shows how. A typical skydiver might be 1.8 m long and 0.4 m wide ($A = 0.72$ m²) with a mass of 75 kg, while a mouse has a mass of perhaps 20 g (0.020 kg) and is 7 cm long and 3 cm wide ($A = 0.07$ m $\times$ 0.03 m $= 0.0021$ m²).

FIGURE 5.24 The cross-section areas of a skydiver and a mouse.

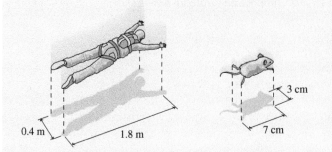

3 cm

7 cm

0.4 m

1.8 m

SOLVE We can rearrange the equation to read

$$v = \sqrt{\frac{4mg}{\rho A}}$$

Using the numbers we estimated as well as the approximate density of air at sea level, we find the following values for the terminal speed:

$$v_{\text{man}} \approx \sqrt{\frac{4(75 \text{ kg})(9.80 \text{ m/s}^2)}{(1.2 \text{ kg/m}^3)(0.72 \text{ m}^2)}} = 60 \text{ m/s}$$

$$v_{\text{mouse}} \approx \sqrt{\frac{4(0.020 \text{ kg})(9.80 \text{ m/s}^2)}{(1.2 \text{ kg/m}^3)(0.0021 \text{ m}^2)}} = 20 \text{ m/s}$$

ASSESS The terminal speed that we calculated for a skydiver is close to what you find if you look up expected speeds for this activity. But how about the mouse? The terminal speed depends on the ratio of the mass to the cross-section area, m/A. Smaller values of this ratio lead to slower terminal speeds. Small animals have smaller values of this ratio and really do experience lower terminal speeds. A mouse, with its very small value of m/A (and thus modest terminal speed), can typically survive a fall from *any* height with no ill effects! Small animals can reduce their terminal speed even further by increasing A. The frog in the photograph at the start of the chapter does this by stretching out its feet, and it experiences a gentle glide to the ground.

Although we've focused our analysis on falling objects, the same ideas apply to objects moving horizontally. If an object is thrown or shot horizontally, $\vec{D}$ causes the object to slow down. An airplane reaches its maximum speed, which is analogous to the terminal speed, when the drag is equal to and opposite the thrust: $D = F_{\text{thrust}}$. The net force is then zero and the plane cannot go any faster.

We will continue to ignore drag unless a problem specifically calls for drag to be considered.

Catalog of Forces Revisited

Before we continue, let's summarize the details of the different forces that we've seen so far, extending the catalog of forces that we presented in ◀ SECTION 4.2.

SYNTHESIS 5.1 A catalog of forces

When solving mechanics problems, you'll often use the directions and details of the most common forces, outlined below.

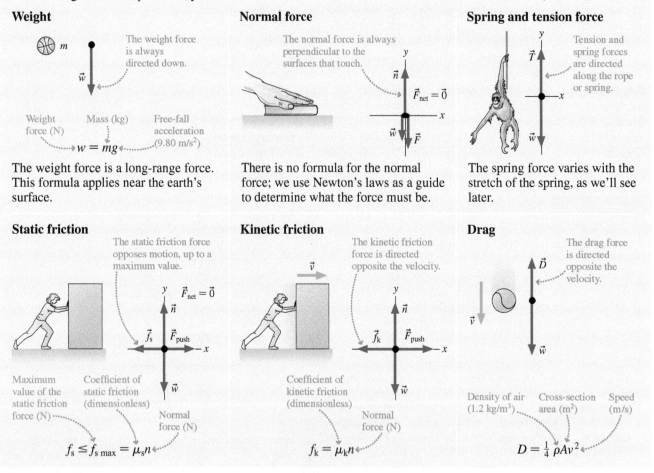

Weight

The weight force is always directed down.

Weight force (N) Mass (kg) Free-fall acceleration (9.80 m/s²)

$w = mg$

The weight force is a long-range force. This formula applies near the earth's surface.

Normal force

The normal force is always perpendicular to the surfaces that touch.

There is no formula for the normal force; we use Newton's laws as a guide to determine what the force must be.

Spring and tension force

Tension and spring forces are directed along the rope or spring.

The spring force varies with the stretch of the spring, as we'll see later.

Static friction

The static friction force opposes motion, up to a maximum value.

Maximum value of the static friction force (N) Coefficient of static friction (dimensionless) Normal force (N)

$f_s \leq f_{s\,\text{max}} = \mu_s n$

Kinetic friction

The kinetic friction force is directed opposite the velocity.

Coefficient of kinetic friction (dimensionless) Normal force (N)

$f_k = \mu_k n$

Drag

The drag force is directed opposite the velocity.

Density of air (1.2 kg/m³) Cross-section area (m²) Speed (m/s)

$D = \frac{1}{4}\rho A v^2$

STOP TO THINK 5.6 The terminal speed of a Styrofoam ball is 15 m/s. Suppose a Styrofoam ball is shot straight down with an initial speed of 30 m/s. Which velocity graph is correct?

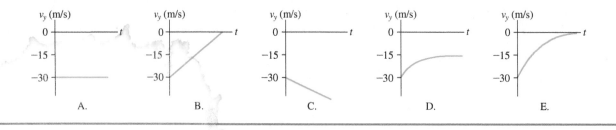

A.   B.   C.   D.   E.

5.7 Interacting Objects

Up to this point we have studied the dynamics of a single object subject to forces exerted on it by other objects. In Example 5.11, for instance, the sofa was acted upon by friction, normal, weight, and pushing forces that came from the floor, the earth, and the person pushing. As we've seen, such problems can be solved by an application of Newton's second law after all the forces have been identified.

But in Chapter 4 we found that real-world motion often involves two or more objects interacting with each other. We further found that forces always come in action/reaction *pairs* that are related by Newton's third law. To remind you, Newton's third law states:

- Every force occurs as one member of an action/reaction pair of forces. The two members of the pair always act on *different* objects.
- The two members of an action/reaction pair point in *opposite* directions and are *equal* in magnitude.

Our goal in this section is to learn how to apply the second *and* third laws to interacting objects.

Objects in Contact

One common way that two objects interact is via direct contact forces between them. Consider, for example, the two blocks being pushed across a frictionless table in FIGURE 5.25. To analyze block A's motion, we need to identify all the forces acting on it and then draw its free-body diagram. We repeat the same steps to analyze the motion of block B. However, the forces on A and B are *not* independent: Forces $\vec{F}_{\text{B on A}}$ acting on block A and $\vec{F}_{\text{A on B}}$ acting on block B are an action/reaction pair and thus have the same magnitude. Furthermore, because the two blocks are in contact, their *accelerations* must be the same, so that $a_{Ax} = a_{Bx} = a_x$. Because the accelerations of both blocks are equal, we can drop the subscripts A and B and call both accelerations a_x.

These observations suggest that we can't solve for the motion of one block without considering the motion of the other block. Solving a motion problem thus means solving two problems in parallel.

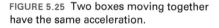

FIGURE 5.25 Two boxes moving together have the same acceleration.

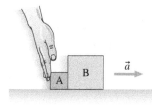

TACTICS
BOX 5.2 **Working with objects in contact** (MP)

When two objects are in contact and their motion is linked, we need to duplicate certain steps in our analysis:

❶ Draw each object separately and prepare a separate force identification diagram for *each* object.
❷ Draw a separate free-body diagram for *each* object.
❸ Write Newton's second law in component form for *each* object.

The two objects in contact exert forces on each other:

❹ Identify the action/reaction pairs of forces. If object A acts on object B with force $\vec{F}_{\text{A on B}}$, then identify the force $\vec{F}_{\text{B on A}}$ that B exerts on A.
❺ Newton's third law says that you can equate the magnitudes of the two forces in each action/reaction pair.

The fact that the objects are in contact simplifies the kinematics:

❻ Objects in contact will have the same acceleration.

Exercises 25, 26, 27 ✎

NOTE ▶ Two steps are especially important when drawing the free-body diagrams. First, draw a *separate* diagram for each object. They need not have the same coordinate system. Second, show only the forces acting *on* that object. The force $\vec{F}_{A \text{ on } B}$ goes on the free-body diagram of object B, but $\vec{F}_{B \text{ on } A}$ goes on the diagram of object A. The two members of an action/reaction pair *always* appear on two different free-body diagrams—*never* on the same diagram. ◀

EXAMPLE 5.15 **Pushing two blocks**

FIGURE 5.26 shows a 5.0 kg block A being pushed with a 3.0 N force. In front of this block is a 10 kg block B; the two blocks move together. What force does block A exert on block B?

FIGURE 5.26 Two blocks are pushed by a hand.

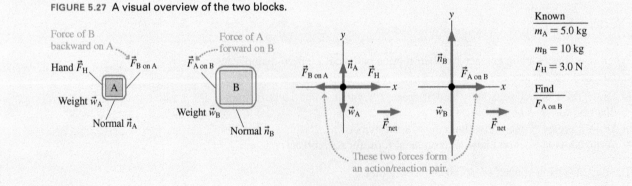

Frictionless surface
5.0 kg 10 kg

PREPARE The visual overview of FIGURE 5.27 lists the known information and identifies $F_{A \text{ on } B}$ as what we're trying to find. Then, following the steps of Tactics Box 5.2, we've drawn *separate* force identification diagrams and *separate* free-body diagrams for the two blocks. Both blocks have a weight force and a normal force, so we've used subscripts A and B to distinguish between them.

The force $\vec{F}_{A \text{ on } B}$ is the contact force that block A exerts on B; it forms an action/reaction pair with the force $\vec{F}_{B \text{ on } A}$ that block B exerts on A. Notice that force $\vec{F}_{A \text{ on } B}$ is drawn acting on block B; it is the force *of* A *on* B. **Force vectors are always drawn on the free-body diagram of the object that *experiences* the force,** not the object exerting the force. Because action/reaction pairs act in opposite directions, force $\vec{F}_{B \text{ on } A}$ pushes backward on block A and appears on A's free-body diagram.

SOLVE We begin by writing Newton's second law in component form for each block. Because the motion is only in the *x*-direction, we need only the *x*-component of the second law. For block A,

$$\sum F_x = (F_H)_x + (F_{B \text{ on } A})_x = m_A a_{Ax}$$

The force components can be "read" from the free-body diagram, where we see $\vec{F}_H$ pointing to the right and $\vec{F}_{B \text{ on } A}$ pointing to the left. Thus

$$F_H - F_{B \text{ on } A} = m_A a_{Ax}$$

For B, we have

$$\sum F_x = (F_{A \text{ on } B})_x = F_{A \text{ on } B} = m_B a_{Bx}$$

We have two additional pieces of information: First, Newton's third law tells us that $F_{B \text{ on } A} = F_{A \text{ on } B}$. Second, the boxes are in contact and must have the same acceleration a_x; that is, $a_{Ax} = a_{Bx} = a_x$. With this information, the two *x*-component equations become

$$F_H - F_{A \text{ on } B} = m_A a_x$$
$$F_{A \text{ on } B} = m_B a_x$$

Our goal is to find $F_{A \text{ on } B}$, so we need to eliminate the unknown acceleration a_x. From the second equation, $a_x = F_{A \text{ on } B}/m_B$. Substituting this into the first equation gives

$$F_H - F_{A \text{ on } B} = \frac{m_A}{m_B} F_{A \text{ on } B}$$

This can be solved for the force of block A on block B, giving

$$F_{A \text{ on } B} = \frac{F_H}{1 + m_A/m_B} = \frac{3.0 \text{ N}}{1 + (5.0 \text{ kg})/(10 \text{ kg})} = \frac{3.0 \text{ N}}{1.5} = 2.0 \text{ N}$$

ASSESS Force F_H accelerates both blocks, a total mass of 15 kg, but force $F_{A \text{ on } B}$ accelerates only block B, with a mass of 10 kg. Thus it makes sense that $F_{A \text{ on } B} < F_H$.

FIGURE 5.27 A visual overview of the two blocks.

Force of B backward on A
Hand $\vec{F}_H$ — $\vec{F}_{B \text{ on } A}$
Weight $\vec{w}_A$
[A]
Normal $\vec{n}_A$

Force of A forward on B
$\vec{F}_{A \text{ on } B}$
[B]
Weight $\vec{w}_B$
Normal $\vec{n}_B$

$\vec{F}_{B \text{ on } A}$ ← $\vec{n}_A$ $\vec{F}_H$ → x
$\vec{w}_A$
$\vec{F}_{\text{net}}$

$\vec{n}_B$
$\vec{F}_{A \text{ on } B}$ → x
$\vec{w}_B$
$\vec{F}_{\text{net}}$

These two forces form an action/reaction pair.

Known
$m_A = 5.0$ kg
$m_B = 10$ kg
$F_H = 3.0$ N
Find
$F_{A \text{ on } B}$

STOP TO THINK 5.7 Boxes P and Q are sliding to the right across a frictionless table. The hand H is slowing them down. The mass of P is larger than the mass of Q. Rank in order, from largest to smallest, the *horizontal* forces on P, Q, and H.

A. $F_{Q \text{ on } H} = F_{H \text{ on } Q} = F_{P \text{ on } Q} = F_{Q \text{ on } P}$
B. $F_{Q \text{ on } H} = F_{H \text{ on } Q} > F_{P \text{ on } Q} = F_{Q \text{ on } P}$
C. $F_{Q \text{ on } H} = F_{H \text{ on } Q} < F_{P \text{ on } Q} = F_{Q \text{ on } P}$
D. $F_{H \text{ on } Q} = F_{H \text{ on } P} > F_{P \text{ on } Q}$

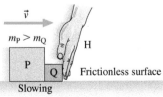

$\vec{v}$
$m_P > m_Q$ H
P Q Frictionless surface
Slowing

5.8 Ropes and Pulleys

Many objects are connected by strings, ropes, cables, and so on. We can learn several important facts about ropes and tension by considering the box being pulled by a rope in FIGURE 5.28. The rope in turn is being pulled by a hand that exerts a force $\vec{F}$ on the rope.

The box is pulled by the rope, so the box's free-body diagram shows a tension force $\vec{T}$. The *rope* is subject to two horizontal forces: the force $\vec{F}$ of the hand on the rope, and the force $\vec{F}_{\text{box on rope}}$ with which the box pulls back on the rope. In problems we'll consider, the mass of a string or rope is significantly less than the mass of the objects it pulls on, so we'll make the approximation—called the **massless string approximation**—that $m_{\text{rope}} = 0$. There is no weight force acting on the string, and so no supporting force is necessary. In this case, there are no forces that act along the vertical axis. $\vec{T}$ and $\vec{F}_{\text{box on rope}}$ form an action/reaction pair, so their magnitudes are equal: $F_{\text{box on rope}} = T$. Newton's second law *for the rope* is thus

$$\sum F_x = F - F_{\text{box on rope}} = F - T = m_{\text{rope}}a_x = 0 \qquad (5.13)$$

We've used the approximation $m_{\text{rope}} = 0$. In this case, we can say that $T = F$. Generally, **the tension in a massless string or rope equals the magnitude of the force pulling on the end of the string or rope.** As a result:

- A massless string or rope "transmits" a force undiminished from one end to the other: If you pull on one end of a rope with force F, the other end of the rope pulls on what it's attached to with a force of the same magnitude F.
- The tension in a massless string or rope is the same from one end to the other.

FIGURE 5.28 A box being pulled by a rope.

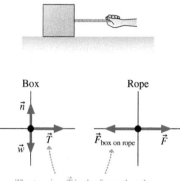

The tension $\vec{T}$ is the force that the rope exerts on the box. Thus $\vec{T}$ and $\vec{F}_{\text{box on rope}}$ are an action/reaction pair and have the same magnitude.

CONCEPTUAL EXAMPLE 5.16   Pulling a rope

FIGURE 5.29a shows a student pulling horizontally with a 100 N force on a rope that is attached to a wall. In FIGURE 5.29b, two students in a tug-of-war pull on opposite ends of a rope with 100 N each. Is the tension in the second rope larger than, smaller than, or the same as that in the first?

FIGURE 5.29 Pulling on a rope. Which produces a larger tension?

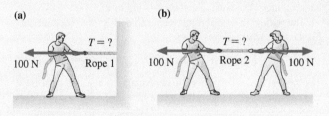

(a)   (b)

REASON Surely pulling on a rope from both ends causes more tension than pulling on one end. Right? Before jumping to

conclusions, let's analyze the situation carefully. We found above that the force pulling on the end of a rope—here, the 100 N force exerted by the student—and the tension in the rope have the same magnitude. Thus, the tension in rope 1 is 100 N, the force with which the student pulls on the rope.

To find the tension in the second rope, consider the force that the *wall* exerts on the *first* rope. The first rope is in equilibrium, so the 100 N force exerted by the student must be balanced by a 100 N force on the rope from the wall. The first rope is being pulled from *both* ends by a 100 N force—the exact same situation as for the second rope, pulled by the students. A rope doesn't care whether it's being pulled on by a wall or by a person, so the tension in the second rope is the *same* as that in the first, or 100 N.

ASSESS This example reinforces what we just learned about ropes: A rope pulls on the objects at each of its ends with a force equal in magnitude to the tension, and the external force applied to each end of the rope and the rope's tension have equal magnitude.

Pulleys

Strings and ropes often pass over pulleys. FIGURE 5.30 on the next page shows a simple situation in which block B drags block A across a table as it falls. As the string moves, static friction between the string and the pulley causes the pulley to turn. If we assume that

- The string *and* the pulley are both massless, and
- There is no friction where the pulley turns on its axle,

FIGURE 5.30 An ideal pulley changes the direction in which a tension force acts, but not its magnitude.

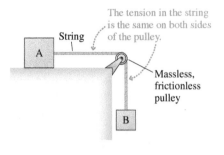

The tension in the string is the same on both sides of the pulley.

String

A

Massless, frictionless pulley

B

then no net force is needed to accelerate the string or turn the pulley. In this case, **the tension in a massless string is unchanged by passing over a massless, frictionless pulley.** We'll assume such an ideal pulley for problems in this chapter.

TACTICS
BOX 5.3   **Working with ropes and pulleys**   (MP)

For massless ropes or strings and massless, frictionless pulleys:

■ If a force pulls on one end of a rope, the tension in the rope equals the magnitude of the pulling force.

■ If two objects are connected by a rope, the tension is the same at both ends.

■ If the rope passes over a pulley, the tension in the rope is unaffected.

Exercises 29–32 ✎

EXAMPLE 5.17   **Placing a leg in traction** BIO

For serious fractures of the leg, the leg may need to have a stretching force applied to it to keep contracting leg muscles from forcing the broken bones together too hard. This is often done using *traction*, an arrangement of a rope, a weight, and pulleys as shown in **FIGURE 5.31**. The rope must make the same angle θ on both sides of the pulley so that the net force of the rope on the pulley is horizontally to the right, but θ can be adjusted to control the amount of traction. The doctor has specified 50 N of traction for this patient, with a 4.2 kg hanging mass. What is the proper angle θ?

FIGURE 5.31 A leg in traction.

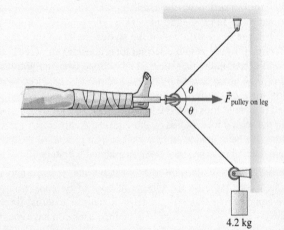

θ

θ

$\vec{F}_{\text{pulley on leg}}$

4.2 kg

PREPARE The pulley attached to the patient's leg is in static equilibrium, so the net force on it must be zero. **FIGURE 5.32** shows a free-body diagram for the pulley, which we'll assume to be frictionless. Forces $\vec{T}_1$ and $\vec{T}_2$ are the tension forces of the rope as it pulls on the pulley. These forces are equal in magnitude for a frictionless pulley, and their combined pull is to the right. This force is balanced by the force $\vec{F}_{\text{leg on pulley}}$ of the patient's leg pulling to the left. The traction force $\vec{F}_{\text{pulley on leg}}$ forms an action/reaction pair with $\vec{F}_{\text{leg on pulley}}$, so 50 N of traction means that $\vec{F}_{\text{leg on pulley}}$ also has a magnitude of 50 N.

FIGURE 5.32 Free-body diagram for the pulley.

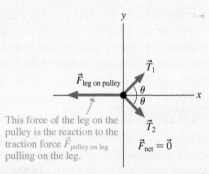

y

$\vec{F}_{\text{leg on pulley}}$

$\vec{T}_1$

θ

θ

x

This force of the leg on the pulley is the reaction to the traction force $\vec{F}_{\text{pulley on leg}}$ pulling on the leg.

$\vec{T}_2$

$\vec{F}_{\text{net}} = \vec{0}$

SOLVE Two important properties of ropes, given in Tactics Box 5.3, are that (1) the tension equals the magnitude of the force pulling on its end and (2) the tension is the same throughout the rope. Thus, if a hanging mass m pulls on the rope with its weight mg, the tension along the entire rope is $T = mg$. For a 4.2 kg hanging mass, the tension is then $T = mg = 41.2$ N.

The pulley, in equilibrium, must satisfy Newton's second law for the case where $\vec{a} = \vec{0}$. Thus

$$\sum F_x = T_{1x} + T_{2x} + (F_{\text{leg on pulley}})_x = ma_x = 0$$

The tension forces both have the same magnitude T, and both are at angle θ from horizontal. The x-component of the leg force is negative because it's directed to the left. Then Newton's law becomes

$$2T\cos\theta - F_{\text{leg on pulley}} = 0$$

so that

$$\cos\theta = \frac{F_{\text{leg on pulley}}}{2T} = \frac{50 \text{ N}}{82.4 \text{ N}} = 0.607$$

$$\theta = \cos^{-1}(0.607) = 53°$$

ASSESS The traction force would approach $2mg = 82$ N if angle θ approached zero because the two tensions would pull in parallel. Conversely, the traction force would approach 0 N if θ approached 90°. Because the desired traction force is roughly halfway between 0 N and 82 N, an angle near 45° is reasonable.

EXAMPLE 5.18 **Lifting a stage set**

A 200 kg set used in a play is stored in the loft above the stage. The rope holding the set passes up and over a pulley, then is tied backstage. The director tells a 100 kg stagehand to lower the set. When he unties the rope, the set falls and the unfortunate man is hoisted into the loft. What is the stagehand's acceleration?

PREPARE FIGURE 5.33 shows the visual overview. The objects of interest are the stagehand M and the set S, for which we've drawn separate free-body diagrams. Assume a massless rope and a massless, frictionless pulley. Tension forces $\vec{T}_S$ and $\vec{T}_M$ are due to a massless rope going over an ideal pulley, so their magnitudes are the same.

FIGURE 5.33 Visual overview for the stagehand and set.

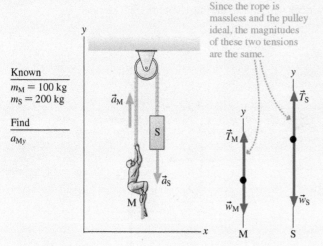

SOLVE From the two free-body diagrams, we can write Newton's second law in component form. For the man we have

$$\sum F_{My} = T_M - w_M = T_M - m_M g = m_M a_{My}$$

For the set we have

$$\sum F_{Sy} = T_S - w_S = T_S - m_S g = m_S a_{Sy}$$

Only the y-equations are needed. Because the stagehand and the set are connected by a rope, the upward distance traveled by one is the *same* as the downward distance traveled by the other. Thus the *magnitudes* of their accelerations must be the same, but, as Figure 5.33 shows, their *directions* are opposite. We can express this mathematically as $a_{Sy} = -a_{My}$. We also know that the two tension forces have equal magnitudes, which we'll call T. Inserting this information into the above equations gives

$$T - m_M g = m_M a_{My}$$
$$T - m_S g = -m_S a_{My}$$

These are simultaneous equations in the two unknowns T and a_{My}. We can solve for T in the first equation to get

$$T = m_M a_{My} + m_M g$$

Inserting this value of T into the second equation then gives

$$m_M a_{My} + m_M g - m_S g = -m_S a_{My}$$

which we can rewrite as

$$(m_S - m_M)g = (m_S + m_M)a_{My}$$

Finally, we can solve for the hapless stagehand's acceleration:

$$a_{My} = \frac{m_S - m_M}{m_S + m_M}g = \left(\frac{100 \text{ kg}}{300 \text{ kg}}\right) \times 9.80 \text{ m/s}^2 = 3.3 \text{ m/s}^2$$

This is also the acceleration with which the set falls. If the rope's tension was needed, we could now find it from $T = m_M a_{My} + m_M g$.

ASSESS If the stagehand weren't holding on, the set would fall with free-fall acceleration g. The stagehand acts as a *counterweight* to reduce the acceleration.

EXAMPLE 5.19 **A not-so-clever bank robbery**

Bank robbers have pushed a 1000 kg safe to a second-story floor-to-ceiling window. They plan to break the window, then lower the safe 3.0 m to their truck. Not being too clever, they stack up 500 kg of furniture, tie a rope between the safe and the furniture, and place the rope over a pulley. Then they push the safe out the window. What is the safe's speed when it hits the truck? The coefficient of kinetic friction between the furniture and the floor is 0.50.

PREPARE The visual overview in FIGURE 5.34 on the next page establishes a coordinate system and defines the symbols that will be needed to calculate the safe's motion. The objects of interest are the safe S and the furniture F, which we will model as particles. We will assume a massless rope and a massless, frictionless pulley. The tension is then the same everywhere in the rope.

SOLVE We can write Newton's second law directly from the free-body diagrams. For the furniture,

$$\sum F_{Fx} = T_F - f_k = T - f_k = m_F a_{Fx}$$
$$\sum F_{Fy} = n - w_F = n - m_F g = 0$$

And for the safe,

$$\sum F_{Sy} = T_S - w_S = T - m_S g = m_S a_{Sy}$$

The safe and the furniture are tied together, so their accelerations have the same magnitude. But as the furniture slides to the right with positive acceleration a_{Fx}, the safe falls in the negative y-direction, so its acceleration a_{Sy} is negative; we can express this

Continued

mathematically as $a_{Fx} = -a_{Sy}$. We also have made use of the fact that $T_S = T_F = T$. We have one additional piece of information, the model of kinetic friction:

$$f_k = \mu_k n = \mu_k m_F g$$

where we used the y-equation of the furniture to deduce that $n = m_F g$. Substitute this result for f_k into the x-equation of the furniture, then rewrite the furniture's x-equation and the safe's y-equation:

$$T - \mu_k m_F g = -m_F a_{Sy}$$

$$T - m_S g = m_S a_{Sy}$$

We have succeeded in reducing our knowledge to two simultaneous equations in the two unknowns a_{Sy} and T. We subtract the second equation from the first to eliminate T:

$$(m_S - \mu_k m_F)g = -(m_S + m_F)a_{Sy}$$

Finally, we can solve for the safe's acceleration:

$$a_{Sy} = -\left(\frac{m_S - \mu_k m_F}{m_S + m_F}\right)g$$

$$= -\frac{1000 \text{ kg} - 0.5(500 \text{ kg})}{1000 \text{ kg} + 500 \text{ kg}} \times 9.80 \text{ m/s}^2 = -4.9 \text{ m/s}^2$$

Now we need to calculate the kinematics of the falling safe. Because the time of the fall is not known or needed, we can use

$$(v_y)_f^2 = (v_y)_i^2 + 2a_{Sy}\,\Delta y = 0 + 2a_{Sy}(y_f - y_i) = -2a_{Sy}y_i$$

$$(v_y)_f = \sqrt{-2a_{Sy}y_i} = \sqrt{-2(-4.9 \text{ m/s}^2)(3.0 \text{ m})} = 5.4 \text{ m/s}$$

The value of $(v_y)_f$ is negative, but we only needed to find the speed, so we took the absolute value. It seems unlikely that the truck will survive the impact of the 1000 kg safe!

FIGURE 5.34 Visual overview of the furniture and falling safe.

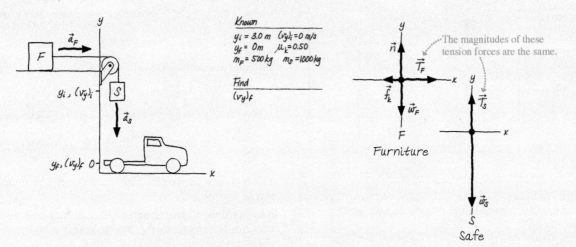

Newton's three laws form the cornerstone of the science of mechanics. These laws allowed scientists to understand many diverse phenomena, from the motion of a raindrop to the orbits of the planets. We will continue to develop Newtonian mechanics in the next few chapters because of its tremendous importance to the physics of everyday life. But it's worth keeping in the back of your mind that Newton's laws aren't the ultimate statement about motion. Later in this text we'll reexamine motion and mechanics from the perspective of Einstein's theory of relativity.

STOP TO THINK 5.8 All three 50 kg blocks are at rest. Is the tension in rope 2 greater than, less than, or equal to the tension in rope 1?

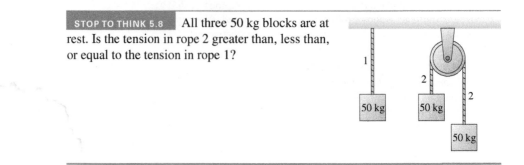

INTEGRATED EXAMPLE 5.20   **Stopping distances**

A 1500 kg car is traveling at a speed of 30 m/s when the driver slams on the brakes and skids to a halt. Determine the stopping distance if the car is traveling up a 10° slope, down a 10° slope, or on a level road.

PREPARE We'll represent the car as a particle and we'll use the model of kinetic friction. We want to solve the problem only once, not three separate times, so we'll leave the slope angle θ unspecified until the end.

FIGURE 5.35 shows the visual overview. We've shown the car sliding uphill, but these representations work equally well for a level or downhill slide if we let θ be zero or negative, respectively. We've used a tilted coordinate system so that the motion is along the x-axis. The car *skids* to a halt, so we've taken the coefficient of *kinetic* friction for rubber on concrete from Table 5.2.

SOLVE Newton's second law and the model of kinetic friction are

$$\sum F_x = n_x + w_x + (f_k)_x$$
$$= 0 - mg\sin\theta - f_k = ma_x$$
$$\sum F_y = n_y + w_y + (f_k)_y$$
$$= n - mg\cos\theta + 0 = ma_y = 0$$

We've written these equations by "reading" the motion diagram and the free-body diagram. Notice that both components of the weight vector $\vec{w}$ are negative. $a_y = 0$ because the motion is entirely along the x-axis.

The second equation gives $n = mg\cos\theta$. Using this in the friction model, we find $f_k = \mu_k mg\cos\theta$. Inserting this result back into the first equation then gives

$$ma_x = -mg\sin\theta - \mu_k mg\cos\theta$$
$$= -mg(\sin\theta + \mu_k\cos\theta)$$
$$a_x = -g(\sin\theta + \mu_k\cos\theta)$$

This is a constant acceleration. Constant-acceleration kinematics gives

$$(v_x)_f^2 = 0 = (v_x)_i^2 + 2a_x(x_f - x_i) = (v_x)_i^2 + 2a_x x_f$$

which we can solve for the stopping distance x_f:

$$x_f = -\frac{(v_x)_i^2}{2a_x} = \frac{(v_x)_i^2}{2g(\sin\theta + \mu_k\cos\theta)}$$

Notice how the minus sign in the expression for a_x canceled the minus sign in the expression for x_f. Evaluating our result at the three different angles gives the stopping distances:

$$x_f = \begin{cases} 48 \text{ m} & \theta = 10° & \text{uphill} \\ 57 \text{ m} & \theta = 0° & \text{level} \\ 75 \text{ m} & \theta = -10° & \text{downhill} \end{cases}$$

The implications are clear about the danger of driving downhill too fast!

ASSESS 30 m/s ≈ 60 mph and 57 m ≈ 180 feet on a level surface. These are similar to the stopping distances you learned when you got your driver's license, so the results seem reasonable. Additional confirmation comes from noting that the expression for a_x becomes $-g\sin\theta$ if $\mu_k = 0$. This is what you learned in Chapter 3 for the acceleration on a frictionless inclined plane.

FIGURE 5.35 Visual overview for a skidding car.

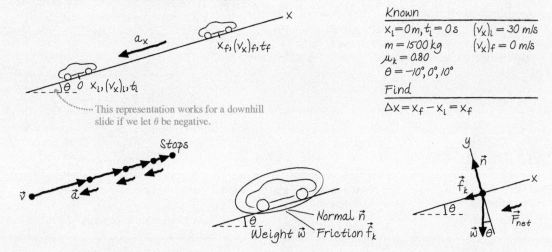

SUMMARY

Goal: To use Newton's laws to solve equilibrium and dynamics problems.

GENERAL STRATEGY

All examples in this chapter follow a three-part strategy. You'll become a better problem solver if you adhere to it as you do the homework problems. The *Dynamics Worksheets* in the *Student Workbook* will help you structure your work in this way.

Equilibrium Problems

Object at rest or moving at constant velocity.

PREPARE Make simplifying assumptions.

- Check that the object is either at rest or moving with constant velocity ($\vec{a} = \vec{0}$).
- Identify forces and show them on a free-body diagram.

SOLVE Use Newton's second law in component form:

$$\sum F_x = ma_x = 0$$
$$\sum F_y = ma_y = 0$$

"Read" the components from the free-body diagram.

ASSESS Is your result reasonable?

Dynamics Problems

Object accelerating.

PREPARE Make simplifying assumptions.
Make a **visual overview:**

- Sketch a pictorial representation.
- Identify known quantities and what the problem is trying to find.
- Identify all forces and show them on a free-body diagram.

SOLVE Use Newton's second law in component form:

$$\sum F_x = ma_x \quad \text{and} \quad \sum F_y = ma_y$$

"Read" the components of the vectors from the free-body diagram. If needed, use kinematics to find positions and velocities.

ASSESS Is your result reasonable?

IMPORTANT CONCEPTS

Specific information about three important forces:

Weight $\vec{w} = (mg, \text{downward})$

Friction $\vec{f}_s = (0 \text{ to } \mu_s n,$ direction as necessary to prevent motion)

$\vec{f}_k = (\mu_k n,$ direction opposite the motion)

$\vec{f}_r = (\mu_r n,$ direction opposite the motion)

Drag $\vec{D} = (\frac{1}{4}\rho A v^2,$ direction opposite the motion) for motion in air

Newton's laws are vector expressions. You must write them out by **components:**

$$(F_{net})_x = \sum F_x = ma_x$$
$$(F_{net})_y = \sum F_y = ma_y$$

For equilibrium problems, $a_x = 0$ and $a_y = 0$.

Objects in Contact

When two objects interact, you need to draw two separate free-body diagrams.

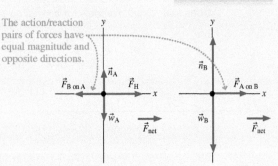

The action/reaction pairs of forces have equal magnitude and opposite directions.

APPLICATIONS

Apparent weight is the magnitude of the contact force supporting an object. It is what a scale would read, and it is your sensation of weight.

Apparent weight equals your true weight $w = mg$ only when the vertical acceleration is zero.

A falling object reaches **terminal speed** when the drag force exactly balances the weight force: $\vec{a} = \vec{0}$.

Strings and pulleys

- A string or rope pulls what it's connected to with a force equal to its tension.
- The tension in a rope is equal to the force pulling on the rope.
- The tension in a massless rope is the same at all points in the rope.
- Tension does not change when a rope passes over a massless, frictionless pulley.

$F_{\text{rope on wall}} = \text{tension}$

$F_{\text{hand on rope}} = \text{tension}$

Problem difficulty is labeled as | (straightforward) to |||| (challenging). Problems labeled INT integrate significant material from earlier chapters; BIO are of biological or medical interest.

For assigned homework and other learning materials, go to MasteringPhysics®

Scan this QR code to launch a Video Tutor Solution that will help you solve problems for this chapter.

QUESTIONS

Conceptual Questions

1. An object is subject to two forces that do not point in opposite directions. Is it possible to choose their magnitudes so that the object is in equilibrium? Explain.
2. Are the objects described here in static equilibrium, dynamic equilibrium, or not in equilibrium at all?
   a. A girder is lifted at constant speed by a crane.
   b. A girder is lowered by a crane. It is slowing down.
   c. You're straining to hold a 200 lb barbell over your head.
   d. A jet plane has reached its cruising speed and altitude.
   e. A rock is falling into the Grand Canyon.
   f. A box in the back of a truck doesn't slide as the truck stops.
3. What forces are acting on you right now? What net force is acting on you right now?
4. Decide whether each of the following is true or false. Give a reason!
   a. The mass of an object depends on its location.
   b. The weight of an object depends on its location.
   c. Mass and weight describe the same thing in different units.
5. An astronaut takes his bathroom scale to the moon and then stands on it. Is the reading of the scale his true weight? Explain.
6. A light block of mass m and a heavy block of mass M are attached to the ends of a rope. A student holds the heavier block and lets the lighter block hang below it, as shown in Figure Q5.6. Then she lets go. Air resistance can be neglected.
   a. What is the tension in the rope while the blocks are falling, before either hits the ground?
   b. Would your answer be different if she had been holding the lighter block initially?

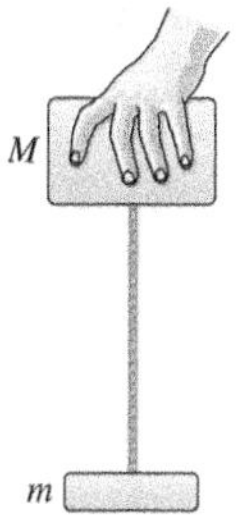

FIGURE Q5.6

7. a. Can the normal force on an object be directed horizontally? If not, why not? If so, provide an example.
   b. Can the normal force on an object be directed downward? If not, why not? If so, provide an example.
8. A ball is thrown straight up. Taking the drag force of air into account, does it take longer for the ball to travel to the top of its motion or for it to fall back down again?
9. You are going sledding with your friends, sliding down a snowy hill. Friction can't be ignored. Riding solo on your sled, you have a certain acceleration. Would the acceleration change if you let a friend ride with you, increasing the mass? Explain.
10. Suppose you are holding a box in front of you and away from your body by squeezing the sides, as shown in Figure Q5.10. Draw a free-body diagram showing all of the forces on the box. What is the force that is holding the box up, the force that is opposite the weight force?

FIGURE Q5.10

11. You are walking up an icy slope. Suddenly your feet slip, and you start to slide backward. Will you slide at a constant speed, or will you accelerate?
12. Three objects move through the air as shown in Figure Q5.12. Rank in order, from largest to smallest, the three drag forces D_1, D_2, and D_3. Some may be equal. Give your answer in the form $A < B = C$ and state your reasoning.

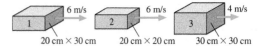
FIGURE Q5.12

13. A skydiver is falling at her terminal speed. Right after she opens her parachute, which has a very large area, what is the direction of the net force on her?
14. Raindrops can fall at different speeds; some fall quite quickly, others quite slowly. Why might this be true?
15. An airplane moves through the air at a constant speed. The engines' thrust applies a force in the direction of motion, and this force is equal in magnitude and opposite in direction to the drag force. Reducing thrust will cause the plane to fly at a slower—but still constant—speed. Explain why this is so.
16. Is it possible for an object to travel in air faster than its terminal speed? If not, why not? If so, explain how this might happen.

For Questions 17 through 20, determine the tension in the rope at the point indicated with a dot.

- All objects are at rest.
- The strings and pulleys are massless, and the pulleys are frictionless.

17.

5 kg

FIGURE Q5.17

18.

5 kg

5 kg

FIGURE Q5.18

19.

5 kg

5 kg

FIGURE Q5.19

20.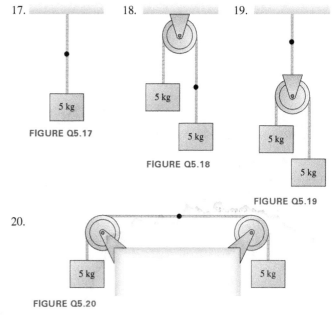

5 kg

5 kg

FIGURE Q5.20

21. In Figure Q5.21, block 2 is moving to the right. There is no friction between the floor and block 2, but there is a friction force between blocks 1 and 2. FIGURE Q5.21 In which direction is the kinetic friction force on block 1? On block 2? Explain.

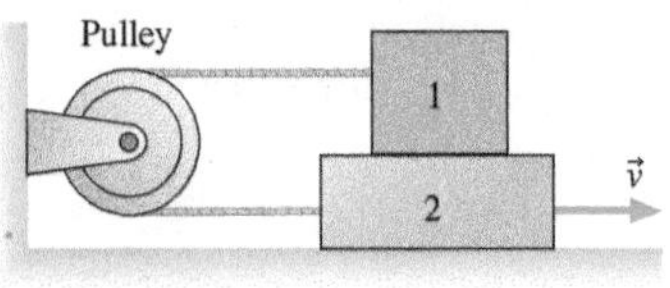

Pulley

1

2

$\vec{v}$

Multiple-Choice Questions

22. ‖ The wood block in Figure Q5.22 is at rest on a wood ramp. In which direction is the static friction force on block 1?
    A. Up the slope.
    B. Down the slope.
    C. The friction force is zero.
    D. There's not enough information to tell.

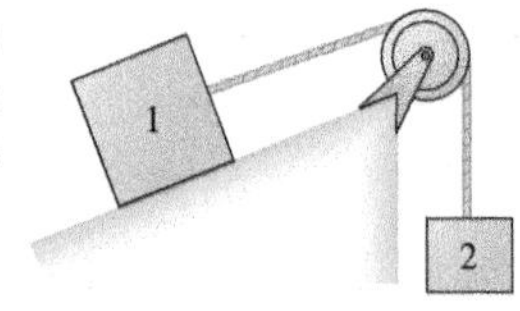

1

2

FIGURE Q5.22

23. ‖ A 2.0 kg ball is suspended by two light strings as shown in Figure Q5.23. What is the tension T in the angled string?

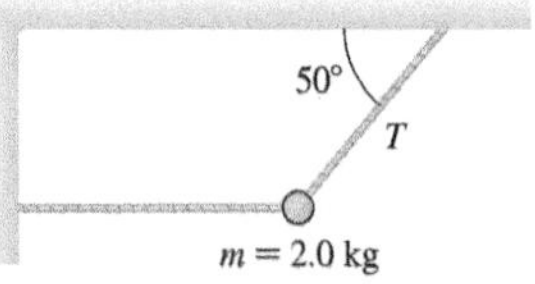

50°

T

$m = 2.0$ kg

FIGURE Q5.23

    A. 9.5 N   B. 15 N
    C. 20 N   D. 26 N   E. 30 N

24. | While standing in a low tunnel, you raise your arms and push against the ceiling with a force of 100 N. Your mass is 70 kg.
    a. What force does the ceiling exert on you?
       A. 10 N   B. 100 N   C. 690 N
       D. 790 N   E. 980 N
    b. What force does the floor exert on you?
       A. 10 N   B. 100 N   C. 690 N
       D. 790 N   E. 980 N

25. | A 5.0 kg dog sits on the floor of an elevator that is accelerating *downward* at 1.20 m/s².
    a. What is the magnitude of the normal force of the elevator floor on the dog?
       A. 34 N   B. 43 N   C. 49 N   D. 55 N   E. 74 N
    b. What is the magnitude of the force of the dog on the elevator floor?
       A. 4.2 N   B. 49 N   C. 55 N   D. 43 N   E. 74 N

26. | A 3.0 kg puck slides due east on a horizontal frictionless surface at a constant speed of 4.5 m/s. Then a force of magnitude 6.0 N, directed due north, is applied for 1.5 s. Afterward,
    a. What is the northward component of the puck's velocity?
       A. 0.50 m/s   B. 2.0 m/s   C. 3.0 m/s
       D. 4.0 m/s   E. 4.5 m/s
    b. What is the speed of the puck?
       A. 4.9 m/s   B. 5.4 m/s   C. 6.2 m/s
       D. 7.5 m/s   E. 11 m/s

27. | Eric has a mass of 60 kg. He is standing on a scale in an elevator that is accelerating downward at 1.7 m/s². What is the approximate reading on the scale?
    A. 0 N   B. 400 N   C. 500 N   D. 600 N

28. ‖ The two blocks in Figure Q5.28 are at rest on frictionless surfaces. What must be the mass of the right block in order that the two blocks remain stationary?
    A. 4.9 kg   B. 6.1 kg   C. 7.9 kg
    D. 9.8 kg   E. 12 kg

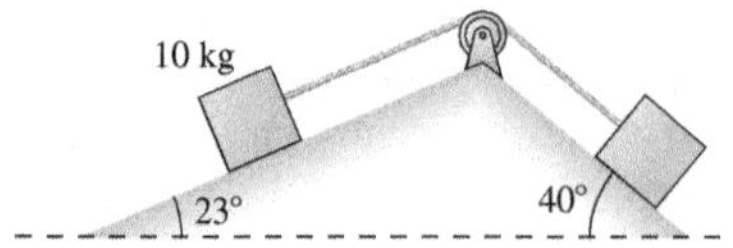

10 kg

23°   40°

FIGURE Q5.28

29. | A football player at practice pushes a 60 kg blocking sled across the field at a constant speed. The coefficient of kinetic friction between the grass and the sled is 0.30. How much force must he apply to the sled?
    A. 18 N   B. 60 N   C. 180 N   D. 600 N

30. | Two football players are pushing a 60 kg blocking sled across the field at a constant speed of 2.0 m/s. The coefficient of kinetic friction between the grass and the sled is 0.30. Once they stop pushing, how far will the sled slide before coming to rest?
    A. 0.20 m   B. 0.68 m   C. 1.0 m   D. 6.6 m

31. ‖ Land Rover ads used to claim that their vehicles could climb a slope of 45°. For this to be possible, what must be the minimum coefficient of static friction between the vehicle's tires and the road?
    A. 0.5   B. 0.7   C. 0.9   D. 1.0

32. ‖ A truck is traveling at 30 m/s on a slippery road. The driver slams on the brakes and the truck starts to skid. If the coefficient of kinetic friction between the tires and the road is 0.20, how far will the truck skid before stopping?
    A. 230 m   B. 300 m   C. 450 m   D. 680 m

PROBLEMS

Section 5.1 Equilibrium

1. | The three ropes in Figure P5.1 are tied to a small, very light ring. Two of the ropes are anchored to walls at right angles, and the third rope pulls as shown. What are T_1 and T_2, the magnitudes of the tension forces in the first two ropes?

2. ‖ The three ropes in Figure P5.2 are tied to a small, very light ring. Two of these ropes are anchored to walls at right angles with the tensions shown in the figure. What are the magnitude and direction of the tension $\vec{T}_3$ in the third rope?

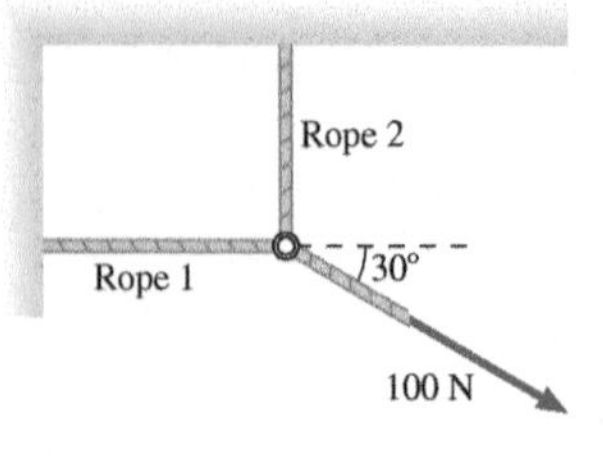

Rope 2

Rope 1

30°

100 N

FIGURE P5.1

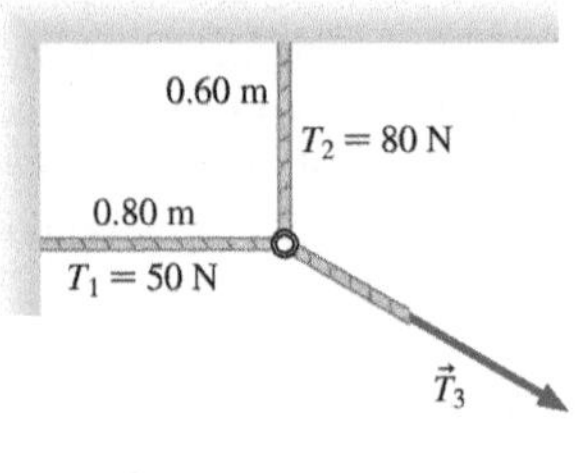

0.60 m

$T_2 = 80$ N

0.80 m

$T_1 = 50$ N

$\vec{T}_3$

FIGURE P5.2

3. ‖‖ A 20 kg loudspeaker is suspended 2.0 m below the ceiling by two cables that are each 30° from vertical. What is the tension in the cables?

4. ‖‖ A construction crew would like to support a 1000 kg steel beam with two angled ropes as shown in Figure P5.4. Their rope can support a maximum tension of 5600 N. Is this rope strong enough to do the job?

Rope 1 — 30° ¦ 30° — Rope 2

FIGURE P5.4

5. ‖‖ When you bend your knee, the
BIO quadriceps muscle is stretched. This increases the tension in the quadriceps tendon attached to your kneecap (patella), which, in turn, increases the tension in the patella tendon that attaches your kneecap to your lower leg bone (tibia). Simultaneously, the end of your upper leg bone (femur) pushes outward on the patella. Figure P5.5 shows how these parts of a knee joint are

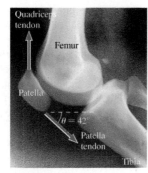

FIGURE P5.5

arranged. What size force does the femur exert on the kneecap if the tendons are oriented as in the figure and the tension in each tendon is 60 N?

6. ‖ An early submersible craft for deep-sea exploration was raised and lowered by a cable from a ship. When the craft was stationary, the tension in the cable was 6000 N. When the craft was lowered or raised at a steady rate, the motion through the water added an 1800 N drag force.
   a. What was the tension in the cable when the craft was being lowered to the seafloor?
   b. What was the tension in the cable when the craft was being raised from the seafloor?

7. ‖ The two angled ropes are used to support the crate in Figure P5.7. The tension in the ropes can have any value up to 1500 N. When the tension exceeds this value, the ropes will break. What is the largest mass the ropes can support?

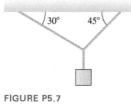

30°   45°

FIGURE P5.7

8. ‖ A 65 kg student is walking on a slackline, a length of webbing stretched between two trees. The line stretches and so has a noticeable sag, as shown in Figure P5.8. At the point where his foot touches the line, the rope applies a tension force in each direction, as shown. What is the tension in the line?

$\vec{T}$ $\vec{T}$
20°   20°

FIGURE P5.8

Section 5.2 Dynamics and Newton's Second Law

9. ‖ A force with x-component F_x acts on a 500 g object as it moves along the x-axis. The object's acceleration graph (a_x versus t) is shown in Figure P5.9. Draw a graph of F_x versus t.

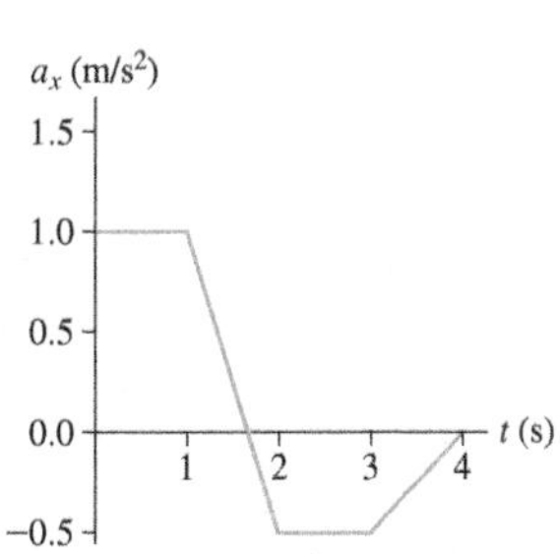

a_x (m/s²)

1.5
1.0
0.5
0.0 ———————— t (s)
      1 2 3 4
−0.5

FIGURE P5.9

10. ‖ The forces in Figure P5.10 are acting on a 2.0 kg object. What is a_x, the x-component of the object's acceleration?

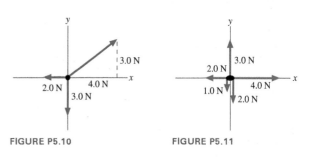

FIGURE P5.10   **FIGURE P5.11**

11. ‖ The forces in Figure P5.11 are acting on a 2.0 kg object. Find the values of a_x and a_y, the x- and y-components of the object's acceleration.

12. ‖ A horizontal rope is tied to a 50 kg box on frictionless ice. What is the tension in the rope if
   a. The box is at rest?
   b. The box moves at a steady 5.0 m/s?
   c. The box has $v_x = 5.0$ m/s and $a_x = 5.0$ m/s²?

13. ‖‖ A crate pushed along the floor with velocity $\vec{v}_i$ slides a distance d after the pushing force is removed.
   a. If the mass of the crate is doubled but the initial velocity is not changed, what distance does the crate slide before stopping? Explain.
   b. If the initial velocity of the crate is doubled to $2\vec{v}_i$ but the mass is not changed, what distance does the crate slide before stopping? Explain.

14. ‖ In a head-on collision, a car stops in 0.10 s from a speed of 14 m/s. The driver has a mass of 70 kg, and is, fortunately, tightly strapped into his seat. What force is applied to the driver by his seat belt during that fraction of a second?

Section 5.3 Mass and Weight

15. ‖ An astronaut's weight on earth is 800 N. What is his weight on Mars, where $g = 3.76$ m/s²?

16. ‖ A woman has a mass of 55.0 kg.
   a. What is her weight on earth?
   b. What are her mass and her weight on the moon, where $g = 1.62$ m/s²?

17. ‖‖ A 75 kg passenger is seated in a cage in the Sling Shot, a carnival ride. Giant bungee cords are stretched as the cage is pulled down, and then they rebound to launch the cage straight up. After the rapid launch, the cords go slack and the cage moves under the influence of gravity alone. What is the rider's apparent weight after the cords have gone slack and the cage is moving upward?

18. ‖ a. How much force does an 80 kg astronaut exert on his chair while sitting at rest on the launch pad?
   b. How much force does the astronaut exert on his chair while accelerating straight up at 10 m/s²?

19. ‖ It takes the elevator in a skyscraper 4.0 s to reach its cruising speed of 10 m/s. A 60 kg passenger gets aboard on the ground floor. What is the passenger's apparent weight
   a. Before the elevator starts moving?
   b. While the elevator is speeding up?
   c. After the elevator reaches its cruising speed?

20. ‖ Riders on the Power Tower are launched skyward with an acceleration of $4g$, after which they experience a period of free fall. What is a 60 kg rider's apparent weight
    a. During the launch?
    b. During the period of free fall?

21. ‖ Zach, whose mass is 80 kg, is in an elevator descending at 10 m/s. The elevator takes 3.0 s to brake to a stop at the first floor.
    a. What is Zach's apparent weight before the elevator starts braking?
    b. What is Zach's apparent weight while the elevator is braking?

22. ‖ A kangaroo carries her 0.51 kg baby in her pouch as she
    BIO bounds across the ground. As she pushes off the ground, she is accelerating upward at 30 m/s². What is the apparent weight of her baby at this instant? By what factor does this exceed her baby's actual weight?

23. ‖‖ Figure P5.23 shows the
    INT velocity graph of a 75 kg passenger in an elevator. What is the passenger's apparent weight at $t = 1.0$ s? At 5.0 s? At 9.0 s?

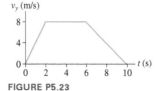

FIGURE P5.23

Section 5.4 Normal Forces

24. ‖ a. A 0.60 kg bullfrog is sitting at rest on a level log. How large is the normal force of the log on the frog?
    b. A second 0.60 kg bullfrog is on a log tilted 30° above horizontal. How large is the normal force of the log on this frog?

25. ‖‖ A 23 kg child goes down a straight slide inclined 38° above horizontal. The child is acted on by his weight, the normal force from the slide, and kinetic friction.
    a. Draw a free-body diagram of the child.
    b. How large is the normal force of the slide on the child?

Section 5.5 Friction

26. ‖ Two workers are sliding a 300 kg crate across the floor. One worker pushes forward on the crate with a force of 380 N while the other pulls in the same direction with a force of 350 N using a rope connected to the crate. Both forces are horizontal, and the crate slides with a constant speed. What is the crate's coefficient of kinetic friction on the floor?

27. ‖‖ A 4000 kg truck is parked on a 7.0° slope. How big is the friction force on the truck?

28. ‖‖ A 1000 kg car traveling at a speed of 40 m/s skids to a halt on wet concrete where $\mu_k = 0.60$. How long are the skid marks?

29. ‖ A stubborn 120 kg pig sits down and refuses to move. To drag the pig to the barn, the exasperated farmer ties a rope around the pig and pulls with his maximum force of 800 N. The coefficients of friction between the pig and the ground are $\mu_s = 0.80$ and $\mu_k = 0.50$. Is the farmer able to move the pig?

30. ‖ It is friction that provides the force for a car to accelerate, so for high-performance cars the factor that limits acceleration isn't the *engine*; it's the *tires*. For typical rubber-on-concrete friction, what is the shortest time in which a car could accelerate from 0 to 60 mph?

31. ‖‖‖ A 10 kg crate is placed on a horizontal conveyor belt. The materials are such that $\mu_s = 0.50$ and $\mu_k = 0.30$.
    a. Draw a free-body diagram showing all the forces on the crate if the conveyer belt runs at constant speed.
    b. Draw a free-body diagram showing all the forces on the crate if the conveyer belt is speeding up.
    c. What is the maximum acceleration the belt can have without the crate slipping?
    d. If the acceleration of the belt exceeds the value determined in part c, what is the acceleration of the crate?

32. ‖ The rolling resistance for steel on steel is quite low; the coefficient of rolling friction is typically $\mu_r = 0.002$. Suppose a 180,000 kg locomotive is rolling at 10 m/s (just over 20 mph) on level rails. If the engineer disengages the engine, how much time will it take the locomotive to coast to a stop? How far will the locomotive move during this time?

33. ‖ What is the minimum downward force on the box in Figure P5.33 that will keep it from slipping? The coefficients of static and kinetic friction between the box and the floor are 0.35 and 0.25, respectively.

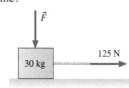

FIGURE P5.33

Section 5.6 Drag

34. ‖ What is the drag force on a 1.6-m-wide, 1.4-m-high car traveling at
    a. 10 m/s ($\approx$ 22 mph)? b. 30 m/s ($\approx$ 65 mph)?

35. ‖‖‖ A 22-cm-diameter bowling ball has a terminal speed of 77 m/s. What is the ball's mass?

36. ‖ Running on a treadmill is slightly easier than running outside
    BIO because there is no drag force to work against. Suppose a 60 kg runner completes a 5.0 km race in 18 minutes. Use the cross-section area estimate of Example 5.14 to determine the drag force on the runner during the race. What is this force as a fraction of the runner's weight?

37. ‖‖‖‖ A 75 kg skydiver can be modeled as a rectangular "box" with dimensions 20 cm × 40 cm × 1.8 m. What is his terminal speed if he falls feet first?

38. ‖ The air is less dense at higher elevations, so skydivers reach a high terminal speed. The highest recorded speed for a skydiver was achieved in a jump from a height of 39,000 m. At this elevation, the density of the air is only 4.3% of the surface density. Use the data from Example 5.14 to estimate the terminal speed of a skydiver at this elevation.

Section 5.7 Interacting Objects

39. ‖‖‖ A 1000 kg car pushes a 2000 kg truck that has a dead battery. When the driver steps on the accelerator, the drive wheels of the car push backward against the ground with a force of 4500 N.
    a. What is the magnitude of the force of the car on the truck?
    b. What is the magnitude of the force of the truck on the car?

40. ‖ A 2200 kg truck has put its front bumper against the rear bumper of a 2400 kg SUV to give it a push. With the engine at full power and good tires on good pavement, the maximum forward force on the truck is 18,000 N.
    a. What is the maximum possible acceleration the truck can give the SUV?
    b. At this acceleration, what is the force of the SUV's bumper on the truck's bumper?

41. ‖‖‖‖ Blocks with masses of 1.0 kg, 2.0 kg, and 3.0 kg are lined up in a row on a frictionless table. All three are pushed forward by a 12 N force applied to the 1.0 kg block. How much force does the 2.0 kg block exert on (a) the 3.0 kg block and (b) the 1.0 kg block?

Section 5.8 Ropes and Pulleys

42. ‖ What is the tension in the rope of Figure P5.42?

43. ‖ A 2.0-m-long, 500 g rope pulls a 10 kg block of ice across a horizontal, frictionless surface. The block accelerates at 2.0 m/s². How much force pulls forward on (a) the block of ice, (b) the rope?

44. ‖ Figure P5.44 shows two 1.00 kg blocks connected by a rope. A second rope hangs beneath the lower block. Both ropes have a mass of 250 g. The entire assembly is accelerated upward at 3.00 m/s² by force $\vec{F}$.
    a. What is F?
    b. What is the tension at the top end of rope 1?
    c. What is the tension at the bottom end of rope 1?
    d. What is the tension at the top end of rope 2?

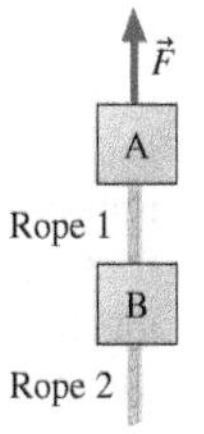

45. ‖ Each of 100 identical blocks sitting on a frictionless surface is connected to the next block by a massless string. The first block is pulled with a force of 100 N.
    a. What is the tension in the string connecting block 100 to block 99?
    b. What is the tension in the string connecting block 50 to block 51?

46. ‖ Two blocks on a frictionless table, A and B, are connected by a massless string. When block A is pulled with a certain force, dragging block B, the tension in the string is 24 N. When block B is pulled by the same force, dragging block A, the tension is 18 N. What is the ratio m_A/m_B of the blocks' masses?

47. ‖ A 500 kg piano is being lowered into position by a crane while two people steady it with ropes pulling to the sides. Bob's rope pulls to the left, 15° below horizontal, with 500 N of tension. Ellen's rope pulls toward the right, 25° below horizontal.
    a. What tension must Ellen maintain in her rope to keep the piano descending vertically at constant speed?
    b. What is the tension in the vertical main cable supporting the piano?

General Problems

48. ‖ Dana has a sports medal suspended by a long ribbon from her rearview mirror. As she accelerates onto the highway, she notices that the medal is hanging at an angle of 10° from the vertical.
    a. Does the medal lean toward or away from the windshield? Explain.
    b. What is her acceleration?

49. ‖ INT Figure P5.49 shows the velocity graph of a 2.0 kg object as it moves along the x-axis. What is the net force acting on this object at $t = 1$ s? At 4 s? At 7 s?

FIGURE P5.49

50. | BIO Your forehead can withstand a force of about 6.0 kN before fracturing, while your cheekbone can only withstand about 1.3 kN.
    a. If a 140 g baseball strikes your head at 30 m/s and stops in 0.0015 s, what is the magnitude of the ball's acceleration?
    b. What is the magnitude of the force that stops the baseball?
    c. What force does the baseball apply to your head? Explain.
    d. Are you in danger of a fracture if the ball hits you in the forehead? In the cheek?

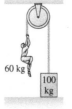

60 kg   100 kg

FIGURE P5.42

$\vec{F}$

A

Rope 1

B

Rope 2

FIGURE P5.44

51. ‖ A 50 kg box hangs from a rope. What is the tension in the rope if
    a. The box is at rest?
    b. The box has $v_y = 5.0$ m/s and is speeding up at 5.0 m/s²?

52. ‖ INT A fisherman has caught a very large, 5.0 kg fish from a dock that is 2.0 m above the water. He is using lightweight fishing line that will break under a tension of 54 N or more. He is eager to get the fish to the dock in the shortest possible time. If the fish is at rest at the water's surface, what's the least amount of time in which the fisherman can raise the fish to the dock without losing it?

53. ‖ A 50 kg box hangs from a rope. What is the tension in the rope if
    a. The box moves up at a steady 5.0 m/s?
    b. The box has $v_y = 5.0$ m/s and is slowing down at 5.0 m/s²?

54. ‖ Riders on the Tower of Doom, an amusement park ride, experience 2.0 s of free fall, after which they are slowed to a stop in 0.50 s. What is a 65 kg rider's apparent weight as the ride is coming to rest? By what factor does this exceed her actual weight?

55. ‖ BIO Seat belts and air bags save lives by reducing the forces exerted on the driver and passengers in an automobile collision. Cars are designed with a "crumple zone" in the front of the car. In the event of an impact, the passenger compartment decelerates over a distance of about 1 m as the front of the car crumples. An occupant restrained by seat belts and air bags decelerates with the car. By contrast, an unrestrained occupant keeps moving forward with no loss of speed (Newton's first law!) until hitting the dashboard or windshield, as we saw in Figure 4.2. These are unyielding surfaces, and the unfortunate occupant then decelerates over a distance of only about 5 mm.
    a. A 60 kg person is in a head-on collision. The car's speed at impact is 15 m/s. Estimate the net force on the person if he or she is wearing a seat belt and if the air bag deploys.
    b. Estimate the net force that ultimately stops the person if he or she is not restrained by a seat belt or air bag.
    c. How do these two forces compare to the person's weight?

56. ‖ INT Elite quarterbacks can throw a football 70 m. To achieve such a throw, a quarterback accelerates the 0.42 kg ball over a distance of approximately 1.0 m.
    a. If you assume that the quarterback throws the ball at the optimal angle, at what speed does he throw the ball?
    b. What force must he apply to the ball to throw it at this speed?

57. ‖ A 20,000 kg rocket has a rocket motor that generates 3.0×10^5 N of thrust.
    a. What is the rocket's initial upward acceleration?
    b. At an altitude of 5.0 km the rocket's acceleration has increased to 6.0 m/s². What mass of fuel has it burned?

58. ‖ You've always wondered about the acceleration of the elevators in the 101-story-tall Empire State Building. One day, while visiting New York, you take your bathroom scale into the elevator and stand on it. The scale reads 150 lb as the door closes. The reading varies between 120 lb and 170 lb as the elevator travels 101 floors.
    a. What is the magnitude of the acceleration as the elevator starts upward?
    b. What is the magnitude of the acceleration as the elevator brakes to a stop?

59. |||| A 23 kg child goes down a straight slide inclined 38° above horizontal. The child is acted on by his weight, the normal force from the slide, kinetic friction, and a horizontal rope exerting a 30 N force as shown in Figure P5.59. How large is the normal force of the slide on the child?

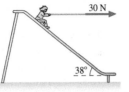

FIGURE P5.59

60. ||| An impala is an African antelope capable of a remarkable
BIO vertical leap. In one recorded leap, a 45 kg impala went into a
INT deep crouch, pushed straight up for 0.21 s, and reached a height of 2.5 m above the ground. To achieve this vertical leap, with what force did the impala push down on the ground? What is the ratio of this force to the antelope's weight?

61. || Josh starts his sled at the top of a 3.0-m-high hill that has a constant slope of 25°. After reaching the bottom, he slides across a horizontal patch of snow. Ignore friction on the hill, but assume that the coefficient of kinetic friction between his sled and the horizontal patch of snow is 0.050. How far from the base of the hill does he end up?

62. || The drag force is an impor-
BIO tant fact of life for the small
INT marine crustaceans called copepods. The drag force for small objects in water is very different from the drag force in air, so the drag equations of this chapter don't apply. However, we can estimate the drag

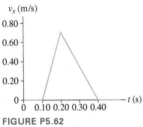

FIGURE P5.62

force from data. The velocity graph of Figure P5.62 shows two phases of the swimming motion of a 1.8 mg copepod. First it swims vigorously, speeding up; then it slows down under the influence of drag. What is the magnitude of the drag force? What is the ratio of the drag force to the copepod's weight?

63. || A wood block, after being given a starting push, slides down a wood ramp at a constant speed. What is the angle of the ramp above horizontal?

64. || Researchers often use *force plates* to measure the forces that
BIO people exert against the floor during movement. A force plate
INT works like a bathroom scale, but it keeps a record of how the reading changes with time. Figure P5.64 shows the data from a force plate as a woman jumps straight up and then lands.
   a. What was the vertical component of her acceleration during push-off?
   b. What was the vertical component of her acceleration while in the air?
   c. What was the vertical component of her acceleration during the landing?
   d. What was her speed as her feet left the force plate?
   e. How high did she jump?

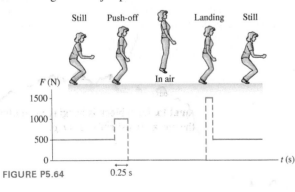

FIGURE P5.64

65. |||| A person with compromised pinch strength in his fingers can only exert a normal force of 6.0 N to either side of a pinch-held object, such as the book shown in Figure P5.65. What is the greatest mass book he can hold onto vertically before it slips out of his fingers? The coefficient of static friction of the surface between the fingers and the book cover is 0.80.

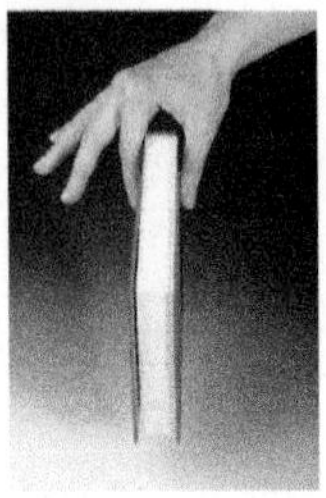

FIGURE P5.65

66. || It's possible for a determined
INT group of people to pull an aircraft. Drag is negligible at low speeds, and the only force impeding motion is the rolling friction of the rubber tires on the concrete runway. In 2000, a team of 60 British police offi-

cers set a world record by pulling a Boeing 747, with a mass of 200,000 kg, a distance of 100 m in 53 s. The plane started at rest. Estimate the force with which each officer pulled on the plane, assuming constant pulling force and constant acceleration.

67. ||| A 1.0 kg wood block is pressed against a vertical wood wall by a 12 N force as shown in Figure P5.67. If the block is initially at rest, will it move upward, move downward, or stay at rest?

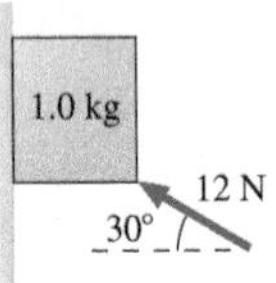

FIGURE P5.67

68. |||| A 2.0 kg wood block is launched up a wooden ramp that is inclined at a 35° angle. The block's initial speed is 10 m/s.
   a. What vertical height does the block reach above its starting point?
   b. What speed does it have when it slides back down to its starting point?

69. ||| Two blocks are at rest on a frictionless incline, as shown in Figure P5.69. What are the tensions in the two strings?

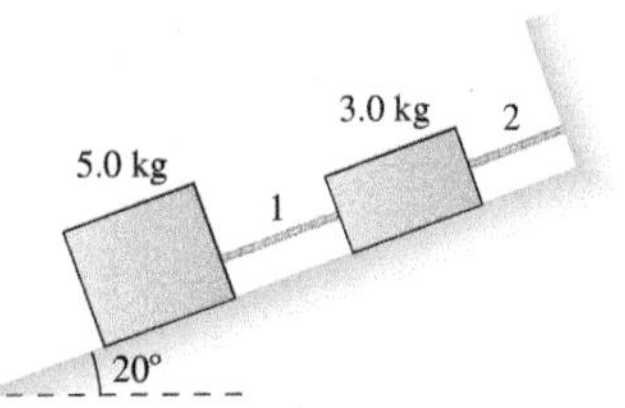

FIGURE P5.69

70. || Running indoors on
BIO a treadmill is slightly easier than running outside because you aren't moving through the air and there is no drag force to oppose your motion. A 60 kg man is running at 4.5 m/s on an indoor treadmill. To experience the same intensity workout as he'd get outdoors, he tilts the treadmill at a slight angle, choosing the angle so that the component of his weight force down the ramp is the same as the missing drag force. What is the necessary angle? Use the cross-section area estimate of Example 5.14 to compute the drag force.

71. || Two identical 2.0 kg blocks are stacked as shown in Figure P5.71. The bottom block is free to slide on a frictionless surface. The coefficient of static friction between the blocks is 0.35. What is the maximum horizontal force that can be applied to the lower block without the upper block slipping?

FIGURE P5.71

72. ‖ A wood block is sliding up a wood ramp. If the angle of the ramp is very steep, the block will reverse direction at some point and slide back down. If the angle of the ramp is shallow, the block will stop when it reaches the highest point of its motion. What is the smallest ramp angle, measured from the horizontal, for which the block will slide back down after reaching its highest point?

73. ‖ A 2.7 g Ping-Pong ball has a diameter of 4.0 cm.
    a. The ball is shot straight up at twice its terminal speed. What is its acceleration immediately after launch?
    b. The ball is shot straight down at twice its terminal speed. What is its acceleration immediately after launch?

74. ‖‖ Two blocks are connected by a string as in Figure P5.74. What is the upper block's acceleration if the coefficient of kinetic friction between the block and the table is 0.20?

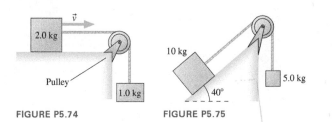

FIGURE P5.74   FIGURE P5.75

75. ‖‖ The ramp in Figure P5.75 is friction-less. If the blocks are released from rest, which way does the 10 kg block slide, and what is the magnitude of its acceleration?

76. ‖ The 100 kg block in Figure P5.76 takes
    [NT] 6.0 s to reach the floor after being released from rest. What is the mass of the block on the left?

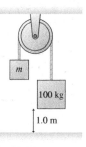

FIGURE P5.76

MCAT-Style Passage Problems

Sliding on the Ice

In the winter sport of curling, players give a 20 kg stone a push across a sheet of ice. The stone moves approximately 40 m before coming to rest. The final position of the stone, in principle, only depends on the initial speed at which it is launched and the force of friction between the ice and the stone, but team members can use brooms to sweep the ice in front of the stone to adjust its speed and trajectory a bit; they must do this without touching the stone. Judicious sweeping can lengthen the travel of the stone by 3 m.

77. | A curler pushes a stone to a speed of 3.0 m/s over a time of 2.0 s. Ignoring the force of friction, how much force must the curler apply to the stone to bring it up to speed?
    A. 3.0 N   B. 15 N   C. 30 N   D. 150 N

78. | The sweepers in a curling competition adjust the trajectory of the stone by
    A. Decreasing the coefficient of friction between the stone and the ice.
    B. Increasing the coefficient of friction between the stone and the ice.
    C. Changing friction from kinetic to static.
    D. Changing friction from static to kinetic.

79. | Suppose the stone is launched with a speed of 3 m/s and travels 40 m before coming to rest. What is the *approximate* magnitude of the friction force on the stone?
    A. 0 N   B. 2 N   C. 20 N   D. 200 N

80. | Suppose the stone's mass is increased to 40 kg, but it is launched at the same 3 m/s. Which one of the following is true?
    A. The stone would now travel a longer distance before coming to rest.
    B. The stone would now travel a shorter distance before coming to rest.
    C. The coefficient of friction would now be greater.
    D. The force of friction would now be greater.

STOP TO THINK ANSWERS

Chapter Preview Stop to Think: B. The elevator is moving at a steady speed, so $a = 0$ and therefore $F_{net} = 0$. There are two forces that act: the downward weight force and the upward tension force. Since $F_{net} = 0$, the magnitudes of these two forces must be equal.

Stop to Think 5.1: A. The ball is stationary, so the net force on it must be zero. Newton's second law for the vertical components of forces is $\sum F_y = T_y - w = 0$. The vertical component of the tension by itself is equal to w, so the tension must be greater than w.

Stop to Think 5.2: A. The lander is descending and slowing. The acceleration vector points upward, and so $\vec{F}_{net}$ points upward. This can be true only if the thrust has a larger magnitude than the weight.

Stop to Think 5.3: D. When you are in the air, there is no contact force supporting you, so your apparent weight is zero: You are weightless.

Stop to Think 5.4: C. The cyclist moves at a constant speed, so the net force is zero. We set the axes parallel to and perpendicular to the slope and compute the components of the weight force as in Figure 5.11. Newton's second law for the vertical components of forces is $\sum F_y = n - w \cos 20° = 0$. For this to be true, it's clear that $n < w$.

Stop to Think 5.5: $f_B > f_C = f_D = f_E > f_A$. Situations C, D, and E are all kinetic friction, which does not depend on either velocity or acceleration. Kinetic friction is less than the maximum static friction that is exerted in B. $f_A = 0$ because no friction is needed to keep the object at rest.

Stop to Think 5.6: D. The ball is shot *down* at 30 m/s, so $v_{0y} = -30$ m/s. This exceeds the terminal speed, so the upward drag force is *greater* than the downward weight force. Thus the ball *slows down* even though it is "falling." It will slow until $v_y = -15$ m/s, the terminal velocity, then maintain that velocity.

Stop to Think 5.7: B. $F_{Q\,on\,H} = F_{H\,on\,Q}$ and $F_{P\,on\,Q} = F_{Q\,on\,P}$ because these are action/reaction pairs. Box Q is slowing down and therefore must have a net force to the left. So from Newton's second law we also know that $F_{H\,on\,Q} > F_{P\,on\,Q}$.

Stop to Think 5.8: Equal to. Each block is hanging in equilibrium, with no net force, so the upward tension force is mg.

6 Circular Motion, Orbits, and Gravity

The horses are rounding a corner on a snowy race course. Why do they lean into the turn this way?

LOOKING AHEAD ➤

Goal: To learn about motion in a circle, including orbital motion under the influence of a gravitational force.

Circular Motion

An object moving in a circle has an acceleration toward the center, so there must be a net force toward the center as well.

How much force does it take to swing the girl in a circle? You'll learn how to solve such problems.

Apparent Forces

The riders feel pushed out. This isn't a real force, though it is often called centrifugal force; it's an **apparent force.**

This apparent force makes the riders "feel heavy." You'll learn to calculate their apparent weight.

Gravity and Orbits

The space station appears to float in space, but gravity is pulling down on it quite forcefully.

You'll learn **Newton's law of gravity**, and you'll see how the force of gravity keeps the station in orbit.

LOOKING BACK ◄

Centripetal Acceleration

In Section 3.8, you learned that an object moving in a circle at a constant speed experiences an acceleration directed toward the center of the circle.

In this chapter, you'll learn how to extend Newton's second law, which relates acceleration to the forces that cause it, to this type of acceleration.

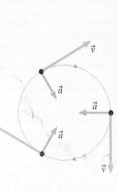

STOP TO THINK

A softball pitcher is throwing a pitch. At the instant shown, the ball is moving in a circular arc at a steady speed. At this instant, the acceleration is

A. Directed up.   B. Directed down.
C. Directed left.   D. Directed right.
E. Zero.

6.1 Uniform Circular Motion

The riders on this carnival ride are going in a circle at a constant speed, a type of motion that we've called uniform circular motion. We saw in ◀ SECTION 3.8 that uniform circular motion requires an acceleration directed toward the center of the circle. This means that there must be a force directed toward the center of the circle. That's the job of the cables, which provide a tension force directed toward the hub of the ride. This force keeps the riders moving in a circle.

In this chapter, we'll look at objects moving in circles or circular arcs. We'll consider the details of the acceleration and the forces that provide this acceleration, combining our discussion and description of motion from Chapters 1–3 with our treatment of Newton's laws and dynamics from Chapters 4 and 5. For now, we consider objects that move at a *constant* speed; circular motion with *changing* speed will wait until Chapter 7.

Uniform circular motion at the fair.

Velocity and Acceleration in Uniform Circular Motion

Although the *speed* of a particle in uniform circular motion is constant, its *velocity* is not constant because the *direction* of the motion is always changing. FIGURE 6.1 reminds you of the details: There is an acceleration at every point in the motion, with the acceleration vector $\vec{a}$ pointing toward the center of the circle. We called this the *centripetal acceleration,* and we showed that for uniform circular motion the acceleration is given by $a = v^2/r$:

$$a = \frac{v^2}{r} \qquad (6.1)$$

Centripetal acceleration for uniform circular motion

QUADRATIC

Acceleration depends on speed but also on distance from the center of the circle.

FIGURE 6.1 Velocity and acceleration for uniform circular motion.

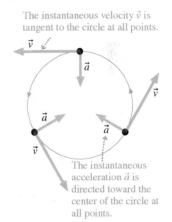

The instantaneous velocity $\vec{v}$ is tangent to the circle at all points.

The instantaneous acceleration $\vec{a}$ is directed toward the center of the circle at all points.

CONCEPTUAL EXAMPLE 6.1 **Rounding a corner**

A car is turning a tight corner at a constant speed. A top view of the motion is shown in FIGURE 6.2. The velocity vector for the car points to the east at the instant shown. What is the direction of the acceleration?

FIGURE 6.2 Top view of a car turning a corner.

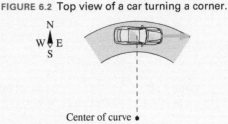

Center of curve

REASON The curve that the car is following is a segment of a circle, so this is an example of uniform circular motion. For uniform circular motion, the acceleration is directed toward the center of the circle, which is to the south.

ASSESS This acceleration is due to a change in direction, not a change in speed. And this matches your experience in a car: If you turn the wheel to the right—as the driver of this car is doing—your car then *changes* its motion toward the right, in the direction of the center of the circle.

NOTE ▶ In the example above, the car is following a curve that is only a *segment* of a circle, not a *full* circle. At the instant shown, though, the motion follows a circular arc. You can have uniform circular motion without completing a full circle. ◀

Period, Frequency, and Speed

It's not necessary to have a full circle to have uniform circular motion, but in most of the cases we'll consider, objects will complete multiple full circles of motion, one after another. Since the motion is uniform, each time around the circle is just a repeat of the one before. The motion is **periodic.**

The time interval it takes an object to go around a circle one time, completing one revolution (abbreviated rev), is called the **period** of the motion. Period is represented by the symbol T.

Rather than specify the time for one revolution, we can specify circular motion by its **frequency,** the number of revolutions per second, for which we use the symbol f. An object with a period of one-half second completes 2 revolutions each second. Similarly, an object can make 10 revolutions in 1 s if its period is one-tenth of a second. This shows that frequency is the inverse of the period:

$$f = \frac{1}{T} \tag{6.2}$$

Although frequency is often expressed as "revolutions per second," *revolutions* are not true units but merely the counting of events. Thus the SI unit of frequency is simply inverse seconds, or s^{-1}. Frequency may also be given in revolutions per minute (rpm) or another time interval, but these usually need to be converted to s^{-1} before calculations are done.

FIGURE 6.3 shows an object moving at a constant speed in a circular path of radius r. We know the time for one revolution—one period T—and we know the distance traveled, so we can write an equation relating the period, the radius, and the speed:

$$v = \frac{2\pi r}{T} \tag{6.3}$$

Given Equation 6.2 relating frequency and period, we can also write this equation as

$$v = 2\pi f r \tag{6.4}$$

We can combine this equation with Equation 6.1 for acceleration to get an expression for the centripetal acceleration in terms of the frequency or the period for the circular motion:

$$a = \frac{v^2}{r} = (2\pi f)^2 r = \left(\frac{2\pi}{T}\right)^2 r \tag{6.5}$$

FIGURE 6.3 Relating frequency and speed.

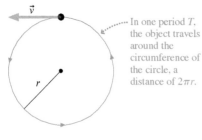

In one period T, the object travels around the circumference of the circle, a distance of $2\pi r$.

EXAMPLE 6.2 **Spinning some tunes**

An audio CD has a diameter of 120 mm and spins at up to 540 rpm. When a CD is spinning at its maximum rate, how much time is required for one revolution? If a speck of dust rides on the outside edge of the disk, how fast is it moving? What is the acceleration?

PREPARE Before we get started, we need to do some unit conversions. The diameter of a CD is given as 120 mm, which is 0.12 m. The radius is 0.060 m. The frequency is given in rpm; we need to convert this to s^{-1}:

$$f = 540 \, \frac{\text{rev}}{\text{min}} \times \frac{1 \, \text{min}}{60 \, \text{s}} = 9.0 \, \frac{\text{rev}}{\text{s}} = 9.0 \, \text{s}^{-1}$$

SOLVE The time for one revolution is the period. This is given by Equation 6.2:

$$T = \frac{1}{f} = \frac{1}{9.0 \, \text{s}^{-1}} = 0.11 \, \text{s}$$

The dust speck is moving in a circle of radius 0.0060 m at a frequency of $9.0 \, \text{s}^{-1}$. We can use Equation 6.4 to find the speed:

$$v = 2\pi f r = 2\pi(9.0 \, \text{s}^{-1})(0.060 \, \text{m}) = 3.4 \, \text{m/s}$$

We can then use Equation 6.5 to find the acceleration:

$$a = (2\pi f)^2 r = (2\pi(9.0 \, \text{s}^{-1}))^2(0.060 \, \text{m}) = 190 \, \text{m/s}^2$$

ASSESS If you've watched a CD spin, you know that it takes much less than a second to go around, so the value for the period seems reasonable. The speed we calculate for the dust speck is nearly 8 mph, but for a point on the edge of the CD to go around so many times in a second, it must be moving pretty fast. And we'd expect that such a high speed in a small circle would lead to a very large acceleration.

EXAMPLE 6.3 **Finding the period of a carnival ride**

In the Quasar carnival ride, passengers travel in a horizontal 5.0-m-radius circle. For safe operation, the maximum sustained acceleration that riders may experience is 20 m/s², approximately twice the free-fall acceleration. What is the period of the ride when it is being operated at the maximum acceleration? How fast are the riders moving when the ride is operated at this period?

PREPARE We will assume that the cars on the ride are in uniform circular motion. The visual overview of **FIGURE 6.4** shows a top view of the motion of the ride.

FIGURE 6.4 Visual overview for the Quasar carnival ride.

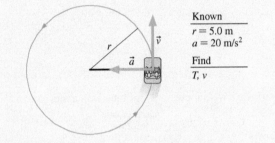

Known
$r = 5.0$ m
$a = 20$ m/s²

Find
T, v

SOLVE Equation 6.5 shows that the acceleration increases with decreasing period: If the riders go around in a shorter time, the acceleration increases. Setting a maximum value for the acceleration means setting a minimum value for the period—as fast as the ride can safely go. We can rearrange Equation 6.5 to find the period in terms of the acceleration. Setting the acceleration equal to the maximum value gives the minimum period:

$$T = 2\pi\sqrt{\frac{r}{a}} = 2\pi\sqrt{\frac{5.0\text{ m}}{20\text{ m/s}^2}} = 3.1\text{ s}$$

We can then use Equation 6.4 to find the speed at which the riders move:

$$v = \frac{2\pi r}{T} = \frac{2\pi(5.0\text{ m})}{3.1\text{ s}} = 10\text{ m/s}$$

ASSESS One rotation in just over 3 seconds seems reasonable for a pretty zippy carnival ride. (The period for this particular ride is actually 3.7 s, so it runs a bit slower than the maximum safe speed.) But in this case we can do a quantitative check on our work. If we use our calculated velocity to find the acceleration using Equation 6.1, we find

$$a = \frac{v^2}{r} = \frac{(10\text{ m/s})^2}{5.0\text{ m}} = 20\text{ m/s}^2$$

This is the acceleration given in the problem statement, so we can have confidence in our work.

STOP TO THINK 6.1 Rank in order, from largest to smallest, the period of the motion of particles A to D.

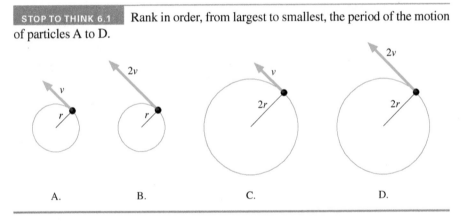

A.   B.   C.   D.

Hurling the heavy hammer Scottish games involve feats of strength. Here, a man is throwing a 30 lb hammer for distance. He starts by swinging the hammer rapidly in a circle. You can see from how he is leaning that he is providing a large force directed toward the center to produce the necessary centripetal acceleration. When the hammer is heading in the right direction, the man lets go. With no force directed toward the center, the hammer will stop going in a circle and fly in the chosen direction across the field.

6.2 Dynamics of Uniform Circular Motion

Riders traveling around on a circular carnival ride are accelerating, as we have seen. Consequently, according to Newton's second law, the riders must have a net *force* acting on them.

We've already determined the acceleration of a particle in uniform circular motion—the centripetal acceleration of Equation 6.1. Newton's second law tells us what the net force must be to cause this acceleration:

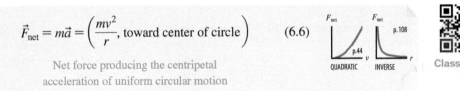

$$\vec{F}_{net} = m\vec{a} = \left(\frac{mv^2}{r}, \text{toward center of circle}\right) \qquad (6.6)$$

Net force producing the centripetal acceleration of uniform circular motion

Class Video

FIGURE 6.5 Net force for circular motion.

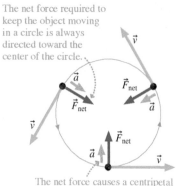

The net force required to keep the object moving in a circle is always directed toward the center of the circle.

The net force causes a centripetal acceleration.

In other words, **a particle of mass *m* moving at constant speed *v* around a circle of radius *r* must always have a net force of magnitude mv^2/r pointing toward the center of the circle**, as in FIGURE 6.5. It is this net force that causes the centripetal acceleration of circular motion. Without such a net force, the particle would move off in a straight line tangent to the circle.

The force described by Equation 6.6 is not a *new* kind of force. The net force will be due to one or more of our familiar forces, such as tension, friction, or the normal force. Equation 6.6 simply tells us how the net force needs to act—how strongly and in which direction—to cause the particle to move with speed *v* in a circle of radius *r*.

In each example of circular motion that we will consider in this chapter, a physical force or a combination of forces directed toward the center produces the necessary acceleration.

CONCEPTUAL EXAMPLE 6.4 **Forces on a car, part I**

Engineers design curves on roads to be segments of circles. They also design dips and peaks in roads to be segments of circles with a radius that depends on expected speeds and other factors. A car is moving at a constant speed and goes into a dip in the road. At the very bottom of the dip, is the normal force of the road on the car greater than, less than, or equal to the car's weight?

REASON FIGURE 6.6 shows a visual overview of the situation. The car is accelerating, even though it is moving at a constant speed, because its direction is changing. When the car is at the bottom of the dip, the center of its circular path is directly above it and so its acceleration vector points straight up. The free-body diagram of Figure 6.6 identifies the only two forces acting on the car as the normal force, pointing upward, and its weight, pointing downward. Which is larger: *n* or *w*?

Because $\vec{a}$ points upward, by Newton's second law there must be a net force on the car that also points upward. In order for this

FIGURE 6.6 Visual overview for the car in a dip.

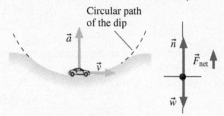

Circular path of the dip

to be the case, the free-body diagram shows that the magnitude of the normal force must be *greater* than the weight.

ASSESS You have probably experienced this situation. As you drive through a dip in the road, you feel "heavier" than normal. As discussed in Section 5.3, this is because your apparent weight—the normal force that supports you—is greater than your true weight.

CONCEPTUAL EXAMPLE 6.5 **Forces on a car, part II**

A car is turning a corner at a constant speed, following a segment of a circle. What force provides the necessary centripetal acceleration?

REASON The car moves along a circular arc at a constant speed—uniform circular motion—for the quarter-circle necessary to complete the turn. We know that the acceleration is directed toward the center of the circle. What force or forces can we identify that provide this acceleration?

Imagine you are driving a car on a frictionless road, such as a very icy road. You would not be able to turn a corner. Turning the steering wheel would be of no use. The car would slide straight ahead, in accordance with both Newton's first law and the experience of anyone who has ever driven on ice! So it must be *friction* that causes the car to turn. The top view of the tire in FIGURE 6.7 shows the force on one of the car's tires as it turns a corner. It must be a *static* friction force, not kinetic, because the tires are not skidding: The points where the tires touch the road are not moving relative to the surface. If you skid, your car won't turn the corner—it will continue in a straight line!

FIGURE 6.7 Top views of a car turning a corner.

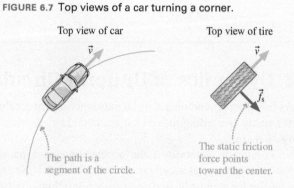

Top view of car

Top view of tire

The path is a segment of the circle.

The static friction force points toward the center.

ASSESS This result agrees with your experience. You know that reduced friction makes it harder to turn corners, and you know as well that skidding on a curve is bad news. So it makes sense that static friction is the force at work.

PROBLEM-SOLVING
STRATEGY 6.1   **Circular dynamics problems**

(MP)

Circular motion involves an acceleration and thus a net force. We can therefore use techniques very similar to those we've already seen for other Newton's second-law problems.

PREPARE Begin your visual overview with a pictorial representation in which you sketch the motion, define symbols, define axes, and identify what the problem is trying to find. There are two common situations:

■ If the motion is in a horizontal plane, like a tabletop, draw the free-body diagram with the circle viewed edge-on, the *x*-axis pointing toward the center of the circle, and the *y*-axis perpendicular to the plane of the circle.

■ If the motion is in a vertical plane, like a Ferris wheel, draw the free-body diagram with the circle viewed face-on, the *x*-axis pointing toward the center of the circle, and the *y*-axis tangent to the circle.

SOLVE Newton's second law for uniform circular motion, $\vec{F}_{net} = (mv^2/r,$ toward center of circle), is a vector equation. Some forces act in the plane of the circle, some act perpendicular to the circle, and some may have components in both directions. In the coordinate system described above, with the *x*-axis pointing toward the center of the circle, Newton's second law is

$$\sum F_x = \frac{mv^2}{r} \quad \text{and} \quad \sum F_y = 0$$

That is, the net force toward the center of the circle has magnitude mv^2/r while the net force perpendicular to the circle is zero. The components of the forces are found directly from the free-body diagram. Depending on the problem, either:

■ Use the net force to determine the speed *v*, then use circular kinematics to find frequencies or other details of the motion.

■ Use circular kinematics to determine the speed *v*, then solve for unknown forces.

ASSESS Make sure your net force points toward the center of the circle. Check that your result has the correct units, is reasonable, and answers the question.

Exercise 9 🖉

EXAMPLE 6.6   **Analyzing the motion of a cart**

An energetic father places his 20 kg child on a 5.0 kg cart to which a 2.0-m-long rope is attached. He then holds the end of the rope and spins the cart and child around in a circle, keeping the rope parallel to the ground. If the tension in the rope is 100 N, how much time does it take for the cart to make one rotation?

PREPARE We proceed according to the steps of Problem-Solving Strategy 6.1. FIGURE 6.8 on the next page shows a visual overview of the problem. The main reason for the pictorial representation on the left is to illustrate the relevant geometry and to define the symbols that will be used. A circular dynamics problem usually does not have starting and ending points like a projectile problem, so subscripts such as x_i or y_f are usually not needed. Here we need to define the cart's speed *v* and the radius *r* of the circle.

The object moving in the circle is the cart plus the child, a total mass of 25 kg; the free-body diagram shows the forces. Because the motion is in a horizontal plane, Problem-Solving Strategy 6.1 tells us to draw the free-body diagram looking at the edge of the circle, with the *x*-axis pointing toward the center of the circle and the *y*-axis perpendicular to the plane of the circle. Three forces are acting on the cart: the weight force $\vec{w}$, the normal force of the ground $\vec{n}$, and the tension force of the rope $\vec{T}$.

Notice that there are two quantities for which we use the symbol *T*: the tension and the period. We will include additional information when necessary to distinguish the two.

SOLVE There is no net force in the *y*-direction, perpendicular to the circle, so $\vec{w}$ and $\vec{n}$ must be equal and opposite. There is a net force in the *x*-direction, toward the center of the circle, as there must be to cause the centripetal acceleration of circular motion.

Continued

FIGURE 6.8 A visual overview of the cart spinning in a circle.

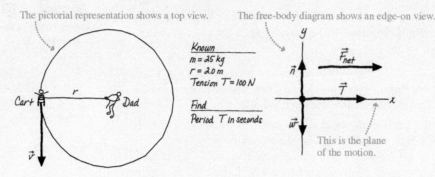

The pictorial representation shows a top view.

The free-body diagram shows an edge-on view.

Known
m = 25 kg
r = 2.0 m
Tension T = 100 N

Find
Period T in seconds

This is the plane of the motion.

Only the tension force has an x-component, so Newton's second law is

$$\sum F_x = T = \frac{mv^2}{r}$$

We know the mass, the radius of the circle, and the tension, so we can solve for v:

$$v = \sqrt{\frac{Tr}{m}} = \sqrt{\frac{(100\ \text{N})(2.0\ \text{m})}{25\ \text{kg}}} = 2.83\ \text{m/s}$$

From this, we can compute the period with a slight rearrangement of Equation 6.3:

$$T = \frac{2\pi r}{v} = \frac{(2\pi)(2.0\ \text{m})}{2.83\ \text{m/s}} = 4.4\ \text{s}$$

ASSESS The speed is about 3 m/s. Because 1 m/s ≈ 2 mph, the child is going about 6 mph. A trip around the circle in just over 4 s at a speed of about 6 mph sounds reasonable. It's a fast ride, but not so fast as to be scary!

EXAMPLE 6.7 **Finding the maximum speed to turn a corner**

What is the maximum speed with which a 1500 kg car can make a turn around a curve of radius 20 m on a level (unbanked) road without sliding? (This radius turn is about what you might expect at a major intersection in a city.)

PREPARE We start with the visual overview in FIGURE 6.9. The car moves along a circular arc at a constant speed—uniform circular motion—during the turn. We saw in Conceptual Example 6.5 that the force that provides the necessary centripetal acceleration is static friction between the tires and the road. The direction of the net force—and thus the static friction force—must point in the direction of the acceleration. With this in mind, the free-body diagram, drawn from behind the car, shows the static friction force pointing toward the center of the circle. Because the motion is in a horizontal plane, we've again chosen an x-axis toward the center of the circle and a y-axis perpendicular to the plane of motion.

FIGURE 6.9 Visual overview of a car turning a corner.

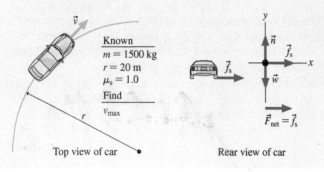

Known
m = 1500 kg
r = 20 m
μ_s = 1.0

Find
v_{max}

Top view of car

Rear view of car

SOLVE The only force in the x-direction, toward the center of the circle, is static friction. Newton's second law along the x-axis is

$$\sum F_x = f_s = \frac{mv^2}{r}$$

The only difference between this example and the preceding one is that the tension force toward the center has been replaced by a static friction force toward the center.

Newton's second law in the y-direction is

$$\sum F_y = n - w = ma_y = 0$$

so that $n = w = mg$.

The net force toward the center of the circle is the force of static friction. Recall from Equation 5.7 in Chapter 5 that static friction has a maximum possible value:

$$f_{s\,max} = \mu_s n = \mu_s mg$$

Because the static friction force has a maximum value, there will be a maximum speed at which a car can turn without sliding. This speed is reached when the static friction force reaches its maximum value $f_{s\,max} = \mu_s mg$. If the car enters the curve at a speed higher than the maximum, static friction cannot provide the necessary centripetal acceleration and the car will slide.

Thus the maximum speed occurs at the maximum value of the force of static friction, or when

$$f_{s\,max} = \frac{mv_{max}^2}{r}$$

Using the known value of $f_{s\,max}$, we find

$$\frac{mv_{max}^2}{r} = f_{s\,max} = \mu_s mg$$

Rearranging, we get

$$v_{max}^2 = \mu_s gr$$

For rubber tires on pavement, we find from Table 5.2 that $\mu_s = 1.0$. We then have

$$v_{max} = \sqrt{\mu_s gr} = \sqrt{(1.0)(9.8 \text{ m/s}^2)(20 \text{ m})} = 14 \text{ m/s}$$

ASSESS 14 m/s $\approx$ 30 mph, which seems like a reasonable upper limit for the speed at which a car can go around a curve without sliding. There are two other things to note about the solution:

- The car's mass canceled out. The maximum speed *does not* depend on the mass of the vehicle, though this may seem surprising.
- The final expression for v_{max} *does* depend on the coefficient of friction and the radius of the turn. Both of these factors make sense. You know, from experience, that the speed at which you can take a turn decreases if μ_s is less (the road is wet or icy) or if r is smaller (the turn is tighter).

Because v_{max} depends on μ_s and because μ_s depends on road conditions, the maximum safe speed through turns can vary dramatically. A car that easily handles a curve in dry weather can suddenly slide out of control when the pavement is wet. Icy conditions are even worse. If you lower the value of the coefficient of friction in Example 6.7 from 1.0 (dry pavement) to 0.1 (icy pavement), the maximum speed for the turn goes down to 4.4 m/s—about 10 mph!

Race cars turn corners at much higher speeds than normal passenger vehicles. One design modification of the *cars* to allow this is the addition of wings, as on the car in FIGURE 6.10. The wings provide an additional force pushing the car *down* onto the pavement by deflecting air upward. This extra downward force increases the normal force, thus increasing the maximum static friction force and making faster turns possible.

There are also design modifications of the *track* to allow race cars to take corners at high speeds. If the track is banked by raising the outside edge of curved sections, the normal force can provide some of the force necessary to produce the centripetal acceleration, as we will see in the next example. The curves on racetracks may be quite sharply banked. Curves on ordinary highways are often banked as well, though at more modest angles suiting the lower speeds.

FIGURE 6.10 Wings on an Indy racer.

A banked turn on a racetrack.

| EXAMPLE 6.8 | **Finding speed on a banked turn** |

A curve on a racetrack of radius 70 m is banked at a 15° angle. At what speed can a car take this curve without assistance from friction?

PREPARE After drawing the pictorial representation in FIGURE 6.11, we use the force identification diagram to find that, given that there

is no friction acting, the only two forces are the normal force and the car's weight. We can then construct the free-body diagram, making sure that we draw the normal force perpendicular to the road's surface.

Even though the car is tilted, it is still moving in a *horizontal* circle. Thus, following Problem-Solving Strategy 6.1, we choose the x-axis to be horizontal and pointing toward the center of the circle.

SOLVE Without friction, $n_x = n\sin\theta$ is the only component of force toward the center of the circle. It is this inward component of the normal force on the car that causes it to turn the corner. Newton's second law is

$$\sum F_x = n\sin\theta = \frac{mv^2}{r}$$
$$\sum F_y = n\cos\theta - w = 0$$

FIGURE 6.11 Visual overview for the car on a banked turn.

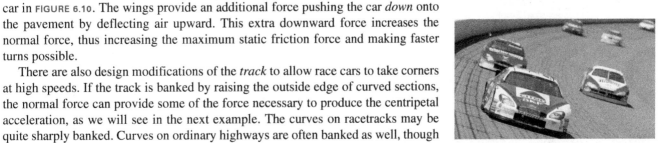

Top view $\vec{v}$ r

Rear view θ r

Weight $\vec{w}$

Normal $\vec{n}$

Known
$r = 70$ m
$\theta = 15°$

Find
v

The normal force is perpendicular to the surface.

Road surface

The free-body diagram is drawn as seen from the rear of the car.

The x-axis points toward the center of the circle.

Continued

where θ is the angle at which the road is banked, and we've assumed that the car is traveling at the correct speed v. From the y-equation,

$$n = \frac{w}{\cos\theta} = \frac{mg}{\cos\theta}$$

Substituting this into the x-equation and solving for v give

$$\left(\frac{mg}{\cos\theta}\right)\sin\theta = mg\tan\theta = \frac{mv^2}{r}$$

$$v = \sqrt{rg\tan\theta} = 14 \text{ m/s}$$

ASSESS This is ≈ 30 mph, a reasonable speed. Only at this exact speed can the turn be negotiated without reliance on friction forces.

FIGURE 6.12 Road forces on a cyclist leaning into a turn.

The friction force provides the necessary centripetal acceleration for cars turning corners, but also for bicycles, horses, and humans. The cyclists in FIGURE 6.12 are going through a tight turn; you can tell by how they lean. The road exerts both a vertical normal force and a horizontal friction force on their tires. A vector sum of these forces points at an angle. The cyclists lean to the side so that the sum of the road forces points along the line of their bikes and their bodies; this keeps them in balance. The horses in the photo that opened the chapter are leaning into the turn for a similar reason.

Maximum Walking Speed BIO

Humans and other two-legged animals have two basic gaits: walking and running. At slow speeds, you walk. When you need to go faster, you run. Why don't you just walk faster? There is an upper limit to the speed of walking, and this limit is set by the physics of circular motion.

Think about the motion of your body as you take a walking stride. You put one foot forward, then push off with your rear foot. Your body pivots over your front foot, and you bring your rear foot forward to take the next stride. As you can see in FIGURE 6.13a, the path that your body takes during this stride is the arc of a circle. **In a walking gait, your body is in circular motion as you pivot on your forward foot.**

A force toward the center of the circle is required for this circular motion, as shown in Figure 6.13. FIGURE 6.13b shows the forces acting on the woman's body during the midpoint of the stride: her weight, directed down, and the normal force of the ground, directed up. Newton's second law for the x-axis is

$$\sum F_x = w - n = \frac{mv^2}{r}$$

Because of her circular motion, the net force must point toward the center of the circle, or, in this case, down. In order for the net force to point down, the normal force must be *less* than her weight. Your body tries to "lift off" as it pivots over your foot, decreasing the normal force exerted on you by the ground. The normal force becomes smaller as you walk faster, but n cannot be less than zero. Thus the maximum possible walking speed v_{max} occurs when $n = 0$. Setting $n = 0$ in Newton's second law gives

$$w = mg = \frac{mv_{max}^2}{r}$$

FIGURE 6.13 Analysis of a walking stride.

(a) Walking stride   During each stride, her hip undergoes circular motion.

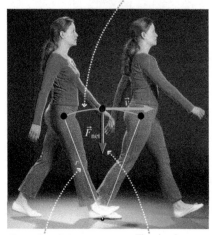

The radius of the circular motion is the length of the leg from the foot to the hip.

The circular motion requires a force directed toward the center of the circle.

(b) Forces in the stride   Side view (same as photo)

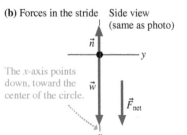

The x-axis points down, toward the center of the circle.

Thus

$$v_{max} = \sqrt{gr} \tag{6.7}$$

The maximum possible walking speed is limited by r, the length of the leg, and g, the free-fall acceleration. This formula is a good approximation of the maximum walking speed for humans and other animals. Giraffes, with their very long legs, can walk at high speeds. Animals such as mice with very short legs have such a low maximum walking speed that they rarely use this gait.

For humans, the length of the leg is approximately 0.7 m, giving $v_{max} \approx 2.6$ m/s ≈ 6 mph. You *can* walk this fast, though it becomes energetically unfavorable to walk at speeds above 4 mph. Most people make a transition to a running gait at about this speed.

STOP TO THINK 6.2 A block on a string spins in a horizontal circle on a frictionless table. Rank in order, from largest to smallest, the tensions T_A to T_E acting on the blocks A to E.

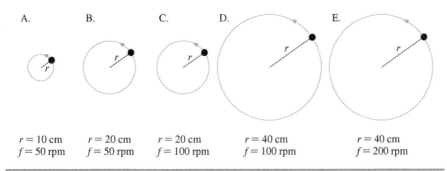

A.	B.	C.	D.	E.
$r = 10$ cm	$r = 20$ cm	$r = 20$ cm	$r = 40$ cm	$r = 40$ cm
$f = 50$ rpm	$f = 50$ rpm	$f = 100$ rpm	$f = 100$ rpm	$f = 200$ rpm

6.3 Apparent Forces in Circular Motion

FIGURE 6.14 shows a carnival ride that spins the riders around inside a large cylinder. The people are "stuck" to the inside wall of the cylinder! As you probably know from experience, the riders *feel* that they are being pushed outward, into the wall. But our analysis has found that an object in circular motion must have an *inward* force to create the centripetal acceleration. How can we explain this apparent difference?

FIGURE 6.14 Inside the Gravitron, a rotating circular room.

Centrifugal Force?

If you are a passenger in a car that turns a corner quickly, you may feel "thrown" by some mysterious force against the door. But is there really such a force? FIGURE 6.15 shows a bird's-eye view of you riding in a car as it makes a left turn. You try to continue moving in a straight line, obeying Newton's first law, when—without having been provoked—the door starts to turn in toward you and so runs into you! You then feel the force of the door because it is now the force of the door, pushing *inward* toward the center of the curve, that is causing you to turn the corner. But you were not "thrown" into the door; the door ran into you.

A "force" that *seems* to push an object to the outside of a circle is called a *centrifugal force*. Despite having a name, there really is no such force. What you feel is your body trying to move ahead in a straight line (which would take you away from the center of the circle) as outside forces act to turn you in a circle. The only real forces, those that appear on free-body diagrams, are the ones pushing inward toward the center. **A centrifugal force will never appear on a free-body diagram and never be included in Newton's laws.**

With this in mind, let's revisit the rotating carnival ride. A person watching from above would see the riders in the cylinder moving in a circle with the walls providing the inward force that causes their centripetal acceleration. The riders *feel* as if they're being pushed outward because their natural tendency to move in a straight line is being resisted by the wall of the cylinder, which keeps getting in the way. But feelings aren't forces. The only actual force is the contact force of the cylinder wall pushing *inward*.

FIGURE 6.15 Bird's-eye view of a passenger in a car turning a corner.

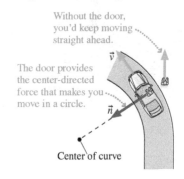

Without the door, you'd keep moving straight ahead.

$\vec{v}$

The door provides the center-directed force that makes you move in a circle.

$\vec{n}$

Center of curve

Apparent Weight in Circular Motion

Imagine swinging a bucket of water over your head. If you swing the bucket quickly, the water stays in. But you'll get a shower if you swing too slowly. Why does the water stay in the bucket? Or think about a roller coaster that does a loop-the-loop. How does the car stay on the track when it's upside down? You might have said that there was a centrifugal force holding the water in the bucket and the car on the track, but we have seen that there really isn't a centrifugal force. Analyzing these questions will tell us a lot about forces in general and circular motion in particular.

FIGURE 6.16 A roller coaster car going around a loop-the-loop.

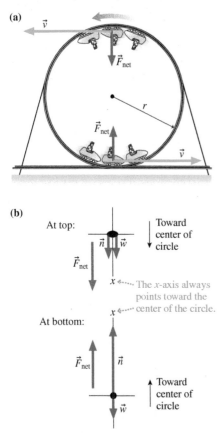

(a)

(b)

At top:

Toward center of circle

$\vec{n}$ $\vec{w}$

$\vec{F}_{net}$

x ····· The x-axis always points toward the center of the circle.

At bottom:

x ·····

$\vec{F}_{net}$ $\vec{n}$

Toward center of circle

$\vec{w}$

When "down" is up BIO You can tell, even with your eyes closed, what direction is down. Organs in your inner ear contain small crystals of calcium carbonate, called *otoliths*. These crystals are supported by a sensitive membrane. Your brain interprets "down" as the opposite of the direction of the normal force of the membrane on the otoliths. At the top of a loop in a roller coaster, this normal force is directed down, so your inner ear tells you that "down" is up. You are upside down, but it doesn't feel that way.

FIGURE 6.16a shows a roller coaster car going around a vertical loop-the-loop of radius r. If you've ever ridden a roller coaster, you know that your sensation of weight changes as you go over the crests and through the dips. To understand why, let's look at the forces on passengers going through the loop. To simplify our analysis, we will assume that the speed of the car stays constant as it moves through the loop.

FIGURE 6.16b shows a passenger's free-body diagram at the top and the bottom of the loop. Let's start by examining the forces on the passenger at the bottom of the loop. The only forces acting on her are her weight $\vec{w}$ and the normal force $\vec{n}$ of the seat pushing up on her. But recall from ◀ SECTION 5.3 that you don't feel the weight force. The force you *feel*, your apparent weight, is the magnitude of the contact force that supports you. Here the seat is supporting the passenger with the normal force $\vec{n}$, so her apparent weight is $w_{app} = n$. Based on our understanding of circular motion, we can say:

- She's moving in a circle, so there must be a net force directed toward the center of the circle—currently directly above her head—to provide the centripetal acceleration.
- The net force points *upward*, so it must be the case that $n > w$.
- Her apparent weight is $w_{app} = n$, so her apparent weight is greater than her true weight ($w_{app} > w$). Thus she "feels heavy" at the bottom of the circle.

This situation is the same as for the car driving through a dip in Conceptual Example 6.4. To analyze the situation quantitatively, we'll apply the steps of Problem-Solving Strategy 6.1. As always, we choose the x-axis to point toward the center of the circle or, in this case, vertically upward. Then Newton's second law is

$$\sum F_x = n_x + w_x = n - w = \frac{mv^2}{r}$$

From this equation, the passenger's apparent weight is

$$w_{app} = n = w + \frac{mv^2}{r} \tag{6.8}$$

Her apparent weight at the bottom is *greater* than her true weight w, which agrees with your experience when you go through a dip or a valley.

Now let's look at the roller coaster car as it crosses the top of the loop. Things are a little trickier here. As Figure 6.16b shows, whereas the normal force of the seat pushes up when the passenger is at the bottom of the circle, it pushes *down* when she is at the top and the seat is above her. It's worth thinking carefully about this diagram to make sure you understand what it is showing.

The passenger is still moving in a circle, so there must be a net force *downward*, toward the center of the circle, to provide her centripetal acceleration. As always, we define the x-axis to be toward the center of the circle, so here the x-axis points vertically downward. Newton's second law gives

$$\sum F_x = n_x + w_x = n + w = \frac{mv^2}{r}$$

Note that w_x is now *positive* because the x-axis is directed downward. We can solve for the passenger's apparent weight:

$$w_{app} = n = \frac{mv^2}{r} - w \tag{6.9}$$

If v is sufficiently large, her apparent weight can exceed the true weight, just as it did at the bottom of the track.

But let's look at what happens if the car goes slower. Notice from Equation 6.9 that, as v decreases, there comes a point when $mv^2/r = w$ and n becomes zero. At that point, the seat is *not* pushing against the passenger at all! Instead, she is able to complete the circle because her weight force alone provides sufficient centripetal acceleration.

The speed for which $n = 0$ is called the *critical speed* v_c. Because for n to be zero we must have $mv_c^2/r = w$, the critical speed is

$$v_c = \sqrt{\frac{rw}{m}} = \sqrt{\frac{rmg}{m}} = \sqrt{gr} \qquad (6.10)$$

What happens if the speed is slower than the critical speed? In this case, Equation 6.9 gives a *negative* value for n if $v < v_c$. But that is physically impossible. The seat can push against the passenger ($n > 0$), but it can't *pull* on her, so the slowest possible speed is the speed for which $n = 0$ at the top. Thus, **the critical speed is the slowest speed at which the car can complete the circle.** If $v < v_c$, the passenger cannot turn the full loop but, instead, will fall from the car as a projectile! (This is why you're always strapped into a roller coaster.)

Water stays in a bucket swung over your head for the same reason. The bottom of the bucket pushes against the water to provide the inward force that causes circular motion. If you swing the bucket too slowly, the force of the bucket on the water drops to zero. At that point, the water leaves the bucket and becomes a projectile following a parabolic trajectory onto your head!

A fast-spinning world Saturn, a gas giant planet composed largely of fluid matter, is quite a bit larger than the earth. It also rotates much more quickly, completing one rotation in just under 11 hours. The rapid rotation decreases the apparent weight at the equator enough to distort the fluid surface. The planet is noticeably out of round, as the red circle shows. The diameter at the equator is 11% greater than the diameter at the poles.

EXAMPLE 6.9   How slow can you go?

A motorcyclist in the Globe of Death, pictured here, rides in a 2.2-m-radius vertical loop. To keep control of the bike, the rider wants the normal force on his tires at the top of the loop to equal or exceed his and the bike's combined weight. What is the minimum speed at which the rider can take the loop?

PREPARE The visual overview for this problem is shown in FIGURE 6.17. At the top of the loop, the normal force of the cage on the tires is a *downward* force. In accordance with Problem-Solving Strategy 6.1, we've chosen the x-axis to point toward the center of the circle.

SOLVE We will consider the forces at the top point of the loop. Because the x-axis points downward, Newton's second law is

$$\sum F_x = w + n = \frac{mv^2}{r}$$

FIGURE 6.17 Riding in a vertical loop around the Globe of Death.

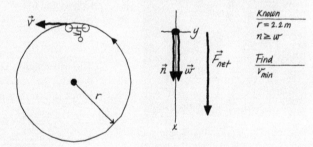

The minimum acceptable speed occurs when $n = w$; thus

$$2w = 2mg = \frac{mv_{min}^2}{r}$$

Solving for the speed, we find

$$v_{min} = \sqrt{2gr} = \sqrt{2(9.8 \text{ m/s}^2)(2.2 \text{ m})} = 6.6 \text{ m/s}$$

ASSESS The minimum speed is ≈ 15 mph, which isn't all that fast; the bikes can easily reach this speed. But normally several bikes are in the globe at one time. The big challenge is to keep all of the riders in the cage moving at this speed in synchrony. The period for the circular motion at this speed is $T = 2\pi r/v \approx 2$ s, leaving little room for error!

Centrifuges BIO

The *centrifuge*, an important biological application of circular motion, is used to separate the components of a liquid that has different densities. Typically these are different types of cells, or the components of cells, suspended in water. You probably know that small particles suspended in water will eventually settle to the bottom. However, the downward motion due to gravity for extremely small objects such as cells is so slow that it could take days or even months for the cells to settle out. It's not practical to wait for biological samples to separate due to gravity alone.

FIGURE 6.18 The operation of a centrifuge.

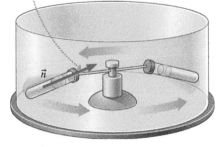

The high centripetal acceleration requires a large normal force, which leads to a large apparent weight.

$\vec{n}$

Human centrifuge BIO If you spin your arm rapidly in a vertical circle, the motion will produce an effect like that in a centrifuge. The motion will assist outbound blood flow in your arteries and retard inbound blood flow in your veins. There will be a buildup of fluid in your hand that you will be able to see (and feel!) quite easily.

The separation would go faster if the force of gravity could be increased. Although we can't change gravity, we can increase the apparent weight of objects in the sample by spinning them very fast, and that is what the centrifuge in FIGURE 6.18 does. The centrifuge produces centripetal accelerations that are thousands of times greater than free-fall acceleration. As the centrifuge effectively increases gravity to thousands of times its normal value, the cells or cell components settle out and separate by density in a matter of minutes or hours.

EXAMPLE 6.10 **Analyzing the ultracentrifuge**

An 18-cm-diameter ultracentrifuge produces an extraordinarily large centripetal acceleration of 250,000g, where g is the free-fall acceleration due to gravity. What is its frequency in rpm? What is the apparent weight of a sample with a mass of 0.0030 kg?

PREPARE The acceleration in SI units is

$$a = 250,000(9.80 \text{ m/s}^2) = 2.45 \times 10^6 \text{ m/s}^2$$

The radius is half the diameter, or $r = 9.0$ cm $= 0.090$ m.

SOLVE We can rearrange Equation 6.5 to find the frequency given the centripetal acceleration:

$$f = \frac{1}{2\pi}\sqrt{\frac{a}{r}} = \frac{1}{2\pi}\sqrt{\frac{2.45 \times 10^6 \text{ m/s}^2}{0.090 \text{ m}}} = 830 \text{ rev/s}$$

Converting to rpm, we find

$$830 \frac{\text{rev}}{\text{s}} \times \frac{60 \text{ s}}{1 \text{ min}} = 50,000 \text{ rpm}$$

The acceleration is so high that every force is negligible except for the force that provides the centripetal acceleration. The net force is simply equal to the inward force, which is also the sample's apparent weight:

$$w_{\text{app}} = F_{\text{net}} = ma = (3.0 \times 10^{-3} \text{ kg})(2.45 \times 10^6 \text{ m/s}^2) = 7.4 \times 10^3 \text{ N}$$

The 3 gram sample has an effective weight of about 1700 pounds!

ASSESS Because the acceleration is 250,000g, the apparent weight is 250,000 times the actual weight. This makes sense, as does the fact that we calculated a very high frequency, which is necessary to give the large acceleration.

STOP TO THINK 6.3   A car is rolling over the top of a hill at constant speed v. At this instant,

A. $n > w$.
B. $n < w$.
C. $n = w$.
D. We can't tell about n without knowing v.

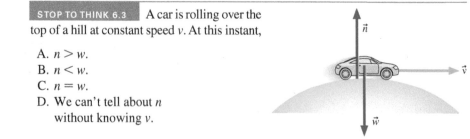

6.4 Circular Orbits and Weightlessness

The International Space Station orbits the earth in a circular path at a speed of over 15,000 miles per hour. What forces act on it? Why does it move in a circle? Before we start considering the physics of orbital motion, let's return, for a moment, to projectile motion. Projectile motion occurs when the only force on an object is gravity. Our analysis of projectiles made an implicit assumption that the earth is flat and that the free-fall acceleration, due to gravity, is everywhere straight down. This is an acceptable approximation for projectiles of limited range, such as baseballs or cannon balls, but there comes a point where we can no longer ignore the curvature of the earth.

Orbital Motion

FIGURE 6.19 shows a perfectly smooth, spherical, airless planet with a vertical tower of height h. A projectile is launched from this tower with initial speed v_i parallel to the ground. If v_i is very small, as in trajectory A, the "flat-earth approximation" is valid and the problem is identical to Example 3.11 in which a dog ran off the end of a dock. The projectile simply falls to the ground along a parabolic trajectory.

As the initial speed v_i is increased, it seems to the projectile that the ground is curving out from beneath it. It is still falling the entire time, always getting closer to the ground, but the distance that the projectile travels before finally reaching the ground—that is, its range—increases because the projectile must "catch up" with the ground that is curving away from it. Trajectories B and C are like this.

If the launch speed v_i is sufficiently large, there comes a point at which the curve of the trajectory and the curve of the earth are parallel. In this case, the projectile "falls" but it never gets any closer to the ground! This is the situation for trajectory D. The projectile returns to the point from which it was launched, at the same speed at which it was launched, making a closed trajectory. Such a closed trajectory around a planet or star is called an **orbit.**

The most important point of this qualitative analysis is that, in the absence of air resistance, **an orbiting projectile is in free fall.** This is, admittedly, a strange idea, but one worth careful thought. An orbiting projectile is really no different from a thrown baseball or a dog jumping off a dock. The only force acting on it is gravity, but its tangential velocity is so great that the curvature of its trajectory matches the curvature of the earth. When this happens, the projectile "falls" under the influence of gravity but never gets any closer to the surface, which curves away beneath it.

When we first studied free fall in Chapter 2, we said that free-fall acceleration is always directed vertically downward. As we see in **FIGURE 6.20**, "downward" really means "toward the center of the earth." For a projectile in orbit, the direction of the force of gravity changes, always pointing toward the center of the earth.

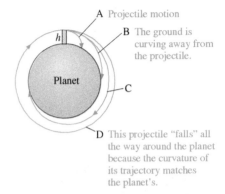

FIGURE 6.19 Projectiles being launched at increasing speeds from height h on a smooth, airless planet.

A Projectile motion

B The ground is curving away from the projectile.

C

D This projectile "falls" all the way around the planet because the curvature of its trajectory matches the planet's.

FIGURE 6.20 The force of gravity is really directed toward the center of the earth.

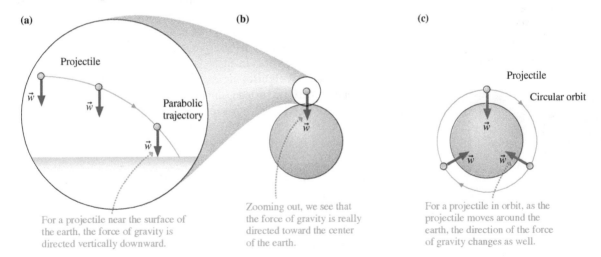

(a)

Projectile

Parabolic trajectory

For a projectile near the surface of the earth, the force of gravity is directed vertically downward.

(b)

Zooming out, we see that the force of gravity is really directed toward the center of the earth.

(c)

Projectile

Circular orbit

For a projectile in orbit, as the projectile moves around the earth, the direction of the force of gravity changes as well.

As you have learned, a force of constant magnitude that always points toward the center of a circle causes the centripetal acceleration of uniform circular motion. Because the only force acting on the orbiting projectile in Figure 6.20 is gravity, and we're assuming the projectile is very near the surface of the earth, we can write

$$a = \frac{F_{net}}{m} = \frac{w}{m} = \frac{mg}{m} = g \qquad (6.11)$$

An object moving in a circle of radius r at speed v_{orbit} will have this centripetal acceleration if

$$a = \frac{(v_{orbit})^2}{r} = g \qquad (6.12)$$

That is, if an object moves parallel to the surface with the speed

$$v_{orbit} = \sqrt{gr} \qquad (6.13)$$

then the free-fall acceleration provides exactly the centripetal acceleration needed for a circular orbit of radius r. An object with any other speed will not follow a circular orbit.

The earth's radius is $r = R_e = 6.37 \times 10^6$ m. The orbital speed of a projectile just skimming the surface of a smooth, airless earth is

$$v_{orbit} = \sqrt{gR_e} = \sqrt{(9.80 \text{ m/s}^2)(6.37 \times 10^6 \text{ m})} = 7900 \text{ m/s} \approx 18,000 \text{ mph}$$

We can use v_{orbit} to calculate the period of the satellite's orbit:

$$T = \frac{2\pi r}{v_{orbit}} = 2\pi\sqrt{\frac{r}{g}} \qquad (6.14)$$

For this earth-skimming orbit, $T = 5065$ s $= 84.4$ min.

Of course, actual satellites must orbit at some height above the surface to be above mountains and trees—and most of the atmosphere, so there is little drag. Most people envision the International Space Station orbiting far above the earth's surface, but the average height is just over 200 miles, giving a value of r that is only 5% greater than the earth's radius—not too far from skimming the surface. At this slightly larger value of r, Equation 6.14 gives $T = 87$ min. In fact, the International Space Station orbits with a period of about 90 min, going around the earth more than 15 times each day.

Weightlessness in Orbit

When we discussed *weightlessness* in ◄ SECTION 5.3, we saw that it occurs during free fall. We asked whether astronauts and their spacecraft are in free fall. We can now give an affirmative answer: They are, indeed, in free fall. They are falling continuously around the earth, under the influence of only the gravitational force, but never getting any closer to the ground because the earth's surface curves beneath them. Weightlessness in space is no different from the weightlessness in a free-falling elevator. **Weightlessness does *not* occur from an absence of weight or an absence of gravity.** Instead, the astronaut, the spacecraft, and everything in it are "weightless" (i.e., their *apparent* weight is zero) because they are all falling together. We know that the free-fall acceleration doesn't depend on mass, so the astronaut and the station follow exactly the same orbit.

The Orbit of the Moon

The moon moves in an orbit around the earth that is approximately circular. The force that holds the moon in its orbit, that provides the necessary centripetal acceleration, is the gravitational attraction of the earth. The moon, like all satellites, is simply "falling" around the earth. But if we use the distance to the moon, $r = 3.84 \times 10^8$ m, in Equation 6.14 to predict the period of the moon's orbit, we get a period of approximately 11 hours. This is clearly wrong; you know that the period of the moon's orbit is about one month. What went wrong?

In using Equation 6.14, we assumed that the free-fall acceleration g is the same at the distance of the moon as it is on or near the earth's surface. But if gravity is the force of the earth pulling on an object, it seems plausible that the size of that force, and thus the size of g, should diminish with increasing distance from the earth. And, indeed, the force of gravity does decrease with distance, in a manner we'll explore in the next section.

Zero apparent weight in space.

Rotating space stations BIO The weightlessness astronauts experience in orbit has serious physiological consequences. Astronauts who spend time in weightless environments lose bone and muscle mass and suffer other adverse effects. One solution is to introduce "artificial gravity." On a space station, the easiest way to do this would be to make the station rotate, producing an apparent weight. The designers of this space station model for the movie *2001: A Space Odyssey* made it rotate for just that reason.

STOP TO THINK 6.4   A satellite is in a low earth orbit. Which of the following changes would increase the orbital period?

A. Increasing the mass of the satellite.
B. Increasing the height of the satellite about the surface.
C. Increasing the value of g.

6.5 Newton's Law of Gravity

Our current understanding of the force of gravity begins with Isaac Newton. The popular image of Newton coming to a key realization about gravity after an apple fell on his head is at least close to the truth: Newton himself said that the "notion of gravitation" came to him as he "sat in a contemplative mood" and "was occasioned by the fall of an apple."

The important notion that came to Newton is this: *Gravity is a universal force that affects all objects in the universe.* The force that causes the fall of an apple is the same force that keeps the moon in orbit. This is something widely accepted now, but at the time this was a revolutionary idea, and there were some important details for Newton to work out—in particular, the way that the force varies with distance.

Gravity Obeys an Inverse-Square Law

Newton proposed that *every* object in the universe attracts *every other* object with a force that has the following properties:

1. The force is inversely proportional to the square of the distance between the objects.
2. The force is directly proportional to the product of the masses of the two objects.

FIGURE 6.21 shows two spherical objects with masses m_1 and m_2 separated by distance r. Each object exerts an attractive force on the other, a force that we call the **gravitational force.** These two forces form an action/reaction pair, so $\vec{F}_{1\,on\,2}$ is equal in magnitude and opposite in direction to $\vec{F}_{2\,on\,1}$. The magnitude of the forces is given by Newton's law of gravity.

FIGURE 6.21 The gravitational forces on masses m_1 and m_2.

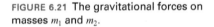

Newton's law of gravity If two objects with masses m_1 and m_2 are a distance r apart, the objects exert attractive forces on each other of magnitude

$$F_{1\,on\,2} = F_{2\,on\,1} = \frac{Gm_1m_2}{r^2} \qquad (6.15)$$

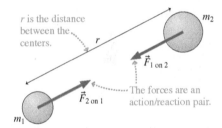

The forces are directed along the line joining the two objects.

The constant G is called the **gravitational constant.** In the SI system of units,

$$G = 6.67 \times 10^{-11} \ \text{N} \cdot \text{m}^2/\text{kg}^2$$

NOTE ▶ Strictly speaking, Newton's law of gravity applies to *particles* with masses m_1 and m_2. However, it can be shown that the law also applies to the force between two spherical objects if r is the distance between their centers. ◄

As the distance r between two objects increases, the gravitational force between them decreases. Because the distance appears squared in the denominator, Newton's law of gravity is what we call an **inverse-square** law. Doubling the distance between two masses causes the force between them to decrease by a factor of 4. This mathematical form is one we will see again, so it is worth our time to explore it in more detail.

Inverse-square relationships

Two quantities have an **inverse-square relationship** if y is inversely proportional to the *square* of x. We write the mathematical relationship as

$$y = \frac{A}{x^2}$$

y is inversely proportional to x^2

Here, A is a constant. This relationship is sometimes written as $y \propto 1/x^2$.

SCALING As the graph shows, inverse-square scaling means, for example:

- If you double x, you decrease y by a factor of 4.
- If you halve x, you increase y by a factor of 4.
- If you increase x by a factor of 3, you decrease y by a factor of 9.
- If you decrease x by a factor of 3, you increase y by a factor of 9.

Generally, **if x increases by a factor of C, y decreases by a factor of C^2.** If x *decreases* by a factor of C, y *increases* by a factor of C^2.

When x is halved, y increases by a factor of 4.

$y = \frac{A}{x^2}$

When x is 1, y is A.

When x is doubled, y is reduced by a factor of 4 (2 squared).

RATIOS For any two values of x—say, x_1 and x_2—we have

$$y_1 = \frac{A}{x_1^2} \quad \text{and} \quad y_2 = \frac{A}{x_2^2}$$

Dividing the y_1-equation by the y_2-equation, we find

$$\frac{y_1}{y_2} = \frac{A/x_1^2}{A/x_2^2} = \frac{A}{x_1^2}\frac{x_2^2}{A} = \frac{x_2^2}{x_1^2}$$

That is, the ratio of y-values is the inverse of the ratio of the squares of the corresponding values of x.

LIMITS As x becomes large, y becomes very small; as x becomes small, y becomes very large.

Exercises 19, 20

CONCEPTUAL EXAMPLE 6.11 **Varying gravitational force**

The gravitational force between two giant lead spheres is 0.010 N when the centers of the spheres are 20 m apart. What is the distance between their centers when the gravitational force between them is 0.160 N?

REASON We can solve this problem without knowing the masses of the two spheres. The key is to consider the ratios of forces and distances. Gravity is an inverse-square relationship;

the force is related to the inverse square of the distance. The force *increases* by a factor of (0.160 N)/(0.010 N) = 16, so the distance must *decrease* by a factor of $\sqrt{16} = 4$. The distance is thus (20 m)/4 = 5.0 m.

ASSESS This type of ratio reasoning is a very good way to get a quick handle on the solution to a problem.

EXAMPLE 6.12 **Gravitational force between two people**

You are seated in your physics class next to another student 0.60 m away. Estimate the magnitude of the gravitational force between you. Assume that you each have a mass of 65 kg.

PREPARE We will model each of you as a sphere. This is not a particularly good model, but it will do for making an estimate. We will take the 0.60 m as the distance between your centers.

SOLVE The gravitational force is given by Equation 6.15:

$$\begin{aligned} F_{(you)\,on\,(other\,student)} &= \frac{Gm_{you}m_{other\,student}}{r^2} \\ &= \frac{(6.67 \times 10^{-11}\ \text{N} \cdot \text{m}^2/\text{kg}^2)(65\ \text{kg})(65\ \text{kg})}{(0.60\ \text{m})^2} \\ &= 7.8 \times 10^{-7}\ \text{N} \end{aligned}$$

ASSESS The force is quite small, roughly the weight of one hair on your head. This seems reasonable; you don't normally sense this attractive force!

There is a gravitational force between all objects in the universe, but the gravitational force between two ordinary-sized objects is extremely small. Only when one (or both) of the masses is exceptionally large does the force of gravity become important. The downward force of the earth on you—your weight—is large because the earth has an enormous mass. And the attraction is mutual: By Newton's third law, you exert an upward force on the earth that is equal to your weight. However, the large mass of the earth makes the *effect* of this force on the earth negligible.

EXAMPLE 6.13 **Gravitational force of the earth on a person**

What is the magnitude of the gravitational force of the earth on a 60 kg person? The earth has mass 5.98×10^{24} kg and radius 6.37×10^6 m.

PREPARE We'll again model the person as a sphere. The distance r in Newton's law of gravity is the distance between the *centers* of the two spheres. The size of the person is negligible compared to the size of the earth, so we can use the earth's radius as r.

SOLVE The force of gravity on the person due to the earth can be computed using Equation 6.15:

$$F_{\text{earth on person}} = \frac{GM_e m}{R_e^2}$$

$$= \frac{(6.67 \times 10^{-11}\ \text{N} \cdot \text{m}^2/\text{kg}^2)(5.98 \times 10^{24}\ \text{kg})(60\ \text{kg})}{(6.37 \times 10^6\ \text{m})^2}$$

$$= 590\ \text{N}$$

ASSESS This force is exactly the same as we would calculate using the formula for the weight force, $w = mg$. This isn't surprising, though. Chapter 5 introduced the weight of an object as simply the "force of gravity" acting on it. Newton's law of gravity is a more fundamental law for calculating the force of gravity, but it's still the same force that we earlier called "weight."

NOTE ▸ We will use uppercase R and M to represent the large mass and radius of a star or planet, as we did in Example 6.13. ◂

Gravity on Other Worlds

The force of gravitational attraction between the earth and you is responsible for your weight. If you traveled to another planet, your *mass* would be the same but your *weight* would vary, as we discussed in Chapter 5. Indeed, when astronauts ventured to the moon, television images showed them walking—and even jumping and skipping—with ease, even though they were wearing life-support systems with a mass greater than 80 kg, a visible reminder that the weight of objects is less on the moon. Let's consider why this is so.

FIGURE 6.22 shows an astronaut on the moon weighing a rock of mass m. When we compute the weight of an object on the surface of the earth, we use the formula $w = mg$. We can do the same calculation for a mass on the moon, as long as we use the value of g on the moon:

$$w = mg_{\text{moon}} \tag{6.16}$$

This is the "little g" perspective. Falling-body experiments on the moon would give the value of g_{moon} as $1.62\ \text{m/s}^2$.

But we can also take a "big G" perspective. The weight of the rock comes from the gravitational attraction of the moon, and we can compute this weight using Equation 6.15. The distance r is the radius of the moon, which we'll call R_{moon}. Thus

$$F_{\text{moon on }m} = \frac{GM_{\text{moon}} m}{R_{\text{moon}}^2} \tag{6.17}$$

Because Equations 6.16 and 6.17 are two names and two expressions for the same force, we can equate the right-hand sides to find that

$$g_{\text{moon}} = \frac{GM_{\text{moon}}}{R_{\text{moon}}^2}$$

We have done this calculation for an object on the moon, but the result is completely general. At the surface of a planet (or a star), the free-fall acceleration g, a consequence of gravity, can be computed as

$$g_{\text{planet}} = \frac{GM_{\text{planet}}}{R_{\text{planet}}^2} \tag{6.18}$$

Free-fall acceleration on the surface of a planet

Variable gravity When we calculated the force of the earth's gravity, we assumed that the earth's shape and composition are uniform. In reality, unevenness in density and other factors create small variations in the earth's gravity, as shown in this image. Red means slightly stronger surface gravity; blue means slightly weaker. These variations are important for scientists who study the earth, but they are small enough that we can ignore them for the computations we'll do in this text.

FIGURE 6.22 An astronaut weighing a mass on the moon.

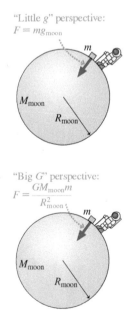

"Little g" perspective:
$F = mg_{\text{moon}}$

"Big G" perspective:
$F = \dfrac{GM_{\text{moon}} m}{R_{\text{moon}}^2}$

Walking on the moon BIO We saw that the maximum walking speed depends on leg length and the value of g. The low lunar gravity makes walking very easy but also makes it very slow: The maximum walking speed on the moon is about 1 m/s—a very gentle stroll! Walking at a reasonable pace was difficult for the Apollo astronauts, but the reduced weight made jumping quite easy. Videos from the surface of the moon often showed the astronauts getting from place to place by hopping or skipping—not for fun, but for speed and efficiency.

If we use values for the mass and the radius of the moon, we compute $g_{moon} = 1.62$ m/s². This means that an object would weigh less on the moon than it would on the earth, where g is 9.80 m/s². A 70 kg astronaut wearing an 80 kg spacesuit would weigh more than 330 lb on the earth but only 54 lb on the moon.

Equation 6.18 gives g at the surface of a planet. More generally, imagine an object at distance $r > R$ from the center of a planet. Its free-fall acceleration at this distance is

$$g = \frac{GM}{r^2} \tag{6.19}$$

This more general result agrees with Equation 6.18 if $r = R$, but it allows us to determine the "local" free-fall acceleration at distances $r > R$. Equation 6.19 expresses Newton's idea that the size of g should decrease as you get farther from the earth.

As you're flying in a jet airplane at a height of about 10 km, the free-fall acceleration is about 0.3% less than on the ground. At the height of the International Space Station, about 300 km, Equation 6.19 gives $g = 8.9$ m/s², about 10% less than the free-fall acceleration on the earth's surface. If you use this slightly smaller value of g in Equation 6.14 for the period of a satellite's orbit, you'll get the correct period of about 90 minutes. This value of g, only slightly less than the ground-level value, emphasizes the point that an object in orbit is not "weightless" due to the absence of gravity, but rather because it is in free fall.

EXAMPLE 6.14 **Finding the speed to orbit Deimos**

Mars has two moons, each much smaller than the earth's moon. The smaller of these two bodies, Deimos, isn't quite spherical, but we can model it as a sphere of radius 6.3 km. Its mass is 1.8×10^{15} kg. At what speed would a projectile move in a very low orbit around Deimos?

SOLVE The free-fall acceleration at the surface of Deimos is small:

$$g_{Deimos} = \frac{GM_{Deimos}}{R^2_{Deimos}}$$

$$= \frac{(6.67 \times 10^{-11} \text{ N} \cdot \text{m}^2/\text{kg}^2)(1.8 \times 10^{15} \text{ kg})}{(6.3 \times 10^3 \text{ m})^2}$$

$$= 0.0030 \text{ m/s}^2$$

Given this, we can use Equation 6.13 to calculate the orbital speed:

$$v_{orbit} = \sqrt{gr} = \sqrt{(0.0030 \text{ m/s}^2)(6.3 \times 10^3 \text{ m})}$$
$$= 4.3 \text{ m/s} \approx 10 \text{ mph}$$

ASSESS This is quite slow. With a good jump, you could easily launch yourself into an orbit around Deimos!

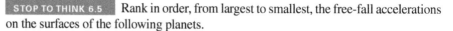

STOP TO THINK 6.5 Rank in order, from largest to smallest, the free-fall accelerations on the surfaces of the following planets.

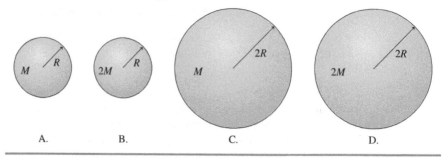

A.   B.   C.   D.

6.6 Gravity and Orbits

The planets of the solar system orbit the sun because the sun's gravitational pull, a force that points toward the center, causes the centripetal acceleration of circular motion. Mercury, the closest planet, experiences the largest acceleration, while Neptune, the most distant, has the smallest.

FIGURE 6.23 shows a large body of mass M, such as the earth or the sun, with a much smaller body of mass m orbiting it. The smaller body is called a **satellite,** even though it may be a planet orbiting the sun. Newton's second law tells us that $F_{M\,\text{on}\,m} = ma$, where $F_{M\,\text{on}\,m}$ is the gravitational force of the large body on the satellite and a is the satellite's acceleration. $F_{M\,\text{on}\,m}$ is given by Equation 6.15, and, because it's moving in a circular orbit, the satellite's acceleration is its centripetal acceleration, mv^2/r. Thus Newton's second law gives

$$F_{M\,\text{on}\,m} = \frac{GMm}{r^2} = ma = \frac{mv^2}{r} \qquad (6.20)$$

Solving for v, we find that the speed of a satellite in a circular orbit is

$$v = \sqrt{\frac{GM}{r}} \qquad (6.21)$$

Speed of a satellite in a circular orbit of radius r
about a star or planet of mass M

A satellite must have this specific speed in order to maintain a circular orbit of radius r about the larger mass M. If the velocity differs from this value, the orbit will become elliptical rather than circular. Notice that the orbital speed does not depend on the satellite's mass m. This is consistent with our previous discoveries that free-fall motion and projectile motion due to gravity are independent of the mass.

For a planet orbiting the sun, the period T is the time to complete one full orbit. The relationship among speed, radius, and period is the same as for any circular motion, $v = 2\pi r/T$. Combining this with the value of v for a circular orbit from Equation 6.21 gives

$$\sqrt{\frac{GM}{r}} = \frac{2\pi r}{T}$$

If we square both sides and rearrange, we find the period of a satellite:

$$T^2 = \left(\frac{4\pi^2}{GM}\right)r^3 \qquad (6.22)$$

Relationship between the orbital period T and radius r for a
satellite in a circular orbit around an object of mass M

In other words, **the square of the period of the orbit is proportional to the cube of the radius of the orbit.**

NOTE ▶ The mass M in Equation 6.22 is the mass of the object at the center of the orbit. ◀

This relationship between radius and period had been deduced from naked-eye observations of planetary motions by the 17th-century astronomer Johannes Kepler. One of Newton's major scientific accomplishments was to use his law of gravity and his laws of motion to prove what Kepler had deduced from observations. Even today, Newton's law of gravity and equations such as Equation 6.22 are essential tools for the NASA engineers who launch probes to other planets in the solar system.

NOTE ▶ The table inside the back cover of this text contains astronomical information about the sun and the planets that will be useful for many of the end-of-chapter problems. Note that planets farther from the sun have longer periods, in agreement with Equation 6.22. ◀

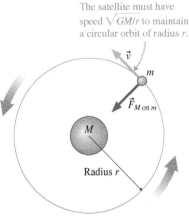

FIGURE 6.23 The orbital motion of a satellite is due to the force of gravity.

EXAMPLE 6.15   **Locating a geostationary satellite**

Communication satellites appear to "hover" over one point on the earth's equator. A satellite that appears to remain stationary as the earth rotates is said to be in a *geostationary orbit*. What is the radius of the orbit of such a satellite?

PREPARE For the satellite to remain stationary with respect to the earth, the satellite's orbital period must be 24 hours; in seconds this is $T = 8.64 \times 10^4$ s.

SOLVE We solve for the radius of the orbit by rearranging Equation 6.22. The mass at the center of the orbit is the earth:

$$r = \left(\frac{GM_e T^2}{4\pi^2}\right)^{\frac{1}{3}}$$

$$= \left(\frac{(6.67 \times 10^{-11} \text{ N} \cdot \text{m}^2/\text{kg}^2)(5.98 \times 10^{24} \text{ kg})(8.64 \times 10^4 \text{ s})^2}{4\pi^2}\right)^{\frac{1}{3}}$$

$$= 4.22 \times 10^7 \text{ m}$$

ASSESS This is a high orbit, and the radius is about 7 times the radius of the earth. Recall that the radius of the International Space Station's orbit is only about 5% larger than that of the earth.

Gravity on a Grand Scale

A spiral galaxy, similar to our Milky Way galaxy.

Although relatively weak, gravity is a long-range force. No matter how far apart two objects may be, there is a gravitational attraction between them. Consequently, gravity is the most ubiquitous force in the universe. It not only keeps your feet on the ground, but also is at work on a much larger scale. The Milky Way galaxy, the collection of stars of which our sun is a part, is held together by gravity. But why doesn't the attractive force of gravity simply pull all of the stars together?

The reason is that all of the stars in the galaxy are in orbit around the center of the galaxy. The gravitational attraction keeps the stars moving in orbits around the center of the galaxy rather than falling inward, much as the planets orbit the sun rather than falling into the sun. In the nearly 5 billion years that our solar system has existed, it has orbited the center of the galaxy approximately 20 times.

The galaxy as a whole doesn't rotate as a fixed object, though. All of the stars in the galaxy are different distances from the galaxy's center, and so orbit with different periods. Stars closer to the center complete their orbits in less time, as we would expect from Equation 6.22. As the stars orbit, their relative positions shift. Stars that are relatively near neighbors now could be on opposite sides of the galaxy at some later time.

The rotation of a *rigid body* like a wheel is much simpler. As a wheel rotates, all of the points keep the same relationship to each other. The rotational dynamics of such rigid bodies is a topic we will take up in the next chapter.

STOP TO THINK 6.6   Each year, the moon gets a little bit farther away from the earth, increasing the radius of its orbit. How does this change affect the length of a month?

A. A month gets longer.
B. A month gets shorter.
C. The length of a month stays the same.

INTEGRATED EXAMPLE 6.16   **A hunter and his sling**

A Stone Age hunter stands on a cliff overlooking a flat plain. He places a 1.0 kg rock in a sling, ties the sling to a 1.0-m-long vine, then swings the rock in a horizontal circle around his head. The plane of the motion is 25 m above the plain below. The tension in the vine increases as the rock goes faster and faster. Suddenly, just as the tension reaches 200 N, the vine snaps. If the rock is moving toward the cliff at this instant, how far out on the plain (from the base of the cliff) will it land?

PREPARE We model the rock as a particle in uniform circular motion. We can use Problem-Solving Strategy 6.1 to analyze this part of the motion. Once the vine breaks, the rock undergoes projectile motion with an initial velocity that is horizontal.

The force identification diagram of **FIGURE 6.24a** shows that the only contact force acting on the rock is the tension in the vine. Because the rock moves in a horizontal circle, you may be tempted to draw a free-body diagram like **FIGURE 6.24b**, where $\vec{T}$ is directed along the *x*-axis. You will quickly run into trouble, however, because in this diagram the net force has a downward *y*-component that would cause the rock to rapidly accelerate downward. But we know that it moves in a horizontal circle and that the net force must point toward the center of the circle. In this free-body diagram, the weight force $\vec{w}$ points straight down and is certainly correct, so the difficulty must be with $\vec{T}$.

FIGURE 6.24 Visual overview of a hunter swinging a rock.

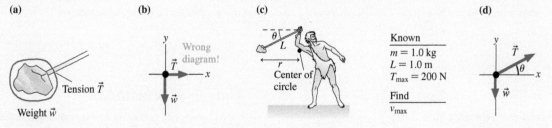

(a)

Tension $\vec{T}$

Weight $\vec{w}$

(b)

Wrong diagram!

$\vec{T}$

x

y

$\vec{w}$

(c)

θ

L

r

Center of circle

(d)

Known
$m = 1.0\text{ kg}$
$L = 1.0\text{ m}$
$T_{max} = 200\text{ N}$

Find
v_{max}

y

$\vec{T}$

θ

x

$\vec{w}$

As an experiment, tie a small weight to a string, swing it over your head, and check the angle of the string. You will discover that the string is not horizontal but, instead, is angled downward. The sketch of **FIGURE 6.24c** labels this angle θ. Notice that the rock moves in a *horizontal* circle, so the center of the circle is not at his hand. The x-axis points horizontally, to the center of the circle, but the tension force is directed along the vine. Thus the correct free-body diagram is the one in **FIGURE 6.24d**.

Once the vine breaks, the visual overview of the situation is shown in **FIGURE 6.25**. The important thing to note here is that the initial x-component of velocity is the speed the rock had an instant before the vine broke.

SOLVE From the free-body diagram of Figure 6.24d, Newton's second law for circular motion is

$$\sum F_x = T\cos\theta = \frac{mv^2}{r}$$

$$\sum F_y = T\sin\theta - mg = 0$$

where θ is the angle of the vine below the horizontal. We can use the y-equation to find the angle of the vine:

$$\sin\theta = \frac{mg}{T}$$

$$\theta = \sin^{-1}\left(\frac{mg}{T}\right) = \sin^{-1}\left(\frac{(1.0\text{ kg})(9.8\text{ m/s}^2)}{200\text{ N}}\right) = 2.81°$$

where we've evaluated the angle at the maximum tension of 200 N. The vine's angle of inclination is small but not zero.

Turning now to the x-equation, we find the rock's speed around the circle is

$$v = \sqrt{\frac{rT\cos\theta}{m}}$$

Be careful! The radius r of the circle is not the length L of the vine. You can see in Figure 6.24c that $r = L\cos\theta$. Thus

$$v = \sqrt{\frac{LT\cos^2\theta}{m}} = \sqrt{\frac{(1.0\text{ m})(200\text{ N})(\cos 2.81°)^2}{1.0\text{ kg}}} = 14.1\text{ m/s}$$

Because this is the horizontal speed of the rock just when the vine breaks, the initial velocity $(v_x)_i$ in the visual overview of the projectile motion, Figure 6.25, must be $(v_x)_i = 14.1\text{ m/s}$. Recall that a projectile has no horizontal acceleration, so the rock's final position is

$$x_f = x_i + (v_x)_i\,\Delta t = 0\text{ m} + (14.1\text{ m/s})\Delta t$$

where Δt is the time the projectile is in the air. We're not given Δt, but we can find it from the vertical motion. For a projectile, the vertical motion is just free-fall motion, so we have

$$y_f = y_i + (v_y)_i\,\Delta t - \frac{1}{2}g(\Delta t)^2$$

The initial height is $y_i = 25\text{ m}$, the final height is $y_f = 0\text{ m}$, and the initial vertical velocity is $(v_y)_i = 0\text{ m/s}$. With these values, we have

$$0\text{ m} = 25\text{ m} + (0\text{ m/s})\Delta t - \frac{1}{2}(9.8\text{ m/s}^2)(\Delta t)^2$$

Solving this for Δt gives

$$\Delta t = \sqrt{\frac{2(25\text{ m})}{9.8\text{ m/s}^2}} = 2.26\text{ s}$$

Now we can use this time to find

$$x_f = 0\text{ m} + (14.1\text{ m/s})(2.26\text{ s}) = 32\text{ m}$$

The rock lands 32 m from the base of the cliff.

ASSESS The circumference of the rock's circle is $2\pi r$, or about 6 m. At a speed of 14.1 m/s, the rock takes roughly half a second to go around once. This seems reasonable. The 32 m distance is about 100 ft, which seems easily attainable from a cliff over 75 feet high.

FIGURE 6.25 Visual overview of the rock in projectile motion.

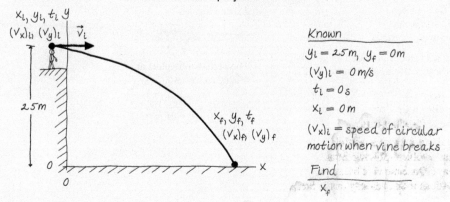

x_i, y_i, t_i
$(v_x)_i\ (v_y)_i$
$\vec{v}_i$
y

25 m

x_f, y_f, t_f
$(v_x)_f\ (v_y)_f$

0

x

0

Known
$y_i = 25\text{ m},\ y_f = 0\text{ m}$
$(v_y)_i = 0\text{ m/s}$
$t_i = 0\text{ s}$
$x_i = 0\text{ m}$

$(v_x)_i =$ speed of circular motion when vine breaks

Find
x_f

SUMMARY

Goal: To learn about motion in a circle, including orbital motion under the influence of a gravitational force.

GENERAL PRINCIPLES

Uniform Circular Motion

An object moving in a circular path is in uniform circular motion if v is constant.

- The speed is constant, but the direction of motion is constantly changing.

- The **centripetal acceleration** is directed toward the center of the circle and has magnitude

$$a = \frac{v^2}{r}$$

- This acceleration requires a net force directed toward the center of the circle. Newton's second law for circular motion is

$$\vec{F}_{net} = m\vec{a} = \left(\frac{mv^2}{r}, \text{ toward center of circle}\right)$$

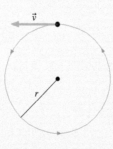

Universal Gravitation

Two objects with masses m_1 and m_2 that are distance r apart exert attractive gravitational forces on each other of magnitude

$$F_{1\,on\,2} = F_{2\,on\,1} = \frac{Gm_1 m_2}{r^2}$$

where the gravitational constant is

$$G = 6.67 \times 10^{-11} \text{ N} \cdot \text{m}^2/\text{kg}^2$$

This is **Newton's law of gravity.** Gravity is an inverse-square law.

IMPORTANT CONCEPTS

Describing circular motion

For an object moving in a circle of radius r at a constant speed v:

- The **period** T is the time to go once around the circle.

  T = time for one revolution

- The **frequency** f is defined as the number of revolutions per second. It is defined in terms of the period:

$$f = \frac{1}{T}$$

- The frequency and period are related to the speed and the radius: $v = 2\pi f r = \frac{2\pi r}{T}$

Planetary gravity

The gravitational attraction between a planet and a mass on the surface depends on the two masses and the distance to the center of the planet.

$$F_{planet\,on\,m} = \frac{GM_{planet}m}{R_{planet}^2}$$

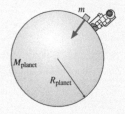

We can use this to define a value of the free-fall acceleration at the surface of a planet:

$$g_{planet} = \frac{GM_{planet}}{R_{planet}^2}$$

APPLICATIONS

Apparent weight and weightlessness

Circular motion requires a net force pointing to the center. The apparent weight $w_{app} = n$ is usually not the same as the true weight w. n must be > 0 for the object to be in contact with a surface.

In orbital motion, the net force is provided by gravity. An astronaut and his spacecraft are both in free fall, so he feels weightless.

Orbital motion

A **satellite** in a circular orbit of radius r around an object of mass M moves at a speed v given by

$$v = \sqrt{\frac{GM}{r}}$$

The period and radius are related as follows:

$$T^2 = \left(\frac{4\pi^2}{GM}\right)r^3$$

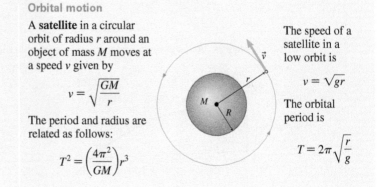

The speed of a satellite in a low orbit is

$$v = \sqrt{gr}$$

The orbital period is

$$T = 2\pi\sqrt{\frac{r}{g}}$$

QUESTIONS

Conceptual Questions

1. A cyclist goes around a level, circular track at constant speed. Do you agree or disagree with the following statement? "Since the cyclist's speed is constant, her acceleration is zero." Explain.
2. In uniform circular motion, which of the following quantities are constant: speed, instantaneous velocity, centripetal acceleration, the magnitude of the net force?
3. A particle moving along a straight line can have nonzero acceleration even when its speed is zero (for instance, a ball in free fall at the top of its path). Can a particle moving in a circle have nonzero *centripetal* acceleration when its speed is zero? If so, give an example. If not, why not?
4. Would having four-wheel drive on a car make it possible to drive faster around corners on an icy road, without slipping, than the same car with two-wheel drive? Explain.
5. Large birds like pheasants often walk short distances. Small birds like chickadees never walk. They either hop or fly. Why might this be?
   BIO
6. When you drive fast on the highway with muddy tires, you can hear the mud flying off the tires into your wheel wells. Why does the mud fly off?
7. A ball on a string moves in a vertical circle as in Figure Q6.7. When the ball is at its lowest point, is the tension in the string greater than, less than, or equal to the ball's weight? Explain. (You may want to include a free-body diagram as part of your explanation.)

FIGURE Q6.7

8. Give an everyday example of circular motion for which the centripetal acceleration is mostly or completely due to a force of the type specified: (a) Static friction. (b) Tension.
9. Give an everyday example of circular motion for which the centripetal acceleration is mostly or completely due to a force of the type specified: (a) Gravity. (b) Normal force.
10. It's been proposed that future space stations create "artificial gravity" by rotating around an axis. (The space station would have to be much larger than the present space station for this to be feasible.)
    a. How would this work? Explain.
    b. Would the artificial gravity be equally effective throughout the space station? If not, where in the space station would the residents want to live and work?
11. A car coasts at a constant speed over a circular hill. Which of the free-body diagrams in Figure Q6.11 is correct? Explain.

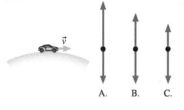

FIGURE Q6.11

A.   B.   C.

12. In Figure Q6.11, at the instant shown, is the apparent weight of the car's driver greater than, less than, or equal to his true weight? Explain.
13. Riding in the back of a pickup truck can be very dangerous. If the truck turns suddenly, the riders can be thrown from the truck bed. Why are the riders ejected from the bed?
14. Playground swings move through an arc of a circle. When you are on a swing, and at the lowest point of your motion, is your apparent weight greater than, less than, or equal to your true weight? Explain.
15. Variation in your apparent weight is desirable when you ride a roller coaster; it makes the ride fun. However, too much variation over a short period of time can be painful. For this reason, the loops of real roller coasters are not simply circles like Figure 6.16a. A typical loop is shown in Figure Q6.15. The radius of the circle that matches the track at the top of the loop is much smaller than that of a matching circle at other places on the track. Explain why this shape gives a more comfortable ride than a circular loop.

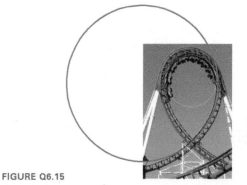

FIGURE Q6.15

16. A small projectile is launched parallel to the ground at height $h = 1$ m with sufficient speed to orbit a completely smooth, airless planet. A bug rides in a small hole inside the projectile. Is the bug weightless? Explain.
17. Why is it impossible for an astronaut inside an orbiting space station to go from one end to the other by walking normally?
18. If every object in the universe feels an attractive gravitational force due to every other object, why don't you feel a pull from someone seated next to you?
19. A mountain climber's weight is slightly less on the top of a tall mountain than at the base, though his mass is the same. Why?
20. Is the earth's gravitational force on the sun larger than, smaller than, or equal to the sun's gravitational force on the earth? Explain.

Multiple-Choice Questions

21. I A ball on a string moves around a complete circle, once a second, on a frictionless, horizontal table. The tension in the string is measured to be 6.0 N. What would the tension be if the ball went around in only half a second?
    A. 1.5 N   B. 3.0 N   C. 12 N   D. 24 N

22. | As seen from above, a car rounds the curved path shown in Figure Q6.22 at a constant speed. Which vector best represents the net force acting on the car?

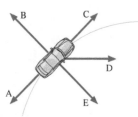

FIGURE Q6.22

23. | As we saw in the chapter, wings on race cars push them into the track. The increased normal force makes large friction forces possible. At one Formula One racetrack, cars turn around a half-circle with diameter 190 m at 68 m/s. For a 610 kg vehicle, the approximate minimum static friction force to complete this turn is
A. 6000 N   B. 15,000 N   C. 18,000 N
D. 24,000 N   E. 30,000 N

24. | Suppose you and a friend, each of mass 60 kg, go to the park and get on a 4.0-m-diameter merry-go-round. You stand on the outside edge of the merry-go-round, while your friend pushes so that it rotates once every 6.0 s. What is the magnitude of the (apparent) outward force that you feel?
A. 7 N   B. 63 N   C. 130 N   D. 260 N

25. | The cylindrical space station in Figure Q6.25, 200 m in diameter, rotates in order to provide artificial gravity of g for the occupants. How much time does the station take to complete one rotation?
A. 3 s   B. 20 s   C. 28 s   D. 32 s

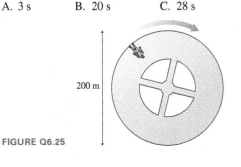

200 m

FIGURE Q6.25

26. || Two cylindrical space stations, the second four times the diameter of the first, rotate so as to provide the same amount of artificial gravity. If the first station makes one rotation in the time T, then the second station makes one rotation in time
A. $T/4$ B. $2T$ C. $4T$ D. $16T$

27. || The radius of Jupiter is 11 times that of earth, and the free-fall acceleration near its surface is 2.5 times that on earth. If we someday put a spacecraft in low Jupiter orbit, its orbital speed will be
A. Greater than that for an earth satellite.
B. The same as that for an earth satellite.
C. Less than that for an earth satellite.

28. | A newly discovered planet has twice the mass and three times the radius of the earth. What is the free-fall acceleration at its surface, in terms of the free-fall acceleration g at the surface of the earth?
A. $\frac{2}{9}g$ B. $\frac{2}{3}g$ C. $\frac{3}{4}g$ D. $\frac{4}{3}g$

29. || Suppose one night the radius of the earth doubled but its mass stayed the same. What would be an approximate new value for the free-fall acceleration at the surface of the earth?
A. 2.5 m/s²   B. 5.0 m/s²   C. 10 m/s²   D. 20 m/s²

30. | Currently, the moon goes around the earth once every 27.3 days. If the moon could be brought into a new circular orbit with a smaller radius, its orbital period would be
A. More than 27.3 days.
B. 27.3 days.
C. Less than 27.3 days.

31. || Two planets orbit a star. You can ignore the gravitational interactions between the planets. Planet 1 has orbital radius r_1 and planet 2 has $r_2 = 4r_1$. Planet 1 orbits with period T_1. Planet 2 orbits with period
A. $T_2 = \frac{1}{2}T_1$ B. $T_2 = 2T_1$ C. $T_2 = 4T_1$ D. $T_2 = 8T_1$

PROBLEMS

Section 6.1 Uniform Circular Motion

1. || A 5.0-m-diameter merry-go-round is turning with a 4.0 s period. What is the speed of a child on the rim?

2. | The blade on a table saw spins at 3450 rpm. Its diameter is 25.0 cm. What is the speed of a tooth on the edge of the blade, in both m/s and mph?

3. | An old-fashioned LP record rotates at $33\frac{1}{3}$ rpm.
   a. What is its frequency, in rev/s?
   b. What is its period, in seconds?

4. | A typical hard disk in a computer spins at 5400 rpm.
   a. What is the frequency, in rev/s?
   b. What is the period, in seconds?

5. || A CD-ROM drive in a computer spins the 12-cm-diameter disks at 10,000 rpm.
   a. What are the disk's period (in s) and frequency (in rev/s)?
   b. What would be the speed of a speck of dust on the outside edge of this disk?
   c. What is the acceleration in units of g that this speck of dust experiences?

6. || The horse on a carousel is 4.0 m from the central axis.
   a. If the carousel rotates at 0.10 rev/s, how long does it take the horse to go around twice?
   b. How fast is a child on the horse going (in m/s)?

7. ||| The radius of the earth's very nearly circular orbit around the sun is 1.50×10^{11} m. Find the magnitude of the earth's (a) velocity and (b) centripetal acceleration as it travels around the sun. Assume a year of 365 days.

8. | Modern wind turbines are larger than they appear, and despite their apparently lazy motion, the speed of the blades tips can be quite high—many times higher than the wind speed. A typical modern turbine has blades 56 m long that spin at 13 rpm. At the tip of a blade, what are (a) the speed and (b) the centripetal acceleration?

9. ‖ Your roommate is working on his bicycle and has the bike upside down. He spins the 60-cm-diameter wheel, and you notice that a pebble stuck in the tread goes by three times every second. What are the pebble's speed and acceleration?

10. ‖ Wind turbines designed for offshore installations are much larger than ones designed for use on land. One company makes a turbine for use on land with blades 40 m long that spin at 17 rpm, and a turbine for use offshore with blades that are twice as long. At what rate should the blades of the offshore turbine rotate if the speed at the tip is to be the same as that of the turbine on land?

11. ‖ To withstand "g-forces" of up to 10g, caused by suddenly pulling out of a steep dive, fighter jet pilots train on a "human centrifuge." 10g is an acceleration of 98 m/s^2. If the length of the centrifuge arm is 12 m, at what speed is the rider moving when she experiences 10g?

12. ‖ A typical running track is an oval with 74-m-diameter half circles at each end. A runner going once around the track covers a distance of 400 m. Suppose a runner, moving at a constant speed, goes once around the track in 1 min 40 s. What is her centripetal acceleration during the turn at each end of the track?

Section 6.2 Dynamics of Uniform Circular Motion

13. ‖‖‖ Figure P6.13 is a bird's-eye view of particles on a string moving in horizontal circles on a tabletop. All are moving at the same speed. Rank in order, from largest to smallest, the tensions T_1 to T_4.

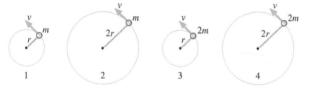

FIGURE P6.13

14. ‖ In short-track speed skating, the track has straight sections and semicircles 16 m in diameter. Assume that a 65 kg skater goes around the turn at a constant 12 m/s.
    a. What is the horizontal force on the skater?
    b. What is the ratio of this force to the skater's weight?

15. ‖‖ A 200 g block on a 50-cm-long string swings in a circle on a horizontal, frictionless table at 75 rpm.
    a. What is the speed of the block?
    b. What is the tension in the string?

16. ‖ A cyclist is rounding a 20-m-radius curve at 12 m/s. What is
INT the minimum possible coefficient of static friction between the bike tires and the ground?

17. ‖ A 1500 kg car drives around a flat 200-m-diameter circular track at 25 m/s. What are the magnitude and direction of the net force on the car? What causes this force?

18. ‖ A fast pitch softball player does a "windmill" pitch, illus-
BIO trated in Figure P6.18, moving her hand through a circular arc to pitch a ball at 70 mph. The 0.19 kg ball is 50 cm from the pivot point at her shoulder. At the lowest point of the circle, the ball has reached its maximum speed.
    a. At the bottom of the circle, just before the ball leaves her hand, what is its centripetal acceleration?
    b. What are the magnitude and direction of the force her hand exerts on the ball at this point?

FIGURE P6.18

19. ‖ A baseball pitching machine works by rotating a light and stiff rigid rod about a horizontal axis until the ball is moving toward the target. Suppose a 144 g baseball is held 85 cm from the axis of rotation and released at the major league pitching speed of 85 mph.
    a. What is the ball's centripetal acceleration just before it is released?
    b. What is the magnitude of the net force that is acting on the ball just before it is released?

20. ‖ A wind turbine has 12,000 kg blades that are 38 m long. The blades spin at 22 rpm. If we model a blade as a point mass at the midpoint of the blade, what is the inward force necessary to provide each blade's centripetal acceleration?

21. ‖ You're driving your pickup truck around a curve with a radius of 20 m. A box in the back of the truck is pressed up against the wall of the truck. How fast must you drive so that the force of the wall on the box equals the weight of the box?

22. ‖ You have seen dogs shake to shed water from their fur. The
BIO motion is complicated, but the fur on a dog's torso rotates back and forth along a roughly circular arc. Water droplets are held to the fur by contact forces, and these forces provide the centripetal acceleration that keeps the droplets moving in a circle, still attached to the fur, if the dog shakes gently. But these contact forces—like static friction—have a maximum possible value. As the dog shakes more vigorously, the contact forces cannot provide sufficient centripetal acceleration and the droplets fly off. A big dog has a torso that is approximately circular, with a radius of 16 cm. At the midpoint of a shake, the dog's fur is moving at a remarkable 2.5 m/s.
    a. What force is required to keep a 10 mg water droplet moving in this circular arc?
    b. What is the ratio of this force to the weight of a droplet?

23. ‖ Gibbons, small Asian apes, move by *brachiation,* swinging
BIO below a handhold to move forward to the next handhold. A 9.0 kg gibbon has an arm length (hand to shoulder) of 0.60 m. We can model its motion as that of a point mass swinging at the end of a 0.60-m-long, massless rod. At the lowest point of its swing, the gibbon is moving at 3.5 m/s. What upward force must a branch provide to support the swinging gibbon?

Section 6.3 Apparent Forces in Circular Motion

24. ‖‖ The passengers in a roller coaster car feel 50% heavier than their true weight as the car goes through a dip with a 30 m radius of curvature. What is the car's speed at the bottom of the dip?

25. ‖ You hold a bucket in one hand. In the bucket is a 500 g rock. You swing the bucket so the rock moves in a vertical circle 2.2 m in diameter. What is the minimum speed the rock must have at the top of the circle if it is to always stay in contact with the bottom of the bucket?

26. ‖ A roller coaster car is going over the top of a 15-m-radius circular rise. At the top of the hill, the passengers "feel light," with an apparent weight only 50% of their true weight. How fast is the coaster moving?

27. ‖‖ As a roller coaster car crosses the top of a 40-m-diameter loop-the-loop, its apparent weight is the same as its true weight. What is the car's speed at the top?

28. ‖ An 80-ft-diameter Ferris wheel rotates once every 24 s. What is the apparent weight of a 70 kg passenger at (a) the lowest point of the circle and (b) at the highest point?

29. ‖‖ A typical laboratory centrifuge rotates at 4000 rpm. Test
BIO tubes have to be placed into a centrifuge very carefully because
INT of the very large accelerations.
   a. What is the acceleration at the end of a test tube that is 10 cm from the axis of rotation?
   b. For comparison, what is the magnitude of the acceleration a test tube would experience if stopped in a 1.0-ms-long encounter with a hard floor after falling from a height of 1.0 m?

Section 6.4 Circular Orbits and Weightlessness

30. ‖‖ A satellite orbiting the moon very near the surface has a period of 110 min. Use this information, together with the radius of the moon from the table on the inside of the back cover, to calculate the free-fall acceleration on the moon's surface.

31. ‖ Spacecraft have been sent to Mars in recent years. Mars is smaller than Earth and has correspondingly weaker surface gravity. On Mars, the free-fall acceleration is only 3.8 m/s². What is the orbital period of a spacecraft in a low orbit near the surface of Mars?

Section 6.5 Newton's Law of Gravity

32. ‖‖ The centers of a 10 kg lead ball and a 100 g lead ball are separated by 10 cm.
   a. What gravitational force does each exert on the other?
   b. What is the ratio of this gravitational force to the weight of the 100 g ball?

33. ‖ The gravitational force of a star on an orbiting planet 1 is F_1. Planet 2, which is twice as massive as planet 1 and orbits at twice the distance from the star, experiences gravitational force F_2. What is the ratio F_2/F_1? You can ignore the gravitational force between the two planets.

34. ‖ The free-fall acceleration at the surface of planet 1 is 20 m/s². The radius and the mass of planet 2 are twice those of planet 1. What is the free-fall acceleration on planet 2?

35. ‖‖ What is the ratio of the sun's gravitational force on you to the earth's gravitational force on you?

36. ‖‖ Suppose the free-fall acceleration at some location on earth was exactly 9.8000 m/s². What would it be at the top of a 1000-m-tall tower at this location? (Give your answer to five significant figures.)

37. ‖ In recent years, astronomers have found planets orbiting nearby stars that are quite different from planets in our solar system. Kepler-12b, has a diameter that is 1.7 times that of Jupiter, but a mass that is only 0.43 that of Jupiter. What is the value of g on this large, but low-density, world?

38. ‖ In recent years, astronomers have found planets orbiting nearby stars that are quite different from planets in our solar system. Kepler-39b, has a diameter that is 1.2 times that of Jupiter, but a mass that is 18 times that of Jupiter. What would be the period of a satellite in a low orbit around this large, dense planet?

39. ‖ a. What is the gravitational force of the sun on the earth?
   b. What is the gravitational force of the moon on the earth?
   c. The moon's force is what percent of the sun's force?

40. ‖ What is the value of g on the surface of Saturn? Explain how such a low value is possible given Saturn's large mass—100 times that of Earth.

41. ‖ What is the free-fall acceleration at the surface of (a) Mars and (b) Jupiter?

Section 6.6 Gravity and Orbits

42. ‖‖ Planet X orbits the star Omega with a "year" that is 200 earth days long. Planet Y circles Omega at four times the distance of Planet X. How long is a year on Planet Y?

43. ‖‖‖ Satellite A orbits a planet with a speed of 10,000 m/s. Satellite B is twice as massive as satellite A and orbits at twice the distance from the center of the planet. What is the speed of satellite B? Assume that both orbits are circular.

44. ‖‖ The International Space Station is in a 250-mile-high orbit. What are the station's orbital period, in minutes, and speed?

45. ‖ The *asteroid belt* circles the sun between the orbits of Mars and Jupiter. One asteroid has a period of 5.0 earth years. What are the asteroid's orbital radius and speed?

46. ‖‖‖ An earth satellite moves in a circular orbit at a speed of 5500 m/s. What is its orbital period?

In recent years, scientists have discovered hundreds of planets orbiting other stars. Some of these planets are in orbits that are similar to that of earth, which orbits the sun ($M_{sun} = 1.99 \times 10^{30}$ kg) at a distance of 1.50×10^{11} m, called 1 *astronomical unit* (1 au). Others have extreme orbits that are much different from anything in our solar system. Problems 47–49 relate to some of these planets that follow circular orbits around other stars.

47. ‖‖ WASP-32b orbits with a period of only 2.7 days a star with a mass that is 1.1 times that of the sun. How many au from the star is this planet?

48. ‖‖ HD 10180g orbits with a period of 600 days at a distance of 1.4 au from its star. What is the ratio of the star's mass to our sun's mass?

49. ‖‖ Kepler-42c orbits at a very close 0.0058 au from a small star with a mass that is 0.13 that of the sun. How long is a "year" on this world?

General Problems

50. ‖ How fast must a plane fly along the earth's equator so that the sun stands still relative to the passengers? In which direction must the plane fly, east to west or west to east? Give your answer in both km/h and mph. The radius of the earth is 6400 km.

51. ‖‖ The car in Figure P6.51 travels at a constant speed along the road shown. Draw vectors showing its acceleration at the three points A, B, and C, or write $\vec{a} = \vec{0}$. The lengths of your vectors should correspond to the magnitudes of the accelerations.

FIGURE P6.51

52. ‖ In the Bohr model of the hydrogen atom, an electron (mass $m = 9.1 \times 10^{-31}$ kg) orbits a proton at a distance of 5.3×10^{-11} m. The proton pulls on the electron with an electric force of 8.2×10^{-8} N. How many revolutions per second does the electron make?

53. ‖ A 75 kg man weighs himself at the north pole and at the equator. Which scale reading is higher? By how much? Assume the earth is a perfect sphere. Explain why the readings differ.

54. ‖ A 1500 kg car takes a 50-m-radius unbanked curve at 15 m/s. What is the size of the friction force on the car?

55. ‖‖ A 500 g ball swings in a vertical circle at the end of a 1.5-m-long string. When the ball is at the bottom of the circle, the tension in the string is 15 N. What is the speed of the ball at that point?

56. ‖ A 5.0 g coin is placed 15 cm from the center of a turntable.
NT The coin has static and kinetic coefficients of friction with the turntable surface of $\mu_s = 0.80$ and $\mu_k = 0.50$. The turntable very slowly speeds up to 60 rpm. Does the coin slide off?

57. ‖‖ A *conical pendulum* is formed by attaching a 500 g ball to a 1.0-m-long string, then allowing the mass to move in a horizontal circle of radius 20 cm. Figure P6.57 shows that the string traces out the surface of a cone, hence the name.
    a. What is the tension in the string?
    b. What is the ball's speed?
    c. What is the period of the ball's orbit?

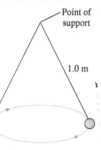

Point of support

1.0 m

FIGURE P6.57

Hint: Determine the horizontal and vertical components of the forces acting on the ball, and use the fact that the vertical component of acceleration is zero since there is no vertical motion.

58. ‖‖ In an old-fashioned amusement park ride, passengers stand inside a 3.0-m-tall, 5.0-m-diameter hollow steel cylinder with their backs against the wall. The cylinder begins to rotate about a vertical axis. Then the floor on which the passengers are standing suddenly drops away! If all goes well, the passengers will "stick" to the wall and not slide. Clothing has a static coefficient of friction against steel in the range 0.60 to 1.0 and a kinetic coefficient in the range 0.40 to 0.70. What is the minimum rotational frequency, in rpm, for which the ride is safe?

59. ‖‖ The 0.20 kg puck on the
NT frictionless, horizontal table in Figure P6.59 is connected by a string through a hole in the table to a hanging 1.20 kg block. With what speed must the puck rotate in a circle of radius 0.50 m if the block is to remain hanging at rest?

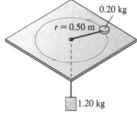

0.20 kg

$r = 0.50$ m

1.20 kg

FIGURE P6.59

60. ‖ While at the county fair, you decide to ride the Ferris wheel. Having eaten too many candy apples and elephant ears, you find the motion somewhat unpleasant. To take your mind off your stomach, you wonder about the motion of the ride. You estimate the radius of the big wheel to be 15 m, and you use your watch to find that each loop around takes 25 s.
    a. What are your speed and magnitude of your acceleration?
    b. What is the ratio of your apparent weight to your true weight at the top of the ride?
    c. What is the ratio of your apparent weight to your true weight at the bottom?

61. ‖ A car drives over the top of a hill that has a radius of 50 m. What maximum speed can the car have without flying off the road at the top of the hill?

62. ‖‖ A 100 g ball on a 60-cm-long string is swung in a vertical
NT circle whose center is 200 cm above the floor. The string suddenly breaks when it is parallel to the ground and the ball is moving upward. The ball reaches a height 600 cm above the floor. What was the tension in the string an instant before it broke?

63. ‖‖ The two identical pucks in Figure P6.63 rotate together on a fric-
NT tionless, horizontal table. They are tied together by strings 1 and 2, each of length l. If they rotate together at a frequency f, what are the tensions in the two strings?

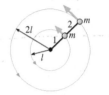

$2l$ 2 ○m

1 ○m

l

FIGURE P6.63

64. ‖‖ The ultracentrifuge is an important tool for separating and
BIO analyzing proteins in biological research. Because of the
NT enormous centripetal accelerations that can be achieved, the apparatus (see Figure 6.18) must be carefully balanced so that each sample is matched by another on the opposite side of the rotor shaft. Failure to do so is a costly mistake, as seen in Figure P6.64. Any difference in mass of the opposing samples will cause a net force in the horizontal plane on the shaft of the rotor. Suppose that a scientist makes a slight error in sample preparation, and one sample has a mass 10 mg greater than the opposing sample. If the samples are 10 cm from the axis of the rotor and the ultracentrifuge spins at 70,000 rpm, what is the magnitude of the net force on the rotor due to the unbalanced samples?

FIGURE P6.64

65. ‖‖ A sensitive gravimeter at a mountain observatory finds that the free-fall acceleration is 0.0075 m/s² less than that at sea level. What is the observatory's altitude?

66. ‖ Suppose we could shrink the earth without changing its mass. At what fraction of its current radius would the free-fall acceleration at the surface be three times its present value?

67. ‖‖ Planet Z is 10,000 km in diameter. The free-fall acceleration on Planet Z is 8.0 m/s².
    a. What is the mass of Planet Z?
    b. What is the free-fall acceleration 10,000 km above Planet Z's north pole?

68. ‖‖ What are the speed and altitude of a geostationary satellite (see Example 6.15) orbiting Mars? Mars rotates on its axis once every 24.8 hours.

69. ‖‖ a. What is the free-fall acceleration on Mars?
BIO   b. Estimate the maximum speed at which an astronaut can walk on the surface of Mars.

70. ‖‖ How long will it take a rock dropped from 2.0 m above the
NT surface of Mars to reach the ground?

71. ▌ A 20 kg sphere is at the origin and a 10 kg sphere is at
INT (x, y) = (20 cm, 0 cm). At what point or points could you place
a small mass such that the net gravitational force on it due to the
spheres is zero?

72. ▌ a. At what height above the earth is the free-fall acceleration
10% of its value at the surface?
   b. What is the speed of a satellite orbiting at that height?

73. ▏ Mars has a small moon, Phobos, that orbits with a period of 7 h
39 min. The radius of Phobos' orbit is 9.4×10^6 m. Use only this
information (and the value of G) to calculate the mass of Mars.

74. ▐ You are the science officer on a visit to a distant solar
system. Prior to landing on a planet you measure its diameter to
be 1.80×10^7 m and its rotation period to be 22.3 h. You have
previously determined that the planet orbits 2.20×10^{11} m from
its star with a period of 402 earth days. Once on the surface
you find that the free-fall acceleration is 12.2 m/s². What are the
masses of (a) the planet and (b) the star?

75. ▐▌ Europa, a satellite of Jupiter,
BIO is believed to have a liquid ocean
of water (with a possibility of
life) beneath its icy surface. In
planning a future mission to
Europa, what is the fastest that
an astronaut with legs of length
0.70 m could walk on the surface
of Europa? Europa is 3100 km in
diameter and has a mass of 4.8×10^{22} kg.

MCAT-Style Passage Problems

Orbiting the Moon

Suppose a spacecraft orbits the moon in a very low, circular orbit, just
a few hundred meters above the lunar surface. The moon has a diameter of 3500 km, and the free-fall acceleration at the surface is 1.6 m/s².

76. ▏ The direction of the net force on the craft is
   A. Away from the surface of the moon.
   B. In the direction of motion.
   C. Toward the center of the moon.
   D. Nonexistent, because the net force is zero.

77. ▏ How fast is this spacecraft moving?
   A. 53 m/s   B. 75 m/s   C. 1700 m/s   D. 2400 m/s

78. ▏ How much time does it take for the spacecraft to complete
one orbit?
   A. 38 min   B. 76 min   C. 110 min   D. 220 min

79. ▏ The material that comprises the side of the moon facing the
earth is actually slightly more dense than the material on the far
side. When the spacecraft is above a more dense area of the sur-
face, the moon's gravitational force on the craft is a bit stronger.
In order to stay in a circular orbit of constant height and speed, the
spacecraft could fire its rockets while passing over the denser area.
The rockets should be fired so as to generate a force on the craft
   A. Away from the surface of the moon.
   B. In the direction of motion.
   C. Toward the center of the moon.
   D. Opposite the direction of motion.

STOP TO THINK ANSWERS

Chapter Preview Stop to Think: D. The ball is in uniform circular
motion. The acceleration is directed toward the center of the circle,
which is to the right at the instant shown.

Stop to Think 6.1: C > D = A > B. Rearranging Equation 6.3
gives $T = \frac{2\pi r}{v}$. For the cases shown, speed is either v or $2v$; the radius
of the circular path is r or $2r$. Going around a circle of radius r at a
speed v takes the same time as going around a circle of radius $2r$ at a
speed $2v$. It's twice the distance at twice the speed.

Stop to Think 6.2: $T_A < T_B < T_C < T_D < T_E$. The tension force
provides the centripetal acceleration, and larger acceleration implies
larger force. So the question reduces to one about acceleration: Rank
the centripetal accelerations for these cases. Equation 6.5 shows that
the acceleration is proportional to the radius of the circle and the
square of the frequency, and so the acceleration increases steadily as
we move from A to E.

Stop to Think 6.3: B. The car is moving in a circle, so there must be
a net force toward the center of the circle. The center of the circle is

below the car, so the net force must point downward. This can be true
only if $w > n$. This makes sense; $n < w$, so the apparent weight is
less than the true weight. The riders in the car "feel light"; if you've
driven over a rise like this, you know that this is what you feel.

Stop to Think 6.4: B. The period of a satellite doesn't depend on
the mass of the satellite. If you increase the height above the surface, you increase the radius of the orbit. This will result in an
increased period. Increasing the value of g would cause the period
to decrease.

Stop to Think 6.5: B > A > D > C. The free-fall acceleration is
proportional to the mass, but inversely proportional to the square of
the radius.

Stop to Think 6.6: A. The length of a month is determined by the
period of the moon's orbit. Equation 6.22 shows that as the moon
gets farther away, the period of the orbit—and thus the length of a
month—increases.

7 Rotational Motion

This cyclist is clearly interested in high speeds and large accelerations. For the largest possible acceleration, she needs to lighten her bike. The most important parts to lighten are the tires and wheels. Why is this?

LOOKING AHEAD ▸

Goal: To understand the physics of rotating objects.

Rotational Kinematics

The spinning roulette wheel isn't going anywhere, but it is moving. This is **rotational motion.**

You'll learn about angular velocity and other quantities we use to describe rotational motion.

Torque

To start something moving, apply a force. To start something rotating, apply a **torque,** as this sailor is doing to the wheel.

You'll see that torque depends on how hard you push and also on *where* you push. A push far from the axle gives a large torque.

Rotational Dynamics

The girl pushes on the outside edge of the merry-go-round, gradually increasing its rotation rate.

You'll learn a version of Newton's second law for rotational motion and use it to solve problems.

LOOKING BACK ◂

Circular Motion

In Chapter 6, you learned to describe *circular* motion in terms of period, frequency, velocity, and centripetal acceleration.

In this chapter, you'll learn to use angular velocity, angular acceleration, and other quantities that describe *rotational* motion.

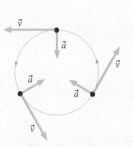

STOP TO THINK

As an audio CD plays, the frequency at which the disk spins changes. At 210 rpm, the speed of a point on the outside edge of the disk is 1.3 m/s. At 420 rpm, the speed of a point on the outside edge is

A. 1.3 m/s   B. 2.6 m/s
C. 3.9 m/s   D. 5.2 m/s

7.1 Describing Circular and Rotational Motion

Moving blades on a wind turbine.

The photo shows a spinning wind turbine. The blades are clearly in motion. As the blades go around, each point on each blade is moving in a circle at a constant speed. But you can see that the outer parts of the blades are more blurred than the parts nearer the hub. This tells us that the outer parts of the blades are moving more rapidly. Different parts of the blades move at different speeds. How do we account for this in our analysis?

In this chapter, we'll consider **rotational motion**, the motion of objects that spin about an axis, like the blade assembly of the wind turbine. We'll introduce some new concepts, but we'll start by returning to a topic we first saw in ◄ SECTION 6.1—the motion of a particle in uniform circular motion, such as the motion of the tip of a wind turbine blade. We'll extend this treatment to describe the motion of the whole system. Doing this will require us to define some new quantities.

Angular Position

FIGURE 7.1 A particle's angular position is described by angle θ.

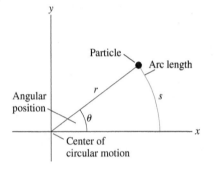

When we describe the motion of a particle as it moves in a circle, it is convenient to use the angle θ from the positive x-axis to describe the particle's location. This is shown in FIGURE 7.1. Because the particle travels in a circle with a fixed radius r, specifying θ completely locates the position of the particle. Thus we call angle θ the **angular position** of the particle.

We define θ to be positive when measured *counterclockwise* from the positive x-axis. An angle measured *clockwise* from the positive x-axis has a negative value. The words "clockwise" and "counterclockwise" in circular motion are analogous, respectively, to "left of the origin" and "right of the origin" in linear motion, which we associate with negative and positive values of x.

Rather than measure angles in degrees, mathematicians and scientists usually measure angle θ in the angular unit of *radians*. In Figure 7.1, we also show the **arc length** s, the distance that the particle has traveled along its circular path. We define the particle's angle θ in **radians** in terms of this arc length and the radius of the circle:

$$\theta \text{ (radians)} = \frac{s}{r} \qquad (7.1)$$

This is a sensible definition of an angle: The farther the particle has traveled around the circle (i.e., the greater s is), the larger the angle θ in radians. The radian, abbreviated rad, is the SI unit of angle. An angle of 1 rad has an arc length s exactly equal to the radius r. An important consequence of Equation 7.1 is that the arc length spanning the angle θ is

$$s = r\theta \qquad (7.2)$$

NOTE ► Equation 7.2 is valid only if θ is measured in radians, not degrees. This very simple relationship between angle and arc length is one of the primary motivations for using radians. ◄

When a particle travels all the way around the circle—completing one *revolution,* abbreviated rev—the arc length it travels is the circle's circumference $2\pi r$. Thus the angle of a full circle is

$$\theta_{\text{full circle}} = \frac{s}{r} = \frac{2\pi r}{r} = 2\pi \text{ rad}$$

We can use this fact to write conversion factors among revolutions, radians, and degrees:

$$1 \text{ rev} = 360° = 2\pi \text{ rad}$$

$$1 \text{ rad} = 1 \text{ rad} \times \frac{360°}{2\pi \text{ rad}} = 57.3°$$

We will often specify angles in degrees, but keep in mind that the SI unit is the radian. You can visualize angles in radians by remembering that 1 rad is just about 60°.

Angular Displacement and Angular Velocity

For the *linear* motion you studied in Chapters 1 and 2, a particle with a larger velocity undergoes a greater displacement in each second than one with a smaller velocity, as FIGURE 7.2a shows. FIGURE 7.2b shows two particles undergoing uniform *circular* motion. The particle on the left is moving slowly around the circle; it has gone only one-quarter of the way around after 5 seconds. The particle on the right is moving much faster around the circle, covering half of the circle in the same 5 seconds. You can see that the particle to the right undergoes twice the **angular displacement** $\Delta\theta$ during each interval as the particle to the left. Its **angular velocity,** the angular displacement through which the particle moves each second, is twice as large.

FIGURE 7.2 Comparing uniform linear and circular motion.

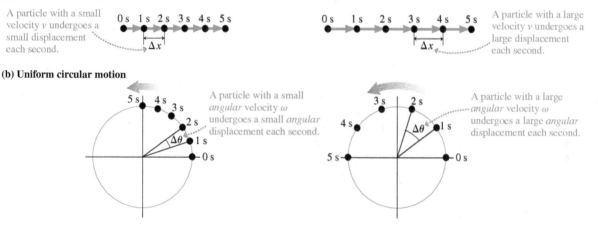

(a) Uniform linear motion

A particle with a small velocity v undergoes a small displacement each second.

A particle with a large velocity v undergoes a large displacement each second.

(b) Uniform circular motion

A particle with a small *angular* velocity ω undergoes a small *angular* displacement each second.

A particle with a large *angular* velocity ω undergoes a large *angular* displacement each second.

In analogy with linear motion, where $v_x = \Delta x / \Delta t$, we thus define angular velocity as

$$\omega = \frac{\text{angular displacement}}{\text{time interval}} = \frac{\Delta\theta}{\Delta t} \qquad (7.3)$$

Angular velocity of a particle in uniform circular motion

The symbol ω is a lowercase Greek omega, *not* an ordinary w. The SI unit of angular velocity is rad/s.

Figure 7.2a shows that the displacement Δx of a particle in uniform linear motion changes by the same amount each second. Similarly, as Figure 7.2b shows, the *angular* displacement $\Delta\theta$ of a particle in uniform *circular* motion changes by the same amount each second. This means that **the angular velocity $\omega = \Delta\theta / \Delta t$ is constant for a particle moving with uniform circular motion.**

EXAMPLE 7.1 **Comparing angular velocities**

Find the angular velocities of the two particles in Figure 7.2b.

PREPARE For uniform circular motion, we can use any angular displacement $\Delta\theta$, as long as we use the corresponding time interval Δt. For each particle, we'll choose the angular displacement corresponding to the motion from $t = 0$ s to $t = 5$ s.

SOLVE The particle on the left travels one-quarter of a full circle during the 5 s time interval. We learned earlier that a full circle corresponds to an angle of 2π rad, so the angular displacement for this particle is $\Delta\theta = (2\pi \text{ rad})/4 = \pi/2 \text{ rad}$. Thus its angular velocity is

$$\omega = \frac{\Delta\theta}{\Delta t} = \frac{\pi/2 \text{ rad}}{5 \text{ s}} = 0.314 \text{ rad/s}$$

The particle on the right travels halfway around the circle, or π rad, in the 5 s interval. Its angular velocity is

$$\omega = \frac{\Delta\theta}{\Delta t} = \frac{\pi \text{ rad}}{5 \text{ s}} = 0.628 \text{ rad/s}$$

ASSESS The speed of the second particle is double that of the first, as it should be. We should also check the scale of the answers. The angular velocity of the particle on the right is 0.628 rad/s, meaning that the particle travels through an angle of 0.628 rad each second. Because 1 rad $\approx 60°$, 0.628 rad is roughly 35°. In Figure 7.2b, the particle on the right appears to move through an angle of about this size during each 1 s time interval, so our answer is reasonable.

FIGURE 7.3 Positive and negative angular velocities.

ω is positive for a counterclockwise rotation.

ω is negative for a clockwise rotation.

Angular velocity, like the velocity v_x of one-dimensional motion, can be positive or negative. The signs for ω noted in FIGURE 7.3 are based on the convention that angles are positive when measured counterclockwise from the positive x-axis.

We've already noted how circular motion is analogous to linear motion, with angular variables replacing linear variables. Thus much of what you learned about linear kinematics and dynamics carries over to circular motion. For example, Equation 2.4 gave us a formula for computing a linear displacement during a time interval:

$$x_f - x_i = \Delta x = v_x \, \Delta t$$

You can see from Equation 7.3 that we can write a similar equation for an angular displacement:

$$\theta_f - \theta_i = \Delta\theta = \omega \, \Delta t \qquad (7.4)$$

Angular displacement for uniform circular motion

For linear motion, we use the term *speed v* when we are not concerned with the direction of motion, *velocity* v_x when we are. For circular motion, we define the **angular speed** to be the absolute value of the angular velocity, so that it's a positive quantity irrespective of the particle's direction of rotation. Although potentially confusing, it is customary to use the symbol ω for angular speed *and* for angular velocity. If the direction of rotation is not important, we will interpret ω to mean angular speed. In kinematic equations, such as Equation 7.4, ω is always the angular velocity, and you need to use a negative value for clockwise rotation.

EXAMPLE 7.2 Kinematics at the roulette wheel

A small steel ball rolls counterclockwise around the inside of a 30.0-cm-diameter roulette wheel, like the one shown in the chapter preview. The ball completes exactly 2 rev in 1.20 s.

a. What is the ball's angular velocity?
b. What is the ball's angular position at $t = 2.00$ s? Assume $\theta_i = 0$.

PREPARE Treat the ball as a particle in uniform circular motion.

SOLVE

a. The ball's angular velocity is $\omega = \Delta\theta/\Delta t$. We know that the ball completes 2 revolutions in 1.20 s and that each revolution corresponds to an angular displacement $\Delta\theta = 2\pi$ rad. Thus

$$\omega = \frac{2(2\pi \text{ rad})}{1.20 \text{ s}} = 10.47 \text{ rad/s}$$

We'll use this value in subsequent calculations, but our final result for the angular velocity should be reported to three significant figures, $\omega = 10.5$ rad/s. Because the rotation direction is counterclockwise, the angular velocity is positive.

b. The ball moves with constant angular velocity, so its angular position is given by Equation 7.4. Thus the ball's angular position at $t = 2.00$ s is

$$\theta_f = \theta_i + \omega \, \Delta t = 0 \text{ rad} + (10.47 \text{ rad/s})(2.00 \text{ s}) = 20.94 \text{ rad}$$

If we're interested in where the ball is in the wheel at $t = 2.00$ s, we can write its angular position as an integer multiple of 2π (representing the number of complete revolutions the ball has made) plus a remainder:

$$\theta_f = 20.94 \text{ rad} = 3.333 \times 2\pi \text{ rad}$$
$$= 3 \times 2\pi \text{ rad} + 0.333 \times 2\pi \text{ rad}$$
$$= 3 \times 2\pi \text{ rad} + 2.09 \text{ rad}$$

In other words, at $t = 2.00$ s, the ball has completed 3 rev and is 2.09 rad $= 120°$ into its fourth revolution. An observer would say that the ball's angular position is $\theta = 120°$.

ASSESS Since the ball completes 2 revolutions in 1.20 s, it seems reasonable that it completes 3.33 revolutions in 2.00 s.

The angular speed ω is closely related to the period T and the frequency f of the motion. If a particle in uniform circular motion moves around a circle once, which by definition takes time T, its angular displacement is $\Delta\theta = 2\pi$ rad. The angular speed is thus

$$\omega = \frac{2\pi \text{ rad}}{T} \qquad (7.5)$$

We can also write the angular speed in terms of the frequency $f = 1/T$:

$$\omega = (2\pi \text{ rad})f \qquad (7.6)$$

where f must be in rev/s.

EXAMPLE 7.3 **Rotations in a car engine**

The crankshaft in your car engine is turning at 3000 rpm. What is the shaft's angular speed?

PREPARE We'll need to convert rpm to rev/s and then use Equation 7.6.

SOLVE We convert rpm to rev/s by

$$\left(3000 \, \frac{\text{rev}}{\text{min}}\right)\left(\frac{1 \, \text{min}}{60 \, \text{s}}\right) = 50.0 \, \text{rev/s}$$

Thus the crankshaft's angular speed is

$$\omega = (2\pi \, \text{rad})f = (2\pi \, \text{rad})(50.0 \, \text{rev/s}) = 314 \, \text{rad/s}$$

Angular-Position and Angular-Velocity Graphs

For the one-dimensional motion you studied in Chapter 3, we found that position- and velocity-versus-time graphs were important and useful representations of motion. We can use the same kinds of graphs to represent angular motion. Let's begin by considering the motion of the roulette ball of Example 7.2. We found that it had angular velocity $\omega = 10.5$ rad/s, meaning that its angular *position* changed by $+10.5$ rad every second. This is exactly analogous to the one-dimensional motion problem of a car driving in a straight line with a velocity of 10.5 m/s, so that its position increases by 10.5 m each second. Using this analogy, we can construct the **angular position-versus-time graph** for the roulette ball shown in FIGURE 7.4.

The angular velocity is given by $\omega = \Delta\theta/\Delta t$. Graphically, this is the *slope* of the angular position-versus-time graph, just as the ordinary velocity is the slope of the position-versus-time graph. Thus we can create an **angular velocity-versus-time graph** by finding the slope of the corresponding angular position-versus-time graph.

FIGURE 7.4 Angular position for the ball on the roulette wheel.

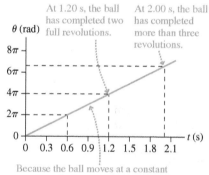

Because the ball moves at a constant angular velocity, a graph of the angular position versus time is a straight line.

EXAMPLE 7.4 **Graphing a bike ride**

Jake rides his bicycle home from campus. FIGURE 7.5 is the angular position-versus-time graph for a small rock stuck in the tread of his tire. First, draw the rock's angular velocity-versus-time graph, using rpm on the vertical axis. Then interpret the graphs with a story about Jake's ride.

PREPARE Angular velocity ω is the slope of the angular position-versus-time graph.

SOLVE We can see that $\omega = 0$ rad/s during the first and last 30 s of Jake's ride because the horizontal segments of the graph have zero slope. Between $t = 30$ s and $t = 150$ s, an interval of 120 s, the rock's angular velocity (the slope of the angular position-versus-time graph) is

$$\omega = \text{slope} = \frac{2500 \, \text{rad} - 0 \, \text{rad}}{120 \, \text{s}} = 20.8 \, \text{rad/s}$$

We need to convert this to rpm:

$$\omega = \left(\frac{20.8 \, \text{rad}}{1 \, \text{s}}\right)\left(\frac{1 \, \text{rev}}{2\pi \, \text{rad}}\right)\left(\frac{60 \, \text{s}}{1 \, \text{min}}\right) = 200 \, \text{rpm}$$

These values have been used to draw the angular velocity-versus-time graph of FIGURE 7.6. It looks like Jake waited 30 s for the light to change, then pedaled so that the bike wheel turned at a constant angular velocity of 200 rpm. 2.0 min later, he quickly braked to a stop for another 30-s-long red light.

ASSESS 200 rpm is about 3 revolutions per second. This seems reasonable for someone cycling at a pretty good clip.

FIGURE 7.5 Angular position-versus-time graph for Jake's bike ride.

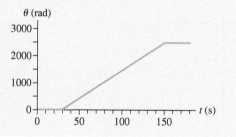

FIGURE 7.6 Angular velocity-versus-time graph for Jake's bike ride.

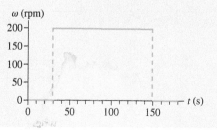

Relating Speed and Angular Speed

Let's think back to the wind turbine example that opened the chapter. Different points on the blades move at different speeds. And you know that points farther from the axis—at larger values of *r*—move at greater speeds. In Chapter 6 we found that the speed of a particle moving with frequency *f* around a circular path of radius *r* is $v = 2\pi fr$. If we combine this result with Equation 7.6 for the angular speed, we find that speed *v* and angular speed ω are related by

$$v = \omega r \qquad (7.7)$$

Relationship between speed and angular speed

NOTE ▶ In Equation 7.7, ω **must be in units of rad/s.** If you are given a frequency in rev/s or rpm, you should convert it to an angular speed in rad/s. ◀

EXAMPLE 7.5 **Finding the speed at two points on a CD**

The diameter of an audio compact disk is 12.0 cm. When the disk is spinning at its maximum rate of 540 rpm, what is the speed of a point (a) at a distance 3.0 cm from the center and (b) at the outside edge of the disk, 6.0 cm from the center?

PREPARE Consider two points A and B on the rotating compact disk in FIGURE 7.7. During one period *T*, the disk rotates once, and both points rotate through the same angle, 2π rad. Thus the angular speed, $\omega = 2\pi/T$, is the same for these two points; in fact, it is the same for all points on the disk. But as they go around one time, the two points move different *distances*. The outer point B goes around a larger circle. The two points thus have different *speeds*. We can solve this problem by first finding the angular speed of the disk and then computing the speeds at the two points.

SOLVE We first convert the frequency of the disk to rev/s:

$$f = \left(540 \frac{\text{rev}}{\text{min}}\right) \times \left(\frac{1 \text{ min}}{60 \text{ s}}\right) = 9.00 \text{ rev/s}$$

We then compute the angular speed using Equation 7.6:

$$\omega = (2\pi \text{ rad})(9.00 \text{ rev/s}) = 56.5 \text{ rad/s}$$

FIGURE 7.7 **The rotation of an audio compact disk.**

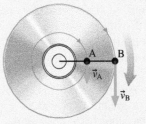

We can now use Equation 7.7 to compute the speeds of points on the disk. At point A, $r = 3.0$ cm $= 0.030$ m, so the speed is

$$v_A = \omega r = (56.5 \text{ rad/s})(0.030 \text{ m}) = 1.7 \text{ m/s}$$

At point B, $r = 6.0$ cm $= 0.060$ m, so the speed at the outside edge is

$$v_B = \omega r = (56.5 \text{ rad/s})(0.060 \text{ m}) = 3.4 \text{ m/s}$$

ASSESS The speeds are a few meters per second, which seems reasonable. The point farther from the center is moving at a higher speed, as we expected.

◀ **Why do clocks go clockwise?** In the northern hemisphere, the rotation of the earth causes the sun to follow a circular arc through the southern sky, rising in the east and setting in the west. The shadow cast by the sun thus sweeps in an arc from west to east, so the shadow on a sundial—the first practical timekeeping device—sweeps around the top of the dial from left to right. Early clockmakers used the same convention, which is how it came to be clockwise.

STOP TO THINK 7.1 Which particle has angular position $5\pi/2$?

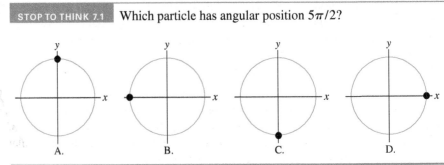

7.2 The Rotation of a Rigid Body

Think about the rotating CD in Example 7.5. Until now, our study of physics has focused almost exclusively on the *particle model*, in which an entire object is represented as a single point in space. The particle model is adequate for understanding motion in a wide variety of situations, but we need to treat the CD as an *extended object*—a system of particles for which the size and shape *do* make a difference and cannot be ignored.

A **rigid body** is an extended object whose size and shape do not change as it moves. For example, a bicycle wheel can be thought of as a rigid body. FIGURE 7.8 shows a rigid body as a collection of atoms held together by the rigid "massless rods" of molecular bonds.

Real molecular bonds are, of course, not perfectly rigid. That's why an object seemingly as rigid as a bicycle wheel can flex and bend. Thus Figure 7.8 is really a simplified *model* of an extended object, the **rigid-body model.** The rigid-body model is a very good approximation for many real objects of practical interest, such as wheels and axles.

FIGURE 7.9 illustrates the three basic types of motion of a rigid body: **translational motion, rotational motion,** and **combination motion.** We've already studied translational motion of a rigid body using the particle model. If a rigid body doesn't rotate, this model is often adequate for describing its motion. The rotational motion of a rigid body will be the main focus of this chapter. We'll also discuss an important case of combination motion—that of a *rolling* object—later in this chapter.

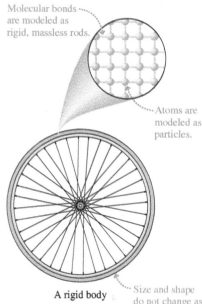

FIGURE 7.8 **The rigid-body model of an extended object.**

Molecular bonds are modeled as rigid, massless rods.

Atoms are modeled as particles.

A rigid body

Size and shape do not change as the object moves.

FIGURE 7.9 Three basic types of motion of a rigid body.

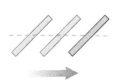  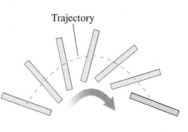

Trajectory

Translational motion:
The object as a whole moves along a trajectory but does not rotate.

Rotational motion:
The object rotates about a fixed point. Every point on the object moves in a circle.

Combination motion:
An object rotates as it moves along a trajectory.

Rotational Motion of a Rigid Body

FIGURE 7.10 shows a wheel rotating on an axle. Notice that as the wheel rotates for a time interval Δt, two points 1 and 2 on the wheel, marked with dots, turn through the *same angle,* even though their distances r from the axis of rotation are different; that is, $\Delta\theta_1 = \Delta\theta_2$ during the time interval Δt. As a consequence, the two points have equal angular velocities: $\omega_1 = \omega_2$. In general, **every point on a rotating rigid body has the same angular velocity.** Because of this, we can refer to the angular velocity ω *of the wheel.* All points move with the same angular velocity, but two points of a rotating object will have different *speeds* if they have different distances from the axis of rotation.

Angular Acceleration

If you push on the edge of a bicycle wheel, it begins to rotate. If you continue to push, it rotates ever faster. Its angular velocity is *changing.* To understand the dynamics of rotating objects, we'll need to be able to describe this case of changing angular velocity—that is, the case of *nonuniform* circular motion.

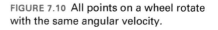

FIGURE 7.10 **All points on a wheel rotate with the same angular velocity.**

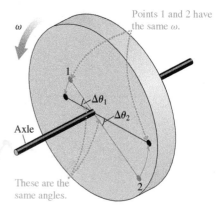

Points 1 and 2 have the same ω.

ω

$\Delta\theta_1$

$\Delta\theta_2$

Axle

These are the same angles.

FIGURE 7.11 A rotating wheel with a changing angular velocity.

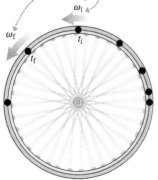

The angular velocity is *changing*, so the wheel has an angular acceleration.

FIGURE 7.12 Determining the sign of the angular acceleration.

α is *positive* when the rigid body is . . .

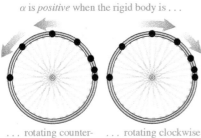

. . . rotating counter-clockwise and speeding up.

. . . rotating clockwise and slowing down.

α is *negative* when the rigid body is . . .

. . . rotating counter-clockwise and slowing down.

. . . rotating clockwise and speeding up.

FIGURE 7.11 shows a bicycle wheel whose angular velocity is changing. The dot represents a particular point on the wheel at successive times. At time t_i the angular velocity is ω_i; at a later time $t_f = t_i + \Delta t$ the angular velocity has changed to ω_f. The change in angular velocity during this time interval is

$$\Delta \omega = \omega_f - \omega_i$$

Recall that in Chapter 2 we defined the *linear* acceleration as

$$a_x = \frac{\Delta v_x}{\Delta t} = \frac{(v_x)_f - (v_x)_i}{\Delta t}$$

By analogy, we now define the **angular acceleration** as

$$\alpha = \frac{\text{change in angular velocity}}{\text{time interval}} = \frac{\Delta \omega}{\Delta t} \qquad (7.8)$$

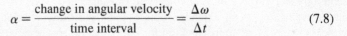

Angular acceleration for a particle in nonuniform circular motion

We use the symbol α (Greek alpha) for angular acceleration. Because the units of ω are rad/s, the units of angular acceleration are (rad/s)/s, or rad/s^2. From Equation 7.8, the sign of α is the same as the sign of $\Delta \omega$. **FIGURE 7.12** shows how to determine the sign of α. Be careful with the sign of α; just as with linear acceleration, positive and negative values of α can't be interpreted as simply "speeding up" and "slowing down." Like ω, the angular acceleration α is the same for every point on a rotating rigid body.

NOTE ▶ Don't confuse the angular acceleration with the centripetal acceleration introduced in Chapter 6. The angular acceleration indicates how rapidly the *angular* velocity is changing. The centripetal acceleration is a vector quantity that points toward the center of a particle's circular path; it is nonzero even if the angular velocity is constant. ◀

Now is a good time to bring together the definitions and equations for both linear and circular motion in order to provide a useful comparison and reference. Putting the variables and equations together this way makes the similarities between the two sets of variables and equations stand out.

SYNTHESIS 7.1 Linear and circular motion

The variables and equations for linear motion have analogs for circular motion.

	Linear motion	**Circular motion**	
Variables	Position (m) ⟶ x	θ ⟵ Angle (rad)	
	Velocity (m/s) $v_x = \dfrac{\Delta x}{\Delta t}$	$\omega = \dfrac{\Delta \theta}{\Delta t}$ Angular velocity (rad/s)	
	Acceleration (m/s^2) $a_x = \dfrac{\Delta v_x}{\Delta t}$	$\alpha = \dfrac{\Delta \omega}{\Delta t}$ Angular acceleration (rad/s^2)	
Equations	Constant velocity	$\Delta x = v \Delta t$	$\Delta \theta = \omega \Delta t$	Constant angular velocity
	Constant acceleration	$\Delta v = a \Delta t$ $\Delta x = v \Delta t + \frac{1}{2} a (\Delta t)^2$	$\Delta \omega = \alpha \Delta t$ $\Delta \theta = \omega_i \Delta t + \frac{1}{2} \alpha (\Delta t)^2$	Constant angular acceleration

EXAMPLE 7.6 **Spinning up a computer disk**

The disk in a computer disk drive spins up from rest to a final angular speed of 5400 rpm in 2.00 s. What is the angular acceleration of the disk? At the end of 2.00 s, how many revolutions has the disk made?

PREPARE The initial angular velocity is $\omega_i = 0$ rad/s. The final angular velocity is $\omega_f = 5400$ rpm. In rad/s this is

$$\omega_f = \frac{5400 \text{ rev}}{\text{min}} \times \frac{1 \text{ min}}{60 \text{ s}} \times \frac{2\pi \text{ rad}}{1 \text{ rev}} = 565 \text{ rad/s}$$

SOLVE This problem is clearly analogous to the linear motion problems we saw in Chapter 2. Something starts at rest and then moves with constant angular acceleration for a fixed interval of time. We can solve this problem by referring to Synthesis 7.1 and using equations that are analogous to the equations we'd use to solve a similar linear motion problem.

First, we use the definition of angular acceleration:

$$\alpha = \frac{\Delta\omega}{\Delta t} = \frac{565 \text{ rad/s} - 0 \text{ rad/s}}{2.00 \text{ s}} = 282.5 \text{ rad/s}^2$$

We've kept an extra significant figure for later calculations, but we'll report our final result as $\alpha = 283$ rad/s^2.

Next, we use the equation for the angular displacement—analogous to the equation for linear displacement—during the period of constant angular acceleration to determine the angle through which the disk moves:

$$\Delta\theta = \omega_i \Delta t + \tfrac{1}{2}\alpha (\Delta t)^2$$
$$= (0 \text{ rad/s})(2.00 \text{ s}) + \tfrac{1}{2}(282.5 \text{ rad/s}^2)(2.00 \text{ s})^2$$
$$= 565 \text{ rad}$$

Each revolution corresponds to an angular displacement of 2π, so we have

$$\text{number of revolutions} = \frac{565 \text{ rad}}{2\pi \text{ rad/revolution}}$$
$$= 90 \text{ revolutions}$$

The disk completes 90 revolutions during the first 2 seconds.

ASSESS The disk spins up to 5400 rpm, which corresponds to 90 rev/s. If the disk spins at full speed for 2.0 s, it will undergo 180 revolutions. But it spins up from rest to this speed over a time of 2.0 s, so it makes sense that it undergoes half this number of revolutions as it gets up to speed.

Graphs for Rotational Motion with Constant Angular Acceleration

In ◀◀ SECTION 2.5, we studied position, velocity, and acceleration graphs for motion with constant acceleration. We can extend that treatment to the current case. Because of the analogies between linear and angular quantities in Synthesis 7.1, the rules for graphing angular variables are identical to those for linear variables. In particular, **the angular velocity is the slope of the angular position-versus-time graph**, and **the angular acceleration is the slope of the angular velocity-versus-time graph.**

EXAMPLE 7.7 **Graphing angular quantities**

FIGURE 7.13 shows the angular velocity-versus-time graph for the propeller of a ship.

a. Describe the motion of the propeller.
b. Draw the angular acceleration graph for the propeller.

FIGURE 7.13 The propeller's angular velocity.

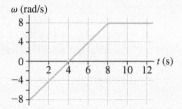

PREPARE The angular acceleration graph is the slope of the angular velocity graph.

SOLVE

a. Initially the propeller has a negative angular velocity, so it is turning clockwise. It slows down until, at $t = 4$ s, it is instantaneously stopped. It then speeds up in the opposite direction until it is turning counterclockwise at a constant angular velocity.

b. The angular acceleration graph is the slope of the angular velocity graph. From $t = 0$ s to $t = 8$ s, the slope is

$$\frac{\Delta\omega}{\Delta t} = \frac{\omega_f - \omega_i}{\Delta t} = \frac{(8.0 \text{ rad/s}) - (-8.0 \text{ rad/s})}{8.0 \text{ s}} = 2.0 \text{ rad/s}^2$$

After $t = 8$ s, the slope is zero, so the angular acceleration is zero. This graph is plotted in FIGURE 7.14.

FIGURE 7.14 Angular acceleration graph for a propeller.

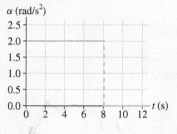

ASSESS A comparison of these graphs with their linear analogs in Figure 2.24 suggests that we're on the right track.

FIGURE 7.15 Uniform and nonuniform circular motion.

(a) Uniform circular motion

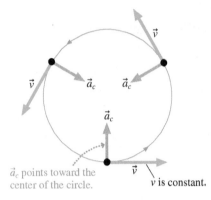

$\vec{a}_c$ points toward the center of the circle.

v is constant.

(b) Nonuniform circular motion

The tangential acceleration $\vec{a}_t$ causes the particle's *speed* to change. There's a tangential acceleration *only* when the particle is speeding up or slowing down.

v is increasing.

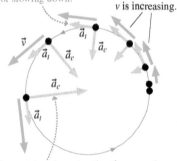

The centripetal acceleration $\vec{a}_c$ causes the particle's *direction* to change. As the particle speeds up, a_c gets larger. Circular motion *always* has a centripetal acceleration.

Tangential Acceleration

As you learned in Chapter 6, and as FIGURE 7.15a reminds you, a particle undergoing uniform circular motion has an acceleration directed inward toward the center of the circle. This centripetal acceleration $\vec{a}_c$ is due to the change in the *direction* of the particle's velocity. Recall that the magnitude of the centripetal acceleration is $a_c = v^2/r = \omega^2 r$.

> NOTE ▶ Centripetal acceleration will now be denoted a_c to distinguish it from tangential acceleration a_t, discussed below. ◀

If a particle is undergoing angular acceleration, its angular speed is changing and, therefore, so is its speed. This means that the particle will have another component to its acceleration. FIGURE 7.15b shows a particle whose speed is increasing as it moves around its circular path. Because the *magnitude* of the velocity is increasing, this second component of the acceleration is directed *tangentially* to the circle, in the same direction as the velocity. This component of acceleration is called the **tangential acceleration.**

The tangential acceleration measures the rate at which the particle's speed around the circle increases. Thus its magnitude is

$$a_t = \frac{\Delta v}{\Delta t}$$

We can relate the tangential acceleration to the *angular* acceleration by using the relationship $v = \omega r$ between the speed of a particle moving in a circle of radius r and its angular velocity ω. We have

$$a_t = \frac{\Delta v}{\Delta t} = \frac{\Delta(\omega r)}{\Delta t} = \frac{\Delta \omega}{\Delta t} r$$

or, because $\alpha = \Delta\omega/\Delta t$ from Equation 7.8,

$$a_t = \alpha r \qquad (7.9)$$

Relationship between tangential and angular acceleration

We've seen that all points on a rotating rigid body have the same angular acceleration. From Equation 7.9, however, the centripetal and tangential accelerations of a point on a rotating object depend on the point's distance r from the axis, so these accelerations are *not* the same for all points.

STOP TO THINK 7.2 A ball on the end of a string swings in a horizontal circle once every second. State whether the magnitude of each of the following quantities is zero, constant (but not zero), or changing.

a. Velocity
b. Angular velocity
c. Centripetal acceleration
d. Angular acceleration
e. Tangential acceleration

FIGURE 7.16 The four forces are the same strength, but they have different effects on the swinging door.

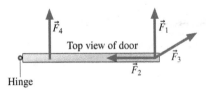

7.3 Torque

We've seen that force is the cause of acceleration. But what about *angular* acceleration? What do Newton's laws tell us about rotational motion? To begin our study of rotational motion, we'll need to find a rotational equivalent of force.

Consider the common experience of pushing open a heavy door. FIGURE 7.16 is a top view of a door that is hinged on the left. Four forces are shown, all of equal strength. Which of these will be most effective at opening the door?

Force $\vec{F}_1$ will open the door, but force $\vec{F}_2$, which pushes straight toward the hinge, will not. Force $\vec{F}_3$ will open the door, but not as easily as $\vec{F}_1$. What about $\vec{F}_4$? It is perpendicular to the door and it has the same magnitude as $\vec{F}_1$, but you know from experience that pushing close to the hinge is not as effective as pushing at the outer edge of the door.

The ability of a force to cause a rotation thus depends on three factors:

1. The magnitude F of the force
2. The distance r from the pivot—the axis about which the object can rotate—to the point at which the force is applied
3. The angle at which the force is applied

We can incorporate these three observations into a single quantity called the **torque** τ (Greek tau). Loosely speaking, τ measures the "effectiveness" of a force at causing an object to rotate about a pivot. **Torque is the rotational equivalent of force.** In Figure 7.16, for instance, the torque τ_1 due to $\vec{F}_1$ is greater than τ_4 due to $\vec{F}_4$.

To make these ideas specific, FIGURE 7.17 shows a force $\vec{F}$ applied at one point of a wrench that's loosening a nut. Figure 7.17 defines the distance r from the pivot to the point at which the force is applied; the **radial line**, the line starting at the pivot and extending through this point; and the angle ϕ (Greek phi) measured from the radial line to the direction of the force.

We saw in Figure 7.16 that force $\vec{F}_1$, which was directed perpendicular to the door, was effective in opening it, but force $\vec{F}_2$, directed toward the hinges, had no effect on its rotation. As shown in FIGURE 7.18, this suggests breaking the force $\vec{F}$ applied to the wrench into two component vectors: $\vec{F}_\perp$ directed perpendicular to the radial line, and $\vec{F}_\parallel$ directed parallel to it. Because $\vec{F}_\parallel$ points either directly toward or away from the pivot, it has no effect on the wrench's rotation, and thus contributes nothing to the torque. Only $\vec{F}_\perp$ tends to cause rotation of the wrench, so it is this component of the force that determines the torque.

NOTE ▶ The perpendicular component $\vec{F}_\perp$ is pronounced "F perpendicular" and the parallel component $\vec{F}_\parallel$ is "F parallel." ◀

We've seen that a force applied at a greater distance r from the pivot has a greater effect on rotation, so we expect a larger value of r to give a greater torque. We also saw that only $\vec{F}_\perp$ contributes to the torque. Both these observations are contained in our first expression for torque:

$$\tau = rF_\perp \qquad (7.10)$$

Torque due to a force with perpendicular component $F_\perp$
acting at a distance r from the pivot

From this equation, we see that the SI units of torque are newton-meters, abbreviated N·m.

FIGURE 7.19 shows an alternative way to calculate torque. The line that is in the direction of the force, and passes through the point at which the force acts, is called the *line of action*. The perpendicular distance from this line to the pivot is called the **moment arm** (or *lever arm*) $r_\perp$. You can see from the figure that $r_\perp = r\sin\phi$. Further, Figure 7.18 showed that $F_\perp = F\sin\phi$. We can then write Equation 7.10 as $\tau = rF\sin\phi = F(r\sin\phi) = Fr_\perp$. Thus an equivalent expression for the torque is

$$\tau = r_\perp F \qquad (7.11)$$

Torque due to a force F with moment arm $r_\perp$

Class Video

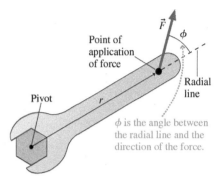
FIGURE 7.17 Force $\vec{F}$ exerts a torque about the pivot point.

Point of application of force

Pivot

$\vec{F}$

ϕ

Radial line

r

ϕ is the angle between the radial line and the direction of the force.

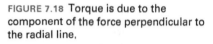
FIGURE 7.18 Torque is due to the component of the force perpendicular to the radial line.

$F_\perp = F\sin\phi$

The component of $\vec{F}$ that is *perpendicular* to the radial line causes a torque.

$\vec{F}_\perp$

$\vec{F}$

ϕ

$\vec{F}_\parallel$

Pivot

r

The *parallel* component does not contribute to the torque.

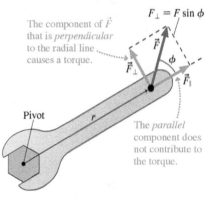

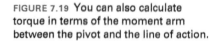
FIGURE 7.19 You can also calculate torque in terms of the moment arm between the pivot and the line of action.

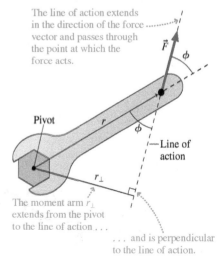
The line of action extends in the direction of the force vector and passes through the point at which the force acts.

$\vec{F}$

ϕ

Pivot

r

ϕ

Line of action

$r_\perp$

The moment arm $r_\perp$ extends from the pivot to the line of action . . .

. . . and is perpendicular to the line of action.

Starting a bike

It is hard to get going if you try to start your bike with the pedal at the highest point. Why is this?

REASON Aided by the weight of the body, the greatest force can be applied to the pedal straight down. But with the pedal at the top, this force is exerted almost directly toward the pivot, causing only a small torque. We could say either that the perpendicular component of the force is small or that the moment arm is small.

ASSESS If you've ever climbed a steep hill while standing on the pedals, you know that you get the greatest forward motion when one pedal is completely forward with the crank parallel to the ground. This gives the maximum possible torque because the force you apply is entirely perpendicular to the radial line, and the moment arm is as long as it can be.

We've seen that Equation 7.10 can be written as $\tau = rF_\perp = r(F\sin\phi)$, and Equation 7.11 as $\tau = r_\perp F = (r\sin\phi)F$. This shows that both methods of calculating torque lead to the same expression for torque—namely:

$$\tau = rF\sin\phi \qquad (7.12)$$

where ϕ is the angle between the radial line and the direction of the force as illustrated in Figure 7.17.

NOTE ▶ Torque differs from force in a very important way. Torque is calculated or measured *about a particular point*. To say that a torque is 20 N · m is meaningless without specifying the point about which the torque is calculated. Torque can be calculated about any point, but its value depends on the point chosen because this choice determines r and ϕ. In practice, we usually calculate torques about a hinge, pivot, or axle. ◀

Equations 7.10–7.12 are three different ways of thinking about—and calculating—the torque due to a force. Equation 7.12 is the most general equation for torque, but Equations 7.10 and 7.11 are generally more useful for practical problem solving. All three equations calculate the *same* torque, and all will give the same value.

◀ **Torque versus speed** To start and stop quickly, the basketball player needs to apply a large torque to her wheel. To make the torque as large as possible, the handrim—the outside wheel that she actually grabs—is almost as big as the wheel itself. The racer needs to move continuously at high speed, so his wheel spins much faster. To allow his hands to keep up, his handrim is much smaller than his chair's wheel, making its linear velocity correspondingly lower. The smaller radius means, however, that the torque he can apply is lower as well.

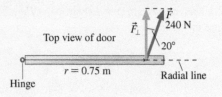

EXAMPLE 7.9 **Torque in opening a door**

Ryan is trying to open a stuck door. He pushes it at a point 0.75 m from the hinges with a 240 N force directed 20° away from being perpendicular to the door. There's a natural pivot point, the hinges. What torque does Ryan exert? How could he exert more torque?

PREPARE In FIGURE 7.20 the radial line is shown drawn from the pivot—the hinge—through the point at which the force $\vec{F}$ is applied.

FIGURE 7.20 Ryan's force exerts a torque on the door.

We see that the component of $\vec{F}$ that is perpendicular to the radial line is $F_\perp = F\cos 20° = 226$ N. The distance from the hinge to the point at which the force is applied is $r = 0.75$ m.

SOLVE We can find the torque on the door from Equation 7.10:

$$\tau = rF_\perp = (0.75 \text{ m})(226 \text{ N}) = 170 \text{ N} \cdot \text{m}$$

The torque depends on how hard Ryan pushes, where he pushes, and at what angle. If he wants to exert more torque, he could push at a point a bit farther out from the hinge, or he could push exactly perpendicular to the door. Or he could simply push harder!

ASSESS As you'll see by doing more problems, 170 N · m is a significant torque, but this makes sense if you are trying to free a stuck door.

Equations 7.10–7.12 give only the magnitude of the torque. But torque, like a force component, has a sign. **A torque that tends to rotate the object in a counterclockwise direction is positive, while a torque that tends to rotate the object in a clockwise direction is negative.** FIGURE 7.21 summarizes the signs. Notice that a force pushing straight toward the pivot or pulling straight out from the pivot exerts *no* torque.

Class Video   Video Tutor Demo

NOTE ▶ When calculating a torque, you must supply the appropriate sign by observing the direction in which the torque acts. ◀

FIGURE 7.21 **Signs and strengths of the torque.**

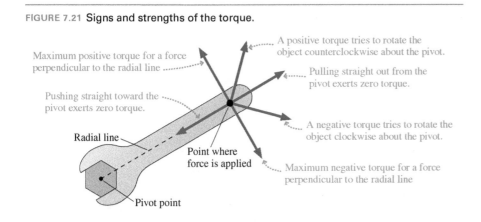

Maximum positive torque for a force perpendicular to the radial line

A positive torque tries to rotate the object counterclockwise about the pivot.

Pulling straight out from the pivot exerts zero torque.

Pushing straight toward the pivot exerts zero torque.

A negative torque tries to rotate the object clockwise about the pivot.

Radial line

Point where force is applied

Maximum negative torque for a force perpendicular to the radial line

Pivot point

STOP TO THINK 7.3 A wheel turns freely on an axle at the center. Given the details noted in Figure 7.21, which one of the forces shown in the figure will provide the largest positive torque on the wheel?

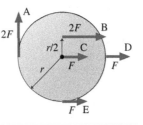

EXAMPLE 7.10 **Calculating the torque on a nut**

Luis uses a 20-cm-long wrench to tighten a nut, turning it clockwise. The wrench handle is tilted 30° above the horizontal, and Luis pulls straight down on the end with a force of 100 N. How much torque does Luis exert on the nut?

PREPARE FIGURE 7.22 shows the situation. The two illustrations correspond to two methods of calculating torque, corresponding to Equations 7.10 and 7.11.

SOLVE According to Equation 7.10, the torque can be calculated as $\tau = rF_\perp$. From Figure 7.22a, we see that the perpendicular component of the force is

$$F_\perp = F\cos 30° = (100 \text{ N})(\cos 30°) = 86.6 \text{ N}$$

FIGURE 7.22 **A wrench being used to turn a nut.**

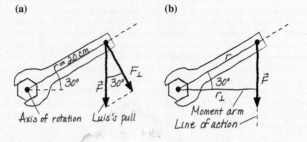

(a)   (b)

Axis of rotation   Luis's pull   Moment arm   Line of action

Continued

This gives a torque of

$$\tau = -rF_\perp = -(0.20\text{ m})(86.6\text{ N}) = -17\text{ N} \cdot \text{m}$$

We put in the minus sign because the torque is negative—it tries to rotate the nut in a *clockwise* direction.

Alternatively, we can use Equation 7.11 to find the torque. Figure 7.22b shows the moment arm $r_\perp$, the perpendicular distance from the pivot to the line of action. From the figure we see that

$$r_\perp = r\cos 30° = (0.20\text{ m})(\cos 30°) \approx 0.173\text{ m}$$

Then the torque is

$$\tau = -r_\perp F = -(0.173\text{ m})(100\text{ N}) = -17\text{ N} \cdot \text{m}$$

Again, we insert the minus sign because the torque acts to give a clockwise rotation.

ASSESS Both of the methods we used gave the same answer for the torque, as should be the case, and this gives us confidence in our results.

Net Torque

FIGURE 7.23 shows the forces acting on the crankset of a bicycle. Forces $\vec{F}_1$ and $\vec{F}_2$ are due to the rider pushing on the pedals, and $\vec{F}_3$ and $\vec{F}_4$ are tension forces from the chain. The crankset is free to rotate about a fixed axle, but the axle prevents it from having any translational motion with respect to the bike frame. It does so by exerting force $\vec{F}_{\text{axle}}$ on the object to balance the other forces and keep $\vec{F}_{\text{net}} = \vec{0}$.

Forces $\vec{F}_1$, $\vec{F}_2$, $\vec{F}_3$, and $\vec{F}_4$ exert torques τ_1, τ_2, τ_3, and τ_4 on the crank (measured about the axle), but $\vec{F}_{\text{axle}}$ does *not* exert a torque because it is applied at the pivot point—the axle—and so has zero moment arm. Thus the *net* torque about the axle is the sum of the torques due to the *applied* forces:

$$\tau_{\text{net}} = \tau_1 + \tau_2 + \tau_3 + \tau_4 + \cdots = \sum \tau \qquad (7.13)$$

FIGURE 7.23 The forces exert a net torque about the pivot point.

The axle exerts a force on the crank to keep $\vec{F}_{\text{net}} = \vec{0}$. This force does not exert a torque.

$\vec{F}_{\text{axle}}$

$\vec{F}_3$

$\vec{F}_1$

Axle

$\vec{F}_2$ $\vec{F}_4$

EXAMPLE 7.11   **Force in turning a capstan**

A capstan is a device used on old sailing ships to raise the anchor. A sailor pushes the long lever, turning the capstan and winding up the anchor rope. If the capstan turns at a constant speed, the net torque on it, as we'll learn later in the chapter, is zero.

Suppose the rope tension due to the weight of the anchor is 1500 N. If the distance from the axis to the point on the lever where the sailor pushes is exactly seven times the radius of the capstan around which the rope is wound, with what force must the sailor push if the net torque on the capstan is to be zero?

PREPARE Shown in **FIGURE 7.24** is a view looking down from above the capstan. The rope pulls with a tension force $\vec{T}$ at distance R from the axis of rotation. The sailor pushes with a force $\vec{F}$ at distance $7R$ from the axis. Both forces are perpendicular to their radial lines, so ϕ in Equation 7.12 is 90°.

FIGURE 7.24 Top view of a sailor turning a capstan.

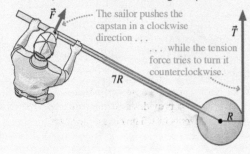

The sailor pushes the capstan in a clockwise direction ...

$\vec{F}$

... while the tension force tries to turn it counterclockwise.

$7R$

$\vec{T}$

R

SOLVE The torque due to the tension in the rope is

$$\tau_T = RT\sin 90° = RT$$

We don't know the capstan radius, so we'll just leave it as R for now. This torque is positive because it tries to turn the capstan counterclockwise. The torque due to the sailor is

$$\tau_S = -(7R)F\sin 90° = -7RF$$

We put the minus sign in because this torque acts in the clockwise (negative) direction. The net torque is zero, so we have $\tau_T + \tau_S = 0$, or

$$RT - 7RF = 0$$

Note that the radius R cancels, leaving

$$F = \frac{T}{7} = \frac{1500\text{ N}}{7} = 210\text{ N}$$

ASSESS 210 N is about 50 lb, a reasonable number. The force the sailor must exert is one-seventh the force the rope exerts: The long lever helps him lift the heavy anchor. In the HMS *Warrior*, built in 1860, it took 200 men turning the capstan to lift the huge anchor that weighed close to 55,000 N!

Note that forces $\vec{F}$ and $\vec{T}$ point in different directions. Their torques depend only on their directions with respect to their own radial lines, not on the directions of the forces with respect to each other. The force the sailor needs to apply remains unchanged as he circles the capstan.

STOP TO THINK 7.4 Two forces act on the wheel shown. What third force, acting at point P, will make the net torque on the wheel zero?

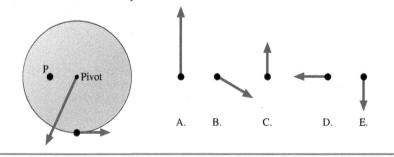

A. B. C. D. E.

7.4 Gravitational Torque and the Center of Gravity

As the gymnast in FIGURE 7.25 pivots around the bar, a torque due to the force of gravity causes her to rotate toward a vertical position. A falling tree and a car hood slamming shut are other examples where gravity exerts a torque on an object. Stationary objects can also experience a torque due to gravity. A diving board experiences a gravitational torque about its fixed end. It doesn't rotate because of a counteracting torque provided by forces from the base at its fixed end.

We've learned how to calculate the torque due to a single force acting on an object. But gravity doesn't act at a single point on an object. It pulls downward on *every particle* that makes up the object, as shown for the gymnast in FIGURE 7.25a, and so each particle experiences a small torque due to the force of gravity that acts on it. The gravitational torque on the object as a whole is then the *net* torque exerted on all the particles. We won't prove it, but the gravitational torque can be calculated by assuming that the net force of gravity—that is, the object's weight $\vec{w}$—acts at a single special point on the object called its **center of gravity** (symbol ☉). Then we can calculate the torque due to gravity by the methods we learned earlier for a single force ($\vec{w}$) acting at a single point (the center of gravity). FIGURE 7.25b shows how we can consider the gymnast's weight as acting at her center of gravity.

FIGURE 7.25 The center of gravity is the point where the weight appears to act.

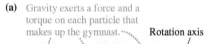

(a) Gravity exerts a force and a torque on each particle that makes up the gymnast. Rotation axis

(b) The weight force provides a torque about the rotation axis.

Center of gravity

$\vec{w}$

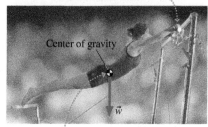

The gymnast responds *as if* her entire weight acts at her center of gravity.

EXAMPLE 7.12 **The torque on a flagpole**

A 3.2 kg flagpole extends from a wall at an angle of 25° from the horizontal. Its center of gravity is 1.6 m from the point where the pole is attached to the wall. What is the gravitational torque on the flagpole about the point of attachment?

PREPARE FIGURE 7.26 shows the situation. For the purpose of calculating torque, we can consider the entire weight of the pole

FIGURE 7.26 Visual overview of the flagpole.

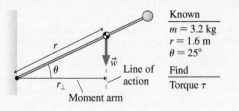

Known
$m = 3.2$ kg
$r = 1.6$ m
$\theta = 25°$

Find
Torque τ

as acting at the center of gravity. Because the moment arm $r_\perp$ is simple to visualize here, we'll use Equation 7.11 for the torque.

SOLVE From Figure 7.26, we see that the moment arm is $r_\perp = (1.6 \text{ m}) \cos 25° = 1.45$ m. Thus the gravitational torque on the flagpole, about the point where it attaches to the wall, is

$$\tau = -r_\perp w = -r_\perp mg = -(1.45 \text{ m})(3.2 \text{ kg})(9.8 \text{ m/s}^2) = -45 \text{ N} \cdot \text{m}$$

We inserted the minus sign because the torque tries to rotate the pole in a clockwise direction.

ASSESS If the pole were attached to the wall by a hinge, the gravitational torque would cause the pole to fall. However, the actual rigid connection provides a counteracting (positive) torque to the pole that prevents this. The net torque is zero.

FIGURE 7.27 Suspending a ruler.

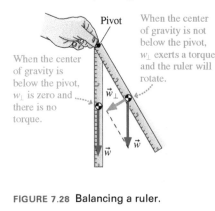

When the center of gravity is below the pivot, $w_\perp$ is zero and there is no torque.

When the center of gravity is not below the pivot, $w_\perp$ exerts a torque and the ruler will rotate.

If you hold a ruler by its end so that it is free to pivot, you know that it will quickly rotate so that it hangs straight down. FIGURE 7.27 explains this result in terms of the center of gravity and gravitational torque. The center of gravity of the ruler lies at its center. If the center of gravity is directly below the pivot, there is no gravitational torque and the ruler will stay put. If you rotate the ruler to the side, the resulting gravitational torque will quickly pull it back until the center of gravity is again below the pivot. We've shown this for a ruler, but this is a general principle: **An object that is free to rotate about a pivot will come to rest with the center of gravity below the pivot point.**

If the center of gravity lies directly *above* the pivot, as in FIGURE 7.28, there is no torque due to the object's weight and it can remain balanced. However, if the object is even slightly displaced to either side, the gravitational torque will no longer be zero and the object will begin to rotate. This question of *balance*—the behavior of an object whose center of gravity lies above the pivot—will be explored in depth in Chapter 8.

FIGURE 7.28 Balancing a ruler.

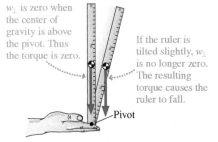

$w_\perp$ is zero when the center of gravity is above the pivot. Thus the torque is zero.

If the ruler is tilted slightly, $w_\perp$ is no longer zero. The resulting torque causes the ruler to fall.

Pivot

Calculating the Position of the Center of Gravity

Because there's no gravitational torque when the center of gravity lies either directly above or directly below the pivot, it must be the case that **the torque due to gravity when the pivot is *at* the center of gravity is zero.** We can use this fact to find a general expression for the position of the center of gravity.

Consider the dumbbell shown in FIGURE 7.29. If we slide the triangular pivot back and forth until the dumbbell balances, the pivot must then be at the center of gravity (at position x_{cg}) and the torque due to gravity must therefore be zero. But we can calculate the gravitational torque directly by calculating and summing the torques about this point due to the two individual weights. Gravity acts on weight 1 with moment arm r_1, so the torque about the pivot at position x_{cg} is

$$\tau_1 = r_1 w_1 = (x_{cg} - x_1)m_1 g$$

Similarly, the torque due to weight 2 is

$$\tau_2 = -r_2 w_2 = -(x_2 - x_{cg})m_2 g$$

FIGURE 7.29 Finding the center of gravity of a dumbbell.

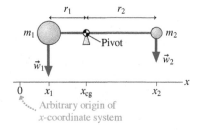

Arbitrary origin of x-coordinate system

This torque is negative because it tends to rotate the dumbbell in a clockwise direction. We've just argued that the net torque must be zero because the pivot is directly under the center of gravity, so

$$\tau_{net} = 0 = \tau_1 + \tau_2 = (x_{cg} - x_1)m_1 g - (x_2 - x_{cg})m_2 g$$

We can solve this equation for the position of the center of gravity x_{cg}:

$$x_{cg} = \frac{x_1 m_1 + x_2 m_2}{m_1 + m_2} \tag{7.14}$$

The following Tactics Box shows how Equation 7.14 can be generalized to find the center of gravity of *any* number of particles. If the particles don't all lie along the x-axis, then we'll also need to find the y-coordinate of the center of gravity.

Video Tutor Demo

TACTICS
BOX 7.1   Finding the center of gravity   (MP)

❶ Choose an origin for your coordinate system. You can choose any convenient point as the origin.

❷ Determine the coordinates (x_1, y_1), (x_2, y_2), (x_3, y_3), . . . for the particles of masses m_1, m_2, m_3, . . ., respectively.

❸ The x-coordinate of the center of gravity is

$$x_{cg} = \frac{x_1 m_1 + x_2 m_2 + x_3 m_3 + \cdots}{m_1 + m_2 + m_3 + \cdots} \tag{7.15}$$

❹ Similarly, the y-coordinate of the center of gravity is

$$y_{cg} = \frac{y_1 m_1 + y_2 m_2 + y_3 m_3 + \cdots}{m_1 + m_2 + m_3 + \cdots} \tag{7.16}$$

Exercises 18–21

Because the center of gravity depends on products such as x_1m_1, objects with large masses count more heavily than objects with small masses. Consequently, **the center of gravity tends to lie closer to the heavier objects or particles** that make up the entire object.

EXAMPLE 7.13 **Where should the dumbbell be lifted?**

A 1.0-m-long dumbbell has a 10 kg mass on the left and a 5.0 kg mass on the right. Find the position of the center of gravity, the point where the dumbbell should be lifted in order to remain balanced.

PREPARE First we sketch the situation as in FIGURE 7.30.

FIGURE 7.30 Finding the center of gravity of the barbell.

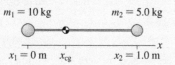

$m_1 = 10$ kg $m_2 = 5.0$ kg

$x_1 = 0$ m x_{cg} $x_2 = 1.0$ m

Next, we can use the steps from Tactics Box 7.1 to find the center of gravity. Let's choose the origin to be at the position of the 10 kg mass on the left, making $x_1 = 0$ m and $x_2 = 1.0$ m.

Because the dumbbell masses lie on the x-axis, the y-coordinate of the center of gravity must also lie on the x-axis. Thus we only need to solve for the x-coordinate of the center of gravity.

SOLVE The x-coordinate of the center of gravity is found from Equation 7.15:

$$x_{cg} = \frac{x_1m_1 + x_2m_2}{m_1 + m_2} = \frac{(0 \text{ m})(10 \text{ kg}) + (1.0 \text{ m})(5.0 \text{ kg})}{10 \text{ kg} + 5.0 \text{ kg}}$$

$$= 0.33 \text{ m}$$

The center of gravity is 0.33 m from the 10 kg mass or, equivalently, 0.17 m left of the center of the bar.

ASSESS The position of the center of gravity is closer to the larger mass. This agrees with our general statement that the center of gravity tends to lie closer to the heavier particles.

The center of gravity of an extended object can often be found by considering the object as made up of pieces, each with mass and center of gravity that are known or can be found. Then the coordinates of the entire object's center of gravity are given by Equations 7.15 and 7.16, with (x_1, y_1), (x_2, y_2), (x_3, y_3), ... the coordinates of the center of gravity of each piece and m_1, m_2, m_3, ... their masses.

This method is widely used in biomechanics and kinesiology to calculate the center of gravity of the human body. FIGURE 7.31 shows how the body can be considered to be made up of segments, each of whose mass and center of gravity have been measured. The numbers shown are appropriate for a man with a total mass of 80 kg. For a given posture the positions of the segments and their centers of gravity can be found, and thus the whole-body center of gravity from Equations 7.15 and 7.16 (and a third equation for the z-coordinate). The next example explores a simplified version of this method.

FIGURE 7.31 Body segment masses and centers of gravity for an 80 kg man.

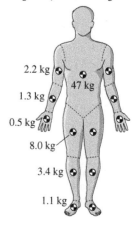

2.2 kg
47 kg
1.3 kg
0.5 kg
8.0 kg
3.4 kg
1.1 kg

EXAMPLE 7.14 **Finding the center of gravity of a gymnast**

A gymnast performing on the rings holds himself in the pike position. FIGURE 7.32 shows how we can consider his body to be made up of two segments whose masses and center-of-gravity positions are shown. The upper segment includes his head, trunk, and arms, while the lower segment consists of his legs. Locate the overall center of gravity of the gymnast.

FIGURE 7.32 Centers of gravity of two segments of a gymnast.

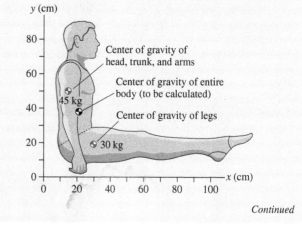

Center of gravity of head, trunk, and arms

Center of gravity of entire body (to be calculated)

Center of gravity of legs

45 kg

30 kg

Continued

PREPARE From Figure 7.32 we can find the x- and y-coordinates of the segment centers of gravity:

$$x_{\text{trunk}} = 15 \text{ cm} \qquad y_{\text{trunk}} = 50 \text{ cm}$$

$$x_{\text{legs}} = 30 \text{ cm} \qquad y_{\text{legs}} = 20 \text{ cm}$$

SOLVE The x- and y-coordinates of the center of gravity are given by Equations 7.15 and 7.16:

$$x_{\text{cg}} = \frac{x_{\text{trunk}} m_{\text{trunk}} + x_{\text{legs}} m_{\text{legs}}}{m_{\text{trunk}} + m_{\text{legs}}}$$

$$= \frac{(15 \text{ cm})(45 \text{ kg}) + (30 \text{ cm})(30 \text{ kg})}{45 \text{ kg} + 30 \text{ kg}} = 21 \text{ cm}$$

and

$$y_{\text{cg}} = \frac{y_{\text{trunk}} m_{\text{trunk}} + y_{\text{legs}} m_{\text{legs}}}{m_{\text{trunk}} + m_{\text{legs}}}$$

$$= \frac{(50 \text{ cm})(45 \text{ kg}) + (20 \text{ cm})(30 \text{ kg})}{45 \text{ kg} + 30 \text{ kg}} = 38 \text{ cm}$$

ASSESS The center-of-gravity position of the entire body, shown in Figure 7.32, is closer to that of the heavier trunk segment than to that of the lighter legs. It also lies along a line connecting the two segment centers of gravity, just as it would for the center of gravity of two point particles. Note also that the gymnast's hands—the pivot point—must lie directly below his center of gravity. Otherwise he would rotate forward or backward.

STOP TO THINK 7.5 The balls are connected by very lightweight rods pivoted at the point indicated by a dot. The rod lengths are all equal except for A, which is twice as long. Rank in order, from least to greatest, the magnitudes of the net gravitational torques about the pivots for arrangements A to D.

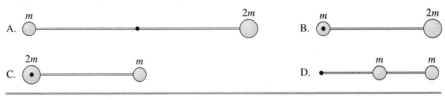

7.5 Rotational Dynamics and Moment of Inertia

In Section 7.3 we asked: What do Newton's laws tell us about rotational motion? We can now answer that question: **A torque causes an angular acceleration.** This is the rotational equivalent of our earlier discovery, for motion along a line, that a force causes an acceleration.

To see where this connection between torque and angular acceleration comes from, let's start by examining a *single particle* subject to a torque. FIGURE 7.33 shows a particle of mass m attached to a lightweight, rigid rod of length r that constrains the particle to move in a circle. The particle is subject to two forces. Because it's moving in a circle, there must be a force—here, the tension $\vec{T}$ from the rod—directed toward the center of the circle. As we learned in Chapter 6, this is the force responsible for changing the *direction* of the particle's velocity. The acceleration associated with this change in the particle's velocity is the centripetal acceleration $\vec{a}_c$.

But the particle in Figure 7.33 is also subject to the force $\vec{F}$ that changes the *speed* of the particle. This force causes a tangential acceleration $\vec{a}_t$. Applying Newton's second law in the direction tangent to the circle gives

$$a_t = \frac{F}{m} \tag{7.17}$$

Now the tangential and angular accelerations are related by $a_t = \alpha r$, so we can rewrite Equation 7.17 as $\alpha r = F/m$, or

$$\alpha = \frac{F}{mr} \tag{7.18}$$

We can now connect this angular acceleration to the torque because force $\vec{F}$, which is perpendicular to the radial line, exerts torque

$$\tau = rF$$

FIGURE 7.33 A tangential force $\vec{F}$ exerts a torque on the particle and causes an angular acceleration.

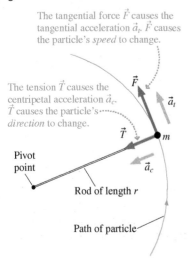

The tangential force $\vec{F}$ causes the tangential acceleration $\vec{a}_t$. $\vec{F}$ causes the particle's *speed* to change.

The tension $\vec{T}$ causes the centripetal acceleration $\vec{a}_c$. $\vec{T}$ causes the particle's *direction* to change.

Pivot point

Rod of length r

Path of particle

With this relationship between F and τ, we can write Equation 7.18 as

$$\alpha = \frac{\tau}{mr^2} \qquad (7.19)$$

Equation 7.19 gives a relationship between the torque on a single particle and its angular acceleration. Now all that remains is to expand this idea from a single particle to an extended object.

Newton's Second Law for Rotational Motion

FIGURE 7.34 shows a rigid body that undergoes rotation about a fixed and unmoving axis. According to the rigid-body model, we can think of the object as consisting of particles with masses $m_1, m_2, m_3, \ldots$ at fixed distances $r_1, r_2, r_3, \ldots$ from the axis. Suppose forces $\vec{F}_1, \vec{F}_2, \vec{F}_3, \ldots$ act on these particles. These forces exert torques around the rotation axis, so the object will undergo an angular acceleration α. Because all the particles that make up the object rotate together, each particle has this *same* angular acceleration α. Rearranging Equation 7.19 slightly, we can write the torques on the particles as

$$\tau_1 = m_1 r_1^2 \alpha \qquad \tau_2 = m_2 r_2^2 \alpha \qquad \tau_3 = m_3 r_3^2 \alpha$$

and so on for every particle in the object. If we add up all these torques, the *net* torque on the object is

$$
\begin{aligned}
\tau_{\text{net}} &= \tau_1 + \tau_2 + \tau_3 + \cdots = m_1 r_1^2 \alpha + m_2 r_2^2 \alpha + m_3 r_3^2 \alpha + \cdots \\
&= \alpha(m_1 r_1^2 + m_2 r_2^2 + m_3 r_3^2 + \cdots) = \alpha \sum m_i r_i^2
\end{aligned}
\qquad (7.20)
$$

By factoring α out of the sum, we're making explicit use of the fact that every particle in a rotating rigid body has the *same* angular acceleration α.

The quantity $\sum mr^2$ in Equation 7.20, which is the proportionality constant between angular acceleration and net torque, is called the object's **moment of inertia** I:

$$I = m_1 r_1^2 + m_2 r_2^2 + m_3 r_3^2 + \cdots = \sum m_i r_i^2 \qquad (7.21)$$

Moment of inertia of a collection of particles

The units of moment of inertia are mass times distance squared, or $kg \cdot m^2$. An object's moment of inertia, like torque, *depends on the axis of rotation*. Once the axis is specified, allowing the values of $r_1, r_2, r_3, \ldots$ to be determined, the moment of inertia *about that axis* can be calculated from Equation 7.21.

NOTE ▶ The word "moment" in "moment of inertia" and "moment arm" has nothing to do with time. It stems from the Latin *momentum*, meaning "motion." ◀

Substituting the moment of inertia into Equation 7.20 puts the final piece of the puzzle into place, giving us the fundamental equation for rigid-body dynamics:

Newton's second law for rotation An object that experiences a net torque τ_{net} about the axis of rotation undergoes an angular acceleration

$$\alpha = \frac{\tau_{\text{net}}}{I} \qquad (7.22)$$

where I is the moment of inertia of the object *about the rotation axis*.

In practice we often write $\tau_{\text{net}} = I\alpha$, but Equation 7.22 better conveys the idea that **a net torque is the cause of angular acceleration.** In the absence of a net torque ($\tau_{\text{net}} = 0$), the object has zero angular acceleration α, so it either does not rotate ($\omega = 0$) or rotates with *constant* angular velocity (ω = constant).

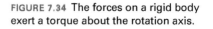
FIGURE 7.34 The forces on a rigid body exert a torque about the rotation axis.

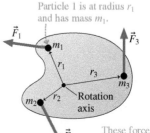
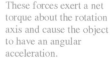

Particle 1 is at radius r_1 and has mass m_1.

These forces exert a net torque about the rotation axis and cause the object to have an angular acceleration.

This large granite ball, with a mass of 8200 kg, floats with nearly zero friction on a thin layer of pressurized water. Even though the girl exerts a large torque on the ball, its angular acceleration is small because of its large moment of inertia.

FIGURE 7.35 Moment of inertia depends on both the mass and how the mass is distributed.

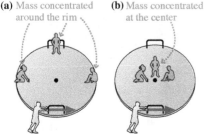

(a) Mass concentrated around the rim

(b) Mass concentrated at the center

Larger moment of inertia, harder to get rotating

Smaller moment of inertia, easier to get rotating

Interpreting the Moment of Inertia

Before rushing to calculate moments of inertia, let's get a better understanding of its meaning. First, notice that **moment of inertia is the rotational equivalent of mass.** It plays the same role in Equation 7.22 as does mass m in the now-familiar $\vec{a} = \vec{F}_{net}/m$. Recall that objects with larger mass have a larger *inertia,* meaning that they're harder to accelerate. Similarly, an object with a larger moment of inertia is harder to get rotating: It takes a larger torque to spin up an object with a larger moment of inertia than an object with a smaller moment of inertia. The fact that "moment of inertia" retains the word "inertia" reminds us of this.

But why does the moment of inertia depend on the distances r from the rotation axis? Think about trying to start a merry-go-round from rest, as shown in FIGURE 7.35. By pushing on the rim of the merry-go-round, you exert a torque on it, and its angular velocity begins to increase. If your friends sit at the rim of the merry-go-round, as in Figure 7.35a, their distances r from the axle are large. Once you've spun up the merry-go-round to a reasonable angular velocity ω, their speeds, given by $v = \omega r$, will be large as well. If your friends sit near the axle, however, as in Figure 7.35b, then r is small and their speeds $v = \omega r$ will be small. In the first case, you've gotten your friends up to a higher speed, and it's no surprise that you will need to push harder to make that happen.

We can express this result using the concept of moment of inertia. In the first case, with large values of r, Equation 7.21 says that the moment of inertia will be large. In the second case, with small values of r, Equation 7.21 gives a small value of moment of inertia. And the larger moment of inertia means you need to apply a larger torque to produce a certain angular acceleration: You'll need to push harder.

Thus an object's moment of inertia depends not only on the object's mass but also on *how the mass is distributed* around the rotation axis. This is well known to bicycle racers. Every time a cyclist accelerates, she has to "spin up" the wheels and tires. The larger the moment of inertia, the more effort it takes and the smaller her acceleration. For this reason, racers use the lightest possible tires, and they put those tires on wheels that have been designed to keep the mass as close as possible to the center without sacrificing the necessary strength and rigidity.

Synthesis 7.1 connected quantities and equations for linear and circular motion. We can now do the same thing for linear and rotational dynamics.

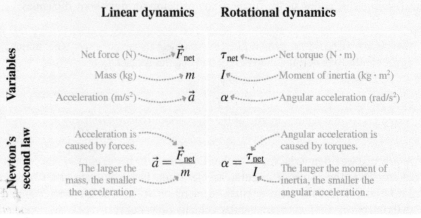

SYNTHESIS 7.2 Linear and rotational dynamics

The variables for linear dynamics have analogs for rotational dynamics. Newton's second law for rotational dynamics is expressed in terms of these variables.

	Linear dynamics	**Rotational dynamics**
Variables	Net force (N) ···· $\vec{F}_{net}$	τ_{net} ···· Net torque (N · m)
	Mass (kg) ···· m	I ···· Moment of inertia (kg · m^2)
	Acceleration (m/s^2) ···· $\vec{a}$	α ···· Angular acceleration (rad/s^2)
Newton's second law	Acceleration is caused by forces. $\vec{a} = \dfrac{\vec{F}_{net}}{m}$ The larger the mass, the smaller the acceleration.	Angular acceleration is caused by torques. $\alpha = \dfrac{\tau_{net}}{I}$ The larger the moment of inertia, the smaller the angular acceleration.

EXAMPLE 7.15 **Calculating the moment of inertia**

Your friend is creating an abstract sculpture that consists of three small, heavy spheres attached by very lightweight 10-cm-long rods as shown in FIGURE 7.36. The spheres have masses $m_1 = 1.0$ kg, $m_2 = 1.5$ kg, and $m_3 = 1.0$ kg. What is the object's moment of inertia if it is rotated about axis A? About axis B?

FIGURE 7.36 Three point particles separated by lightweight rods.

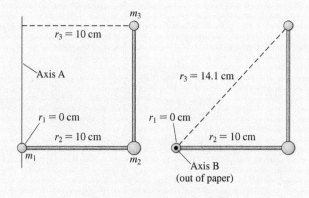

PREPARE We'll use Equation 7.21 for the moment of inertia:

$$I = m_1 r_1^2 + m_2 r_2^2 + m_3 r_3^2$$

In this expression, r_1, r_2, and r_3 are the distances of each particle from the axis of rotation, so they depend on the axis chosen. Particle 1 lies on both axes, so $r_1 = 0$ cm in both cases. Particle 2 lies

10 cm (0.10 m) from both axes. Particle 3 is 10 cm from axis A but farther from axis B. We can find r_3 for axis B by using the Pythagorean theorem, which gives $r_3 = 14.1$ cm. These distances are indicated in the figure.

SOLVE For each axis, we can prepare a table of the values of r, m, and mr^2 for each particle, then add the values of mr^2. For axis A we have

Particle	r	m	mr^2
1	0 m	1.0 kg	0 kg $\cdot$ m^2
2	0.10 m	1.5 kg	0.015 kg $\cdot$ m^2
3	0.10 m	1.0 kg	0.010 kg $\cdot$ m^2
			$I_A = 0.025$ kg $\cdot$ m^2

For axis B we have

Particle	r	m	mr^2
1	0 m	1.0 kg	0 kg $\cdot$ m^2
2	0.10 m	1.5 kg	0.015 kg $\cdot$ m^2
3	0.141 m	1.0 kg	0.020 kg $\cdot$ m^2
			$I_B = 0.035$ kg $\cdot$ m^2

ASSESS We've already noted that the moment of inertia of an object is higher when its mass is distributed farther from the axis of rotation. Here, m_3 is farther from axis B than from axis A, leading to a higher moment of inertia about that axis.

▶ **Novel golf clubs** In recent years, manufacturers have introduced golf putters with heads that have high moments of inertia. When the putter hits the ball, the ball—by Newton's third law—exerts a force on the putter and thus exerts a torque that causes the head of the putter to rotate around the shaft. A large moment of inertia of the head will keep the resulting angular acceleration small, thus reducing unwanted rotation and allowing a truer putt.

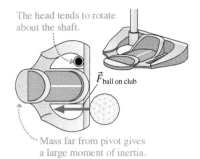

The head tends to rotate about the shaft.

$\vec{F}_{\text{ball on club}}$

Mass far from pivot gives a large moment of inertia.

The Moments of Inertia of Common Shapes

Newton's second law for rotational motion is easy to write, but we can't make use of it without knowing an object's moment of inertia. Unlike mass, we can't measure moment of inertia by putting an object on a scale. And although we can guess that the center of gravity of a symmetrical object is at the physical center of the object, we can *not* guess the moment of inertia of even a simple object.

For an object consisting of only a few point particles connected by massless rods, we can use Equation 7.21 to directly calculate I. But such an object is pretty unrealistic. All real objects are made up of solid material that is itself composed of countless atoms. To calculate the moment of inertia of even a simple object requires integral calculus and is beyond the scope of this text. A short list of common moments of inertia is given in Table 7.1 on the next page. We use a capital M for the total mass of an extended object.

We can make some general observations about the moments of inertia in Table 7.1. For instance, the cylindrical hoop is composed of particles that are all the same distance R from the axis. Thus each particle of mass m makes the *same* contribution mR^2

TABLE 7.1 Moments of inertia of objects with uniform density and total mass M

Object and axis	Picture	I	Object and axis	Picture	I
Thin rod (of any cross section), about center		$\frac{1}{12}ML^2$	Cylinder or disk, about center		$\frac{1}{2}MR^2$
Thin rod (of any cross section), about end		$\frac{1}{3}ML^2$	Cylindrical hoop, about center		MR^2
Plane or slab, about center		$\frac{1}{12}Ma^2$	Solid sphere, about diameter		$\frac{2}{5}MR^2$
Plane or slab, about edge		$\frac{1}{3}Ma^2$	Spherical shell, about diameter		$\frac{2}{3}MR^2$

to the hoop's moment of inertia. Adding up all these contributions gives

$$I = m_1R^2 + m_2R^2 + m_3R^2 + \cdots = (m_1 + m_2 + m_3 + \cdots)R^2 = MR^2$$

as given in the table. The solid cylinder of the same mass and radius has a *lower* moment of inertia than the hoop because much of the cylinder's mass is nearer its center. In the same way we can see why a slab rotated about its center has a lower moment of inertia than the same slab rotated about its edge: In the latter case, some of the mass is twice as far from the axis as the farthest mass in the former case. Those particles contribute *four times* as much to the moment of inertia, leading to an overall larger moment of inertia for the slab rotated about its edge.

STOP TO THINK 7.6 Four very lightweight disks of equal radii each have three identical heavy marbles glued to them as shown. Rank in order, from largest to smallest, the moments of inertia of the disks about the indicated axis.

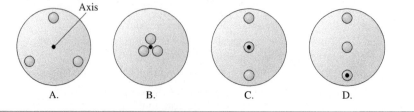

7.6 Using Newton's Second Law for Rotation

In this section we'll look at several examples of rotational dynamics for rigid bodies that rotate about a *fixed axis*. The restriction to a fixed axis avoids complications that arise for an object undergoing a combination of rotational and translational motion.

PROBLEM-SOLVING
STRATEGY 7.1 **Rotational dynamics problems**

We can use a problem-solving strategy for rotational dynamics that is very similar to the strategy for linear dynamics in Chapter 5.

PREPARE Model the object as a simple shape. Draw a pictorial representation to clarify the situation, define coordinates and symbols, and list known information.

- Identify the axis about which the object rotates.
- Identify the forces and determine their distance from the axis.
- Calculate the torques caused by the forces, and find the signs of the torques.

SOLVE The mathematical representation is based on Newton's second law for rotational motion:

$$\tau_{net} = I\alpha \quad \text{or} \quad \alpha = \frac{\tau_{net}}{I}$$

- Find the moment of inertia either by direct calculation using Equation 7.21 or from Table 7.1 for common shapes of objects.
- Use rotational kinematics to find angular positions and velocities.

ASSESS Check that your result has the correct units, is reasonable, and answers the question.

Exercise 31

EXAMPLE 7.16 **Angular acceleration of a falling pole**

In the caber toss, a contest of strength and skill that is part of Scottish games, contestants toss a heavy uniform pole, landing it on its end. A 5.9-m-tall pole with a mass of 79 kg has just landed on its end. It is tipped by 25° from the vertical and is starting to rotate about the end that touches the ground. Estimate the angular acceleration.

PREPARE The situation is shown in FIGURE 7.37, where we define our symbols and list the known information. Two forces are acting on the pole: the pole's weight $\vec{w}$, which acts at the center of gravity, and the force of the ground on the pole (not shown). This second force exerts no torque because it acts at the axis of rotation. The torque on the pole is thus due only to gravity. From the figure we see that this torque tends to rotate the pole in a counterclockwise direction, so the torque is positive.

SOLVE We'll model the pole as a uniform thin rod rotating about one end. Its center of gravity is at its center, a distance $L/2$ from the axis. You can see from the figure that the perpendicular component of $\vec{w}$ is $w_\perp = w\sin\theta$. Thus the torque due to gravity is

$$\tau_{net} = \left(\frac{L}{2}\right)w_\perp = \left(\frac{L}{2}\right)w\sin\theta = \frac{mgL}{2}\sin\theta$$

From Table 7.1, the moment of inertia of a thin rod rotated about its end is $I = \frac{1}{3}mL^2$. Thus, from Newton's second law for rotational motion, the angular acceleration is

FIGURE 7.37 A falling pole undergoes an angular acceleration due to a gravitational torque.

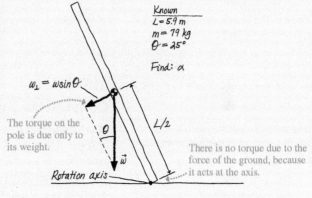

$$\alpha = \frac{\tau_{net}}{I} = \frac{\frac{1}{2}mgL\sin\theta}{\frac{1}{3}mL^2} = \frac{3g\sin\theta}{2L}$$

$$= \frac{3(9.8 \text{ m/s}^2)\sin 25°}{2(5.9 \text{ m})} = 1.1 \text{ rad/s}^2$$

ASSESS The final result for the angular acceleration did not depend on the mass, as we might expect given the analogy with free-fall problems. And the final value for the angular acceleration is quite modest. This is reasonable: You can see that the angular acceleration is inversely proportional to the length of the pole, and it's a long pole. The modest value of angular acceleration is fortunate—the caber is pretty heavy, and folks need some time to get out of the way when it topples!

Balancing a meter stick

You can easily balance a meter stick or a baseball bat on your palm. (Try it!) But it's nearly impossible to balance a pencil this way. Why?

REASON Suppose you've managed to balance a vertical stick on your palm, but then it starts to fall. You'll need to quickly adjust your hand to bring the stick back into balance. As Example 7.16 showed, the angular acceleration α of a thin rod is *inversely proportional* to L. Thus a long object like a meter stick topples much more slowly than a short one like a pencil. Your reaction time is

fast enough to correct for a slowly falling meter stick but not for a rapidly falling pencil.

ASSESS If we double the length of a rod, its mass doubles and its center of gravity is twice as high, so the gravitational torque τ on it is four times as much. But because a rod's moment of inertia is $I = \frac{1}{3}ML^2$, the longer rod's moment of inertia will be *eight* times greater, so the angular acceleration will be only half as large. This gives you more time to move to put the support back under the center of gravity.

EXAMPLE 7.18   **Starting an airplane engine**

The engine in a small airplane is specified to have a torque of 500 N·m. This engine drives a 2.0-m-long, 40 kg single-blade propeller. On startup, how long does it take the propeller to reach 2000 rpm?

PREPARE The propeller can be modeled as a rod that rotates about its center. The engine exerts a torque on the propeller. FIGURE 7.38 shows the propeller and the rotation axis.

SOLVE The moment of inertia of a rod rotating about its center is found in Table 7.1:

$$I = \tfrac{1}{12}ML^2 = \tfrac{1}{12}(40\ \text{kg})(2.0\ \text{m})^2 = 13.3\ \text{kg} \cdot \text{m}^2$$

FIGURE 7.38 A rotating airplane propeller.

The torque from the engine rotates the propeller.

$M = 40$ kg

$L = 2.0$ m

Axis

The 500 N·m torque of the engine causes an angular acceleration of

$$\alpha = \frac{\tau}{I} = \frac{500\ \text{N} \cdot \text{m}}{13.3\ \text{kg} \cdot \text{m}^2} = 37.5\ \text{rad/s}^2$$

The time needed to reach $\omega_f = 2000$ rpm = 33.3 rev/s = 209 rad/s is

$$\Delta t = \frac{\Delta \omega}{\alpha} = \frac{\omega_f - \omega_i}{\alpha} = \frac{209\ \text{rad/s} - 0\ \text{rad/s}}{37.5\ \text{rad/s}^2} = 5.6\ \text{s}$$

ASSESS We've assumed a constant angular acceleration, which is reasonable for the first few seconds while the propeller is still turning slowly. Eventually, air resistance and friction will cause opposing torques and the angular acceleration will decrease. At full speed, the negative torque due to air resistance and friction cancels the torque of the engine. Then $\tau_{net} = 0$ and the propeller turns at *constant* angular velocity with no angular acceleration.

Constraints Due to Ropes and Pulleys

FIGURE 7.39 The rope's motion must match the motion of the rim of the pulley.

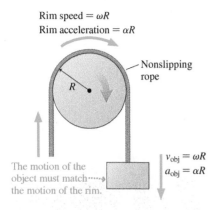

Rim speed = ωR
Rim acceleration = αR

Nonslipping rope

R

The motion of the object must match the motion of the rim.

$v_{obj} = \omega R$
$a_{obj} = \alpha R$

Many important applications of rotational dynamics involve objects that are attached to ropes that pass over pulleys. FIGURE 7.39 shows a rope passing over a pulley and connected to an object in linear motion. If the pulley turns *without the rope slipping on it*, then the rope's speed v_{rope} must exactly match the speed of the rim of the pulley, which is $v_{rim} = \omega R$. If the pulley has an angular acceleration, the rope's acceleration a_{rope} must match the *tangential* acceleration of the rim of the pulley, $a_t = \alpha R$.

The object attached to the other end of the rope has the same speed and acceleration as the rope. Consequently, the object must obey the constraints

$$v_{obj} = \omega R$$
$$a_{obj} = \alpha R \tag{7.23}$$

Motion constraints for an object connected to a pulley of radius R by a nonslipping rope

These constraints are similar to the acceleration constraints introduced in Chapter 5 for two objects connected by a string or rope.

> **NOTE** ▶ The constraints are given as magnitudes. Specific problems will require you to specify signs that depend on the direction of motion and on the choice of coordinate system. ◀

EXAMPLE 7.19 **Time for a bucket to fall**

Josh has just raised a 2.5 kg bucket of water using a well's winch when he accidentally lets go of the handle. The winch consists of a rope wrapped around a 3.0 kg, 4.0-cm-diameter cylinder, which rotates on an axle through the center. The bucket is released from rest 4.0 m above the water level of the well. How long does it take to reach the water?

PREPARE Assume the rope is massless and does not slip. **FIGURE 7.40a** gives a visual overview of the falling bucket. **FIGURE 7.40b** shows the free-body diagrams for the cylinder and the bucket. The rope tension exerts an upward force on the bucket and a downward force on the outer edge of the cylinder. The rope is massless, so these two tension forces have equal magnitudes, which we'll call T.

SOLVE Newton's second law applied to the linear motion of the bucket is

$$ma_y = T - mg$$

where, as usual, the y-axis points upward. What about the cylinder? There is a normal force $\vec{n}$ on the cylinder due to the axle and the weight of the cylinder $\vec{w}_c$. However, neither of these forces exerts a torque because each passes through the rotation axis. The only torque comes from the rope tension. The moment arm for the tension is $r_\perp = R$, and the torque is positive because the rope turns the cylinder counterclockwise. Thus $\tau_{\text{rope}} = TR$, and Newton's second law for the rotational motion is

$$\alpha = \frac{\tau_{\text{net}}}{I} = \frac{TR}{\frac{1}{2}MR^2} = \frac{2T}{MR}$$

The moment of inertia of a cylinder rotating about a center axis was taken from Table 7.1.

The last piece of information we need is the constraint due to the fact that the rope doesn't slip. Equation 7.23 relates only the magnitudes of the linear and angular accelerations, but in this problem α is positive (counterclockwise acceleration), while a_y is negative (downward acceleration). Hence

$$a_y = -\alpha R$$

Using α from the cylinder's equation in the constraint, we find

$$a_y = -\alpha R = -\frac{2T}{MR}R = -\frac{2T}{M}$$

FIGURE 7.40 Visual overview of a falling bucket.

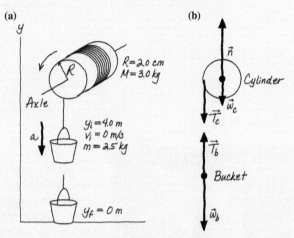

Thus the tension is $T = -\frac{1}{2}Ma_y$. If we use this value of the tension in the bucket's equation, we can solve for the acceleration:

$$ma_y = -\frac{1}{2}Ma_y - mg$$

$$a_y = -\frac{g}{(1 + M/2m)} = -6.13 \text{ m/s}^2$$

The time to fall through $\Delta y = y_f - y_i = -4.0$ m is found from kinematics:

$$\Delta y = \frac{1}{2}a_y(\Delta t)^2$$

$$\Delta t = \sqrt{\frac{2\Delta y}{a_y}} = \sqrt{\frac{2(-4.0 \text{ m})}{-6.13 \text{ m/s}^2}} = 1.1 \text{ s}$$

ASSESS The expression for the acceleration gives $a_y = -g$ if $M = 0$. This makes sense because the bucket would be in free fall if there were no cylinder. When the cylinder has mass, the downward force of gravity on the bucket has to accelerate the bucket and spin the cylinder. Consequently, the acceleration is reduced and the bucket takes longer to fall.

7.7 Rolling Motion

Rolling is a *combination motion* in which an object rotates about an axis that is moving along a straight-line trajectory. For example, **FIGURE 7.41** is a time-exposure photo of a rolling wheel with one lightbulb on the axis and a second lightbulb at the edge. The axis light moves straight ahead, but the edge light follows a curve called a *cycloid*. Let's see if we can understand this interesting motion. We'll consider only objects that roll without slipping.

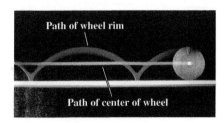

▶ **FIGURE 7.41** The trajectories of the center of a wheel and of a point on the rim are seen in a time-exposure photograph.

FIGURE 7.42 An object rolling through one revolution.

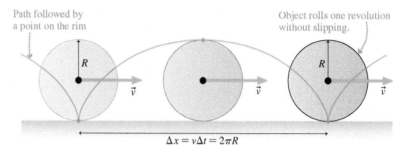

To understand rolling motion, consider FIGURE 7.42, which shows a round object—a wheel or a sphere—that rolls forward, *without slipping*, exactly one revolution. The point initially at the bottom follows the blue curve to the top and then back to the bottom. The overall position of the object is measured by the position x of the object's center. Because the object doesn't slip, in one revolution the center moves forward exactly one circumference, so that $\Delta x = 2\pi R$. The time for the object to turn one revolution is its period T, so we can compute the speed of the object's center as

$$v = \frac{\Delta x}{T} = \frac{2\pi R}{T} \qquad (7.24)$$

But $2\pi/T$ is the angular velocity ω, as you learned in Chapter 6, which leads to

$$v = \omega R \qquad (7.25)$$

Equation 7.25 is the **rolling constraint,** the basic link between translation and rotation for objects that roll without slipping.

We can find the velocity for any point on a rolling object by adding the velocity of that point when the object is in pure translation, without rolling, to the velocity of the point when the object is in pure rotation, without translating. FIGURE 7.43 shows how the velocity vectors at the top, center, and bottom of a rotating wheel are found in this way.

FIGURE 7.43 Rolling is a combination of translation and rotation.

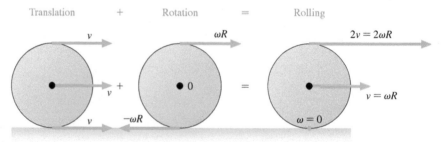

Thus the point at the top of the wheel has a forward speed of v due to its translational motion plus a forward speed of $\omega R = v$ due to its rotational motion. The speed of a point at the top of a wheel is then $2v = 2\omega R$, or *twice* the speed of its center. On the other hand, the point at the bottom of the wheel, where it touches the ground, still has a forward speed of v due to its translational motion. But its velocity due to rotation points *backward* with a magnitude of $\omega R = v$. Adding these, we find that the velocity of this lowest point is *zero*. In other words, **the point on the bottom of a rolling object is instantaneously at rest.**

Although this seems surprising, it is really what we mean by "rolling without slipping." If the bottom point had a velocity, it would be moving horizontally relative to the surface. In other words, it would be slipping or sliding across the surface. To roll without slipping, the bottom point, the point touching the surface, must be at rest.

Ancient movers The great stone *moai* of Easter Island were moved as far as 16 km from a quarry to their final positions. Archeologists believe that one possible method of moving these 14 ton statues was to place them on rollers. One disadvantage of this method is that the statues, placed on top of the rollers, move twice as fast as the rollers themselves. Thus rollers are continuously left behind and have to be carried back to the front and reinserted. Sadly, the indiscriminate cutting of trees for moving *moai* may have hastened the demise of this island civilization.

EXAMPLE 7.20   **Rotating your tires**

The diameter of your tires is 0.60 m. You take a 60 mile trip at a speed of 45 mph.

a. During this trip, what was your tires' angular speed?
b. How many times did they revolve?

PREPARE The angular speed is related to the speed of a wheel's center by Equation 7.25: $v = \omega R$. Because the center of the wheel turns on an axle fixed to the car, the speed v of the wheel's center is the same as that of the car. We prepare by converting the car's speed to SI units:

$$v = (45 \text{ mph}) \times \left(0.447 \frac{\text{m/s}}{\text{mph}} \right) = 20 \text{ m/s}$$

Once we know the angular speed, we can find the number of times the tires turned from the rotational-kinematic equation $\Delta\theta = \omega \Delta t$. We'll need to find the time traveled Δt from $v = \Delta x / \Delta t$.

SOLVE a. From Equation 7.25 we have

$$\omega = \frac{v}{R} = \frac{20 \text{ m/s}}{0.30 \text{ m}} = 67 \text{ rad/s}$$

b. The time of the trip is

$$\Delta t = \frac{\Delta x}{v} = \frac{60 \text{ mi}}{45 \text{ mi/h}} = 1.33 \text{ h} \times \frac{3600 \text{ s}}{1 \text{ h}} = 4800 \text{ s}$$

Thus the total angle through which the tires turn is

$$\Delta\theta = \omega \Delta t = (67 \text{ rad/s})(4800 \text{ s}) = 3.2 \times 10^5 \text{ rad}$$

Because each turn of the wheel is 2π rad, the number of turns is

$$\frac{3.2 \times 10^5 \text{ rad}}{2\pi \text{ rad}} = 51{,}000 \text{ turns}$$

ASSESS You probably know from seeing tires on passing cars that a tire rotates several times a second at 45 mph. Because there are 3600 s in an hour, and your 60 mile trip at 45 mph is going to take over an hour—say, ≈ 5000 s—you would expect the tire to make many thousands of revolutions. So 51,000 turns seems to be a reasonable answer. You can see that your tires rotate roughly a thousand times per mile. During the lifetime of a tire, about 50,000 miles, it will rotate about 50 million times!

STOP TO THINK 7.7   A wheel rolls without slipping. Which is the correct velocity vector for point P on the wheel?

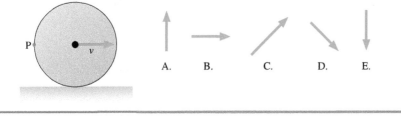

A.   B.   C.   D.   E.

INTEGRATED EXAMPLE 7.21   **Spinning a gyroscope**

A gyroscope is a top-like toy consisting of a heavy ring attached by light spokes to a central axle. The axle and ring are free to turn on bearings. To get the gyroscope spinning, a 30-cm-long string is wrapped around the 2.0-mm-diameter axle, then pulled with a constant force of 5.0 N. If the ring's diameter is 5.0 cm and its mass is 30 g, at what rate is it spinning, in rpm, once the string is completely unwound?

PREPARE Because the ring is heavy compared to the spokes and the axle, we'll model it as a cylindrical hoop, taking its moment of inertia from Table 7.1 to be $I = MR^2$. FIGURE 7.44 on the next page shows a visual overview of the problem. Two points are worth noting. First, Tactics Box 5.3 tells us that the tension in the string has the same magnitude as the force that pulls on the string, so the tension is $T = 5.0$ N. Second, it is a good

idea to convert all the known quantities in the problem statement to SI units, and to collect them all in one place, as we have done in the visual overview of Figure 7.44. Here, radius R is half the 5.0 cm ring diameter, and radius r is half the 2.0 mm axle diameter.

We are asked at what rate the ring is spinning when the string is unwound. This is a question about the ring's final *angular velocity*, which we've labeled ω_f. We've assumed that the initial angular velocity is $\omega_i = 0$ rad/s. Because the angular velocity is changing, the ring must have an angular acceleration that, as we know, is caused by a torque. So a good strategy will be to find the torque on the ring, from which we can find its angular acceleration and, using kinematics, the final angular velocity.

SOLVE The torque on the ring is due to the tension in the string. Because the string—and the line of action of the tension—is tangent to the axle, the moment arm of the tension force is the radius r of the axle. Thus $\tau = r_\perp T = rT$. Now we can apply Newton's

Continued

FIGURE 7.44 Visual overview of a gyroscope being spun.

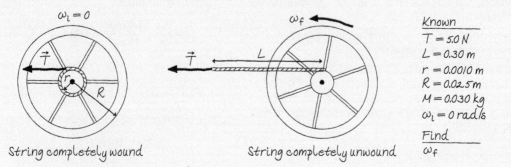

String completely wound

String completely unwound

$\omega_i = 0$

ω_f

$\vec{T}$

L

$\vec{T}$

r

R

Known
$T = 5.0\ N$
$L = 0.30\ m$
$r = 0.0010\ m$
$R = 0.025\ m$
$M = 0.030\ kg$
$\omega_i = 0\ rad/s$

Find
ω_f

second law for rotational motion, Equation 7.22, to find the angular acceleration:

$$\alpha = \frac{\tau_{net}}{I} = \frac{rT}{MR^2} = \frac{(0.0010\ \text{m})(5.0\ \text{N})}{(0.030\ \text{kg})(0.025\ \text{m})^2} = 267\ \text{rad/s}^2$$

We next use constant-angular-acceleration kinematics to find the final angular velocity. For the equation $\Delta\theta = \omega_i \Delta t + \frac{1}{2}\alpha \Delta t^2$ in Synthesis 7.1, we know α and ω_i, and we should be able to find $\Delta\theta$ from the length of string unwound, but we don't know Δt. For the equation $\Delta\omega = \omega_f - \omega_i = \alpha \Delta t$, we know α and ω_i, and ω_f is what we want to find, but again we don't know Δt. To find an equation that doesn't contain Δt, we first write

$$\Delta t = \frac{\omega_f - \omega_i}{\alpha}$$

from the second kinematic equation. Inserting this value for Δt into the first equation gives

$$\Delta\theta = \omega_i \frac{\omega_f - \omega_i}{\alpha} + \frac{1}{2}\alpha \left(\frac{\omega_f - \omega_i}{\alpha} \right)^2$$

which can be simplified to

$$\omega_f^2 = \omega_i^2 + 2\alpha \Delta\theta$$

This equation, which is the rotational analog of the linear motion Equation 2.13, will allow us to find ω_f once $\Delta\theta$ is known.

FIGURE 7.45 shows how to find $\Delta\theta$. As a segment of string of length s unwinds, the axle turns through an angle (based on the definition of radian measure) $\theta = s/r$. Thus as the whole string, of

FIGURE 7.45 Relating the angle turned to the length of string unwound.

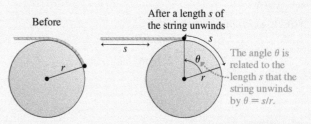

Before

After a length s of the string unwinds

r

s

θ

r

s

The angle θ is related to the length s that the string unwinds by $\theta = s/r$.

length L, unwinds, the axle (and the ring) turns through an angular displacement

$$\Delta\theta = \frac{L}{r} = \frac{0.30\ \text{m}}{0.0010\ \text{m}} = 300\ \text{rad}$$

Now we can use our kinematic equation to find that

$$\omega_f^2 = \omega_i^2 + 2\alpha \Delta\theta = (0\ \text{rad/s})^2 + 2(267\ \text{rad/s}^2)(300\ \text{rad})$$
$$= 160{,}000\ (\text{rad/s})^2$$

from which we find that $\omega_f = 400\ \text{rad/s}$. Converting rad/s to rpm, we find that the gyroscope ring is spinning at

$$400\ \text{rad/s} = \left(\frac{400\ \text{rad}}{\text{s}} \right) \left(\frac{60\ \text{s}}{1\ \text{min}} \right) \left(\frac{1\ \text{rev}}{2\pi\ \text{rad}} \right) = 3800\ \text{rpm}$$

ASSESS This is fast, about the speed of your car engine when it's on the highway, but if you've ever played with a gyroscope or a string-wound top, you know you can really get it spinning fast.

SUMMARY

Goal: To understand the physics of rotating objects.

GENERAL PRINCIPLES

Newton's Second Law for Rotational Motion

If a net torque τ_{net} acts on an object, the object will experience an angular acceleration given by $\alpha = \tau_{net}/I$, where I is the object's moment of inertia about the rotation axis.

This law is analogous to Newton's second law for linear motion, $\vec{a} = \vec{F}_{net}/m$.

IMPORTANT CONCEPTS

Describing circular motion

We define new variables for circular motion. By convention, counterclockwise is positive.

Angular displacement: $\Delta\theta = \theta_f - \theta_i$

Angular velocity: $\omega = \dfrac{\Delta\theta}{\Delta t}$

Angular acceleration: $\alpha = \dfrac{\Delta\omega}{\Delta t}$

Angles are measured in radians:
$$1 \text{ rev} = 360° = 2\pi \text{ rad}$$
The angular velocity depends on the frequency and period:

$$\omega = \frac{2\pi}{T} = 2\pi f$$

Relating linear and circular motion quantities

Linear and angular speeds are related by: $v = \omega r$

If the particle's speed is increasing, it will also have a tangential acceleration $\vec{a}_t$ directed tangent to the circle and an angular acceleration α.

Angular and tangential accelerations are related by: $a_t = \alpha r$

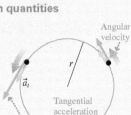

The **moment of inertia** is the rotational equivalent of mass. For an object made up of particles of masses $m_1, m_2, \ldots$ at distances $r_1, r_2, \ldots$ from the axis, the moment of inertia is

$$I = m_1 r_1^2 + m_2 r_2^2 + m_3 r_3^2 + \cdots = mr^2$$

Torque

A force causes an object to undergo a linear acceleration, a torque causes an object to undergo an angular acceleration.

There are two interpretations of torque:

Interpretation 1: $\tau = rF_\perp$ Interpretation 2: $\tau = r_\perp F$

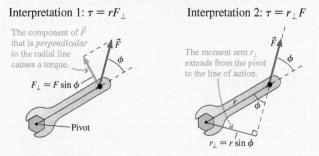

The component of $\vec{F}$ that is *perpendicular* to the radial line causes a torque.

$F_\perp = F \sin\phi$

Pivot

The moment arm $r_\perp$ extends from the pivot to the line of action.

$r_\perp = r \sin\phi$

Both interpretations give the same expression for the magnitude of the torque:

$$\tau = rF \sin\phi$$

Center of gravity

The **center of gravity** of an object is the point at which gravity can be considered to be acting.

Gravity acts on each particle that makes up the object.   The object responds *as if* its entire weight acts at the center of gravity.

The **position of the center of gravity** depends on the distance $x_1, x_2, \ldots$ of each particle of mass $m_1, m_2, \ldots$ from the origin:

$$x_{cg} = \frac{x_1 m_1 + x_2 m_2 + x_3 m_3 + \cdots}{m_1 + m_2 + m_3 + \cdots}$$

APPLICATIONS

Moments of inertia of common shapes

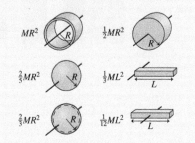

MR^2 $\frac{1}{2}MR^2$

$\frac{2}{5}MR^2$ $\frac{1}{3}ML^2$

$\frac{2}{3}MR^2$ $\frac{1}{12}ML^2$

Rotation about a fixed axis

When a net torque is applied to an object that rotates about a fixed axis, the object will undergo an **angular acceleration** given by

$$\alpha = \frac{\tau_{net}}{I}$$

If a rope unwinds from a pulley of radius R, the linear motion of an object tied to the rope is related to the angular motion of the pulley by

$$a_{obj} = \alpha R \qquad v_{obj} = \omega R$$

Rolling motion

For an object that rolls without slipping,

$$v = \omega R$$

The velocity of a point at the top of the object is twice that of the center.

QUESTIONS

Conceptual Questions

1. The batter in a baseball game hits a home run. As he circles the bases, is his angular velocity positive or negative?
2. Viewed from somewhere in space above the north pole, would a point on the earth's equator have a positive or negative angular velocity due to the earth's rotation?
3. Figure Q7.3 shows four pulleys, each with a heavy and a light block strung over it. The blocks' velocities are shown. What are the signs (+ or −) of the angular velocity and angular acceleration of the pulley in each case?

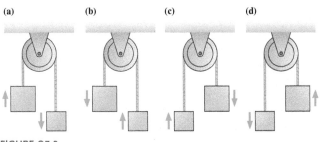

(a)   **(b)**   **(c)**   **(d)**

FIGURE Q7.3

4. If you are using a wrench to loosen a very stubborn nut, you can make the job easier by using a "cheater pipe." This is a piece of pipe that slides over the handle of the wrench, as shown in Figure Q7.4, making it effectively much longer. Explain why this would help you loosen the nut.

FIGURE Q7.4

5. Five forces are applied to a door, as seen from above in Figure Q7.5. For each force, is the torque about the hinge positive, negative, or zero?

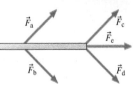

FIGURE Q7.5

6. A screwdriver with a very thick handle requires less force to operate than one with a very skinny handle. Explain why this is so.
7. If you have ever driven a truck, you likely found that it had a steering wheel with a larger diameter than that of a passenger car. Why is this?
8. A common type of door stop is a wedge made of rubber. Is such a stop more effective when jammed under the door near or far from the hinges? Why?

9. A student gives a steady push to a ball at the end of a massless, rigid rod for 1 s, causing the ball to rotate clockwise in a horizontal circle as shown in Figure Q7.9. The rod's pivot is frictionless. Sketch a graph of the ball's angular velocity as a function of time for the first 3 s of the ball's motion. You won't be able to include numbers on the vertical axis, but your graph should have the correct sign and the correct shape.

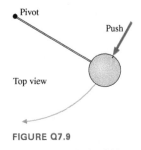

FIGURE Q7.9

10. You can use a simple technique to find the center of gravity of an irregular shape. Figure Q7.10 shows a cardboard cutout of the outline of the continental United States. The map is suspended from pivot 1 and allowed to hang freely; then a blue vertical line is drawn. The map is then suspended from pivot 2, hangs freely, and the red vertical line is drawn. The center of gravity lies at the intersection of the two lines. Explain how this technique works.

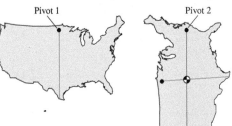

FIGURE Q7.10

11. The two ends of the dumbbell shown in Figure Q7.11 are made of the same material. Is the dumbbell's center of gravity at point 1, 2, or 3? Explain.

FIGURE Q7.11

12. If you grasp a hammer by its lightweight handle and wave it back and forth, and then grasp it by its much heavier head and wave it back and forth, as in the figure, you'll find that you can wave the hammer much more rapidly in the second case, when you grasp it by the head. Explain why this is so.

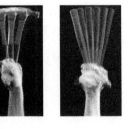

13. Suppose you have two identical-looking metal spheres of the same size and the same mass. One of them is solid; the other is hollow. Describe a simple test that you could do to determine which is which.
14. The moment of inertia of a uniform rod about an axis through its center is $ML^2/12$. The moment of inertia about an axis at one end is $ML^2/3$. Explain *why* the moment of inertia is larger about the end than about the center.

15. The wheel in Figure Q7.15 is rolling to the right without slipping. Rank in order, from fastest to slowest, the *speeds* of the points labeled 1 through 5. Explain your reasoning.

16. With care, it's possible to walk on top of a barrel as it rolls. It is much easier to do this if the barrel is full than if it is empty. Explain why this is so.

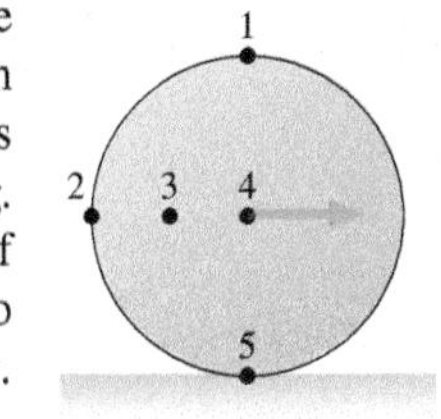

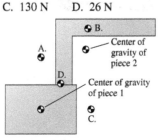

FIGURE Q7.15

Multiple-Choice Questions

17. | A nut needs to be tightened with a wrench. Which force shown in Figure Q7.17 will apply the greatest torque to the nut?

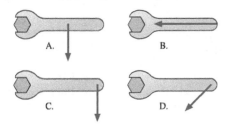

A.
B.
C.
D.

FIGURE Q7.17

18. | Suppose a bolt on your car engine needs to be tightened to a torque of 20 N · m. You are using a 15-cm-long wrench, and you apply a force at the very end in the direction that produces maximum torque. What force should you apply?
    A. 1300 N   B. 260 N   C. 130 N   D. 26 N

19. | A machine part is made up of two pieces, with centers of gravity shown in Figure Q7.19. Which point could be the center of gravity of the entire part?

A.
B.
Center of gravity of piece 2
D.
Center of gravity of piece 1
C.

FIGURE Q7.19

20. ‖ A typical compact disk has a mass of 15 g and a diameter of 120 mm. What is its moment of inertia about an axis through its center, perpendicular to the disk?
    A. 2.7×10^{-5} kg · m^2 B. 5.4×10^{-5} kg · m^2
    C. 1.1×10^{-4} kg · m^2 D. 2.2×10^{-4} kg · m^2

21. ‖ Suppose manufacturers increase the size of compact disks so that they are made of the same material and have the same thickness as a current disk but have twice the diameter. By what factor will the moment of inertia increase?
    A. 2   B. 4   C. 8   D. 16

22. | Two horizontal rods are each held up by vertical strings tied to their ends. Rod 1 has length L and mass M; rod 2 has length $2L$ and mass $2M$. Each rod then has one of its supporting strings cut, causing the rod to begin pivoting about the end that is still tied up. Which rod has a larger initial angular acceleration?
    A. Rod 1   B. Rod 2
    C. The initial angular acceleration is the same for both.

23. | A baseball bat has a thick, heavy barrel and a thin, light handle. If you want to hold a baseball bat on your palm so that it balances vertically, you should
    A. Put the end of the handle in your palm, with the barrel up.
    B. Put the end of the barrel in your palm, with the handle up.
    C. The bat will be equally easy to balance in either configuration.

24. | A particle undergoing circular motion in the *xy*-plane stops on the positive *y*-axis. Which of the following does *not* describe its angular position?
    A. $\pi/2$ rad   B. π rad   C. $5\pi/2$ rad   D. $-3\pi/2$ rad

Questions 25 through 27 concern a classic figure-skating jump called the axel. A skater starts the jump moving forward as shown in Figure Q7.25, leaps into the air, and turns one-and-a-half revolutions before landing. The typical skater is in the air for about 0.5 s, and the skater's hands are located about 0.8 m from the rotation axis.

FIGURE Q7.25

25. ‖ What is the approximate angular speed of the skater during the leap?
    A. 2 rad/s   B. 6 rad/s
    C. 9 rad/s   D. 20 rad/s

26. | The skater's arms are fully extended during the jump. What is the approximate centripetal acceleration of the skater's hand?
    A. 10 m/s^2 B. 30 m/s^2 C. 300 m/s^2 D. 450 m/s^2

27. | What is the approximate speed of the skater's hand?
    A. 1 m/s   B. 3 m/s   C. 9 m/s   D. 15 m/s

PROBLEMS

Section 7.1 Describing Circular and Rotational Motion

1. ‖ What is the angular position in radians of the minute hand of a clock at (a) 5:00, (b) 7:15, and (c) 3:35?

2. | A child on a merry-go-round takes 3.0 s to go around once. What is his angular displacement during a 1.0 s time interval?

3. ‖ What is the angular speed of the tip of the minute hand on a clock, in rad/s?

4. ‖ An old-fashioned vinyl record rotates on a turntable at 45 rpm. What are (a) the angular speed in rad/s and (b) the period of the motion?

5. ‖ The earth's radius is about 4000 miles. Kampala, the capital of Uganda, and Singapore are both nearly on the equator. The distance between them is 5000 miles.
    a. Through what angle do you turn, relative to the earth, if you fly from Kampala to Singapore? Give your answer in both radians and degrees.
    b. The flight from Kampala to Singapore takes 9 hours. What is the plane's angular speed relative to the earth?

6. ‖ A Ferris wheel rotates at an angular velocity of 0.036 rad/s. At $t = 0$ min, your friend Seth is at the very top of the ride. What is Seth's angular position at $t = 3.0$ min, measured counterclockwise from the top? Give your answer as an angle in degrees between 0° and 360°.

7. ⦀ A turntable rotates counterclockwise at 78 rpm. A speck of dust on the turntable is at $\theta = 0.45$ rad at $t = 0$ s. What is the angle of the speck at $t = 8.0$ s? Your answer should be between 0 and 2π rad.

8. ‖ A fast-moving superhero in a comic book runs around a circular, 70-m-diameter track five and a half times (ending up directly opposite her starting point) in 3.0 s. What is her angular speed, in rad/s?

9. ‖ Figure P7.9 shows the angular position of a potter's wheel.
   a. What is the angular displacement of the wheel between $t = 5$ s and $t = 15$ s?
   b. What is the angular velocity of the wheel at $t = 15$ s?

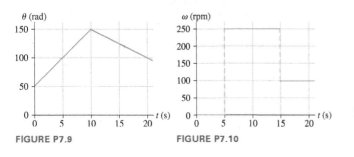

FIGURE P7.9   FIGURE P7.10

10. ‖ The angular velocity (in rpm) of the blade of a blender is given in Figure P7.10.
    a. If $\theta = 0$ rad at $t = 0$ s, what is the blade's angular position at $t = 20$ s?
    b. At what time has the blade completed 10 full revolutions?

Section 7.2 The Rotation of a Rigid Body

11. ⎸ The 1.00-cm-long second hand on a watch rotates smoothly.
    a. What is its angular velocity?
    b. What is the speed of the tip of the hand?

12. ‖ The earth's radius is 6.37×10^6 m; it rotates once every 24 hours.
    a. What is the earth's angular speed?
    b. Viewed from a point above the north pole, is the angular velocity positive or negative?
    c. What is the speed of a point on the equator?
    d. What is the speed of a point on the earth's surface halfway between the equator and the pole? (Hint: What is the radius of the circle in which the point moves?)

13. ⦀ To throw a discus, the thrower holds it with a fully outstretched arm. Starting from rest, he begins to turn with a constant angular acceleration, releasing the discus after making one complete revolution. The diameter of the circle in which the discus moves is about 1.8 m. If the thrower takes 1.0 s to complete one revolution, starting from rest, what will be the speed of the discus at release?

14. ⦀ A computer hard disk starts from rest, then speeds up with an angular acceleration of 190 rad/s² until it reaches its final angular speed of 7200 rpm. How many revolutions has the disk made 10.0 s after it starts up?

15. ⦀ The crankshaft in a race car goes from rest to 3000 rpm in 2.0 s.
    a. What is the crankshaft's angular acceleration?
    b. How many revolutions does it make while reaching 3000 rpm?

Section 7.3 Torque

16. ⎸ Reconsider the situation in Example 7.10. If Luis pulls straight down on the end of a wrench that is in the same orientation but is 35 cm long, rather than 20 cm, what force must he apply to exert the same torque?

17. ⦀ Balls are attached to light rods and can move in horizontal circles as shown in Figure P7.17. Rank in order, from smallest to largest, the torques τ_1 to τ_4 about the centers of the circles. Explain.

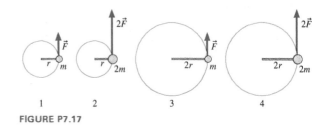

FIGURE P7.17

18. ⦀ Six forces, each of magnitude either F or $2F$, are applied to a door as seen from above in Figure P7.18. Rank in order, from smallest to largest, the six torques τ_1 to τ_6 about the hinge.

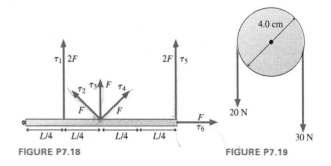

FIGURE P7.18   FIGURE P7.19

19. ⎸ What is the net torque about the axle on the pulley in Figure P7.19?

20. ⦀ The tune-up specifications of a car call for the spark plugs to be tightened to a torque of 38 N·m. You plan to tighten the plugs by pulling on the end of a 25-cm-long wrench. Because of the cramped space under the hood, you'll need to pull at an angle of 120° with respect to the wrench shaft. With what force must you pull?

21. ⦀ A professor's office door is 0.91 m wide, 2.0 m high, and 4.0 cm thick; has a mass of 25 kg; and pivots on frictionless hinges. A "door closer" is attached to the door and the top of the door frame. When the door is open and at rest, the door closer exerts a torque of 5.2 N·m. What is the least force that you need to apply to the door to hold it open?

22. ‖ In Figure P7.22, force $\vec{F}_2$ acts half as far from the pivot as $\vec{F}_1$. What magnitude of $\vec{F}_2$ causes the net torque on the rod to be zero?

FIGURE P7.22

23. ⦀ Tom and Jerry both push on the 3.00-m-diameter merry-go-round shown in Figure P7.23.
    a. If Tom pushes with a force of 50.0 N and Jerry pushes with a force of 35.0 N, what is the net torque on the merry-go-round?

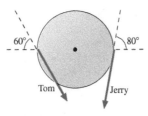

FIGURE P7.23

   b. What is the net torque if Jerry reverses the direction he pushes by 180° without changing the magnitude of his force?

24. ‖ What is the net torque on the bar shown in Figure P7.24, about the axis indicated by the dot?

FIGURE P7.24

FIGURE P7.25

25. ‖ What is the net torque on the bar shown in Figure P7.25, about the axis indicated by the dot?

26. ‖ What is the net torque on the bar shown in Figure P7.26, about the axis indicated by the dot?

FIGURE P7.26

Section 7.4 Gravitational Torque and the Center of Gravity

27. ‖ A 1.7-m-long barbell has a 20 kg weight on its right end and a 35 kg weight on its left end.
    a. If you ignore the weight of the bar itself, how far from the left end of the barbell is the center of gravity?
    b. Where is the center of gravity if the 8.0 kg mass of the barbell itself is taken into account?

28. ‖ Three identical coins lie on three corners of a square 10.0 cm on a side, as shown in Figure P7.28. Determine the x- and y-coordinates of the center of gravity of the group of three coins.

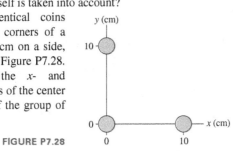

FIGURE P7.28

29. ⏐ Hold your arm outstretched so that it is horizontal. Estimate the mass of your arm and the position of its center of gravity. What is the gravitational torque on your arm in this position, computed around the shoulder joint?

30. ⦀ A solid cylinder sits on top of a solid cube as shown in Figure P7.30. How far above the table's surface is the center of gravity of the combined object?

FIGURE P7.30

31. ‖ The 2.0 kg, uniform, horizontal rod in Figure P7.31 is seen from the side. What is the gravitational torque about the point shown?

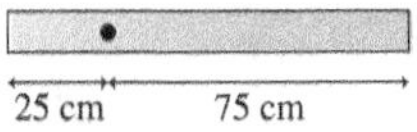

FIGURE P7.31

32. ⦀ A 4.00-m-long, 500 kg steel beam extends horizontally from the point where it has been bolted to the framework of a new building under construction. A 70.0 kg construction worker stands at the far end of the beam. What is the magnitude of the gravitational torque about the point where the beam is bolted into place?

33. ⦀ An athlete at the gym holds a 3.0 kg steel ball in his hand. His
    BIO arm is 70 cm long and has a mass of 4.0 kg. What is the magnitude of the gravitational torque about his shoulder if he holds his arm
    a. Straight out to his side, parallel to the floor?
    b. Straight, but 45° below horizontal?

34. ⦀ The 2.0-m-long, 15 kg beam in Figure P7.34 is hinged at its left end. It is "falling" (rotating clockwise, under the influence of gravity), and the figure shows its position at three different times. What is the gravitational torque on the beam about an axis through the hinged end when the beam is at the
    a. Upper position?
    b. Middle position?
    c. Lower position?

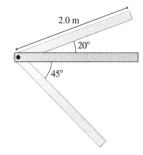

FIGURE P7.34

35. ⦀⦀ Two thin beams are joined end-to-end as shown in Figure P7.35 to make a single object. The left beam is 10.0 kg and 1.00 m long and the right one is 40.0 kg and 2.00 m long.
    a. How far from the left end of the left beam is the center of gravity of the object?
    b. What is the gravitational torque on the object about an axis through its left end? The object is seen from the side.

FIGURE P7.35

36. ⦀ Figure P7.36 shows two thin beams joined at right angles. The vertical beam is 15.0 kg and 1.00 m long and the horizontal beam is 25.0 kg and 2.00 m long.

FIGURE P7.36

   a. Find the center of gravity of the two joined beams. Express your answer in the form (x, y), taking the origin at the corner where the beams join.
   b. Calculate the gravitational torque on the joined beams about an axis through the corner. The beams are seen from the side.

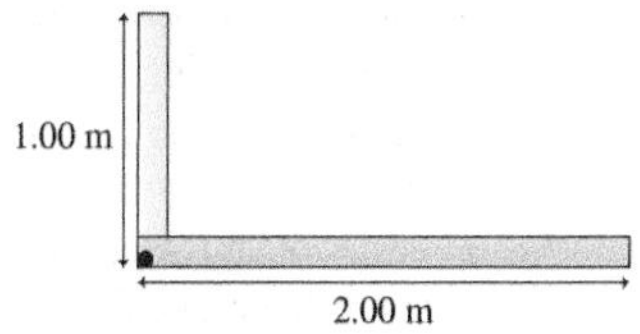

Section 7.5 Rotational Dynamics and Moment of Inertia

37. ⦀ A regulation table tennis ball is a thin spherical shell 40 mm in diameter with a mass of 2.7 g. What is its moment of inertia about an axis that passes through its center?

38. ‖ Three pairs of balls are connected by very light rods as shown in Figure P7.38. Rank in order, from smallest to largest, the moments of inertia I_1, I_2, and I_3 about axes through the centers of the rods.

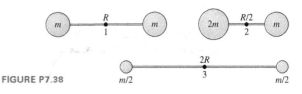

FIGURE P7.38

39. ‖ A playground toy has four seats, each 5.0 kg, attached to very light, 1.5-m-long rods, as seen from above in Figure P7.39. If two children, with masses of 15 kg and 20 kg, sit in seats opposite one another, what is the moment of inertia about the rotation axis?

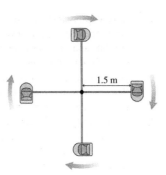

FIGURE P7.39

40. ‖‖ A solid cylinder with a radius of 4.0 cm has the same mass as a solid sphere of radius R. If the cylinder and sphere have the same moment of inertia about their centers, what is the sphere's radius?

41. ‖ A bicycle rim has a diameter of 0.65 m and a moment of inertia, measured about its center, of 0.19 kg·m². What is the mass of the rim?

Section 7.6 Using Newton's Second Law for Rotation

42. ‖‖ a. What is the moment of inertia of the door in Problem 21?
    b. If you let go of the open door, what is its angular acceleration immediately afterward?

43. ‖ A small grinding wheel has a moment of inertia of 4.0×10^{-5} kg·m². What net torque must be applied to the wheel for its angular acceleration to be 150 rad/s²?

44. ‖ While sitting in a swivel chair, you push against the floor with your heel to make the chair spin. The 7.0 N frictional force is applied at a point 40 cm from the chair's rotation axis, in the direction that causes the greatest angular acceleration. If that angular acceleration is 1.8 rad/s², what is the total moment of inertia about the axis of you and the chair?

45. ‖ An object's moment of inertia is 2.0 kg·m². Its angular velocity is increasing at the rate of 4.0 rad/s per second. What is the net torque on the object?

46. ‖‖ A 200 g, 20-cm-diameter plastic disk is spun on an axle through its center by an electric motor. What torque must the motor supply to take the disk from 0 to 1800 rpm in 4.0 s?

47. ‖‖ The 2.5 kg object shown in Figure P7.47 has a moment of inertia about the rotation axis of 0.085 kg·m². The rotation axis is horizontal. When released, what will be the object's initial angular acceleration?

FIGURE P7.47   FIGURE P7.48

48. ‖ A frictionless pulley, which can be modeled as a 0.80 kg solid cylinder with a 0.30 m radius, has a rope going over it, as shown in Figure P7.48. The tension in the rope is 10 N on one side and 12 N on the other. What is the angular acceleration of the pulley?

49. ‖‖ If you lift the front wheel of a poorly maintained bicycle off the ground and then start it spinning at 0.72 rev/s, friction in the bearings causes the wheel to stop in just 12 s. If the moment of inertia of the wheel about its axle is 0.30 kg·m², what is the magnitude of the frictional torque?

50. ‖ On page 207 there is a photograph of a girl pushing on a large stone sphere. The sphere has a mass of 8200 kg and a radius of 90 cm. Suppose that she pushes on the sphere tangent to its surface with a steady force of 50 N and that the pressured water provides a frictionless support. How long will it take her to rotate the sphere one time, starting from rest?

51. ‖‖‖ A toy top with a spool of diameter 5.0 cm has a moment of inertia of 3.0×10^{-5} kg·m² about its rotation axis. To get the top spinning, its string is pulled with a tension of 0.30 N. How long does it take for the top to complete the first five revolutions? The string is long enough that it is wrapped around the top more than five turns.

52. ‖‖‖ A 1.5 kg block and a 2.5 kg block are attached to opposite ends of a light rope. The rope hangs over a solid, frictionless pulley that is 30 cm in diameter and has a mass of 0.75 kg. When the blocks are released, what is the acceleration of the lighter block?

Section 7.7 Rolling Motion

53. ‖‖ A bicycle with 0.80-m-diameter tires is coasting on a level road at 5.6 m/s. A small blue dot has been painted on the tread of the rear tire.
    a. What is the angular speed of the tires?
    b. What is the speed of the blue dot when it is 0.80 m above the road?
    c. What is the speed of the blue dot when it is 0.40 m above the road?

54. ‖ A 2.0-m-long slab of concrete is supported by rollers, as shown in Figure P7.54. If the slab is pushed to the right, it will move off the supporting rollers, one by one, on the left side. How far can the slab be moved before its center of gravity is to the right of the contact point with the rightmost roller—at which point the slab begins to tip?

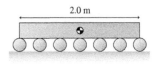

FIGURE P7.54

General Problems

55. ‖ Figure P7.55 shows the angular position-versus-time graph for a particle moving in a circle.
    a. Write a description of the particle's motion.
    b. Draw the angular velocity-versus-time graph.

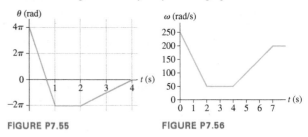

FIGURE P7.55   FIGURE P7.56

56. ‖ The graph in Figure P7.56 shows the angular velocity of the crankshaft in a car. Draw a graph of the angular acceleration versus time. Include appropriate numerical scales on both axes.

57. ‖ A car with 58-cm-diameter tires accelerates uniformly from rest to 20 m/s in 10 s. How many times does each tire rotate?

58. ‖‖‖ The cable lifting an elevator is wrapped around a 1.0-m-diameter cylinder that is turned by the elevator's motor. The elevator is moving upward at a speed of 1.6 m/s. It then slows to a stop as the cylinder makes one complete turn at constant angular acceleration. How long does it take for the elevator to stop?

59. ⦀ The 20-cm-diameter disk in Figure P7.59 can rotate on an axle through its center. What is the net torque about the axle?

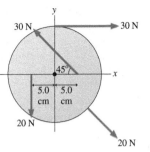

FIGURE P7.59

60. ⦀ A combination lock has a 1.0-cm-diameter knob that is part of the dial you turn to unlock the lock. To turn that knob, you grip it between your thumb and forefinger with a force of 0.60 N as you twist your wrist. Suppose the coefficient of static friction between the knob and your fingers is only 0.12 because some oil accidentally got onto the knob. What is the most torque you can exert on the knob without having it slip between your fingers?

61. ⦀ A 70 kg man's arm, including the hand, can be modeled as a 75-cm-long uniform cylinder with a mass of 3.5 kg. In raising both his arms, from hanging down to straight up, by how much does he raise his center of gravity?

62. ⦀ The three masses shown in Figure P7.62 are connected by massless, rigid rods.
   a. Find the coordinates of the center of gravity.
   b. Find the moment of inertia about an axis that passes through mass A and is perpendicular to the page.
   c. Find the moment of inertia about an axis that passes through masses B and C.

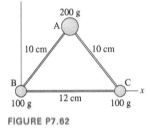

FIGURE P7.62

63. ⦀ A reasonable estimate of the moment of inertia of an ice skater spinning with her arms at her sides can be made by modeling most of her body as a uniform cylinder. Suppose the skater has a mass of 64 kg. One-eighth of that mass is in her arms, which are 60 cm long and 20 cm from the vertical axis about which she rotates. The rest of her mass is approximately in the form of a 20-cm-radius cylinder.
   a. Estimate the skater's moment of inertia to two significant figures.
   b. If she were to hold her arms outward, rather than at her sides, would her moment of inertia increase, decrease, or remain unchanged? Explain.

64. ⦀ Starting from rest, a 12-cm-diameter compact disk takes 3.0 s to reach its operating angular velocity of 2000 rpm. Assume that the angular acceleration is constant. The disk's moment of inertia is 2.5×10^{-5} kg·m².
   a. How much torque is applied to the disk?
   b. How many revolutions does it make before reaching full speed?

65. ⦀ The ropes in Figure P7.65 are each wrapped around a cylinder, and the two cylinders are fastened together. The smaller cylinder has a diameter of 10 cm and a mass of 5.0 kg; the larger cylinder has a diameter of 20 cm and a mass of 20 kg. What is the angular acceleration of the cylinders? Assume that the cylinders turn on a frictionless axle.

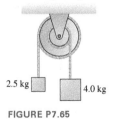

FIGURE P7.65

66. ⦀ Flywheels are large, massive wheels used to store energy. They can be spun up slowly, then the wheel's energy can be released quickly to accomplish a task that demands high power. An industrial flywheel has a 1.5 m diameter and a mass of 250 kg. A motor spins up the flywheel with a constant torque of 50 N·m. How long does it take the flywheel to reach top angular speed of 1200 rpm?

67. ⦀ A 1.0 kg ball and a 2.0 kg ball are connected by a 1.0-m-long rigid, massless rod. The rod and balls are rotating clockwise about their center of gravity at 20 rpm. What torque will bring the balls to a halt in 5.0 s?

68. ⦀ A 1.5 kg block is connected by a rope across a 50-cm-diameter, 2.0 kg, frictionless pulley, as shown in Figure P7.68. A constant 10 N tension is applied to the other end of the rope. Starting from rest, how long does it take the block to move 30 cm?

FIGURE P7.68

69. ⦀ The two blocks in Figure P7.69 are connected by a massless rope that passes over a pulley. The pulley is 12 cm in diameter and has a mass of 2.0 kg. As the pulley turns, friction at the axle exerts a torque of magnitude 0.50 N·m. If the blocks are released from rest, how long does it take the 4.0 kg block to reach the floor?

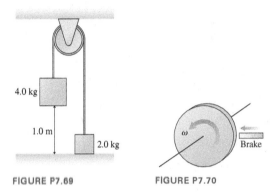

FIGURE P7.69   **FIGURE P7.70**

70. ⦀ The 2.0 kg, 30-cm-diameter disk in Figure P7.70 is spinning at 300 rpm. How much friction force must the brake apply to the rim to bring the disk to a halt in 3.0 s?

71. ⦀ A tradesman sharpens a knife by pushing it with a constant force against the rim of a grindstone. The 30-cm-diameter stone is spinning at 200 rpm and has a mass of 28 kg. The coefficient of kinetic friction between the knife and the stone is 0.20. If the stone slows steadily to 180 rpm in 10 s of grinding, what is the force with which the man presses the knife against the stone?

MCAT-Style Passage Problems

The Bunchberry BIO

The bunchberry flower has the fastest-moving parts ever seen in a plant. Initially, the stamens are held by the petals in a bent position, storing energy like a coiled spring. As the petals release, the tips of the stamens fly up and quickly release a burst of pollen.

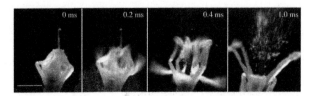

Figure P7.72 shows the details of the motion. The tips of the stamens act like a catapult, flipping through a 60° angle; the times on the earlier photos show that this happens in just 0.30 ms. We can model a stamen tip as a 1.0-mm-long, 10 μg rigid rod with a 10 μg anther sac at one end and a pivot point at the opposite end. Though an oversimplification, we will model the motion by assuming the angular acceleration is constant throughout the motion.

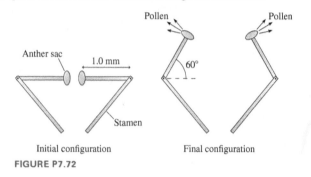

FIGURE P7.72

72. | What is the angular acceleration of the anther sac during the motion?
   A. 3.5×10^3 rad/s²   B. 7.0×10^3 rad/s²
   C. 1.2×10^7 rad/s²   D. 2.3×10^7 rad/s²
73. | What is the speed of the anther sac as it releases its pollen?
   A. 3.5 m/s   B. 7.0 m/s   C. 10 m/s   D. 14 m/s
74. ‖ How large is the "straightening torque"? (You can omit gravitational forces from your calculation; the gravitational torque is much less than this.)
   A. 2.3×10^{-7} N·m   B. 3.1×10^{-7} N·m
   C. 2.3×10^{-5} N·m   D. 3.1×10^{-5} N·m

The Illusion of Flight

The grand jeté is a classic ballet maneuver in which a dancer executes a horizontal leap while moving her arms and legs up and then down. At the center of the leap, the arms and legs are gracefully extended, as we see in Figure P7.75a. The goal of the leap is to create the illusion of flight. As the dancer moves through the air, he or she is in free fall. In Chapter 3, we saw that this leads to projectile motion. But what part of the dancer follows the usual parabolic path? It won't come as a surprise to learn that it's the center of gravity. But

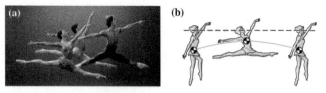

FIGURE P7.75

when you watch a dancer leap through the air, you don't watch her center of gravity, you watch her head. If the translational motion of her head is horizontal—not parabolic—this creates the illusion that she is flying through the air, held up by unseen forces.

Figure P7.75b illustrates how the dancer creates this illusion. While in the air, she changes the position of her center of gravity relative to her body by moving her arms and legs up, then down. Her center of gravity moves in a parabolic path, but her head moves in a straight line. It's not flight, but it will appear that way, at least for a moment.

75. | To perform this maneuver, the dancer relies on the fact that the position of her center of gravity
   A. Is near the center of the torso.
   B. Is determined by the positions of her arms and legs.
   C. Moves in a horizontal path.
   D. Is outside of her body.
76. | Suppose you wish to make a vertical leap with the goal of getting your head as high as possible above the ground. At the top of your leap, your arms should be
   A. Held at your sides.
   B. Raised above your head.
   C. Outstretched, away from your body.
77. | When the dancer is in the air, is there a gravitational torque on her? Take the dancer's rotation axis to be through her center of gravity.
   A. Yes, there is a gravitational torque.
   B. No, there is not a gravitational torque.
   C. It depends on the positions of her arms and legs.
78. | In addition to changing her center of gravity, a dancer may change her moment of inertia. Consider her moment of inertia about a vertical axis through the center of her body. When she raises her arms and legs, this
   A. Increases her moment of inertia.
   B. Decreases her moment of inertia.
   C. Does not change her moment of inertia.

STOP TO THINK ANSWERS

Chapter Preview Stop to Think: B. The speed is proportional to the frequency. Doubling the frequency means doubling the speed.

Stop to Think 7.1: A. Because $5\pi/2$ rad $= 2\pi$ rad $+ \pi/2$ rad, the particle's position is one complete revolution (2π rad) plus an extra $\pi/2$ rad. This extra $\pi/2$ rad puts the particle at position A.

Stop to Think 7.2: a. constant (but not zero), b. constant (but not zero), c. constant (but not zero), d. zero, e. zero. The angular velocity ω is constant. Thus the magnitudes of the velocity $v = \omega r$ and the centripetal acceleration $a_c = \omega^2 r$ are constant. This also means that the ball's angular acceleration α and tangential acceleration $a_t = \alpha r$ are both zero.

Stop to Think 7.3: E. Forces D and C act at or in line with the pivot and provide no torque. Of the others, only E tries to rotate the wheel counterclockwise, so it is the only choice that gives a positive nonzero torque.

Stop to Think 7.4: A. The force acting at the axis exerts no torque. Thus the third force needs to exert an equal but opposite torque to that exerted by the force acting at the rim. Force A, which has twice the magnitude but acts at half the distance from the axis, does so.

Stop to Think 7.5: $\tau_B > \tau_D > \tau_A = \tau_C$. The torques are $\tau_B = 2mgL$, $\tau_D = \frac{3}{2}mgL$, and $\tau_A = \tau_C = mgL$, where L is the length of the rod in B.

Stop to Think 7.6: $I_D > I_A > I_C > I_B$. The moments of inertia are $I_B \approx 0$, $I_C = 2mr^2$, $I_A = 3mr^2$, and $I_D = mr^2 + m(2r)^2 = 5mr^2$.

Stop to Think 7.7: C. The velocity of P is the vector sum of $\vec{v}$ directed to the right and an upward velocity of the same magnitude due to the rotation of the wheel.

8 Equilibrium and Elasticity

How does a dancer balance so gracefully *en pointe*? And how does her foot withstand the great stresses concentrated on her toes? In this chapter we'll find answers to both these questions.

LOOKING AHEAD ▸

Goal: To learn about the static equilibrium of extended objects, and the basic properties of springs and elastic materials.

Static Equilibrium

As this cyclist balances on his back tire, the net force *and* the net torque on him must be zero.

You'll learn to analyze objects that are in **static equilibrium**.

Springs

When a rider takes a seat, the spring is compressed and exerts a **restoring force**, pushing upward.

You'll learn to solve problems involving stretched and compressed springs.

Properties of Materials

All materials have some "give"; if you pull on them, they stretch, and at some point they break.

Is this spider silk as strong as steel? You'll learn to think about what the question means, and how to answer it.

LOOKING BACK ◂

Torque

In Chapter 7, you learned to calculate the torque on an object due to an applied force.

In this chapter, we'll extend our analysis to consider objects with many forces—and many torques—that act on them.

STOP TO THINK

An old-fashioned tire swing exerts a force on the branch and a torque about the point where the branch meets the trunk. If you hang the swing closer to the trunk, this will _____ the force and _____ the torque.

A. increase, increase
B. not change, increase
C. not change, not change
D. not change, decrease
E. decrease, not change
F. decrease, decrease

8.1 Torque and Static Equilibrium

FIGURE 8.1 A block with no net force acting on it may still be out of equilibrium.

(a) When the net force on a particle is zero, the particle is in static equilibrium.

(b) Both the net force and the net torque are zero, so the block is in static equilibrium.

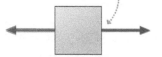

(c) The net force is still zero, but the net torque is *not* zero. The block is not in equilibrium.

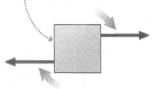

We have now spent several chapters studying motion and its causes. In many disciplines, it is just as important to understand the conditions under which objects do *not* move. Buildings and dams must be designed such that they remain motionless, even when huge forces act on them. And joints in the body must sustain large forces when the body is supporting heavy loads, as in holding or carrying heavy objects.

Recall from ◄ **SECTION 5.1** that an object at rest is in *static equilibrium*. As long as an object can be modeled as a *particle*, the condition necessary for static equilibrium is that the net force $\vec{F}_{net}$ on the particle is zero, as in **FIGURE 8.1a**, where the two forces applied to the particle balance and the particle can remain at rest.

But in Chapter 7 we moved beyond the particle model to study extended objects that can rotate. Consider, for example, the block in **FIGURE 8.1b**. In this case the two forces act along the same line, the net force is zero, and the block is in equilibrium. But what about the block in **FIGURE 8.1c**? The net force is still zero, but this time the block begins to rotate because the two forces exert a net *torque*. For an extended object, $\vec{F}_{net} = \vec{0}$ is not by itself enough to ensure static equilibrium. There is a second condition for static equilibrium of an extended object: The net torque τ_{net} on the object must also be zero.

If we write the net force in component form, the conditions for static equilibrium of an extended object are

$$\left.\begin{array}{l} \sum F_x = 0 \\ \sum F_y = 0 \end{array}\right\} \quad \text{No net force}$$

$$\sum \tau = 0 \quad \} \quad \text{No net torque} \tag{8.1}$$

Conditions for static equilibrium of an extended object

EXAMPLE 8.1   **Finding the force from the biceps tendon** BIO

In the joints of the human body, muscles usually attach quite near the joint. This means that muscle and tendon forces are much larger than applied forces. In weightlifting, where the applied forces are large, holding the body in static equilibrium requires muscle forces that are quite large indeed. In the *strict curl* event, a standing athlete lifts a barbell by moving only his forearms, which pivot at the elbow. The record weight lifted in the strict curl is over 200 pounds (about 900 N). **FIGURE 8.2** shows the arm bones and the main lifting muscle when the forearm is horizontal. The distance from the tendon to the elbow joint is 4.0 cm, and from the barbell to the elbow 35 cm.

a. What is the tension in the tendon connecting the biceps muscle to the bone while a 900 N barbell is held stationary in this position?
b. What is the force exerted by the elbow on the forearm bones?

PREPARE FIGURE 8.3 shows a simplified model of the arm and the forces acting on the forearm. $\vec{F}_t$ is the tension force due to the tendon, $\vec{F}_b$ is the downward force of the barbell, and $\vec{F}_e$ is the force of the elbow joint on the forearm. As a simplification, we've ignored the weight of the arm itself because it is so much less than the weight of the barbell. Because $\vec{F}_t$ and $\vec{F}_b$ have no *x*-component, neither can $\vec{F}_e$. If it did, the net force in the *x*-direction would not be zero, and the forearm could not be in equilibrium. Because each arm supports half the weight of the barbell, the magnitude of the barbell force is $F_b = 450$ N.

FIGURE 8.2 An arm holding a barbell.

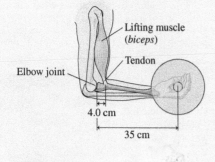

Lifting muscle (*biceps*)

Tendon

Elbow joint

4.0 cm

35 cm

FIGURE 8.3 Visual overview of holding a barbell.

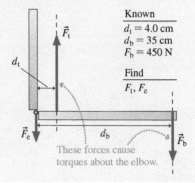

$\vec{F}_t$

d_t

Known
$d_t = 4.0$ cm
$d_b = 35$ cm
$F_b = 450$ N

Find

F_t, F_e

$\vec{F}_e$

d_b

$\vec{F}_b$

These forces cause torques about the elbow.

SOLVE a. For the forearm to be in static equilibrium, the net force and net torque on it must both be zero. Setting the net force to zero gives

$$\sum F_y = F_t - F_e - F_b = 0$$

We don't know either of the forces F_t and F_e, nor does the force equation give us enough information to find them. But the fact that in static equilibrium the torque also must be zero gives us the extra information that we need.

Recall that the torque must be calculated about a particular point. Here, a natural choice is the elbow joint, about which the forearm can pivot. Given this pivot, we can calculate the torque due to each of the three forces in terms of their magnitudes F and moment arms $r_\perp$ as $\tau = r_\perp F$. The moment arm is the perpendicular distance between the pivot and the "line of action" along which the force is applied. Figure 8.3 shows that the moment arms for $\vec{F}_t$ and $\vec{F}_b$ are the distances d_t and d_b, respectively, measured along the beam representing the forearm. The moment arm for $\vec{F}_e$ is zero, because this force acts directly at the pivot. Thus we have

$$\tau_{net} = F_e \times 0 + F_t d_t - F_b d_b = 0$$

The tension in the tendon tries to rotate the arm counterclockwise, so it produces a positive torque. The torque due to the barbell, which tries to rotate the arm in a clockwise direction, is negative. The magnitudes of the two torques must be equal: The magnitude of the torque trying to rotate the arm counterclockwise equals the magnitude of the torque trying to rotate the arm clockwise:

$$F_t d_t = F_b d_b$$

We can solve this equation for the force in the tendon:

$$F_t = F_b \frac{d_b}{d_t} = (450\ \text{N}) \frac{35\ \text{cm}}{4.0\ \text{cm}} = 3940\ \text{N}$$

We've kept an extra significant figure to avoid rounding error in the next step, but we'll report the result of part a as $F_t = 3900\ \text{N}$.

b. Next, we use the force equation to find the force in the elbow joint:

$$F_e = F_t - F_b = 3940\ \text{N} - 450\ \text{N} = 3500\ \text{N}$$

ASSESS This large value for F_t makes sense: The short distance d_t from the tendon to the elbow joint means that the force supplied by the biceps has to be very large to counter the torque generated by a force applied at the opposite end of the forearm. It also makes sense that balancing this large upward force requires a large downward force in the elbow joint.

STOP TO THINK 8.1 Which of these objects is in static equilibrium?

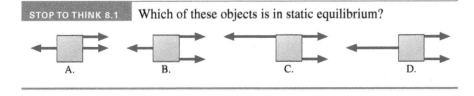

A.   B.   C.   D.

Choosing the Pivot Point

In Example 8.1, we calculated the net torque using the elbow joint as the axis of rotation or pivot point. But we learned in Chapter 7 that the torque depends on which point is chosen as the pivot point. Was there something special about our choice of the elbow joint?

Consider the hammer shown in FIGURE 8.4, supported on a pegboard by two pegs A and B. Because the hammer is in static equilibrium, the net torque around the pivot at peg A must be zero: The clockwise torque due to the weight $\vec{w}$ is exactly balanced by the counterclockwise torque due to the force $\vec{n}_B$ of peg B. (Recall that the torque due to $\vec{n}_A$ is zero because here $\vec{n}_A$ acts at the pivot A.) But if instead we take B as the pivot, the net torque is still zero. The counterclockwise torque due to $\vec{w}$ (with a large force but small moment arm) balances the clockwise torque due to $\vec{n}_A$ (with a small force but large moment arm). Indeed, **for an object in static equilibrium, the net torque about *every* point must be zero.** This means you can pick *any* point you wish as a pivot point for calculating the torque.

Although any choice of a pivot point will work, some choices are better because they simplify the calculations. Often, there is a "natural" axis of rotation in the problem, an axis about which rotation *would* occur if the object were not in static equilibrium. Example 8.1 is of this type, with the elbow joint as a natural axis of rotation.

In many problems, there are forces that are unknown or poorly specified. Choosing to put the pivot point where such a force acts greatly simplifies the solution. For

FIGURE 8.4 A hammer resting on two pegs.

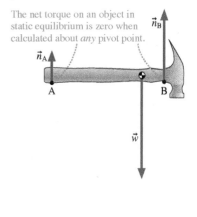

The net torque on an object in static equilibrium is zero when calculated about *any* pivot point.

FIGURE 8.5 Choosing the pivot for a woman rappelling down a rock wall.

The torque due to $\vec{F}$ about this point is zero. This makes this point a good choice as the pivot.

instance, the woman in FIGURE 8.5 is in equilibrium as she rests on the rock wall. The force of the wall on her feet is a mix of normal forces and friction forces; the direction isn't well specified. The directions of the other two forces are well known: The tension force points along the rope, and the weight force points down. So a good choice of pivot point is where the woman's foot contacts the wall because this choice eliminates the torque due to the force of the wall on her foot, which isn't as well known as the other forces.

PROBLEM-SOLVING STRATEGY 8.1   **Static equilibrium problems**

If an object is in static equilibrium, we can use the fact that there is no net force and no net torque as a basis for solving problems.

PREPARE Model the object as a simple shape. Draw a visual overview that shows all forces and distances. List known information.

- Pick an axis or pivot about which the torques will be calculated.
- Determine the torque about this pivot point due to each force acting on the object. The torques due to any forces acting *at* the pivot are zero.
- Determine the sign of each torque about this pivot point.

SOLVE The mathematical steps are based on the conditions:

$$\vec{F}_{net} = \vec{0} \qquad \text{and} \qquad \tau_{net} = 0$$

- Write equations for $\sum F_x = 0$, $\sum F_y = 0$, and $\sum \tau = 0$.
- Solve the resulting equations.

ASSESS Check that your result is reasonable and answers the question.

EXAMPLE 8.2   **Forces on a board resting on sawhorses**

A board weighing 100 N sits across two sawhorses, as shown in FIGURE 8.6. What are the magnitudes of the normal forces of the sawhorses acting on the board?

FIGURE 8.6 A board sitting on two sawhorses.

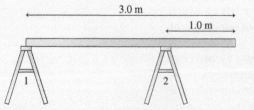

PREPARE The board and the forces acting on it are shown in FIGURE 8.7. $\vec{n}_1$ and $\vec{n}_2$ are the normal forces on the board due to the sawhorses, and $\vec{w}$ is the weight of the board acting at the center of gravity.

As discussed above, a good choice for the pivot is a point at which an unknown force acts because that force contributes nothing to the torque. Either the point where $\vec{n}_1$ acts or the point where $\vec{n}_2$ acts will work; let's choose the left end of the board, where $\vec{n}_1$ acts, for this example. With this choice of pivot point, the moment arm for $\vec{w}$ is $d_1 = 1.5$ m, half the board's length. Because $\vec{w}$ tends to rotate the board clockwise, its torque is negative. The moment arm for $\vec{n}_2$ is the distance of the second sawhorse from the pivot

point, which is $d_2 = 2.0$ m. This force tends to rotate the board counterclockwise, so it exerts a positive torque.

FIGURE 8.7 Visual overview of a board on two sawhorses.

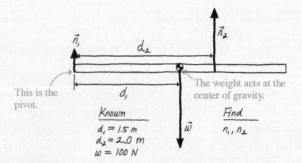

SOLVE The board is in static equilibrium, so the net force $\vec{F}_{net}$ and the net torque τ_{net} must both be zero. The forces have only y-components, so the force equation is

$$\sum F_y = n_1 - w + n_2 = 0$$

As we've seen, the gravitational torque is negative and the torque from the upward force from the sawhorse is positive, so the torque equation is

$$\tau_{net} = -wd_1 + n_2 d_2 = 0$$

We now have two simultaneous equations with the two unknowns n_1 and n_2. We can solve for n_2 in the torque equation and then substitute that result into the force equation. From the torque equation,

$$n_2 = \frac{d_1 w}{d_2} = \frac{(1.5 \text{ m})(100 \text{ N})}{2.0 \text{ m}} = 75 \text{ N}$$

The force equation is then $n_1 - 100 \text{ N} + 75 \text{ N} = 0$, which we can solve for n_1:

$$n_1 = w - n_2 = 100 \text{ N} - 75 \text{ N} = 25 \text{ N}$$

ASSESS It seems reasonable that $n_2 > n_1$ because more of the board sits over the right sawhorse.

The center of gravity of a uniform board is easy to see—it's right at the midpoint—but how about the center of gravity of your body? The location of your center of gravity depends on how you are posed, but even for the simple case of standing straight with your arms at your side, you can't simply assume that the center of gravity is at the midpoint of your body; you need to *measure* the position. The next example illustrates the use of a *reaction board* and a scale to make this determination. This is a standard measurement in biomechanics.

EXAMPLE 8.3 | **Finding the center of gravity of the human body** BIO

A woman weighing 600 N lies on a 2.5-m-long, 60 N reaction board with her feet over the pivot. The scale on the right reads 250 N. What is the distance d from the woman's feet to her center of gravity?

PREPARE The forces and distances in the problem are shown in FIGURE 8.8. We'll consider the board and woman as a single object. We've assumed that the board is uniform, so its center of gravity is at its midpoint. To eliminate the unknown magnitude of

FIGURE 8.8 Visual overview of the reaction board and woman.

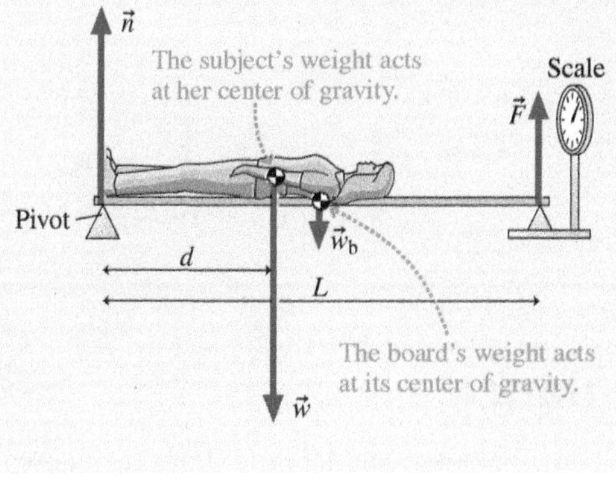

Known
$w = 600$ N
$w_b = 60$ N
$L = 2.5$ m
$F = 250$ N

Find
d

$\vec{n}$ from the torque equation, we'll choose the pivot to be the left end of the board. The torque due to $\vec{F}$ is positive, and those due to $\vec{w}$ and $\vec{w}_b$ are negative.

SOLVE Because the board and woman are in static equilibrium, the net force and net torque on them must be zero. The force equation reads

$$\sum F_y = n - w_b - w + F = 0$$

and the torque equation gives

$$\sum \tau = -\frac{L}{2} w_b - dw + LF = 0$$

In this case, the force equation isn't needed because we can solve the torque equation for d:

$$d = \frac{LF - \frac{1}{2} L w_b}{w} = \frac{(2.5 \text{ m})(250 \text{ N}) - \frac{1}{2}(2.5 \text{ m})(60 \text{ N})}{600 \text{ N}}$$

$$= 0.92 \text{ m}$$

ASSESS If the woman is 5′ 6″ (1.68 m) tall, her center of gravity is $(0.92 \text{ m})/(1.68 \text{ m}) = 55\%$ of her height, or a little more than halfway up her body. This seems reasonable and this is, in fact, a typical value for women.

EXAMPLE 8.4 | **Will the ladder slip?**

A 3.0-m-long ladder leans against a wall at an angle of 60° with respect to the floor. What is the minimum value of μ_s, the coefficient of static friction with the ground, that will prevent the ladder from slipping? Assume that friction between the ladder and the wall is negligible.

PREPARE The ladder is a rigid rod of length L. To not slip, both the net force and net torque on the ladder must be zero. FIGURE 8.9 on the next page shows the ladder and the forces acting on it. We are asked to find the necessary coefficient of static friction. First, we'll solve for the magnitudes of the static friction force and the

normal force. Then we can use these values to determine the necessary value of the coefficient of friction. These forces both act at the bottom corner of the ladder, so even though we are interested in these forces, this is a good choice for the pivot point because two of the forces that act provide no torque, which simplifies the solution. With this choice of pivot, the weight of the ladder, acting at the center of gravity, exerts torque $d_1 w$ and the force of the wall exerts torque $-d_2 n_2$. The signs are based on the observation that $\vec{w}$ would cause the ladder to rotate counterclockwise, while $\vec{n}_2$ would cause it to rotate clockwise.

Continued

FIGURE 8.9 Visual overview of a ladder in static equilibrium.

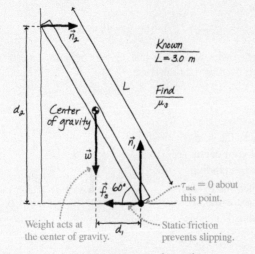

Known
$L = 3.0$ m

Find
μ_s

$\vec{n}_2$

d_2 — Center of gravity

$\vec{w}$

$\vec{n}_1$

$\vec{f}_s$ 60°

$\tau_{net} = 0$ about this point.

Weight acts at the center of gravity.

Static friction prevents slipping.

d_1

SOLVE The x- and y-components of $\vec{F}_{net} = \vec{0}$ are

$$\sum F_x = n_2 - f_s = 0$$
$$\sum F_y = n_1 - w = n_1 - Mg = 0$$

The torque about the bottom corner is

$$\tau_{net} = d_1 w - d_2 n_2 = \frac{1}{2}(L\cos 60°)Mg - (L\sin 60°)n_2 = 0$$

Altogether, we have three equations with the three unknowns n_1, n_2, and f_s. If we solve the third equation for n_2,

$$n_2 = \frac{\frac{1}{2}(L\cos 60°)Mg}{L\sin 60°} = \frac{Mg}{2\tan 60°}$$

we can then substitute this into the first equation to find

$$f_s = \frac{Mg}{2\tan 60°}$$

Our model of static friction is $f_s \leq f_{s\,max} = \mu_s n_1$. We can find n_1 from the second equation: $n_1 = Mg$. From this, the model of friction tells us that

$$f_s \leq \mu_s Mg$$

Comparing these two expressions for f_s, we see that μ_s must obey

$$\mu_s \geq \frac{1}{2\tan 60°} = 0.29$$

Thus the minimum value of the coefficient of static friction is 0.29.

ASSESS You know from experience that you can lean a ladder or other object against a wall if the ground is "rough," but it slips if the surface is too smooth. 0.29 is a "medium" value for the coefficient of static friction, which is reasonable.

STOP TO THINK 8.2 A beam with a pivot on its left end is suspended from a rope. In which direction is the force of the pivot on the beam?

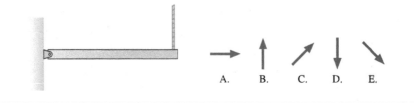

A.   B.   C.   D.   E.

8.2 Stability and Balance

Class Video

If you tilt a box up on one edge by a small amount and let go, it falls back down. If you tilt it too much, it falls over. And if you tilt it "just right," you can get the box to balance on its edge. What determines these three possible outcomes?

FIGURE 8.10 illustrates the idea with a car, but the results are general and apply in many situations. An extended object, whether it's a car, a box, or a person, has a *base of support* on which it rests when in static equilibrium. If you tilt the object, one edge of the base of support becomes a pivot point. As long as the object's center of gravity remains over the base of support, torque due to gravity will rotate the object back toward its stable equilibrium position; we say that the object is **stable.** This is the situation in Figure 8.10b.

A *critical angle* θ_c is reached when the center of gravity is directly over the pivot point, as in Figure 8.10c. This is the point of balance, with no net torque. If the car continues to tip, the center of gravity moves outside the base of support, as in Figure 8.10d. Now, the gravitational torque causes a rotation in the opposite direction and the car rolls over; it is **unstable.** If an accident (or taking a corner too fast) causes a vehicle to pivot up onto two wheels, it will roll back to an upright position as long as $\theta < \theta_c$, but it will roll over if $\theta > \theta_c$.

FIGURE 8.10 A car—or any object—will fall over when tilted too far.

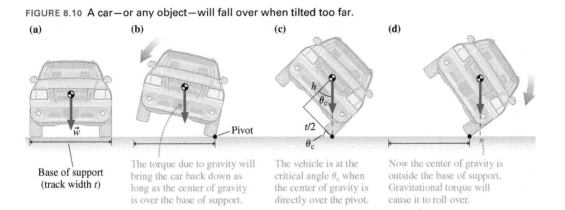

(a)   (b)   (c)   (d)

Base of support
(track width t)

The torque due to gravity will bring the car back down as long as the center of gravity is over the base of support.

The vehicle is at the critical angle θ_c when the center of gravity is directly over the pivot.

Now the center of gravity is outside the base of support. Gravitational torque will cause it to roll over.

For vehicles, the distance between the tires—the base of support—is called the track width t. The height of the center of gravity is h. You can see from Figure 8.10c that the ratio of the two, the height-to-width ratio, determines when the critical angle is reached. It's the height-to-width ratio that's important, not the absolute height of the center of gravity. FIGURE 8.11 compares a passenger car and a sport utility vehicle (SUV). For the passenger car, with $h/t \approx 0.33$, the critical angle is $\theta_c \approx 57°$. But for the SUV, with $h/t \approx 0.47$, the critical angle is $\theta_c \approx 47°$. Loading an SUV with cargo further raises the center of gravity, especially if the roof rack is used, reducing θ_c even more. Various automobile safety groups have determined that a vehicle with $\theta_c > 50°$ is unlikely to roll over in an accident. A rollover becomes increasingly likely as θ_c is reduced below this threshold.

The same argument we made for tilted vehicles can be made for any object, leading to the general rule that **a wider base of support and/or a lower center of gravity improves stability.**

FIGURE 8.11 Compared to a passenger car, an SUV has a high center of gravity relative to its width.

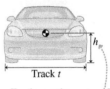

Track t

For the car the center-of-gravity height h is 33% of t.

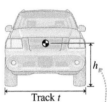

Track t

For the SUV, the center-of-gravity height h is 47% of t.

CONCEPTUAL EXAMPLE 8.5   How far to walk the plank?

A cat walks along a plank that extends out from a table. If the cat walks too far out on the plank, the plank will begin to tilt. What determines when this happens?

REASON An object is stable if its center of gravity lies over its base of support, and unstable otherwise. Let's take the cat and the plank to be one combined object whose center of gravity lies along a line between the cat's center of gravity and that of the plank.

In FIGURE 8.12a, when the cat is near the left end of the plank, the combined center of gravity is over the base of support and the plank is stable. As the cat moves to the right, he reaches a point where the combined center of gravity is directly over the edge of the table, as shown in FIGURE 8.12b. If the cat takes one more step, the cat and plank will become unstable and the plank will begin to tilt.

FIGURE 8.12 Changing stability as a cat walks on a plank.

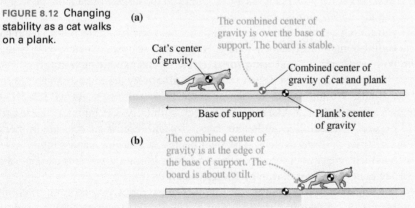

(a)

Cat's center of gravity

The combined center of gravity is over the base of support. The board is stable.

Combined center of gravity of cat and plank

Base of support

Plank's center of gravity

(b)

The combined center of gravity is at the edge of the base of support. The board is about to tilt.

ASSESS Because the plank's center of gravity must be to the left of the edge for it to be stable by itself, the cat can actually walk a short distance out onto the unsupported part of the plank before it starts to tilt. The heavier the plank is, the farther the cat can walk.

Balancing a soda can Try to balance a soda can—full or empty—on the narrow bevel at the bottom. It can't be done because, either full or empty, the center of gravity is near the center of the can. If the can is tilted enough to sit on the bevel, the center of gravity lies far outside this small base of support. But if you put about 2 ounces (60 ml) of water in an empty can, the center of gravity will be right over the bevel and the can will balance.

Stability and Balance of the Human Body BIO

FIGURE 8.13 Standing on tiptoes.

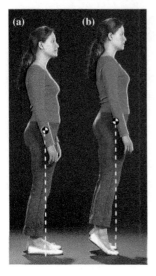

The human body is remarkable for its ability to constantly adjust its stance to remain stable on just two points of support. In walking, running, or even the simple act of rising from a chair, the position of the body's center of gravity is constantly changing. To maintain stability, we unconsciously adjust the positions of our arms and legs to keep our center of gravity over our base of support.

A simple example of how the body naturally realigns its center of gravity is found in the act of standing up on tiptoes. FIGURE 8.13a shows the body in its normal standing position. Notice that the center of gravity is well centered over the base of support (the feet), ensuring stability. If the person were now to stand on tiptoes *without* otherwise adjusting the body position, her center of gravity would fall behind the base of support, which is now the balls of the feet, and she would fall backward. To prevent this, as shown in FIGURE 8.13b, the body naturally leans forward, regaining stability by moving the center of gravity over the balls of the feet. Try this: Stand facing a wall with your toes touching the base of the wall. Now try standing on your toes. Your body can't move forward to keep your center of gravity over your toes, so you can't do it!

The chapter opened with a photo of a dancer who is performing a very delicate balancing act, adjusting the positions of her torso, head, arms, and legs so that her center of gravity is above a very small point of support. Now that you understand the basics of balance, you can see what a remarkable feat this is!

STOP TO THINK 8.3   Rank in order, from least stable to most stable, the three objects shown in the figure. The positions of their centers of gravity are marked. (For the centers of gravity to be positioned like this, the objects must have a nonuniform composition.)

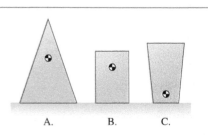

8.3 Springs and Hooke's Law

Elasticity in action A golf ball compresses quite a bit when struck. The restoring force that pushes the ball back into its original shape helps launch the ball off the face of the club, making for a longer drive.

We have assumed that objects in equilibrium maintain their shape as forces and torques are applied to them. In reality this is an oversimplification. Every solid object stretches, compresses, or deforms when a force acts on it. This change is easy to see when you press on a green twig on a tree, but even the largest branch on the tree will bend slightly under your weight.

If you stretch a rubber band, there is a force that tries to pull the rubber band back to its equilibrium, or unstretched, length. A force that restores a system to an equilibrium position is called a **restoring force.** Systems that exhibit such restoring forces are called **elastic.** The most basic examples of **elasticity** are things like springs and rubber bands. We introduced the spring force in ◄ SECTION 4.2. As we saw, if you stretch a spring, a tension-like force pulls back. Similarly, a compressed spring tries to re-expand to its equilibrium length. Elasticity and restoring forces are properties of much stiffer systems as well. The steel beams of a bridge bend slightly as you drive your car over it, but they are restored to equilibrium after your car passes by. Your leg bones flex a bit during each step you take.

When no forces act on a spring to compress or extend it, it will rest at its equilibrium length. If you then stretch the spring, it pulls back. If you compress the spring, it pushes back, as illustrated in FIGURE 8.14a. In general, the spring force always points in the direction opposite the displacement from equilibrium. How hard the spring pulls back depends on how much it is stretched, as shown in FIGURE 8.14b. FIGURE 8.14c is a graph of real data for a spring, showing the magnitude of the spring force as the stretch of the spring is varied. You can see that **the spring force is *proportional* to the displacement of the end of the spring.** This is a *linear relationship,* and the slope k of the line is the proportionality constant:

$$F_{sp} = k \, \Delta x \tag{8.2}$$

Compressing or stretching the spring twice as far results in a restoring force that is twice as large.

FIGURE 8.14 The restoring force for a spring.

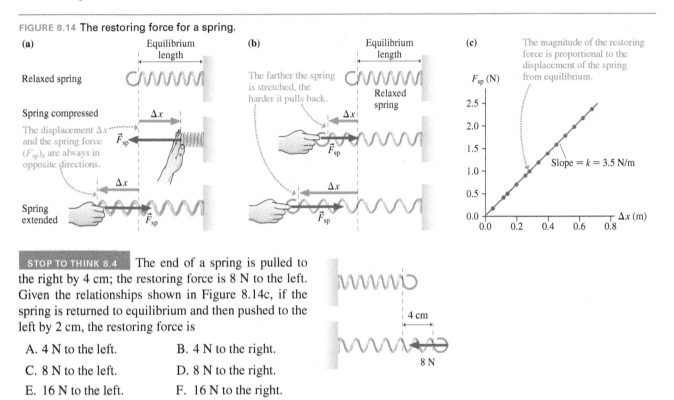

STOP TO THINK 8.4   The end of a spring is pulled to the right by 4 cm; the restoring force is 8 N to the left. Given the relationships shown in Figure 8.14c, if the spring is returned to equilibrium and then pushed to the left by 2 cm, the restoring force is

A. 4 N to the left.   B. 4 N to the right.

C. 8 N to the left.   D. 8 N to the right.

E. 16 N to the left.   F. 16 N to the right.

The value of the constant k in Equation 8.2 depends on the spring. We call k the **spring constant**; it has units of N/m. The spring constant k is a property that characterizes a spring, just as the mass m characterizes a particle. If k is large, it takes a large pull to cause a significant stretch, and we call the spring a "stiff" spring. If k is small, we can stretch the spring with very little force, and we call it a "soft" spring. Every spring has its own unique value of k. The spring constant for the data in Figure 8.14c can be determined from the slope of the straight line to be $k = 3.5$ N/m.

As Figure 8.14a shows, if the spring is compressed, Δx is positive and, because $\vec{F}_{sp}$ points to the left, its component $(F_{sp})_x$ is negative. If the spring is stretched, however, Δx is negative and, because $\vec{F}_{sp}$ points to the right, its component $(F_{sp})_x$ is positive. If we rewrite Equation 8.2 in terms of the *component* of the spring force, we get the most general form of the relationship between the restoring force and the displacement of the end of a spring, which is known as **Hooke's law**:

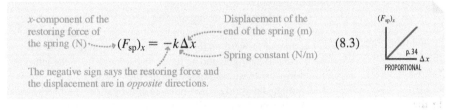

x-component of the restoring force of the spring (N) $\longrightarrow (F_{sp})_x = -k \, \Delta x$ Displacement of the end of the spring (m) $\cdots\cdots$ Spring constant (N/m) (8.3)

The negative sign says the restoring force and the displacement are in *opposite* directions.

For motion in the vertical (y) direction, Hooke's law is $(F_{sp})_y = -k \, \Delta y$.

Hooke's law is not a true "law of nature" in the sense that Newton's laws are. It is actually just a model of a restoring force. It works extremely well for some springs, such as the one in Figure 8.14c, but less well for others. Hooke's law will fail for any spring if it is compressed or stretched too far, as we'll see in the next section.

NOTE ▶ Just as we used massless strings, we will adopt the idealization of a *massless spring*. Though not a perfect description, it is a good approximation if the mass attached to a spring is much greater than the mass of the spring itself. ◀

EXAMPLE 8.6 Weighing a fish

A scale used to weigh fish consists of a spring hung from a support. The spring's equilibrium length is 10.0 cm. When a 4.0 kg fish is suspended from the end of the spring, it stretches to a length of 12.4 cm.

a. What is the spring constant k for this spring?
b. If an 8.0 kg fish is suspended from the spring, what will be the length of the spring?

PREPARE The visual overview in FIGURE 8.15 shows the details for the first part of the problem. The fish hangs in static equilibrium, so the net force in the y-direction and the net torque must be zero.

FIGURE 8.15 Visual overview of a mass suspended from a spring.

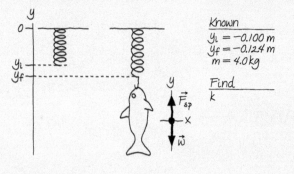

SOLVE a. Because the fish is in static equilibrium, we have

$$\sum F_y = (F_{sp})_y + w_y = -k\,\Delta y - mg = 0$$

so that $k = -mg/\Delta y$. (The net torque is zero because the fish's center of gravity comes to rest directly under the pivot point of the hook.) From Figure 8.15, the displacement of the spring from equilibrium is $\Delta y = y_f - y_i = (-0.124\ \text{m}) - (-0.100\ \text{m}) = -0.024\ \text{m}$. This displacement is *negative* because the fish moves in the $-y$-direction. We can now solve for the spring constant:

$$k = -\frac{mg}{\Delta y} = -\frac{(4.0\ \text{kg})(9.8\ \text{m/s}^2)}{-0.024\ \text{m}} = 1600\ \text{N/m}$$

b. The restoring force is proportional to the displacement of the spring from its equilibrium length. If we double the mass (and thus the weight) of the fish, the displacement of the end of the spring will double as well, to $\Delta y = -0.048\ \text{m}$. Thus the spring will be 0.048 m longer, so its new length is $0.100\ \text{m} + 0.048\ \text{m} = 0.148\ \text{m} = 14.8\ \text{cm}$.

ASSESS The spring doesn't stretch very much when a 4.0 kg mass is hung from it. A large spring constant of 1600 N/m thus seems reasonable for this stiff spring.

EXAMPLE 8.7 When does the block slip?

FIGURE 8.16 shows a spring attached to a 2.0 kg block. The other end of the spring is pulled by a motorized toy train that moves forward at 5.0 cm/s. The spring constant is 50 N/m, and the coefficient of static friction between the block and the surface is 0.60. The spring is at its equilibrium length at $t = 0$ s when the train starts to move. When does the block slip?

FIGURE 8.16 A toy train stretches the spring until the block slips.

5.0 cm/s

2.0 kg

PREPARE We model the block as a particle and the spring as a massless spring. FIGURE 8.17 is a free-body diagram for the block. We convert the speed of the train into m/s: $v = 0.050\ \text{m/s}$.

SOLVE Recall that the tension in a massless string pulls equally at *both* ends of the string. The same is true for the spring force: It

FIGURE 8.17 Free-body diagram for the block.

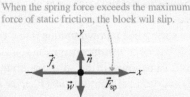

When the spring force exceeds the maximum force of static friction, the block will slip.

pulls (or pushes) equally at *both* ends. Imagine holding a rubber band with your left hand and stretching it with your right hand. Your left hand feels the pulling force, even though it was the right end of the rubber band that moved.

This is the key to solving the problem. As the right end of the spring moves, stretching the spring, the spring pulls backward on the train *and* forward on the block with equal strength. The train is moving to the right, and so the spring force pulls to the left on the train—as we would expect. But the block is at the other end of the spring; the spring force pulls to the right on the block, as shown in Figure 8.17. As the spring stretches, the static friction force on the block increases in magnitude to keep the block at rest. The block is in static equilibrium, so

$$\sum F_x = (F_{sp})_x + (f_s)_x = F_{sp} - f_s = 0$$

where F_{sp} is the magnitude of the spring force. This magnitude is $F_{sp} = k \Delta x$, where $\Delta x = vt$ is the distance the train has moved. Thus

$$f_s = F_{sp} = k \Delta x$$

The block slips when the static friction force reaches its maximum value $f_{s\,max} = \mu_s n = \mu_s mg$. This occurs when the train has

moved a distance

$$\Delta x = \frac{f_{s\,max}}{k} = \frac{\mu_s mg}{k} = \frac{(0.60)(2.0\ \text{kg})(9.8\ \text{m/s}^2)}{50\ \text{N/m}} = 0.235\ \text{m}$$

The time at which the block slips is

$$t = \frac{\Delta x}{v} = \frac{0.235\ \text{m}}{0.050\ \text{m/s}} = 4.7\ \text{s}$$

ASSESS The result of about 5 s seems reasonable for a slowly moving toy train to stretch the spring enough for the block to slip.

STOP TO THINK 8.5 A 1.0 kg weight is suspended from a spring, stretching it by 5.0 cm. How much does the spring stretch if the 1.0 kg weight is replaced by a 3.0 kg weight?

A. 5.0 cm   B. 10.0 cm   C. 15.0 cm   D. 20.0 cm

8.4 Stretching and Compressing Materials

In Chapter 4 we noted that we could model most solid materials as being made of particle-like atoms connected by spring-like bonds. We can model a steel rod this way, as illustrated in FIGURE 8.18a. The spring-like bonds between the atoms in steel are quite stiff, but they can be stretched or compressed, meaning that even a steel rod is elastic. If you pull on the end of a steel rod, as in Figure 8.18a, you will slightly stretch the bonds between the particles that make it up, and the rod itself will stretch. The stretched bonds pull back on your hand with a restoring force that causes the rod to return to its original length when released. In this sense, the entire rod acts like a very stiff spring. As is the case for a spring, a restoring force is also produced by compressing the rod.

In FIGURE 8.18b, real data for a 1.0-m-long, 1.0-cm-diameter steel rod show that, just as for a spring, the restoring force is proportional to the change in length. However, the *scale* of the stretch of the rod and the restoring force is much different from that for a spring. It would take a force of 16,000 N to stretch the rod by only 1 mm, corresponding to a spring constant of 1.6×10^7 N/m! Steel is elastic, but under normal forces, it experiences only very small changes in dimension. Materials of this sort are called **rigid.**

The behavior of other materials, such as the rubber in a rubber band, can be quite different. A rubber band can be stretched quite far—several times its equilibrium length—with a very small force, and then snaps back to its original shape when released. Materials that show large deformations with small forces are called **pliant.**

A rod's spring constant depends on several factors, as shown in FIGURE 8.19. First, we expect that a thick rod, with a large cross-section area A, will be more difficult to stretch than a thinner rod. Second, a rod with a long length L will be easier to stretch by a given amount than a short rod (think of trying to stretch a rope by 1 cm—this would be easy to do for a 10-m-long rope, but it would be pretty hard for a piece of rope only 10 cm long). Finally, the stiffness of the rod will depend on the material that it's made of. Experiments bear out these observations, and it is found that the spring constant of the rod can be written as

$$k = \frac{YA}{L} \qquad (8.4)$$

where the constant Y is called **Young's modulus.** Young's modulus is a property of the *material* from which the rod is made—it does not depend on shape or size.

FIGURE 8.18 Stretching a steel rod.

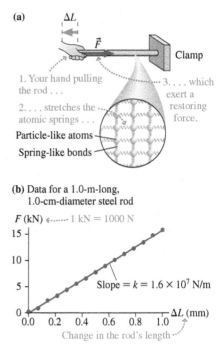

(a)

1. Your hand pulling the rod . . .

2. . . . stretches the atomic springs . . .

3. . . . which exert a restoring force.

Particle-like atoms

Spring-like bonds

Clamp

(b) Data for a 1.0-m-long, 1.0-cm-diameter steel rod

F (kN) ⟵ 1 kN = 1000 N

Slope = $k = 1.6 \times 10^7$ N/m

ΔL (mm)

Change in the rod's length

FIGURE 8.19 A rod stretched by length ΔL.

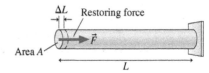

ΔL Restoring force

Area A

L

TABLE 8.1 Young's modulus for rigid materials

Material	Young's modulus (10^{10} N/m^2)
Cast iron	20
Steel	20
Silicon	13
Copper	11
Aluminum	7
Glass	7
Concrete	3
Wood (Douglas Fir)	1

From Equation 8.2, the magnitude of the restoring force for a spring is related to the change in its length as $F_{sp} = k\,\Delta x$. Writing the change in the length of a rod as ΔL, as shown in Figure 8.19, we can use Equation 8.4 to write the restoring force F of a rod as

$$F = \frac{YA}{L}\,\Delta L \qquad (8.5)$$

Equation 8.5 applies both to elongation (stretching) and to compression.

It's useful to rearrange Equation 8.5 in terms of two new ratios, the *stress* and the *strain*:

The ratio of force to cross-section area is called **stress**. $\qquad \frac{F}{A} = Y\left(\frac{\Delta L}{L}\right) \qquad$ The ratio of the change in length to the original length is called **strain**. $\qquad (8.6)$

The unit of stress is N/m^2. If the stress is due to stretching, we call it a **tensile stress**. The strain is the fractional change in the rod's length. If the rod's length changes by 1%, the strain is 0.01. Because strain is dimensionless, Young's modulus Y has the same units as stress. Table 8.1 gives values of Young's modulus for several rigid materials. Large values of Y characterize materials that are stiff. "Softer" materials have smaller values of Y.

EXAMPLE 8.8 **Finding the stretch of a wire**

A *Foucault pendulum* in a physics department (used to prove that the earth rotates) consists of a 120 kg steel ball that swings at the end of a 6.0-m-long steel cable. The cable has a diameter of 2.5 mm. When the ball was first hung from the cable, by how much did the cable stretch?

PREPARE The amount by which the cable stretches depends on the elasticity of the steel cable. Young's modulus for steel is given in Table 8.1 as $Y = 20 \times 10^{10} \text{ N/m}^2$.

SOLVE Equation 8.6 relates the stretch of the cable ΔL to the restoring force F and to the properties of the cable. Rearranging terms, we find that the cable stretches by

$$\Delta L = \frac{LF}{AY}$$

The cross-section area of the cable is

$$A = \pi r^2 = \pi(0.00125 \text{ m})^2 = 4.91 \times 10^{-6} \text{ m}^2$$

The restoring force of the cable is equal to the ball's weight:

$$F = w = mg = (120 \text{ kg})(9.8 \text{ m/s}^2) = 1180 \text{ N}$$

The change in length is thus

$$\Delta L = \frac{(6.0 \text{ m})(1180 \text{ N})}{(4.91 \times 10^{-6} \text{ m}^2)(20 \times 10^{10} \text{ N/m}^2)}$$
$$= 0.0072 \text{ m} = 7.2 \text{ mm}$$

ASSESS If you've ever strung a guitar with steel strings, you know that the strings stretch several millimeters with the force you can apply by turning the tuning pegs. So a stretch of 7 mm under a 120 kg load seems reasonable.

FIGURE 8.20 Stretch data for a steel rod.

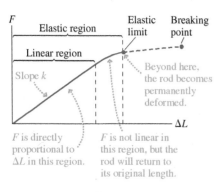

F is directly proportional to ΔL in this region.

F is not linear in this region, but the rod will return to its original length.

Beyond the Elastic Limit

In the previous section, we found that if we stretch a rod by a small amount ΔL, it will pull back with a restoring force F, according to Equation 8.5. But if we continue to stretch the rod, this simple linear relationship between ΔL and F will eventually break down. **FIGURE 8.20** is a graph of the rod's restoring force from the start of the stretch until the rod finally breaks.

As you can see, the graph has a *linear region*, the region where F and ΔL are proportional to each other, obeying Hooke's law: $F = k\,\Delta L$. **As long as the stretch stays within the linear region, a solid rod acts like a spring and obeys Hooke's law.**

How far can you stretch the rod before damaging it? As long as the stretch is less than the **elastic limit,** the rod will return to its initial length L when the force is removed. The elastic limit is the end of the **elastic region.** Stretching the rod beyond the elastic limit will permanently deform it, and the rod won't return to its original length. Finally, at a certain point the rod will reach a breaking point, where it will snap in two. The force that causes the rod to break depends on the area: A thicker rod can sustain a larger force. For a rod or cable of a particular material, we can determine an *ultimate stress*, also known as the **tensile strength,** the largest stress that the material can sustain before breaking:

TABLE 8.2 Tensile strengths of rigid materials

Material	Tensile strength (N/m²)
Polypropylene	20×10^6
Glass	60×10^6
Cast iron	150×10^6
Aluminum	400×10^6
Steel	1000×10^6

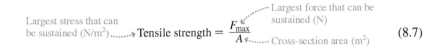

$$\text{Largest stress that can be sustained (N/m}^2) \dashrightarrow \text{Tensile strength} = \frac{F_{max}}{A} \qquad (8.7)$$

Largest force that can be sustained (N)

Cross-section area (m²)

Table 8.2 lists values of tensile strength for rigid materials. When we speak of the *strength* of a material, we are referring to its tensile strength.

EXAMPLE 8.9 **Breaking a pendulum cable**

After a late night of studying physics, several 80 kg students decide it would be fun to swing on the Foucault pendulum of Example 8.8. What's the maximum number of students that the pendulum cable could support?

PREPARE The tensile strength, given for steel in Table 8.2 as $1000 \times 10^6 \text{ N/m}^2$, or $1.0 \times 10^9 \text{ N/m}^2$, is the largest stress the cable can sustain. Because the stress in the cable is F/A, we can find the maximum force F_{max} the cable can supply before it fails.

SOLVE We have

$$F_{max} = A(1.0 \times 10^9 \text{ N/m}^2)$$

From Example 8.8, the diameter of the cable is 2.5 mm, so its radius is 0.00125 m. Thus

$$F_{max} = \left(\pi(0.00125 \text{ m})^2\right)(1.0 \times 10^9 \text{ N/m}^2) = 4.9 \times 10^3 \text{ N}$$

This force is the weight of the heaviest mass the cable can support: $w = m_{max}g$. The maximum mass that can be supported is

$$m_{max} = \frac{F_{max}}{g} = 500 \text{ kg}$$

The ball has a mass of 120 kg, leaving 380 kg for the students. Four students have a mass of 320 kg, which is less than this value. But five students, totaling 400 kg, would cause the cable to break.

ASSESS Steel has a very large tensile strength, so it's reasonable that this very narrow wire can still support 4900 N $\approx$ 1100 lb.

Biological Materials BIO

Suppose we take equal lengths of spider silk and steel wire, stretch each, and measure the restoring force of each until it breaks. The graph of stress versus strain might appear as in **FIGURE 8.21**.

The spider silk is certainly less stiff: For a given stress, the silk will stretch about 100 times farther than steel. Interestingly, though, spider silk and steel eventually fail at approximately the same stress. In this sense, spider silk is "as strong as steel." Many pliant biological materials share this combination of low stiffness and large tensile strength. These materials can undergo significant deformations without failing. Tendons, the walls of arteries, and the web of a spider are all quite strong but nonetheless capable of significant stretch.

Most bones in your body are made of two different kinds of bony material: dense and rigid cortical (or compact) bone on the outside, and porous, flexible cancellous (or spongy) bone on the inside. **FIGURE 8.22** on the next page shows a cross section of a typical bone. Cortical and cancellous bones have very different values of Young's

FIGURE 8.21 Stress-versus-strain graphs for steel and spider silk.

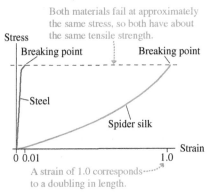

Both materials fail at approximately the same stress, so both have about the same tensile strength.

A strain of 1.0 corresponds to a doubling in length.

FIGURE 8.22 Cross section of a long bone.

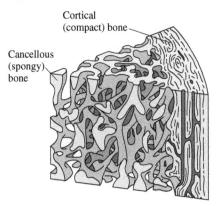

Cortical
(compact) bone

Cancellous
(spongy)
bone

modulus. Young's modulus for cortical bone approaches that of concrete, so it is very rigid with little ability to stretch or compress. In contrast, cancellous bone has a much lower Young's modulus. Consequently, the elastic properties of bones can be well modeled as those of a hollow cylinder.

The structure of bones in birds actually approximates a hollow cylinder quite well. FIGURE 8.23 shows that a typical bone is a thin-walled tube of cortical bone with a tenuous structure of cancellous bone inside. Most of a cylinder's rigidity comes from the material near its surface. A hollow cylinder retains most of the rigidity of a solid one, but it is much lighter. Bird bones carry this idea to its extreme.

FIGURE 8.23 Section of a bone from a bird.

TABLE 8.3 Young's modulus for biological materials

Material	Young's modulus $(10^{10}\ \text{N/m}^2)$
Tooth enamel	6
Cortical bone	1.6
Cancellous bone	0.02–0.3
Spider silk	0.2
Tendon	0.15
Cartilage	0.0001
Blood vessel (aorta)	0.00005

TABLE 8.4 Tensile strengths of biological materials

Material	Tensile strength (N/m^2)
Cancellous bone	5×10^6
Cortical bone	100×10^6
Tendon	100×10^6
Spider silk	1000×10^6

Table 8.3 gives values of Young's modulus for biological materials. Note the large difference between pliant and rigid materials. Table 8.4 shows the tensile strengths for biological materials. The values in Table 8.4 are for static forces—forces applied for a long time in a testing machine. Bone can withstand significantly greater stresses if the forces are applied for only a very short period of time.

EXAMPLE 8.10   **Finding the compression of a bone** BIO

The femur, the long bone in the thigh, can be modeled as a tube of cortical bone for most of its length. A 70 kg person has a femur with a cross-section area (of the cortical bone) of $4.8 \times 10^{-4}\ \text{m}^2$, a typical value.

a. If this person supports his entire weight on one leg, what fraction of the tensile strength of the bone does this stress represent?
b. By what fraction of its length does the femur shorten?

PREPARE The stress on the femur is F/A. Here F, the force compressing the femur, is the person's weight, so $F = mg$. The fractional change $\Delta L/L$ in the femur is the strain, which we can find using Equation 8.6, taking the value of Young's modulus for cortical bone from Table 8.3.

SOLVE

a. The person's weight is $mg = (70\ \text{kg})(9.8\ \text{m/s}^2) = 690\ \text{N}$. The resulting stress on the femur is

$$\frac{F}{A} = \frac{690\ \text{N}}{4.8 \times 10^{-4}\ \text{m}^2} = 1.4 \times 10^6\ \text{N/m}^2$$

A stress of $1.4 \times 10^6\ \text{N/m}^2$ is 1.4% of the tensile strength of cortical bone given in Table 8.4.

b. We can compute the strain as

$$\frac{\Delta L}{L} = \left(\frac{1}{Y}\right)\frac{F}{A} = \left(\frac{1}{1.6 \times 10^{10}\ \text{N/m}^2}\right)(1.4 \times 10^6\ \text{N/m}^2) = 8.8 \times 10^{-5} \approx 0.0001$$

The femur compression is $\Delta L \approx 0.0001L$, or $\approx 0.01\%$ of its length. (The femur is far from a uniform structure, so we've expressed the answer as an approximate result to one significant figure.)

ASSESS It makes sense that, under ordinary standing conditions, the stress on the femur is only a percent or so of the maximum value it can sustain.

The dancer in the chapter-opening photo stands *en pointe,* balanced delicately on the tip of her shoe with her entire weight supported on a very small area. The stress on the bones in her toes is very large, but it is still much less than the tensile strength of bone.

STOP TO THINK 8.6   A 10 kg mass is hung from a 1-m-long cable, causing the cable to stretch by 2 mm. Suppose a 10 kg mass is hung from a 2 m length of the same cable. By how much does the cable stretch?

A. 0.5 mm   B. 1 mm   C. 2 mm   D. 3 mm   E. 4 mm

INTEGRATED EXAMPLE 8.11   **Elevator cable stretch**

The steel cables that hold elevators stretch only a very small fraction of their length, but in a tall building this small fractional change can add up to a noticeable stretch. This example uses realistic numbers for such an elevator to make this point. The 2300 kg car of a high-speed elevator in a tall building is supported by six 1.27-cm-diameter cables. Young's modulus for the cables is 10×10^{10} N/m², a typical value for steel cables with multiple strands. When the elevator is on the bottom floor, the cables rise 90 m up the shaft to the motor above.

On a busy morning, the elevator is on the bottom floor and fills up with 20 people who have a total mass of 1500 kg. The elevator then accelerates upward at 2.3 m/s² until it reaches its cruising speed. How much do the cables stretch due to the weight of the car alone? How much additional stretch occurs when the passengers are in the car? And, what is the total stretch of the cables while the elevator is accelerating? In all cases, you can ignore the mass of the cables.

PREPARE   We can compute the stretch of the cables by rewriting Equation 8.6, the equation relating stress and strain, to get

$$\Delta L = \frac{LF}{YA}$$

Here L is the length of the cables, F is the restoring force exerted by the cables, and Y is Young's modulus, which we're given. Although there are six cables, we can imagine them combined into one cable with a cross-section area A six times that of each individual cable. Each cable has radius 0.00635 m and cross-section area $\pi r^2 = 1.27 \times 10^{-4}$ m². Multiplying the area by 6 gives a total cross-section area $A = 7.62 \times 10^{-4}$ m².

FIGURE 8.24 shows the details. The restoring force exerted by the cable is just the tension in the cable. For the first two questions, the cable supports the elevator in static equilibrium; for the third, there is a net force because the elevator is accelerating upward.

SOLVE   When the elevator is at rest, the net force is zero and so the restoring force of the cable—the tension—is equal to the weight suspended from it. For the first two questions, the forces are

$$F_1 = m_{\text{car}}g = (2300 \text{ kg})(9.8 \text{ m/s}^2) = 22,500 \text{ N}$$

$$F_2 = m_{\text{car+passengers}}g = (2300\text{kg} + 1500\text{kg})(9.8\text{m/s}^2) = 37,200\text{N}$$

These forces stretch the cable by

$$\Delta L_1 = \frac{(90 \text{ m})(22,500 \text{ N})}{(10 \times 10^{10} \text{ N/m}^2)(7.62 \times 10^{-4} \text{ m}^2)} = 0.027 \text{ m} = 2.7 \text{ cm}$$

$$\Delta L_2 = \frac{(90 \text{ m})(37,200 \text{ N})}{(10 \times 10^{10} \text{ N/m}^2)(7.62 \times 10^{-4} \text{ m}^2)} = 0.044 \text{ m} = 4.4 \text{ cm}$$

FIGURE 8.24 Details of the cable stretch and the forces acting on the elevator.

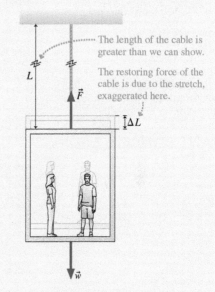

The length of the cable is greater than we can show.

The restoring force of the cable is due to the stretch, exaggerated here.

The additional stretch when the passengers board is

$$4.4 \text{ cm} - 2.7 \text{ cm} = 1.7 \text{ cm}$$

When the elevator is accelerating upward, the tension in the cables must increase. Newton's second law for the vertical motion is

$$\Sigma F_y = F - w = ma_y$$

The restoring force, or tension, is thus

$$F = w + ma_y = mg + ma_y = (3800 \text{ kg})(9.8 \text{ m/s}^2 + 2.3 \text{ m/s}^2)$$

$$= 46,000 \text{ N}$$

Thus right at the start of the motion, when the full 90 m of cable is still deployed, we find that the cables are stretched by

$$\Delta L = \frac{(90 \text{ m})(46,000 \text{ N})}{(10 \times 10^{10} \text{ N/m}^2)(7.62 \times 10^{-4} \text{ m}^2)} = 0.054 \text{ m} = 5.4 \text{ cm}$$

ASSESS   When the passengers enter the car, the cable stretches by 1.7 cm, or about two-thirds of an inch. This is large enough to notice (as we might expect given the problem statement) but not large enough to cause concern to the passengers. The total stretch for a fully loaded elevator accelerating upward is 5.4 cm, just greater than 2 inches. This is not unreasonable for cables that are nearly as long as a football field—the fractional change in length is still quite small.

SUMMARY

Goal: To learn about the static equilibrium of extended objects, and the basic properties of springs and elastic materials.

GENERAL PRINCIPLES

Static Equilibrium

An object in **static equilibrium** must have no net force on it and no net torque. Mathematically, we express this as

$$\sum F_x = 0$$

$$\sum F_y = 0$$

$$\sum \tau = 0$$

Since the net torque is zero about *any* point, the pivot point for calculating the torque can be chosen at any convenient location.

Springs and Hooke's Law

When a spring is stretched or compressed, it exerts a force proportional to the change Δx in its length but in the opposite direction. This is known as **Hooke's law:**

$$(F_{sp})_x = -k\,\Delta x$$

The constant of proportionality k is called the **spring constant.** It is larger for a "stiff" spring.

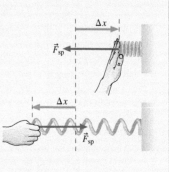

IMPORTANT CONCEPTS

Stability

An object is **stable** if its center of gravity is over its base of support; otherwise, it is **unstable.**

If an object is tipped, it will reach the limit of its stability when its center of gravity is over the edge of the base. This defines the **critical angle** θ_c.

Greater stability is possible with a lower center of gravity or a broader base of support.

Stable   Unstable

Base of support

This object is at its critical angle.   This object has a wider base of support and hence a larger critical angle.   This object has a lower center of gravity, so its critical angle is larger too.

θ_c θ_c θ_c

Elastic materials and Young's modulus

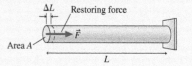

ΔL Restoring force

Area A $\vec{F}$

L

A solid rod illustrates how materials respond when stretched or compressed.

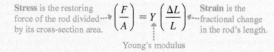

Stress is the restoring force of the rod divided by its cross-section area. $\left(\dfrac{F}{A}\right) = Y\left(\dfrac{\Delta L}{L}\right)$ **Strain** is the fractional change in the rod's length.

Young's modulus

This equation can also be written as

This is the "spring constant" k for the rod. $F = \left(\dfrac{YA}{L}\right)\Delta L$

showing that a rod obeys Hooke's law and acts like a very stiff spring.

APPLICATIONS

Forces in the body

Muscles and tendons apply the forces and torques needed to maintain static equilibrium. These forces may be quite large.

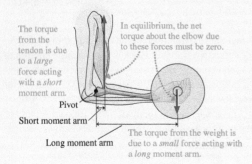

The torque from the tendon is due to a *large* force acting with a *short* moment arm.

In equilibrium, the net torque about the elbow due to these forces must be zero.

Pivot

Short moment arm

Long moment arm

The torque from the weight is due to a *small* force acting with a *long* moment arm.

The elastic limit and beyond

If a rod or other object is not stretched too far, when released it will return to its original shape.

If stretched too far, an object will permanently deform and finally break. The stress at which an object breaks is its **tensile stress.**

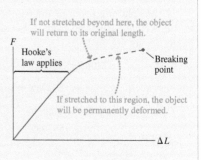

If not stretched beyond here, the object will return to its original length.

F

Hooke's law applies

Breaking point

If stretched to this region, the object will be permanently deformed.

ΔL

Problem difficulty is labeled as| (straightforward) to ||||| (challenging). Problems labeled INT integrate significant material from earlier chapters; BIO are of biological or medical interest.

 For assigned homework and other learning materials, go to Mastering Physics®

Scan this QR code to launch a Video Tutor Solution that will help you solve problems for this chapter.

QUESTIONS

Conceptual Questions

1. An object is acted upon by two (and only two) forces that are of equal magnitude and oppositely directed. Is the object necessarily in static equilibrium?
2. Sketch a force acting at point P in Figure Q8.2 that would make the rod be in static equilibrium. Is there only one such force?
3. Could a ladder on a level floor lean against a wall in static equilibrium if there were no friction forces? Explain.
4. If you are using a rope to raise a tall mast, attaching the rope to the middle of the mast as in Figure Q8.4a gives a very small torque about the base of the mast when the mast is at a shallow angle. You can get a larger torque by adding a pole with a pulley on top, as in Figure Q8.4b. Draw a diagram showing all of the forces acting on the mast and explain why, for the same tension in the rope, adding this pole increases the torque on the mast.

FIGURE Q8.2

(a) **(b)**

FIGURE Q8.4

5. As divers stand on tiptoes on the edge of a diving platform, in preparation for a high dive, as shown in Figure Q8.5, they usually extend their arms in front of them. Why do they do this?
6. Where are the centers of gravity of the two people doing the classic yoga poses shown in Figure Q8.6?

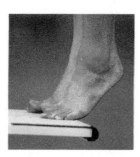

FIGURE Q8.5

(a) **(b)**

FIGURE Q8.6

7. You must lean quite far forward as you rise from a chair (try it!). Explain why.
8. A spring exerts a 10 N force after being stretched by 1 cm from its equilibrium length. By how much will the spring force *increase* if the spring is stretched from 4 cm away from equilibrium to 5 cm from equilibrium?
9. The left end of a spring is attached to a wall. When Bob pulls on the right end with a 200 N force, he stretches the spring by 20 cm. The same spring is then used for a tug-of-war between Bob and Carlos. Each pulls on his end of the spring with a 200 N force.
   a. How far does Bob's end of the spring move? Explain.
   b. How far does Carlos's end of the spring move? Explain.
10. A spring is attached to the floor and pulled straight up by a string. The string's tension is measured. The graph in Figure Q8.10 shows the tension in the spring as a function of the spring's length L.
   a. Does this spring obey Hooke's law? Explain.
   b. If it does, what is the spring constant?

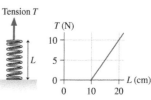

FIGURE Q8.10

11. A typical mattress has a network of springs that provide support. If you sit on a mattress, the springs compress. A heavier person compresses the springs more than a lighter person. Use the properties of springs and spring forces to explain why this is so.
12. Take a spring and cut it in half to make two springs. Is the spring constant of these smaller springs larger than, smaller than, or the same as the spring constant of the original spring? Explain.
13. A wire is stretched right to its breaking point by a 5000 N force. A longer wire made of the same material has the same diameter. Is the force that will stretch it right to its breaking point larger than, smaller than, or equal to 5000 N? Explain.
14. Steel nails are rigid and unbending. Steel wool is soft and squishy. How would you account for this difference?

Multiple-Choice Questions

15. | The rod in Figure Q8.15 pivots around an axle at the left end. With forces applied as noted, the object
   A. Will rotate counterclockwise.
   B. Is in static equilibrium.
   C. Will rotate clockwise.

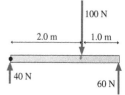

FIGURE Q8.15

16. ‖ Two children hold opposite ends of a lightweight, 1.8-m-long horizontal pole with a water bucket hanging from it. The older child supports twice as much weight as the younger child. How far is the bucket from the older child?
    A. 0.3 m   B. 0.6 m
    C. 0.9 m   D. 1.2 m

17. ‖ The uniform rod in Figure Q8.17 has a weight of 14.0 N. What is the magnitude of the normal force exerted on the rod by the surface?
    A. 7 N   B. 14 N
    C. 20 N   D. 28 N

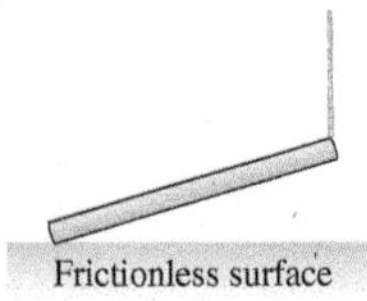

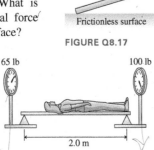

Frictionless surface

FIGURE Q8.17

18. ‖ A student lies on a very light, rigid board with a scale under each end. Her feet are directly over one scale, and her body is positioned as shown in Figure Q8.18. The two scales read the values shown in the figure. What is the student's weight?
    A. 65 lb   B. 75 lb   C. 100 lb   D. 165 lb

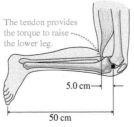

65 lb   100 lb

2.0 m

FIGURE Q8.18

19. ‖ For the student in Figure Q8.18, approximately how far from her feet is her center of gravity?
    A. 0.6 m   B. 0.8 m   C. 1.0 m   D. 1.2 m

Questions 20 through 22 use the information in the following paragraph and figure.

Suppose you stand on one foot while holding your other leg up behind you. Your muscles will have to apply a force to hold your leg in this raised position. We can model this situation as in Figure Q8.20. The leg pivots at the knee joint, and the force to hold the leg up is provided by a tendon attached to the lower leg as shown. Assume that the lower leg

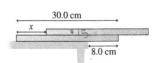

The tendon provides the torque to raise the lower leg.

5.0 cm

50 cm

FIGURE Q8.20

and the foot together have a combined mass of 4.0 kg, and that their combined center of gravity is at the center of the lower leg.

20. ‖ How much force must the tendon exert to keep the leg in this
    BIO position?
    A. 40 N   B. 200 N   C. 400 N   D. 1000 N

21. ‖ As you hold your leg in this position, the upper leg exerts a
    BIO force on the lower leg at the knee joint. What is the direction of this force?
    A. Up   B. Down   C. Right   D. Left

22. ‖ What is the magnitude of the force of the upper leg on the
    BIO lower leg at the knee joint?
    A. 40 N   B. 160 N   C. 200 N   D. 240 N

23. ‖ You have a heavy piece of equipment hanging from a 1.0-mm-diameter wire. Your supervisor asks that the length of the wire be doubled without changing how far the wire stretches. What diameter must the new wire have?
    A. 1.0 mm B. 1.4 mm C. 2.0 mm D. 4.0 mm

24. ‖ A 30.0-cm-long board is placed on a table such that its right end hangs over the edge by 8.0 cm. A second identical board is stacked on top of the first, as shown in Figure Q8.24. What is the largest that the distance x can be before both boards topple over?
    A. 4.0 cm   B. 8.0 cm
    C. 14 cm   D. 15 cm

30.0 cm

x

8.0 cm

FIGURE Q8.24

25. ‖ A 20 kg block resting on the floor is to be raised by a 5.0-mm-diameter rope with Young's modulus $1.5 \times 10^9 \, \text{N/m}^2$. The rope goes up and over a pulley that is 2.5 m above the floor. You hold the rope at a point 1.5 m above the floor and pull hard enough to lift the block just off the floor. By how much does the rope stretch?
    A. 3.0 mm   B. 6.3 mm
    C. 9.3 mm   D. 13 mm

PROBLEMS

Section 8.1 Torque and Static Equilibrium

1. ‖ A 64 kg student stands on a very light, rigid board that rests on a bathroom scale at each end, as shown in Figure P8.1. What is the reading on each of the scales?

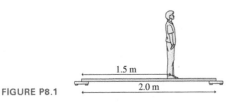

1.5 m

2.0 m

FIGURE P8.1

2. ‖ Suppose the student in Figure P8.1 is 54 kg, and the board being stood on has a 10 kg mass. What is the reading on each of the scales?

3. ‖ How close to the right edge of the 56 kg picnic table shown in Figure P8.3 can a 70 kg man stand without the table tipping over? **Hint:** When the table is just about to tip, what is the force of the ground on the table's left leg?

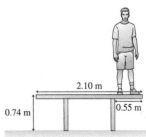

2.10 m

0.74 m

0.55 m

FIGURE P8.3

4. ‖ In Figure P8.4, a 70 kg man walks out on a 10 kg beam that rests on, but is not attached to, two supports. When the beam just starts to tip, what is the force exerted on the beam by the right support?

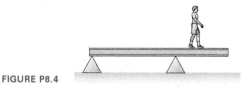

FIGURE P8.4

5. ‖‖ You're carrying a 3.6-m-long, 25 kg pole to a construction site when you decide to stop for a rest. You place one end of the pole on a fence post and hold the other end of the pole 35 cm from its tip. How much force must you exert to keep the pole motionless in a horizontal position?

6. ‖ A typical horse weighs 5000 N.
BIO The distance between the front and rear hooves and the distance from the rear hooves to the center of mass for a typical horse are shown in Figure P8.6. What fraction of the horse's weight is borne by the front hooves?

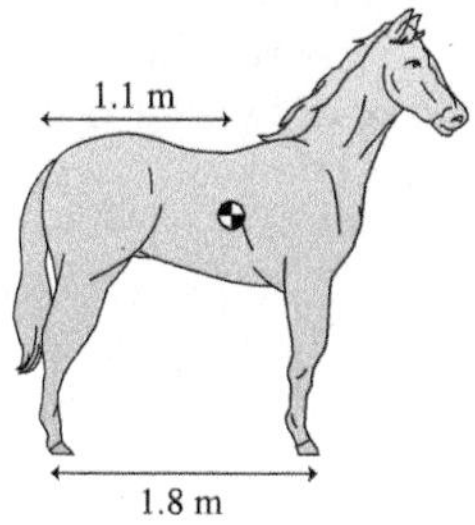

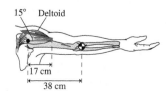

FIGURE P8.6

7. ‖‖ How much torque must the pin exert to keep the rod in Figure P8.7 from rotating? Calculate this torque about an axis that passes through the point where the pin enters the rod and is perpendicular to the plane of the figure.

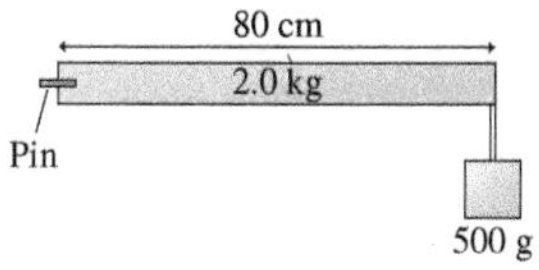

80 cm
2.0 kg
Pin
500 g

FIGURE P8.7

8. ‖ A vendor hangs an 8.0 kg sign in front of his shop with a cable held away from the building by a lightweight pole. The pole is free to pivot about the end where it touches the wall, as shown in Figure P8.8. What is the tension in the cable?

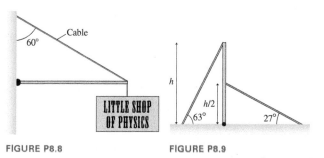

FIGURE P8.8   **FIGURE P8.9**

9. ‖ Figure P8.9 shows a vertical pole of height h that can rotate about a hinge at the bottom. The pole is held in position by two wires under tension. What is the ratio of the tension in the left wire to the tension in the right wire?

10. ‖ Consider the procedure for measuring a woman's center of gravity given in Example 8.3. The 600 N woman is in place on the board, with the scale reading 250 N. She now extends her arms upward, in front of her body. This raises her center of gravity relative to her feet by 4.0 cm. What is the new reading on the scale?

11. ‖‖ The two objects in Figure P8.11 are balanced on the pivot. What is distance d?

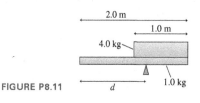

FIGURE P8.11

12. ‖ If you hold your arm outstretched with palm upward, as in
BIO Figure P8.12, the force to keep your arm from falling comes from your deltoid muscle. The arm of a typical person has mass 4.0 kg and the distances and angles shown in the figure.
   a. What force must the deltoid muscle provide to keep the arm in this position?
   b. By what factor does this force exceed the weight of the arm?

15° Deltoid
17 cm
38 cm

FIGURE P8.12

13. ‖‖ A 60 kg diver stands at the end of a 30 kg spring-board, as shown in Figure P8.13. The board is attached to a hinge at the left end but simply rests on the right support. What is the magnitude of the vertical force exerted by the hinge on the board?

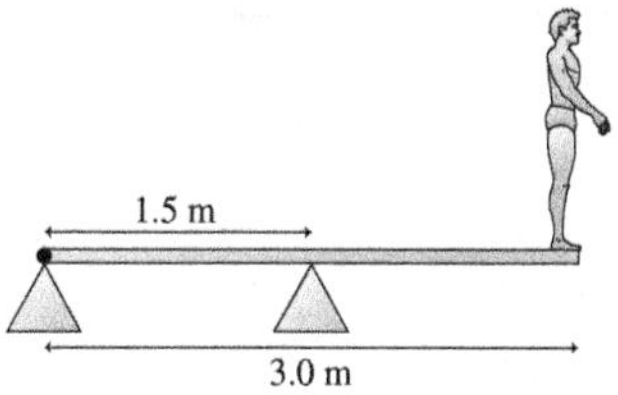

1.5 m
3.0 m

FIGURE P8.13

14. ‖ Hold your upper arm ver-
BIO tical and your lower arm horizontal with your hand palm-down on a table, as shown in Figure P8.14. If you now push down on the table, you'll feel that your triceps muscle has contracted and is trying to pivot your lower arm about the elbow joint. If a person with the arm dimensions shown pushes down hard with a 90 N force (about 20 lb), what force must the triceps muscle provide? You can ignore the mass of the arm and hand in your calculation.

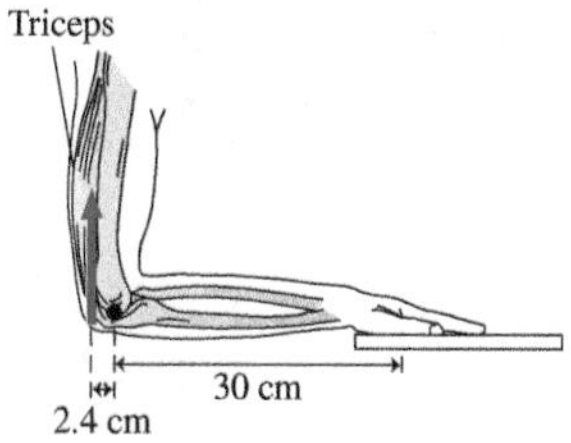

Triceps
30 cm
2.4 cm

FIGURE P8.14

15. ‖ A uniform beam of length 1.0 m and mass 10 kg is attached to a wall by a cable, as shown in Figure P8.15. The beam is free to pivot at the point where it attaches to the wall. What is the tension in the cable?

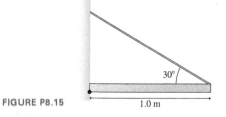

30°
1.0 m

FIGURE P8.15

16. ⫶ The towers holding small wind turbines are often raised and lowered for easy servicing of the turbine. Figure P8.16 shows a 1000 kg wind turbine mounted on the end of a 24-m-long, 700 kg tower that connects to a support column at a pivot. A piston connected 3.0 m from the pivot applies the force needed to raise or lower the tower. At the instant shown, the wind turbine is being raised at a very slow, constant speed. What magnitude force is the piston applying?

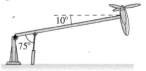

FIGURE P8.16

Section 8.2 Stability and Balance

17. ⎮ A standard four-drawer filing cabinet is 52 inches high and 15 inches wide. If it is evenly loaded, the center of gravity is at the center of the cabinet. A worker is tilting a filing cabinet to the side to clean under it. To what angle can he tilt the cabinet before it tips over?

18. ⎮ The stability of a vehicle is often rated by the *static stability factor*, which is one-half the track width divided by the height of the center of gravity above the road. A typical SUV has a static stability factor of 1.2. What is the critical angle?

19. ⫽ You want to slowly push a stiff board across a 20 cm gap between two tabletops that are at the same height. If you apply only a horizontal force, what is the minimum-length board that won't tilt down into the gap before reaching the other side?

20. ⎮ A magazine rack has a center of gravity 16 cm above the floor, as shown in Figure P8.20. Through what maximum angle, in degrees, can the rack be tilted without falling over?

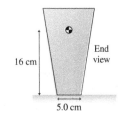

FIGURE P8.20

21. ⫽ A car manufacturer claims that you can drive its new vehicle across a hill with a 47° slope before the vehicle starts to tip. If the vehicle is 2.0 m wide, how high is its center of gravity?

22. ⫽ A thin 2.00 kg box rests on a 6.00 kg board that hangs over the end of a table, as shown in Figure P8.22. How far can the center of the box be from the end of the table before the board begins to tilt?

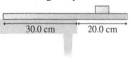

FIGURE P8.22

Section 8.3 Springs and Hooke's Law

23. ⎮ One end of a spring is attached to a wall. A 25 N pull on the other end causes the spring to stretch by 3.0 cm. What is the spring constant?

24. ⎮ An orthodontic spring, connected between the upper and lower BIO jaws, is adjusted to provide no force with the mouth open. When the patient closes her mouth, however, the spring compresses by 6.0 mm. What force is exerted if the spring constant is 160 N/m?

25. ⎮ Experiments using "optical tweezers" measure the elasticity BIO of individual DNA molecules. For small enough changes in length, the elasticity has the same form as that of a spring. A DNA molecule is anchored at one end, then a force of 1.5 nN $(1.5 \times 10^{-9} \text{ N})$ pulls on the other end, causing the molecule to stretch by 5.0 nm $(5.0 \times 10^{-9} \text{ m})$. What is the spring constant of that DNA molecule?

26. ⫽ A spring has an unstretched length of 10 cm. It exerts a restoring force F when stretched to a length of 11 cm.
    a. For what total stretched length of the spring is its restoring force $3F$?
    b. At what compressed length is the restoring force $2F$?

27. ⫽ One end of a 10-cm-long spring is attached to the ceiling. When a 2.0 kg mass is hung from the other end, the spring stretches to a length of 15 cm.
    a. What is the spring constant?
    b. How long is the spring when a 3.0 kg mass is suspended from it?

28. ⫽ A spring stretches 5.0 cm when a 0.20 kg block is hung from it. If a 0.70 kg block replaces the 0.20 kg block, how far does the spring stretch?

29. ⫽ A 1.2 kg block is hung from a vertical spring, causing the spring to stretch by 2.4 cm. How much farther will the spring stretch if a 0.60 kg block is added to the 1.2 kg block?

30. ⎮ You need to make a spring scale to measure the mass of objects hung from it. You want each 1.0 cm length along the scale to correspond to a mass difference of 0.10 kg. What should be the value of the spring constant?

Section 8.4 Stretching and Compressing Materials

31. ⫽ A force stretches a wire by 1.0 mm.
    a. A second wire of the same material has the same cross section and twice the length. How far will it be stretched by the same force?
    b. A third wire of the same material has the same length and twice the diameter as the first. How far will it be stretched by the same force?

32. ⫼ What hanging mass will stretch a 2.0-m-long, 0.50-mm-diameter steel wire by 1.0 mm?

33. ⫼ How much force does it take to stretch a 10-m-long, 1.0-cm-diameter steel cable by 5.0 mm?

34. ⫼ An 80-cm-long, 1.0-mm-diameter steel guitar string must be tightened to a tension of 2.0 kN by turning the tuning screws. By how much is the string stretched?

35. ⫽ A student is testing a 1.0 m length of 2.5-mm-diameter steel wire.
    a. How much force is required to stretch this wire by 1.0 mm?
    b. What length of 5.0-mm-diameter wire would be stretched by 1.0 mm by this force?

36. ⫼ A 1.2-m-long steel rod with a diameter of 0.50 cm hangs vertically from the ceiling. An auto engine weighing 4.7 kN is hung from the rod. By how much does the rod stretch?

37. ⫼ A mineshaft has an ore elevator hung from a single braided cable of diameter 2.5 cm. Young's modulus of the cable is $10 \times 10^{10} \text{ N/m}^2$. When the cable is fully extended, the end of the cable is 800 m below the support. How much does the fully extended cable stretch when 1000 kg of ore is loaded into the elevator?

38. ‖ The normal force of the ground on the foot can reach three
BIO times a runner's body weight when the foot strikes the pavement. By what amount does the 52-cm-long femur of an 80 kg runner compress at this moment? The cross-section area of the bone of the femur can be taken as 5.2×10^{-4} m^2.

39. ‖‖ A three-legged wooden bar stool made out of solid Douglas fir has legs that are 2.0 cm in diameter. When a 75 kg man sits on the stool, by what percent does the length of the legs decrease? Assume, for simplicity, that the stool's legs are vertical and that each bears the same load.

40. ‖‖ A 3.0-m-tall, 50-cm-diameter concrete column supports a 200,000 kg load. By how much is the column compressed?

41. ‖ A glass optical fiber in a communications system has a diameter of 9.0 μm.
   a. What maximum force could this fiber support without breaking?
   b. Assume that the fiber stretches in a linear fashion until the instant it breaks. By how much will a 10-m-long fiber have stretched when it is at the breaking point?

42. ‖ The Achilles tendon connects the muscles in your calf to the
BIO back of your foot. When you are sprinting, your Achilles tendon alternately stretches, as you bring your weight down onto your forward foot, and contracts to push you off the ground. A 70 kg runner has an Achilles tendon that is 15 cm long and has a cross-section area of 110 mm^2, typical values for a person of this size.
   a. By how much will the runner's Achilles tendon stretch if the maximum force on it is 8.0 times his weight, a typical value while running?
   b. What fraction of the tendon's length does this correspond to?

General Problems

43. ‖‖ A 3.0-m-long rigid beam with a mass of 100 kg is supported at each end, as shown in Figure P8.43. An 80 kg student stands 2.0 m from support 1. How much upward force does each support exert on the beam?

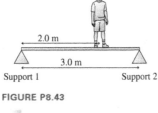

FIGURE P8.43

44. ‖ An 80 kg construction worker sits down 2.0 m from the end of a 1450 kg steel beam to eat his lunch, as shown in Figure P8.44. The cable supporting the beam is rated at 15,000 N. Should the worker be worried?

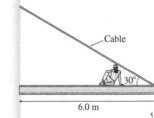

FIGURE P8.44

45. ‖‖ Using the information in Figure 8.2, calculate the tension in
BIO the biceps tendon if the hand is holding a 10 kg ball while the forearm is held 45° below horizontal.

46. ‖‖ A woman weighing 580 N does a
BIO pushup from her knees, as shown in Figure P8.46. What are the normal forces of the floor on (a) each of her hands and (b) each of her knees?

FIGURE P8.46

47. ‖‖‖ When you bend over, a series of large muscles, the erector
BIO spinae, pull on your spine to hold you up. Figure P8.47 shows a simplified model of the spine as a rod of length L that pivots at its lower end. In this model, the center of gravity of the 320 N weight of the upper torso is at the center of the spine. The 160 N weight of the head and arms acts at the top of the spine. The erector spinae muscles are modeled as a single muscle that acts at an 12° angle to the spine. Suppose the person in Figure P8.47 bends over to an angle of 30° from the horizontal.
   a. What is the tension in the erector muscle?
   Hint: Align your x-axis with the axis of the spine.
   b. A force from the pelvic girdle acts on the base of the spine. What is the component of this force in the direction of the spine? (This large force is the cause of many back injuries).

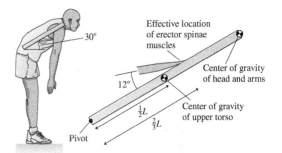

FIGURE P8.47

48. ‖‖ A man is attempting to raise a 7.5-m-long, 28 kg flagpole that has a hinge at the base by pulling on a rope attached to the top of the pole, as shown in Figure P8.48. With what force does the man have to pull on the rope to hold the pole motionless in this position?

FIGURE P8.48

49. ‖‖‖ A 40 kg, 5.0-m-long beam is supported by, but not attached to, the two posts in Figure P8.49. A 20 kg boy starts walking along the beam. How close can he get to the right end of the beam without it tipping?

FIGURE P8.49   **FIGURE P8.50**

50. ‖‖ The wheel of mass m in Figure P8.50 is pulled on by a horizontal force applied at its center. The wheel is touching a curb whose height is half the wheel's radius. What is the minimum force required to just raise the wheel off the ground?

51. ‖ A 5.0 kg mass hanging from a spring scale is slowly lowered onto a vertical spring, as shown in Figure P8.51. The scale reads in newtons.
   a. What does the spring scale read just before the mass touches the lower spring?
   b. The scale reads 20 N when the lower spring has been compressed by 2.0 cm. What is the value of the spring constant for the lower spring?
   c. At what compression distance will the scale read zero?

FIGURE P8.51

52. ‖ Two identical, side-by-side springs with spring constant 240 N/m support a 2.00 kg hanging box. Each spring supports the same weight. By how much is each spring stretched?

53. ‖ Two springs have the same equilibrium length but different spring constants. They are arranged as shown in Figure P8.53, then a block is pushed against them, compressing both by 1.00 cm. With what net force do they push back on the block?

$k_1 = 12.0$ N/cm

$k_2 = 5.4$ N/cm

FIGURE P8.53

54. ‖ Figure P8.54 shows two springs attached to a block that can slide on a frictionless surface. In the block's equilibrium position, the left spring is compressed by 2.0 cm.

$k_1 = 10$ N/m $k_2 = 20$ N/m

FIGURE P8.54

   a. By how much is the right spring compressed?
   b. What is the net force on the block if it is moved 15 cm to the right of its equilibrium position?

55. ‖ A 60 kg student is standing atop a spring in an elevator that is accelerating upward at 3.0 m/s². The spring constant is 2.5×10^3 N/m. By how much is the spring compressed?

56. ‖ A 25 kg child bounces on a pogo stick. The pogo stick has a spring with spring constant 2.0×10^4 N/m. When the child makes a nice big bounce, she finds that at the bottom of the bounce she is accelerating *upward* at 9.8 m/s². How much is the spring compressed?

57. ‖ Figure P8.57 shows a lightweight plank supported at its right end by a 7.0-mm-diameter rope with a tensile strength of 6.0×10^7 N/m².

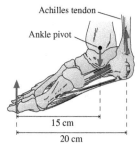

3.5 m

FIGURE P8.57

   a. What is the maximum force that the rope can support?
   b. What is the greatest distance, measured from the pivot, that the center of gravity of an 800 kg piece of heavy machinery can be placed without snapping the rope?

58. ‖ In the hammer throw, an athlete spins a heavy mass in a circle at the end of a cable before releasing it for distance. For male athletes, the "hammer" is a mass of 7.3 kg at the end of a 1.2 m cable, which is typically a 3.0-mm-diameter steel cable. A world-class thrower can get the hammer up to a speed of 29 m/s. If an athlete swings the mass in a horizontal circle centered on the handle he uses to hold the cable
   a. What is the tension in the cable?
   b. How much does the cable stretch?

59. ‖ There is a disk of cartilage between each pair of vertebrae in your spine. Suppose a disk is 0.50 cm thick and 4.0 cm in diameter. If this disk supports half the weight of a 65 kg person, by what fraction of its thickness does the disk compress?

60. ‖ In Example 8.1, the tension in the biceps tendon for a person doing a strict curl of a 900 N barbell was found to be 3900 N. What fraction does this represent of the maximum possible tension the biceps tendon can support? You can assume a typical cross-section area of 130 mm².

61. ‖ Larger animals have sturdier bones than smaller animals. A mouse's skeleton is only a few percent of its body weight, compared to 16% for an elephant. To see why this must be so, recall, from Example 8.10, that the stress on the femur for a man standing on one leg is 1.4% of the bone's tensile strength. Suppose we scale this man up by a factor of 10 in all dimensions, keeping the same body proportions. Use the data for Example 8.10 to compute the following.
   a. Both the inside and outside diameter of the femur, the region of cortical bone, will increase by a factor of 10. What will be the new cross-section area?
   b. The man's body will increase by a factor of 10 in each dimension. What will be his new mass?
   c. If the scaled-up man now stands on one leg, what fraction of the tensile strength is the stress on the femur?

62. ‖ Orb spiders make silk with a typical diameter of 0.15 mm.
   a. A typical large orb spider has a mass of 0.50 g. If this spider suspends itself from a single 12-cm-long strand of silk, by how much will the silk stretch?
   b. What is the maximum weight that a single thread of this silk could support?

MCAT-Style Passage Problems

Standing on Tiptoes BIO

When you stand on your tiptoes, your feet pivot about your ankle. As shown in Figure P8.63, the forces on your foot are an upward force on your toes from the floor, a downward force on your ankle from the lower leg bone, and an upward force on the heel of your foot from your Achilles tendon. Suppose a 60 kg woman stands on tiptoes with the sole of her foot making a 25° angle with the floor. Assume that each foot supports half her weight.

Achilles tendon

Ankle pivot

15 cm

20 cm

FIGURE P8.63

63. ‖ What is the upward force of the floor on the toes of one foot?
   A. 140 N   B. 290 N
   C. 420 N   D. 590 N

64. ‖ What upward force does the Achilles tendon exert on the heel of her foot?
   A. 290 N   B. 420 N
   C. 590 N   D. 880 N

65. ‖ The tension in the Achilles tendon will cause it to stretch. If the Achilles tendon is 15 cm long and has a cross-section area of 110 mm², by how much will it stretch under this force?
   A. 0.2 mm   B. 0.8 mm
   C. 2.3 mm   D. 5.2 mm

Chapter Preview Stop to Think: D. The tension in the rope is equal to the weight suspended from it. The force does not change with the position at which the rope is attached. But the torque about the point where the branch meets the trunk depends on both this force and the distance from the pivot point. As the distance decreases, so does the torque.

Stop to Think 8.1: D. Only object D has both zero net force and zero net torque.

Stop to Think 8.2: B. The tension in the rope and the weight have no horizontal component. To make the net force zero, the force due to the pivot must also have no horizontal component, so we know it points either up or down. Now consider the torque about the point where the rope is attached. The tension provides no torque. The weight exerts a counterclockwise torque. To make the net torque zero, the pivot force must exert a *clockwise* torque, which it can do only if it points *up*.

Stop to Think 8.3: B, A, C. The critical angle θ_c, shown in the figure, measures how far the object can be tipped before falling. B has the smallest critical angle, followed by A, then C.

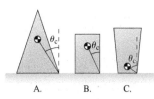

Stop to Think 8.4: B. The spring is now compressed, so the restoring force will be to the right. The end of the spring is displaced by half of what it was in the earlier case, so the restoring force will be half what it was as well.

Stop to Think 8.5: C. The restoring force of the spring is proportional to the stretch. Increasing the restoring force by a factor of 3 requires increasing the stretch by a factor of 3.

Stop to Think 8.6: E. The cables have the same diameter, and the force is the same, so the stress is the same in both cases. This means that the strain, $\Delta L/L$, is the same. The 2 m cable will experience twice the change in length of the 1 m cable.

Force and Motion

The goal of Part I has been to discover the connection between force and motion. We started with kinematics, the mathematical description of motion; then we proceeded to dynamics, the explanation of motion in terms of forces. We then used these descriptions to analyze and explain motions ranging from the motion of the moon about the earth to the forces in your elbow when you lift a weight. Newton's three laws of motion formed the basis of all of our explanations.

The table below is a *knowledge structure* for force and motion. The knowledge structure does not represent everything you have learned over the past eight chapters. It's a summary of the "big picture," outlining the basic goals, the general principles, and the primary applications of the part of the textbook we have just finished. When you are immersed in a chapter, it may be hard to see the connections among all

of the different topics. Before we move on to new topics, we will finish each part of the text with a knowledge structure to make these connections clear.

Work through the knowledge structure from top to bottom. First are the goals and general principles. There aren't that many general principles, but we can use them along with the general problem-solving strategy to solve a wide range of problems. Once you recognize a problem as a dynamics problem, you immediately know to start with Newton's laws. You can then determine the category of motion and apply Newton's second law in the appropriate form. The kinematic equations for that category of motion then allow you to reach the solution you seek. These equations and other detailed information from the chapters are summarized in the bottom section.

KNOWLEDGE STRUCTURE I Force and Motion

BASIC GOALS	How can we describe motion? How does an object respond to a force? How do systems interact? What is the nature of the force of gravity? How can we analyze the motion and deformation of extended objects?

GENERAL PRINCIPLES	**Newton's first law**	An object with no forces acting on it will remain at rest or move in a straight line at a constant speed.
	Newton's second law	$\vec{F}_{net} = m\vec{a}$
	Newton's third law	$\vec{F}_{A\,on\,B} = -\vec{F}_{B\,on\,A}$
	Newton's law of gravity	$F_{1\,on\,2} = F_{2\,on\,1} = \dfrac{Gm_1 m_2}{r^2}$

BASIC PROBLEM-SOLVING STRATEGY

Use Newton's second law for each particle or system. Use Newton's third law to equate the magnitudes of the two members of an action/reaction pair.

Types of forces:
$$\vec{w} = (mg, \text{downward})$$
$$\vec{f}_k = (\mu_k n, \text{opposite motion})$$
$$(F_{sp})_x = -k\,\Delta x$$

Linear and projectile motion:
$$\left.\begin{array}{l} \sum F_x = ma_x \\ \sum F_y = 0 \end{array}\right\} \text{ or } \left\{\begin{array}{l} \sum F_x = 0 \\ \sum F_y = ma_y \end{array}\right.$$

Circular motion:
The force is directed to the center:
$$\vec{F}_{net} = \left(\frac{mv^2}{r}, \text{toward center of circle}\right)$$

Rigid-body motion:
When a torque is exerted on an object with moment of inertia I,
$$\tau_{net} = I\alpha$$

Equilibrium:
For an object at rest,
$$\sum F_x = 0 \qquad \sum \tau = 0$$
$$\sum F_y = 0$$

Linear and projectile kinematics

Uniform motion: $\qquad x_f = x_i + v_x\,\Delta t$
($a_x = 0$, v_x = constant)

Constant acceleration: $\quad (v_x)_f = (v_x)_i + a_x\,\Delta t$
(a_x = constant)
$$x_f = x_i + (v_x)_i\,\Delta t + \tfrac{1}{2}a_x(\Delta t)^2$$
$$(v_x)_f^2 = (v_x)_i^2 + 2a_x\,\Delta x$$

Projectile motion:
Projectile motion is uniform horizontal motion and constant-acceleration vertical motion with $a_y = -g$.

Velocity is the slope of the position-versus-time graph.
Acceleration is the slope of the velocity-versus-time graph.

Circular kinematics

Uniform circular motion:
$$f = \frac{1}{T} \qquad\qquad \omega = 2\pi f$$
$$v = \frac{2\pi r}{T} = \omega r \qquad a = \frac{v^2}{r} = \omega^2 r$$

Rigid bodies

Torque $\tau = rF_\perp = r_\perp F$

Center of gravity $x_{cg} = \dfrac{x_1 m_1 + x_2 m_2 + \cdots}{m_1 + m_2 + \cdots}$

Moment of inertia $I = \sum mr^2$

Dark Matter and the Structure of the Universe

The idea that the earth exerts a gravitational force on us is something we now accept without questioning. But when Isaac Newton developed this idea to show that the gravitational force also holds the moon in its orbit, it was a remarkable, ground-breaking insight. It changed the way that we look at the universe we live in.

Newton's laws of motion and gravity are tools that allow us to continue Newton's quest to better understand our place in the cosmos. But it sometimes seems that the more we learn, the more we realize how little we actually know and understand.

Here's an example. Advances in astronomy over the past 100 years have given us great insight into the structure of the universe. But everything our telescopes can see appears to be only a small fraction of what is out there. Approximately 80% of the mass in the universe is *dark matter*—matter that gives off no light or other radiation that we can detect. Everything that we have ever seen through a telescope is merely the tip of the cosmic iceberg.

What is this dark matter? Black holes? Neutrinos? Some form of exotic particle? We simply aren't sure. It could be any of these, or all of them—or something entirely different that no one has yet dreamed of. You might wonder how we know that such matter exists if no one has seen it. Even though we can't directly observe dark matter, we see its effects. And you now know enough physics to understand why.

Whatever dark matter is, it has mass, and so it has gravity. This picture of the Andromeda galaxy shows a typical spiral galaxy structure: a dense collection of stars in the center surrounded by a disk of stars and other matter. This is the shape of our own Milky Way galaxy.

The spiral Andromeda galaxy.

This structure is reminiscent of the structure of the solar system: a dense mass (the sun) in the center surrounded by a disk of other matter (the planets, asteroids, and comets). The sun's gravity keeps the planets in their orbits, but the planets would fall into the sun unless they were in constant motion around it. The same is true of a spiral galaxy; everything in the galaxy orbits its center. Our solar system orbits the center of our galaxy with a period of about 200 million years.

The orbital speed of an object depends on the mass that pulls on it. If you analyze our sun's motion about the center of the Milky Way, or the motion of stars in the Andromeda galaxy about its center, you find that the orbits are much faster than they should be, based on how many stars we see. There must be some other mass present.

There's another problem with the orbital motion of stars around the center of their galaxies. We know that the orbital speeds of planets decrease with distance from the sun; Neptune orbits at a much slower speed than the earth. We might expect something similar for galaxies: Stars farther from the center should orbit at reduced speeds. But they don't. As we measure outward from the center of the galaxy, the orbital speed stays about the same—even as we get to the edge of the visible disk. There must be some other mass—the invisible dark matter—exerting a gravitational force on the stars. This dark matter, which far outweighs the matter we can see, seems to form a halo around the centers of galaxies, providing the gravitational force necessary to produce the observed rotation. Other observations of the motions of galaxies with respect to each other verify this basic idea.

On a cosmic scale, the picture is even stranger. The universe is currently expanding. The mutual gravitational attraction of all matter—regular and dark—in the universe should slow this expansion. But recent observations of the speeds of distant galaxies imply that the expansion of the universe is accelerating, so there must be yet another component to the universe, something that "pushes out." The best explanation at present is that the acceleration is caused by *dark energy*. The nature of dark matter isn't known, but the nature of dark energy is even more mysterious. If current theories hold, it's the most abundant stuff in the universe. And we don't know what it is.

This sort of mystery is what drives scientific investigation. It's what drove Newton to wonder about the connection between the fall of an apple and the motion of the moon, and what drove investigators to develop all of the techniques and theories you will learn about in the coming chapters.

The following questions are related to the passage "Dark Matter and the Structure of the Universe" on the previous page.

1. As noted in the passage, our solar system orbits the center of the Milky Way galaxy in about 200 million years. If there were no dark matter in our galaxy, this period would be
   A. Longer.
   B. The same.
   C. Shorter.

2. Saturn is approximately 10 times as far away from the sun as the earth. This means that its orbital acceleration is _____ that of the earth.
   A. 1/10
   B. 1/100
   C. 1/1000
   D. 1/10,000

3. Saturn is approximately 10 times as far away from the sun as the earth. If dark matter changed the orbital properties of the planets so that Saturn had the same orbital speed as the earth, Saturn's orbital acceleration would be _____ that of the earth.
   A. 1/10
   B. 1/100
   C. 1/1000
   D. 1/10,000

4. Which of the following might you expect to be an additional consequence of the fact that galaxies contain more mass than expected?
   A. The gravitational force between galaxies is greater than expected.
   B. Galaxies appear less bright than expected.
   C. Galaxies are farther away than expected.
   D. There are more galaxies than expected.

The following passages and associated questions are based on the material of Part I.

Animal Athletes BIO

Different animals have very different capacities for running. A horse can maintain a top speed of 20 m/s for a long distance but has a maximum acceleration of only 6.0 m/s², half what a good human sprinter can achieve with a block to push against. Greyhounds, dogs especially bred for feats of running, have a top speed of 17 m/s, but their acceleration is much greater than that of the horse. Greyhounds are particularly adept at turning corners at a run.

FIGURE I.1

5. If a horse starts from rest and accelerates at the maximum value until reaching its top speed, how much time elapses, to the nearest second?
   A. 1 s   B. 2 s
   C. 3 s   D. 4 s

6. If a horse starts from rest and accelerates at the maximum value until reaching its top speed, how far does it run, to the nearest 10 m?
   A. 40 m   B. 30 m
   C. 20 m   D. 10 m

7. A greyhound on a racetrack turns a corner at a constant speed of 15 m/s with an acceleration of 7.1 m/s². What is the radius of the turn?
   A. 40 m   B. 30 m
   C. 20 m   D. 10 m

8. A human sprinter of mass 70 kg starts a run at the maximum possible acceleration, pushing backward against a block set in the track. What is the force of his foot on the block?
   A. 1500 N   B. 840 N
   C. 690 N   D. 420 N

9. In the photograph of the greyhounds in Figure I.1, what is the direction of the net force on each dog?
   A. Up
   B. Down
   C. Left, toward the outside of the turn
   D. Right, toward the inside of the turn

Sticky Liquids BIO

The drag force on an object moving in a liquid is quite different from that in air. Drag forces in air are largely the result of the object having to push the air out of its way as it moves. For an object moving slowly through a liquid, however, the drag force is mostly due to the *viscosity* of the liquid, a measure of how much resistance to flow the fluid has. Honey, which drizzles slowly out of its container, has a much higher viscosity than water, which flows fairly freely.

The *viscous drag* force in a liquid depends on the shape of the object, but there is a simple result called *Stokes's law* for the drag on a sphere. The drag force on a sphere of radius r moving at speed v through a fluid with viscosity η is

$$\vec{D} = (6\pi\eta rv, \text{ direction opposite motion})$$

At small scales, viscous drag becomes very important. To a paramecium (Figure I.2), a single-celled animal that can propel itself through water with fine hairs on its body, swimming through water feels like swimming through honey would to you. We can model a paramecium as a sphere of diameter 50 μm, with a mass of 6.5×10^{-11} kg. Water has a viscosity of 0.0010 N·s/m².

FIGURE I.2

10. A paramecium swimming at a constant speed of 0.25 mm/s ceases propelling itself and slows to a stop. At the instant it stops swimming, what is the magnitude of its acceleration?
    A. 0.2g B. 0.5g
    C. 2g D. 5g

11. If the acceleration of the paramecium in Problem 10 were to stay constant as it came to rest, approximately how far would it travel before stopping?
    A. 0.02 μm   B. 0.2 μm
    C. 2 μm   D. 20 μm

12. If the paramecium doubles its swimming speed, how does this change the drag force?
    A. The drag force decreases by a factor of 2.
    B. The drag force is unaffected.
    C. The drag force increases by a factor of 2.
    D. The drag force increases by a factor of 4.

13. You can test the viscosity of a liquid by dropping a steel sphere into it and measuring the speed at which it sinks. For viscous fluids, the sphere will rapidly reach a terminal speed. At this terminal speed, the net force on the sphere is
    A. Directed downward.
    B. Zero.
    C. Directed upward.

Pulling Out of a Dive BIO

Falcons are excellent fliers that can reach very high speeds by diving nearly straight down. To pull out of such a dive, a falcon extends its wings and flies through a circular arc that redirects its motion. The forces on the falcon that control its motion are its weight and an upward lift force—like an airplane—due to the air flowing over its wings. At the bottom of the arc, as in Figure I.3, a falcon can easily achieve an acceleration of 15 m/s².

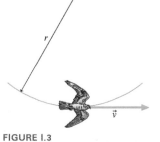

FIGURE I.3

14. At the bottom of the arc, as in Figure I.3, what is the direction of the net force on the falcon?
    A. To the left, opposite the motion
    B. To the right, in the direction of the motion
    C. Up
    D. Down
    E. The net force is zero.
15. Suppose the falcon weighs 8.0 N and is turning with an acceleration of 15 m/s² at the lowest point of the arc. What is the magnitude of the upward lift force at this instant?
    A. 8.0 N   B. 12 N
    C. 16 N   D. 20 N
16. A falcon starts from rest, does a free-fall dive from a height of 30 m, and then pulls out by flying in a circular arc of radius 50 m. Which segment of the motion has a higher acceleration?
    A. The free-fall dive
    B. The circular arc
    C. The two accelerations are equal.

Bending Beams

If you bend a rod down, it compresses the lower side of the rod and stretches the top, resulting in a restoring force. Figure I.4 shows a

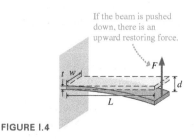

FIGURE I.4

beam of length L, width w, and thickness t fixed at one end and free to move at the other. Deflecting the end of the beam causes a restoring force F at the end of the beam. The magnitude of the restoring force F depends on the dimensions of the beam, the Young's modulus Y for the material, and the deflection d. For small values of the deflection, the restoring force is

$$F = \left[\frac{Ywt^3}{4L^3} \right] d$$

This is similar to the formula for the restoring force of a spring, with the quantity in brackets playing the role of the spring constant k.

When a 70 kg man stands on the end of a springboard (a type of diving board), the board deflects by 4.0 cm.

17. If a 35 kg child stands at the end of the board, the deflection is
    A. 1.0 cm.   B. 2.0 cm.
    C. 3.0 cm.   D. 4.0 cm.
18. A 70 kg man jumps up and lands on the end of the board, deflecting it by 12 cm. At this instant, what is the approximate magnitude of the upward force the board exerts on his feet?
    A. 700 N   B. 1400 N
    C. 2100 N   D. 2800 N
19. If the board is replaced by one that is half the length but otherwise identical, how much will it deflect when a 70 kg man stands on the end?
    A. 0.50 cm   B. 1.0 cm
    C. 2.0 cm   D. 4.0 cm

Additional Integrated Problems

20. You go to the playground and slide down the slide, a 3.0-m-long ramp at an angle of 40° with respect to horizontal. The pants that you've worn aren't very slippery; the coefficient of kinetic friction between your pants and the slide is $\mu_k = 0.45$. A friend gives you a very slight push to get you started. How long does it take you to reach the bottom of the slide?
21. If you stand on a scale at the equator, the scale will read slightly less than your true weight due to your circular motion with the rotation of the earth.
    a. Draw a free-body diagram to show why this is so.
    b. By how much is the scale reading reduced for a person with a true weight of 800 N?
22. Dolphins and other sea creatures can leap to great heights by swimming straight up and exiting the water at a high speed. A 210 kg dolphin leaps straight up to a height of 7.0 m. When the dolphin reenters the water, drag from the water brings it to a stop in 1.5 m. Assuming that the force of the water on the dolphin stays constant as it slows down,
    a. How much time does it take for the dolphin to come to rest?
    b. What is the force of the water on the dolphin as it is coming to rest?

Conservation Laws

The kestrel is pulling in its wings to begin a steep dive, in which it can achieve a speed of 60 mph. How does the bird achieve such a speed, and why does this speed help the kestrel catch its prey? Such questions are best answered by considering the conservation of energy and momentum.

Why Some Things Stay the Same

Part I of this textbook was about *change*. Simple observations show us that most things in the world around us are changing. Even so, there are some things that *don't* change even as everything else is changing around them. Our emphasis in Part II will be on things that stay the same.

Consider, for example, a strong, sealed box in which you have replaced all the air with a mixture of hydrogen and oxygen. The mass of the box plus the gases inside is 600.0 g. Now, suppose you use a spark to ignite the hydrogen and oxygen. As you know, this is an explosive reaction, with the hydrogen and oxygen combining to create water—and quite a bang. But the strong box contains the explosion and all of its products.

What is the mass of the box after the reaction? The gas inside the box is different now, but a careful measurement would reveal that the mass hasn't changed—it's still 600.0 g! We say that the mass is *conserved*. Of course, this is true only if the box has stayed sealed. For conservation of mass to apply, the system must be *closed*.

Conservation Laws

A closed system of interacting particles has another remarkable property. Each system is characterized by a certain number, and no matter how complex the interactions, the value of this number never changes. This number is called the *energy* of the system, and the fact that it never changes is called the *law of conservation of energy*. It is, perhaps, the single most important physical law ever discovered.

The law of conservation of energy is much more general than Newton's laws. Energy can be converted to many different forms, and, in all cases, the total energy stays the same:

- Gasoline, diesel, and jet engines convert the energy of a fuel into the mechanical energy of moving pistons, wheels, and gears.
- A solar cell converts the electromagnetic energy of light into electrical energy.
- An organism converts the chemical energy of food into a variety of other forms of energy, including kinetic energy, sound energy, and thermal energy.

Energy will be *the* most important concept throughout the remainder of this textbook, and much of Part II will focus on understanding what energy is and how it is used.

But energy is not the only conserved quantity. We will begin Part II with the study of two other quantities that are conserved in a closed system: *momentum* and *angular momentum*. Their conservation will help us understand a wide range of physical processes, from the forces when two rams butt heads to the graceful spins of ice skaters.

Conservation laws will give us a new and different *perspective* on motion. Some situations are most easily analyzed from the perspective of Newton's laws, but others make much more sense when analyzed from a conservation-law perspective. An important goal of Part II is to learn which perspective is best for a given problem.

9 Momentum

Male rams butt heads at high speeds in a ritual to assert their dominance. How can the force of this collision be minimized so as to avoid damage to their brains?

LOOKING AHEAD ►

Goal: To learn about impulse, momentum, and a new problem-solving strategy based on conservation laws.

Impulse

This golf club delivers an **impulse** to the ball as the club strikes it.

You'll learn that a longer-lasting, stronger force delivers a greater impulse to an object.

Momentum and Impulse

The impulse delivered by the player's head *changes* the ball's **momentum**.

You'll learn how to calculate this momentum change using the **impulse-momentum theorem**.

Conservation of Momentum

The momentum of these pool balls before and after they collide is the *same*—it is **conserved**.

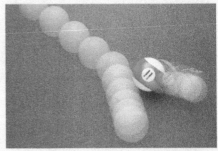

You'll learn a powerful new *before-and-after* problem-solving strategy using this **law of conservation of momentum**.

LOOKING BACK ◄

Newton's Third Law

In Section 4.7 you learned about Newton's third law. In this chapter, you'll apply this law in order to understand the conservation of momentum.

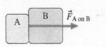

Newton's third law states that the force that object B exerts on A has *equal magnitude* but is *directed opposite to* the force that A exerts on B.

A hammer hits a nail. The force of the nail on the hammer is

A. Greater than the force of the hammer on the nail.
B. Less than the force of the hammer on the nail.
C. Equal to the force of the hammer on the nail.
D. Zero.

9.1 Impulse

A **collision** is a short-duration interaction between two objects. The collision between a tennis ball and a racket, or your foot and a soccer ball, may seem instantaneous to your eye, but that is a limitation of your perception. The sequence of high-speed photos of a soccer kick shown in FIGURE 9.1 reveals that the ball is compressed as the foot begins its contact. It takes time to compress the ball, and more time for the ball to re-expand as it leaves the foot.

FIGURE 9.1 A sequence of high-speed photos of a soccer ball being kicked.

The duration of a collision depends on the materials from which the objects are made, but 1 to 10 ms (0.001 to 0.010 s) is typical. This is the time during which the two objects are in contact with each other. The harder the objects, the shorter the contact time. A collision between two steel balls lasts less than 1 ms, while that between your foot and a soccer ball might last 10 ms.

Let's begin our discussion by again considering the soccer kick shown in Figure 9.1. As the foot and the ball just come into contact, as shown in the left frame, the ball is just beginning to compress. By the middle frame of Figure 9.1, the ball has sped up and become greatly compressed. Finally, as shown in the right frame, the ball, now moving very fast, is again only slightly compressed.

The amount by which the ball is compressed is a measure of the magnitude of the force the foot exerts on the ball; more compression indicates a greater force. If we were to graph this force versus time, it would look something like FIGURE 9.2. The force is zero until the foot first contacts the ball, rises quickly to a maximum value, and then falls back to zero as the ball leaves the foot. Thus there is a well-defined duration Δt of the force. A large force like this exerted during a short interval of time is called an **impulsive force**. The forces of a hammer on a nail and of a bat on a baseball are other examples of impulsive forces.

A harder kick (i.e., a taller force curve) or a kick of longer duration (a wider force curve) causes the ball to leave the kicker's foot with a higher speed; that is, the *effect* of the kick is larger. Now a taller or wider force-versus-time curve has a larger *area* between the curve and the axis (i.e., the area "under" the force curve is larger), so we can say that **the effect of an impulsive force is proportional to the area under the force-versus-time curve.** This area, shown in FIGURE 9.3a, is called the **impulse** J of the force.

Impulsive forces can be complex, and the shape of the force-versus-time graph often changes in a complicated way. Consequently, it is often useful to think of the collision in terms of an *average* force F_{avg}. As FIGURE 9.3b shows, F_{avg} is defined to be the constant force that has the same duration Δt and the same area under the force curve as the real force. You can see from the figure that the area under the force curve can be written simply as $F_{avg} \Delta t$. Thus

$$\text{impulse } J = \text{area under the force curve} = F_{avg} \Delta t \qquad (9.1)$$

Impulse due to a force acting for a duration Δt

From Equation 9.1 we see that impulse has units of $N \cdot s$, but $N \cdot s$ are equivalent to $kg \cdot m/s$. We'll see shortly why the latter are the preferred units for impulse.

FIGURE 9.2 **The force on a soccer ball changes rapidly.**

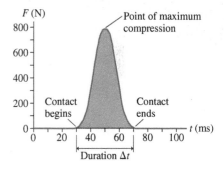
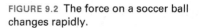

FIGURE 9.3 **Looking at the impulse graphically.**

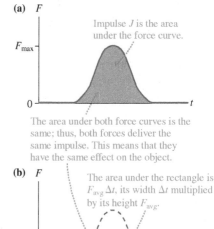

So far, we've been assuming the force is directed along a coordinate axis, such as the x-axis. In this case impulse is a *signed* quantity—it can be positive or negative. A positive impulse results from an average force directed in the positive x-direction (that is, F_{avg} is positive), while a negative impulse is due to a force directed in the negative x-direction (F_{avg} is negative). More generally, the impulse is a *vector* quantity pointing in the direction of the average force vector:

$$\vec{J} = \vec{F}_{avg}\,\Delta t \qquad\qquad (9.2)$$

EXAMPLE 9.1   Finding the impulse on a bouncing ball

A rubber ball experiences the force shown in FIGURE 9.4 as it bounces off the floor.

a. What is the impulse on the ball?
b. What is the average force on the ball?

PREPARE The impulse is the area under the force curve. Here the shape of the graph is triangular, so we'll need to use the fact that the area of a triangle is $\frac{1}{2} \times$ height $\times$ base.

FIGURE 9.4 The force of the floor on a bouncing ball.

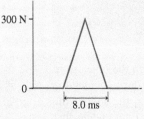

SOLVE a. The impulse is

$$J = \tfrac{1}{2}(300\ \text{N})(0.0080\ \text{s}) = 1.2\ \text{N}\cdot\text{s} = 1.2\ \text{kg}\cdot\text{m/s}$$

b. From Equation 9.1, $J = F_{avg}\,\Delta t$, we can find the average force that would give this same impulse:

$$F_{avg} = \frac{J}{\Delta t} = \frac{1.2\ \text{N}\cdot\text{s}}{0.0080\ \text{s}} = 150\ \text{N}$$

ASSESS In this particular example, the average value of the force is half the maximum value. This is not surprising for a triangular force because the area of a triangle is *half* the base times the height.

STOP TO THINK 9.1   Graph A is the force-versus-time graph for a hockey stick hitting a 160 g puck. Graph B is the force-versus-time graph for a golf club hitting a 46 g golf ball. Which force delivers the greater impulse?

A. Force A
B. Force B
C. Both forces deliver the same impulse.

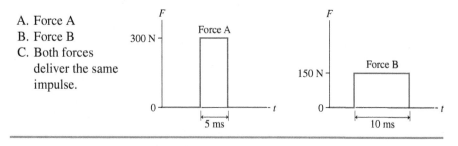

9.2 Momentum and the Impulse-Momentum Theorem

We've noted that the effect of an impulsive force depends on the impulse delivered to the object. The effect also depends on the object's mass. Our experience tells us that giving a kick to a heavy object will change its velocity much less than giving the same kick to a light object. We want now to find a quantitative relationship for impulse, mass, and velocity change.

Consider the puck of mass m in FIGURE 9.5, sliding with an initial velocity $\vec{v}_i$. It is struck by a hockey stick that delivers an impulse $\vec{J} = \vec{F}_{avg}\,\Delta t$ to the puck. After the impulse, the puck leaves the stick with a final velocity $\vec{v}_f$. How is this final velocity related to the initial velocity?

From Newton's second law, the average acceleration of the puck during the time the stick is in contact with it is

$$\vec{a}_{avg} = \frac{\vec{F}_{avg}}{m} \qquad\qquad (9.3)$$

FIGURE 9.5 The stick exerts an impulse on the puck, changing its speed.

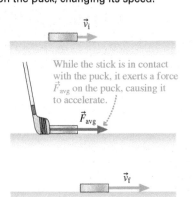

While the stick is in contact with the puck, it exerts a force $\vec{F}_{avg}$ on the puck, causing it to accelerate.

The average acceleration is related to the change in the velocity by

$$\vec{a}_{avg} = \frac{\Delta \vec{v}}{\Delta t} = \frac{\vec{v}_f - \vec{v}_i}{\Delta t} \tag{9.4}$$

Combining Equations 9.3 and 9.4, we have

$$\frac{\vec{F}_{avg}}{m} = \vec{a}_{avg} = \frac{\vec{v}_f - \vec{v}_i}{\Delta t}$$

or, rearranging,

$$\vec{F}_{avg} \, \Delta t = m\vec{v}_f - m\vec{v}_i \tag{9.5}$$

We recognize the left side of this equation as the impulse $\vec{J}$. The right side is the *change* in the quantity $m\vec{v}$. This quantity, the product of the object's mass and velocity, is called the **momentum** of the object. The symbol for momentum is $\vec{p}$:

$$\vec{p} = m\vec{v} \tag{9.6}$$

Momentum of an object of mass m and velocity $\vec{v}$

From Equation 9.6, the units of momentum are those of mass times velocity, or kg · m/s. We noted above that kg · m/s are the preferred units of impulse. Now we see that the reason for that preference is to match the units of momentum.

FIGURE 9.6 shows that the momentum $\vec{p}$ is a *vector* quantity that points in the same direction as the velocity vector $\vec{v}$. Like any vector, $\vec{p}$ can be decomposed into x- and y-components. Equation 9.6, which is a vector equation, is a shorthand way to write the two equations

$$\begin{aligned} p_x &= mv_x \\ p_y &= mv_y \end{aligned} \tag{9.7}$$

NOTE ▶ One of the most common errors in momentum problems is failure to use the correct signs. The momentum component p_x has the same sign as v_x. Just like velocity, momentum is positive for a particle moving to the right (on the x-axis) or up (on the y-axis), but *negative* for a particle moving to the left or down. ◀

The *magnitude* of an object's momentum is simply the product of the object's mass and speed, or $p = mv$. A heavy, fast-moving object will have a great deal of momentum, while a light, slow-moving object will have very little. Two objects with very different masses can have similar momenta if their speeds are very different as well. Table 9.1 gives some typical values of the momenta (the plural of *momentum*) of various moving objects. You can see that the momenta of a bullet and a fastball are similar. The momentum of a moving car is almost a billion times greater than that of a falling raindrop.

The Impulse-Momentum Theorem

We can now write Equation 9.5 in terms of impulse and momentum:

$$\vec{J} = \vec{p}_f - \vec{p}_i = \Delta \vec{p} \tag{9.8}$$

Impulse-momentum theorem

where $\vec{p}_i = m\vec{v}_i$ is the object's initial momentum, $\vec{p}_f = m\vec{v}_f$ is its final momentum after the impulse, and $\Delta \vec{p} = \vec{p}_f - \vec{p}_i$ is the *change* in its momentum. This expression is known as the **impulse-momentum theorem**. It states that **an impulse delivered to an object causes the object's momentum to change.** That is, the *effect* of an impulsive force is to change the object's momentum from $\vec{p}_i$ to

$$\vec{p}_f = \vec{p}_i + \vec{J} \tag{9.9}$$

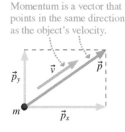

Class Video

FIGURE 9.6 A particle's momentum vector $\vec{p}$ can be decomposed into x- and y-components.

Momentum is a vector that points in the same direction as the object's velocity.

TABLE 9.1 Some typical momenta (approximate)

Object	Mass (kg)	Speed (m/s)	Momentum (kg · m/s)
Falling raindrop	2×10^{-5}	5	10^{-4}
Bullet	0.004	500	2
Pitched baseball	0.15	40	6
Running person	70	3	200
Car on highway	1000	30	3×10^4

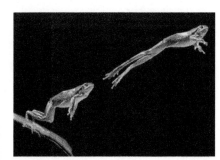

Legging it BIO A frog making a jump wants to gain as much momentum as possible before leaving the ground. This means that he wants the greatest impulse $J = F_{avg} \, \Delta t$ delivered to him by the ground. There is a maximum force that muscles can exert, limiting F_{avg}. But the time interval Δt over which the force is exerted can be greatly increased by having long legs. Many animals that are good jumpers have particularly long legs.

Equation 9.8 can also be written in terms of its x- and y-components as

$$J_x = \Delta p_x = (p_x)_f - (p_x)_i = m(v_x)_f - m(v_x)_i$$
$$J_y = \Delta p_y = (p_y)_f - (p_y)_i = m(v_y)_f - m(v_y)_i$$

(9.10)

The impulse-momentum theorem is illustrated by two examples in FIGURE 9.7. In the first, the putter strikes the ball, exerting a force on it and delivering an impulse $\vec{J} = \vec{F}_{avg} \, \Delta t$. Notice that the direction of the impulse is the same as that of the force. Because $\vec{p}_i = \vec{0}$ in this situation, we can use the impulse-momentum theorem to find that the ball leaves the putter with momentum $\vec{p}_f = \vec{p}_i + \vec{J} = \vec{J}$. This is shown graphically in Figure 9.7a.

NOTE ▶ You can think of the putter as changing the ball's momentum by transferring momentum to it as an impulse. Thus we say the putter *delivers* an impulse to the ball, and the ball *receives* an impulse from the putter. ◀

The soccer player in Figure 9.7b presents a more complicated case. Here, the initial momentum of the ball is directed downward to the left. The impulse delivered to it by the player's head, upward to the right, is strong enough to reverse the ball's motion and send it off in a new direction. The graphical addition of vectors in Figure 9.7b again shows that $\vec{p}_f = \vec{p}_i + \vec{J}$.

FIGURE 9.7 **Impulse causes a *change* in momentum.**

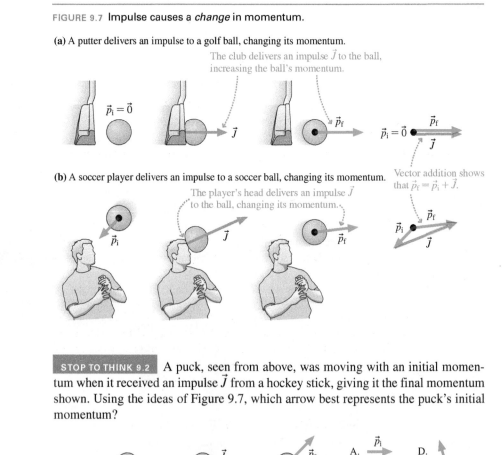

(a) A putter delivers an impulse to a golf ball, changing its momentum.

(b) A soccer player delivers an impulse to a soccer ball, changing its momentum.

STOP TO THINK 9.2   A puck, seen from above, was moving with an initial momentum when it received an impulse $\vec{J}$ from a hockey stick, giving it the final momentum shown. Using the ideas of Figure 9.7, which arrow best represents the puck's initial momentum?

EXAMPLE 9.2   **Calculating the change in momentum**

A ball of mass $m = 0.25$ kg rolling to the right at 1.3 m/s strikes a wall and rebounds to the left at 1.1 m/s. What is the change in the ball's momentum? What is the impulse delivered to it by the wall?

PREPARE A visual overview of the ball bouncing is shown in FIGURE 9.8. This is a new kind of visual overview, one in which we show the situation "before" and "after" the interaction. We'll have more to say about before-and-after pictures in the next section. The ball is moving along the x-axis, so we'll write the momentum in component form, as in Equation 9.7. The change

FIGURE 9.8 Visual overview for a ball bouncing off a wall.

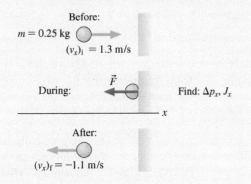

in momentum is then the difference between the final and initial values of the momentum. By the impulse-momentum theorem, the impulse is equal to this change in momentum.

SOLVE The x-component of the initial momentum is

$$(p_x)_i = m(v_x)_i = (0.25 \text{ kg})(1.3 \text{ m/s}) = 0.325 \text{ kg} \cdot \text{m/s}$$

The y-component of the momentum is zero both before and after the bounce. After the ball rebounds, the x-component is

$$(p_x)_f = m(v_x)_f = (0.25 \text{ kg})(-1.1 \text{ m/s}) = -0.275 \text{ kg} \cdot \text{m/s}$$

It is particularly important to notice that the x-component of the momentum, like that of the velocity, is negative. This indicates that the ball is moving to the *left*. The change in momentum is

$$\Delta p_x = (p_x)_f - (p_x)_i = (-0.275 \text{ kg} \cdot \text{m/s}) - (0.325 \text{ kg} \cdot \text{m/s})$$
$$= -0.60 \text{ kg} \cdot \text{m/s}$$

By the impulse-momentum theorem, the impulse delivered to the ball by the wall is equal to this change, so

$$J_x = \Delta p_x = -0.60 \text{ kg} \cdot \text{m/s}$$

ASSESS The impulse is negative, indicating that the force causing the impulse is pointing to the left, which makes sense.

SYNTHESIS 9.1 **Momentum and impulse**

A moving object has momentum. A force acting on an object delivers an *impulse* that changes the object's momentum.

The **momentum** of an object is the product of its mass and its velocity.

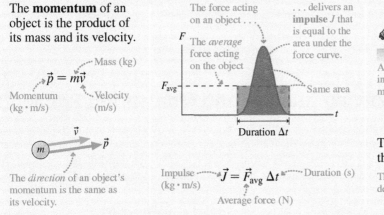

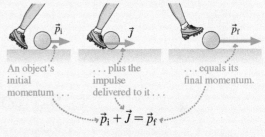

$$\vec{p}_i + \vec{J} = \vec{p}_f$$

This relationship can also be written in terms of the *change* in the object's momentum:

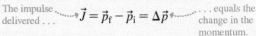

An interesting application of the impulse-momentum theorem is to the question of how to slow down a fast-moving object in the gentlest possible way. For instance, a car is headed for a collision with a bridge abutment. How can this crash be made survivable? How do the rams in the chapter-opening photo avoid injury when they collide?

In these examples, the object has momentum $\vec{p}_i$ just before impact and zero momentum after (i.e., $\vec{p}_f = \vec{0}$). The impulse-momentum theorem tells us that

$$\vec{J} = \vec{F}_{avg} \Delta t = \Delta \vec{p} = \vec{p}_f - \vec{p}_i = -\vec{p}_i$$

or

$$\vec{F}_{avg} = -\frac{\vec{p}_i}{\Delta t} \qquad (9.11)$$

That is, the average force needed to stop an object is *inversely proportional* to the duration Δt of the collision. **If the duration of the collision can be increased, the**

A spiny cushion BIO The spines of a hedgehog obviously help protect it from predators. But they serve another function as well. If a hedgehog falls from a tree—a not uncommon occurrence—it simply rolls itself into a ball before it lands. Its thick spines then cushion the blow by increasing the time it takes for the animal to come to rest. Indeed, hedgehogs have been observed to fall out of trees on purpose to get to the ground!

FIGURE 9.9 The total momentum of three pool balls.

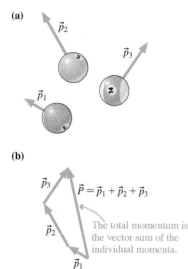

(a)

(b)

force of the impact will be decreased. This is the principle used in most impact-lessening techniques.

For example, obstacles such as bridge abutments are made safer by placing a line of water-filled barrels in front of them. In case of a collision, the time it takes for the car to plow through these barrels is much longer than the time it would take the car to stop if it hit the abutment head-on. The force on the car (and on the driver from his or her seat belt) is greatly reduced by the longer-duration collision with the barrels.

The butting rams shown in the photo at the beginning of this chapter also have adaptations of this kind that allow them to collide at high speeds without injury to their brains. The cranium has a double wall to prevent skull injuries, and there is a thick spongy mass that increases the time it takes for the brain to come to rest upon impact, again reducing the magnitude of the force on the brain.

Total Momentum

If we have more than one object moving—a *system* of particles—then the system as a whole has an overall momentum. The **total momentum** $\vec{P}$ (note the capital P) of a system of particles is the vector sum of the momenta of the individual particles:

$$\vec{P} = \vec{p}_1 + \vec{p}_2 + \vec{p}_3 + \cdots$$

FIGURE 9.9 shows how the momentum vectors of three moving pool balls are graphically added to find the total momentum. The concept of total momentum will be of key importance when we discuss the conservation law for momentum in Section 9.4.

STOP TO THINK 9.3 The cart's change of momentum is

A. $-30 \text{ kg} \cdot \text{m/s}$
B. $-20 \text{ kg} \cdot \text{m/s}$
C. $-10 \text{ kg} \cdot \text{m/s}$
D. $10 \text{ kg} \cdot \text{m/s}$
E. $20 \text{ kg} \cdot \text{m/s}$
F. $30 \text{ kg} \cdot \text{m/s}$

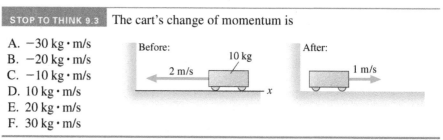

9.3 Solving Impulse and Momentum Problems

Visual overviews have become an important problem-solving tool. The visual overviews and free-body diagrams that you learned to draw in Chapters 1–8 were oriented toward the use of Newton's laws and a subsequent kinematical analysis. Now we are interested in making a connection between "before" and "after."

TACTICS BOX 9.1 Drawing a before-and-after visual overview (MP)

❶ **Sketch the situation.** Use two drawings, labeled "Before" and "After," to show the objects *immediately before* they interact and again *immediately after* they interact.

❷ **Establish a coordinate system.** Select your axes to match the motion.

❸ **Define symbols.** Define symbols for the masses and for the velocities before and after the interaction. Position and time are not needed.

❹ **List known information.** List the values of quantities known from the problem statement or that can be found quickly with simple geometry or unit conversions. Before-and-after pictures are usually simpler than the pictures you used for dynamics problems, so listing known information on the sketch is often adequate.

❺ **Identify the desired unknowns.** What quantity or quantities will allow you to answer the question? These should have been defined as symbols in step 3.

Exercises 9–11

EXAMPLE 9.3 **Force in hitting a baseball**

A 150 g baseball is thrown with a speed of 20 m/s. It is hit straight back toward the pitcher at a speed of 40 m/s. The impulsive force of the bat on the ball has the shape shown in FIGURE 9.10. What is the *maximum* force F_{max} that the bat exerts on the ball? What is the *average* force that the bat exerts on the ball?

FIGURE 9.10 The interaction force between the baseball and the bat.

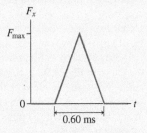

FIGURE 9.11 A before-and-after visual overview.

❶ Draw the before-and-after pictures.

Before:

❷ Establish a coordinate system.

$(v_x)_i = -20$ m/s
$m = 0.15$ kg

❸ Define symbols.
❹ List known information.

After:

$(v_x)_f = 40$ m/s

Find: F_{max} and F_{avg}

❺ Identify desired unknowns.

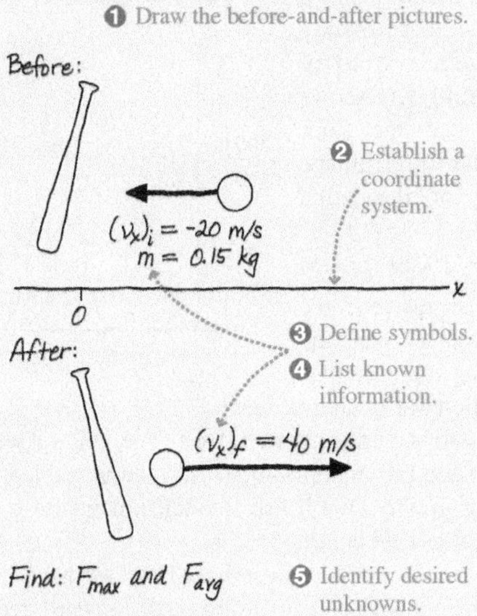

PREPARE We can model the interaction as a collision. FIGURE 9.11 is a before-and-after visual overview in which the steps from Tactics Box 9.1 are explicitly noted. Because F_x is positive (a force to the right), we know the ball was initially moving toward the left and is hit back toward the right. Thus we converted the statements about *speeds* into information about *velocities*, with $(v_x)_i$ negative.

SOLVE In the last several chapters we've started the mathematical solution with Newton's second law. Now we want to use the impulse-momentum theorem:

$$\Delta p_x = J_x = \text{area under the force curve}$$

We know the velocities before and after the collision, so we can find the change in the ball's momentum:

$$\Delta p_x = m(v_x)_f - m(v_x)_i = (0.15 \text{ kg})(40 \text{ m/s} - (-20 \text{ m/s}))$$
$$= 9.0 \text{ kg} \cdot \text{m/s}$$

The force curve is a triangle with height F_{max} and width 0.60 ms. As in Example 9.1, the area under the curve is

$$J_x = \text{area} = \tfrac{1}{2} \times F_{max} \times (6.0 \times 10^{-4} \text{ s})$$
$$= (F_{max})(3.0 \times 10^{-4})$$

According to the impulse-momentum theorem, $\Delta p_x = J_x$, so we have

$$9.0 \text{ kg} \cdot \text{m/s} = (F_{max})(3.0 \times 10^{-4} \text{ s})$$

Thus the *maximum* force is

$$F_{max} = \frac{9.0 \text{ kg} \cdot \text{m/s}}{3.0 \times 10^{-4} \text{ s}} = 30{,}000 \text{ N}$$

Using Equation 9.1, we find that the *average* force, which depends on the collision duration $\Delta t = 6.0 \times 10^{-4}$ s, has the smaller value:

$$F_{avg} = \frac{J_x}{\Delta t} = \frac{\Delta p_x}{\Delta t} = \frac{9.0 \text{ kg} \cdot \text{m/s}}{6.0 \times 10^{-4} \text{ s}} = 15{,}000 \text{ N}$$

ASSESS F_{max} is a large force, but quite typical of the impulsive forces during collisions.

The Impulse Approximation

When two objects interact during a collision or other brief interaction, such as that between the bat and ball of Example 9.3, the forces *between* them are generally quite large. Other forces may also act on the interacting objects, but usually these forces are *much* smaller than the interaction forces. In Example 9.3, for example, the 1.5 N weight of the ball is vastly less than the 30,000 N force of the bat on the ball. We can reasonably ignore these small forces *during* the brief time of the impulsive force. Doing so is called the **impulse approximation.**

When we use the impulse approximation, $(p_x)_i$ and $(p_x)_f$—and $(v_x)_i$ and $(v_x)_f$—are the momenta (and velocities) *immediately* before and *immediately* after the collision. For example, the velocities in Example 9.3 are those of the ball just before and after it collides with the bat. We could then do a follow-up problem, including weight and drag, to find the ball's speed a second later as the second baseman catches it.

EXAMPLE 9.4   **Height of a bouncing ball**

A 100 g rubber ball is thrown straight down onto a hard floor so that it strikes the floor with a speed of 11 m/s. FIGURE 9.12 shows the force that the floor exerts on the ball. Estimate the height of the ball's bounce.

PREPARE The ball experiences an impulsive force while in contact with the floor. Using the impulse approximation, we'll ignore the ball's weight during these 5.0 ms. The ball's rise after the bounce is free-fall motion—that is, motion subject to only the force of gravity. We'll use free-fall kinematics to describe the motion after the bounce.

FIGURE 9.13 is a visual overview. Here we have a two-part problem, an impulsive collision followed by upward free fall. The overview thus shows the ball just before the collision, where we label its velocity as v_{1y}; just after the collision, where its velocity

FIGURE 9.12 The force of the floor on a bouncing rubber ball.

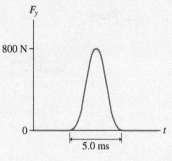

FIGURE 9.13 Before-and-after visual overview for a bouncing ball.

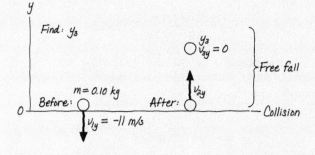

is v_{2y}; and at the highest point of its rising free fall, where its velocity is $v_{3y} = 0$.

SOLVE The impulse-momentum theorem tells us that $J_y = \Delta p_y = p_{2y} - p_{1y}$, so that $p_{2y} = p_{1y} + J_y$. The initial momentum, just before the collision, is $p_{1y} = mv_{1y} = (0.10 \text{ kg})(-11 \text{ m/s}) = -1.1 \text{ kg} \cdot \text{m/s}$.

Next, we need to find the impulse J_y, which is the area under the curve in Figure 9.12. Because the force is given as a smooth curve, we'll have to *estimate* this area. Recall that the area can be written as $F_{avg} \Delta t$. From the curve, we might estimate F_{avg} to be about 400 N, or half the maximum value of the force. With this estimate we have

$$J_y = \text{area under the force curve} \approx (400 \text{ N}) \times (0.0050 \text{ s})$$
$$= 2.0 \text{ N} \cdot \text{s} = 2.0 \text{ kg} \cdot \text{m/s}$$

Thus

$$p_{2y} = p_{1y} + J_y = (-1.1 \text{ kg} \cdot \text{m/s}) + 2.0 \text{ kg} \cdot \text{m/s}$$
$$= 0.9 \text{ kg} \cdot \text{m/s}$$

and the post-collision velocity is

$$v_{2y} = \frac{p_{2y}}{m} = \frac{0.9 \text{ kg} \cdot \text{m/s}}{0.10 \text{ kg}} = 9 \text{ m/s}$$

The rebound speed is less than the impact speed, as expected. Finally, we can use free-fall kinematics to find

$$v_{3y}^2 = 0 = v_{2y}^2 - 2g \Delta y = v_{2y}^2 - 2gy_3$$
$$y_3 = \frac{v_{2y}^2}{2g} = \frac{(9 \text{ m/s})^2}{2(9.8 \text{ m/s}^2)} = 4 \text{ m}$$

We estimate that the ball bounces to a height of 4 m.

ASSESS This is a reasonable height for a rubber ball thrown down quite hard.

STOP TO THINK 9.4   A 10 g rubber ball and a 10 g clay ball are each thrown at a wall with equal speeds. The rubber ball bounces; the clay ball sticks. Which ball receives the greater impulse from the wall?

A. The clay ball receives a greater impulse because it sticks.
B. The rubber ball receives a greater impulse because it bounces.
C. They receive equal impulses because they have equal momenta.
D. Neither receives an impulse because the wall doesn't move.

9.4 Conservation of Momentum

The impulse-momentum theorem was derived from Newton's second law and is really just an alternative way of looking at that law. It is used in the context of single-particle dynamics, much as we used Newton's law in Chapters 4–7.

However, consider two objects, such as the rams shown in the opening photo of this chapter, that interact during the brief moment of a collision. During a collision, two objects exert forces on each other that vary in a complex way. We usually don't even know the magnitudes of these forces. Using Newton's second law alone to predict the outcome of such a collision would thus be a daunting challenge. However, by using Newton's *third* law in the language of impulse and momentum, we'll find that it's possible to describe the *outcome* of a collision—the final speeds and directions of the colliding objects—in a simple way. Newton's third law will lead us to one of the most important conservation laws in physics.

FIGURE 9.14 shows two balls initially headed toward each other. The balls collide, then bounce apart. The forces during the collision, when the balls are interacting, form an action/reaction pair $\vec{F}_{1\,on\,2}$ and $\vec{F}_{2\,on\,1}$. For now, we'll continue to assume that the motion is one dimensional along the x-axis.

During the collision, the impulse J_{2x} delivered to ball 2 by ball 1 is the average value of $\vec{F}_{1\,on\,2}$ multiplied by the collision time Δt. Likewise, the impulse J_{1x} delivered to ball 1 by ball 2 is the average value of $\vec{F}_{2\,on\,1}$ multiplied by Δt. Because $\vec{F}_{1\,on\,2}$ and $\vec{F}_{2\,on\,1}$ form an action/reaction pair, they have equal magnitudes but opposite directions. As a result, the two impulses J_{1x} and J_{2x} are also equal in magnitude but opposite in sign, so that $J_{1x} = -J_{2x}$.

According to the impulse-momentum theorem, the change in the momentum of ball 1 is $\Delta p_{1x} = J_{1x}$ and the change in the momentum of ball 2 is $\Delta p_{2x} = J_{2x}$. Because $J_{1x} = -J_{2x}$, the change in the momentum of ball 1 is equal in magnitude but opposite in sign to the change in momentum of ball 2. If ball 1's momentum increases by a certain amount during the collision, ball 2's momentum will *decrease* by exactly the same amount. This implies that the total momentum $P_x = p_{1x} + p_{2x}$ of the two balls is *unchanged* by the collision; that is,

$$(P_x)_f = (P_x)_i \tag{9.12}$$

Because it doesn't change during the collision, we say that the x-component of total momentum is *conserved*. Equation 9.12 is our first example of a *conservation law*.

Law of Conservation of Momentum

The same arguments just presented for the two colliding balls can be extended to systems containing any number of objects. FIGURE 9.15 shows the idea. Each pair of particles in the system (the boundary of which is denoted by the red line) interacts via forces that are an action/reaction pair. Exactly as for the two-particle collision, the change in momentum of particle 2 due to the force from particle 3 is equal in magnitude, but opposite in direction, to the change in particle 3's momentum due to particle 2. The *net* change in the momentum of these two particles due to their interaction forces is thus zero. The same argument holds for every pair, with the result that, no matter how complicated the forces between the particles, **there is no change in the** *total* **momentum $\vec{P}$ of the system.** The total momentum of the system remains constant: It is *conserved.*

Figure 9.15 showed particles interacting only with other particles inside the system. Forces that act only between particles within the system are called **internal forces.** As we've just seen, **the total momentum of a system subject to only internal forces is conserved.**

Most systems are also subject to forces from agents outside the system. These forces are called **external forces.** For example, the system consisting of a student on a skateboard is subject to three external forces—the normal force of the ground on the skateboard, the force of gravity on the student, and the force of gravity on the board. How do external forces affect the momentum of a system of particles?

In FIGURE 9.16 on the next page we show the same three-particle system of Figure 9.15, but now with *external* forces acting on the three particles. These external forces *can* change the momentum of the system. During a time interval Δt, for instance, the external force $\vec{F}_{ext\,on\,1}$ acting on particle 1 changes its momentum, according to the impulse-momentum theorem, by $\Delta \vec{p}_1 = (\vec{F}_{ext\,on\,1})\Delta t$. The

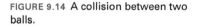

FIGURE 9.14 A collision between two balls.

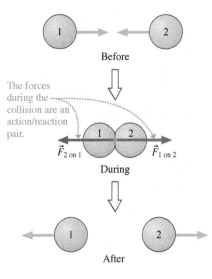

FIGURE 9.15 A system of three particles.

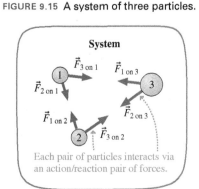

FIGURE 9.16 A system of particles subject to external forces.

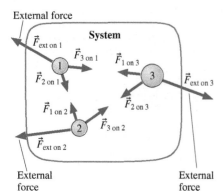

momenta of the other two particles change similarly. Thus the change in the total momentum is

$$\Delta \vec{P} = \Delta \vec{p}_1 + \Delta \vec{p}_2 + \Delta \vec{p}_3$$
$$= (\vec{F}_{\text{ext on 1}} \Delta t) + (\vec{F}_{\text{ext on 2}} \Delta t) + (\vec{F}_{\text{ext on 3}} \Delta t) \qquad (9.13)$$
$$= (\vec{F}_{\text{ext on 1}} + \vec{F}_{\text{ext on 2}} + \vec{F}_{\text{ext on 3}}) \Delta t$$
$$= \vec{F}_{\text{net}} \Delta t$$

where $\vec{F}_{\text{net}}$ is the net force due to *external forces*.

Equation 9.13 has a very important implication in the case where the net force on a system is zero: If $\vec{F}_{\text{net}} = \vec{0}$, the *total* momentum $\vec{P}$ of the system does not change. The total momentum remains constant, *regardless* of whatever interactions are going on *inside* the system.

Earlier, we found that a system's total momentum is conserved when the system has no external forces acting on it. Now we've found that the system's total momentum is also conserved when the net external force acting on it is zero. With no net external force acting that can change its momentum, we call a system with $\vec{F}_{\text{net}} = \vec{0}$ an **isolated system.**

The importance of these results is sufficient to elevate them to a law of nature, alongside Newton's laws.

> **Law of conservation of momentum** The total momentum $\vec{P}$ of an isolated system is a constant. Interactions within the system do not change the system's total momentum.

NOTE ▶ It is worth emphasizing the critical role of Newton's third law in the derivation of Equation 9.13. The law of conservation of momentum is a direct consequence of the fact that interactions within an isolated system are action/reaction pairs. ◀

Mathematically, the law of conservation of momentum for an isolated system is

$$\vec{P}_{\text{f}} = \vec{P}_{\text{i}} \qquad (9.14)$$

Law of conservation of momentum for an isolated system

The total momentum after an interaction is equal to the total momentum before the interaction. Because Equation 9.14 is a vector equation, the equality is true for each of the components of the momentum vector; that is,

$$
\begin{array}{l}
x\text{-component} \longrightarrow \overbrace{(p_{1x})_{\text{f}} + (p_{2x})_{\text{f}} + (p_{3x})_{\text{f}} + \cdots}^{\text{Final momentum}} = \overbrace{(p_{1x})_{\text{i}} + (p_{2x})_{\text{i}} + (p_{3x})_{\text{i}} + \cdots}^{\text{Initial momentum}} \\
\qquad\qquad\quad \text{Particle 1 Particle 2 Particle 3} \qquad\qquad\qquad\qquad (9.15) \\
y\text{-component} \longrightarrow (p_{1y})_{\text{f}} + (p_{2y})_{\text{f}} + (p_{3y})_{\text{f}} + \cdots = (p_{1y})_{\text{i}} + (p_{2y})_{\text{i}} + (p_{3y})_{\text{i}} + \cdots
\end{array}
$$

EXAMPLE 9.5 **Speed of ice skaters pushing off**

Two ice skaters, Sandra and David, stand facing each other on frictionless ice. Sandra has a mass of 45 kg, David a mass of 80 kg. They then push off from each other. After the push, Sandra moves off at a speed of 2.2 m/s. What is David's speed?

PREPARE The two skaters interact with each other, but they form an isolated system because, for each skater, the upward normal force of the ice balances their downward weight force to make $\vec{F}_{\text{net}} = \vec{0}$. Thus the total momentum of the system of the two skaters is conserved.

FIGURE 9.17 shows a before-and-after visual overview for the two skaters. The total momentum before they push off is $\vec{P}_i = \vec{0}$ because both skaters are at rest. Consequently, the total momentum will still be $\vec{0}$ *after* they push off.

FIGURE 9.17 Before-and-after visual overview for two skaters pushing off from each other.

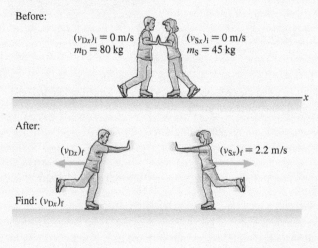

Before:

$(v_{Dx})_i = 0$ m/s
$m_D = 80$ kg

$(v_{Sx})_i = 0$ m/s
$m_S = 45$ kg

After:

$(v_{Dx})_f$

$(v_{Sx})_f = 2.2$ m/s

Find: $(v_{Dx})_f$

SOLVE Since the motion is only in the *x*-direction, we'll need to consider only *x*-components of momentum. We write Sandra's initial momentum as $(p_{Sx})_i = m_S(v_{Sx})_i$, where m_S is her mass and $(v_{Sx})_i$ her initial velocity. Similarly, we write David's initial momentum as $(p_{Dx})_i = m_D(v_{Dx})_i$. Both these momenta are zero because both skaters are initially at rest.

We can now apply the mathematical statement of momentum conservation, Equation 9.15. Writing the final momentum of Sandra as $m_S(v_{Sx})_f$ and that of David as $m_D(v_{Dx})_f$, we have

$$\underbrace{m_S(v_{Sx})_f + m_D(v_{Dx})_f}_{\text{The skaters' final momentum ...}} = \underbrace{m_S(v_{Sx})_i + m_D(v_{Dx})_i}_{\text{... equals their initial momentum ....}} = \underbrace{0}_{\substack{\text{... which} \\ \text{was zero.}}}$$

Solving for $(v_{Dx})_f$, we find

$$(v_{Dx})_f = -\frac{m_S}{m_D}(v_{Sx})_f = -\frac{45 \text{ kg}}{80 \text{ kg}} \times 2.2 \text{ m/s} = -1.2 \text{ m/s}$$

David moves backward with a *speed* of 1.2 m/s.

Notice that we didn't need to know any details about the force between David and Sandra in order to find David's final speed. Conservation of momentum *mandates* this result.

ASSESS It seems reasonable that Sandra, whose mass is less than David's, would have the greater final speed.

PROBLEM-SOLVING STRATEGY 9.1 Conservation of momentum problems (MP)

We can use the law of conservation of momentum to relate the momenta and velocities of objects *after* an interaction to their values *before* the interaction.

PREPARE Clearly define the *system*.

- If possible, choose a system that is isolated ($\vec{F}_{net} = \vec{0}$) or within which the interactions are sufficiently short and intense that you can ignore external forces for the duration of the interaction (the impulse approximation). Momentum is then conserved.
- If it's not possible to choose an isolated system, try to divide the problem into parts such that momentum is conserved during one segment of the motion. Other segments of the motion can be analyzed using Newton's laws or, as you'll learn in Chapter 10, conservation of energy.

Following Tactics Box 9.1, draw a before-and-after visual overview. Define symbols that will be used in the problem, list known values, and identify what you're trying to find.

SOLVE The mathematical representation is based on the law of conservation of momentum, Equations 9.15. Because we generally want to solve for the velocities of objects, we usually use Equations 9.15 in the equivalent form

$$m_1(v_{1x})_f + m_2(v_{2x})_f + \cdots = m_1(v_{1x})_i + m_2(v_{2x})_i + \cdots$$
$$m_1(v_{1y})_f + m_2(v_{2y})_f + \cdots = m_1(v_{1y})_i + m_2(v_{2y})_i + \cdots$$

ASSESS Check that your result has the correct units, is reasonable, and answers the question.

EXAMPLE 9.6   Getaway speed of a cart

Bob is running from the police and thinks he can make a faster getaway by jumping on a stationary cart in front of him. He runs toward the cart, jumps on, and rolls along the horizontal street. Bob has a mass of 75 kg and the cart's mass is 25 kg. If Bob's speed is 4.0 m/s when he jumps onto the cart, what is the cart's speed after Bob jumps on?

PREPARE When Bob lands on and sticks to the cart, a "collision" occurs between Bob and the cart. If we take Bob and the cart together to be the system, the forces involved in this collision—friction forces between Bob's feet and the cart—are internal forces. Because the normal force balances the weight of both Bob and the cart, the net external force on the system is zero, so the total momentum of Bob + cart is conserved: It is the same before and after the collision.

The visual overview in **FIGURE 9.18** shows the important point that Bob and the cart move together after he lands on the cart, so $(v_x)_f$ is their common final velocity.

SOLVE To solve for the final velocity of Bob and the cart, we'll use conservation of momentum: $(P_x)_f = (P_x)_i$. In terms of the individual momenta, we have

$$(P_x)_i = m_B (v_{Bx})_i + m_C \underbrace{(v_{Cx})_i}_{0 \text{ m/s}} = m_B (v_{Bx})_i$$

$$(P_x)_f = m_B (v_x)_f + m_C (v_x)_f = (m_B + m_C)(v_x)_f$$

FIGURE 9.18 Before-and-after visual overview of Bob and the cart.

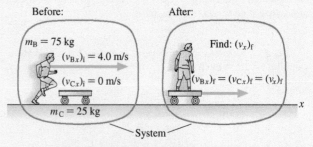

In the second equation, we've used the fact that both Bob and the cart travel at the common velocity of $(v_x)_f$. Equating the final and initial total momenta gives

$$(m_B + m_C)(v_x)_f = m_B (v_{Bx})_i$$

Solving this for $(v_x)_f$, we find

$$(v_x)_f = \frac{m_B}{m_B + m_C}(v_{Bx})_i = \frac{75 \text{ kg}}{100 \text{ kg}} \times 4.0 \text{ m/s} = 3.0 \text{ m/s}$$

The cart's speed is 3.0 m/s immediately after Bob jumps on.

ASSESS It makes sense that Bob has *lost* speed because he had to share his initial momentum with the cart. Not a good way to make a getaway!

Notice how easy this was! No forces, no kinematic equations, no simultaneous equations. Why didn't we think of this before? Although conservation laws are indeed powerful, they can answer only certain questions. Had we wanted to know how far Bob slid across the cart before sticking to it, how long the slide took, or what the cart's acceleration was during the collision, we would not have been able to answer such questions on the basis of the conservation law. There is a price to pay for finding a simple connection between before and after, and that price is the loss of information about the details of the interaction. If we are satisfied with knowing only about before and after, then conservation laws are a simple and straightforward way to proceed. But many problems *do* require us to understand the interaction, and for these there is no avoiding Newton's laws and all they entail.

It Depends on the System

The first step in the problem-solving strategy asks you to clearly define the *system*. This is worth emphasizing because many problem-solving errors arise from trying to apply momentum conservation to an inappropriate system. **The goal is to choose a system whose momentum will be conserved.** Even then, it is the *total* momentum of the system that is conserved, not the momenta of the individual particles within the system.

In Example 9.6, we chose the system to be Bob and the cart. Why this choice? Let's see what would happen if we had chosen the system to be Bob alone, as shown in **FIGURE 9.19**. As the free-body diagram shows, as Bob lands on the cart, there are three forces acting on him: the normal force $\vec{n}$ of the cart on Bob, his weight $\vec{w}$, and a friction force $\vec{f}_{C \text{ on } B}$ of the cart on Bob. This last force is subtle. We know that Bob's feet must exert a rightward-directed friction force $\vec{f}_{B \text{ on } C}$ *on the cart* as he lands; it is this friction force that causes the cart to speed up. By Newton's third law, then, the cart exerts a leftward-directed force $\vec{f}_{C \text{ on } B}$ on Bob.

The free-body diagram of Figure 9.19 then shows that there is a net force on Bob directed to the left. Thus the system consisting of Bob alone is *not* isolated, and Bob's momentum will not be conserved. Indeed, we know that Bob slows down after landing on the cart, so that his momentum clearly *decreases*.

FIGURE 9.19 An analysis of the system consisting of Bob alone.

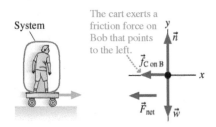

If we had chosen the cart to be the system, the unbalanced rightward force $\vec{f}_{\text{B on C}}$ of Bob on the cart would also lead to a nonzero net force. Thus the cart's momentum would not be conserved; in fact, we know it *increases* because the cart speeds up.

Only by choosing the system to be Bob and the cart *together* is the net force on the system zero and the total momentum conserved. The momentum lost by Bob is gained by the cart, so the total momentum of the two is unchanged.

Explosions

An **explosion,** where the particles of the system move apart after a brief, intense interaction, is the opposite of a collision. The explosive forces, which could be from an expanding spring or from expanding hot gases, are *internal* forces. If the system is isolated, its total momentum during the explosion will be conserved.

Video Tutor
Demo

EXAMPLE 9.7 **Recoil speed of a rifle**

A 30 g ball is fired from a 1.2 kg spring-loaded toy rifle with a speed of 15 m/s. What is the recoil speed of the rifle?

PREPARE As the ball moves down the barrel, there are complicated forces exerted on the ball and on the rifle. However, if we take the system to be the ball + rifle, these are *internal* forces that do not change the total momentum.

The *external* forces of the rifle's and ball's weights are balanced by the external force exerted by the person holding the rifle, so $\vec{F}_{\text{net}} = \vec{0}$. This is an isolated system and the law of conservation of momentum applies.

FIGURE 9.20 shows a visual overview before and after the ball is fired. We'll assume the ball is fired in the +x-direction.

SOLVE The x-component of the total momentum is $P_x = p_{Bx} + p_{Rx}$. Everything is at rest before the trigger is pulled, so the initial momentum is zero. After the trigger is pulled, the internal force of the spring pushes the ball down the barrel *and* pushes the rifle backward. Conservation of momentum gives

$$(P_x)_f = m_B(v_{Bx})_f + m_R(v_{Rx})_f = (P_x)_i = 0$$

Solving for the rifle's velocity, we find

$$(v_{Rx})_f = -\frac{m_B}{m_R}(v_{Bx})_f = -\frac{0.030 \text{ kg}}{1.2 \text{ kg}} \times 15 \text{ m/s} = -0.38 \text{ m/s}$$

FIGURE 9.20 Before-and-after visual overview for a toy rifle.

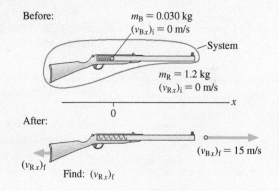

The minus sign indicates that the rifle's recoil is to the left. The recoil *speed* is 0.38 m/s.

ASSESS Real rifles fire their bullets at much higher velocities, and their recoil is correspondingly higher. Shooters need to brace themselves against the "kick" of the rifle back against their shoulder.

We would not know where to begin to solve a problem such as this using Newton's laws. But Example 9.7 is a simple problem when approached from the before-and-after perspective of a conservation law. The selection of ball + rifle as the "system" was the critical step. For momentum conservation to be a useful principle, we had to select a system in which the complicated forces due to the spring and to friction were all internal forces. The rifle by itself is *not* an isolated system, so its momentum is *not* conserved.

Much the same reasoning explains how a rocket or jet aircraft accelerates. FIGURE 9.21 shows a rocket with a parcel of fuel on board. Burning converts the fuel to hot gases that are expelled from the rocket motor. If we choose rocket + gases to be the system, then the burning and expulsion are internal forces. In deep space there are no other forces, so the total momentum of the rocket + gases system must be conserved. The rocket gains forward velocity and momentum as the exhaust gases are shot out the back, but the *total* momentum of the system remains zero.

FIGURE 9.21 Rocket propulsion is an example of conservation of momentum.

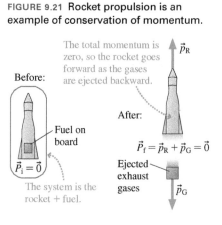

Squid propulsion BIO Squids use a form of "rocket propulsion" to make quick movements to escape enemies or catch prey. The squid draws in water through a pair of valves in its outer sheath, or mantle, and then quickly expels the water through a funnel, propelling the squid backward.

Many people find it hard to understand how a rocket can accelerate in the vacuum of space because there is nothing to "push against." Thinking in terms of momentum, you can see that the rocket does not push against anything *external*, but only against the gases that it pushes out the back. In return, in accordance with Newton's third law, the gases push forward on the rocket.

> **STOP TO THINK 9.5** An explosion in a rigid pipe shoots three balls out of its ends. A 6 g ball comes out the right end. A 4 g ball comes out the left end with twice the speed of the 6 g ball. From which end, left or right, does the third ball emerge?

9.5 Inelastic Collisions

Collisions can have different possible outcomes. A rubber ball dropped on the floor bounces—it's *elastic*—but a ball of clay sticks to the floor without bouncing; we call such a collision *inelastic*. A golf club hitting a golf ball causes the ball to rebound away from the club (elastic), but a bullet striking a block of wood becomes embedded in the block (inelastic).

A collision in which the two objects stick together and move with a common final velocity is called a **perfectly inelastic collision.** The clay hitting the floor and the bullet embedding itself in the wood are examples of perfectly inelastic collisions. Other examples include railroad cars coupling together upon impact and darts hitting a dart board. FIGURE 9.22 emphasizes the fact that the two objects have a common final velocity after they collide. (We have drawn the combined object moving to the right, but it could have ended up moving to the left, depending on the objects' masses and initial velocities.)

In other collisions, the two objects bounce apart. We've looked at some examples of these kinds of collisions, but a full analysis requires some ideas about energy. We will learn more about collisions and energy in Chapter 10.

FIGURE 9.22 A perfectly inelastic collision.

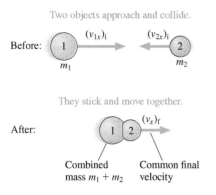

> **EXAMPLE 9.8** **A perfectly inelastic collision of railroad cars**
>
> In assembling a train from several railroad cars, two of the cars, with masses 2.0×10^4 kg and 4.0×10^4 kg, are rolled toward each other. When they meet, they couple and stick together. The lighter car has an initial speed of 1.5 m/s; the collision causes it to reverse direction at 0.25 m/s. What was the initial speed of the heavier car?
>
> **PREPARE** We model the cars as particles and define the two cars as the system. This is an isolated system, so its total momentum is conserved in the collision. The cars stick together, so this is a perfectly inelastic collision.
>
> FIGURE 9.23 Before-and-after visual overview for two train cars colliding on a track.
>
>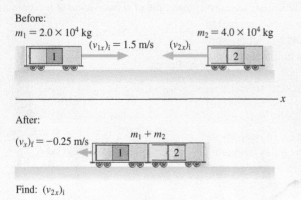
>
> Find: $(v_{2x})_i$
>
> FIGURE 9.23 shows a visual overview. We've chosen to let the 2.0×10^4 kg car (car 1) start out moving to the right, so $(v_{1x})_i$ is a positive 1.5 m/s. The cars move to the left after the collision, so their common final velocity is $(v_x)_f = -0.25$ m/s. You can see that velocity $(v_{2x})_i$ must be negative in order to "turn around" both cars.
>
> **SOLVE** The law of conservation of momentum, $(P_x)_f = (P_x)_i$, is
>
> $$(m_1 + m_2)(v_x)_f = m_1(v_{1x})_i + m_2(v_{2x})_i$$
>
> where we made use of the fact that the combined mass $m_1 + m_2$ moves together after the collision. We can easily solve for the initial velocity of the 4.0×10^4 kg car:
>
> $$(v_{2x})_i = \frac{(m_1 + m_2)(v_x)_f - m_1(v_{1x})_i}{m_2}$$
>
> $$= \frac{(6.0 \times 10^4 \text{ kg})(-0.25 \text{ m/s}) - (2.0 \times 10^4 \text{ kg})(1.5 \text{ m/s})}{4.0 \times 10^4 \text{ kg}}$$
>
> $$= -1.1 \text{ m/s}$$
>
> The negative sign, which we anticipated, indicates that the heavier car started out moving to the left. The initial *speed* of the car, which we were asked to find, is 1.1 m/s.
>
> **ASSESS** The key step in solving inelastic collision problems is that both objects move after the collision with the same velocity. You should thus choose a single symbol (here, $(v_x)_f$) for this common velocity.

STOP TO THINK 9.6 The two particles shown collide and stick together. After the collision, the combined particles

A. Move to the right as shown.
B. Move to the left.
C. Are at rest.

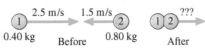

9.6 Momentum and Collisions in Two Dimensions

Our examples thus far have been confined to motion along a one-dimensional axis. Many practical examples of momentum conservation involve motion in a plane. The total momentum $\vec{P}$ is the *vector* sum of the momenta $\vec{p} = m\vec{v}$ of the individual particles. Consequently, as Equations 9.15 showed, momentum is conserved only if each component of $\vec{P}$ is conserved:

$$(p_{1x})_f + (p_{2x})_f + (p_{3x})_f + \cdots = (p_{1x})_i + (p_{2x})_i + (p_{3x})_i + \cdots$$

$$(p_{1y})_f + (p_{2y})_f + (p_{3y})_f + \cdots = (p_{1y})_i + (p_{2y})_i + (p_{3y})_i + \cdots$$

The steps of Problem-Solving Strategy 9.1 still apply for momentum conservation in two dimensions.

Collisions and explosions often involve motion in two dimensions.

EXAMPLE 9.9 **Analyzing a peregrine falcon strike** BIO

Peregrine falcons often grab their prey from above while both falcon and prey are in flight. A falcon, flying at 18 m/s, swoops down at a 45° angle from behind a pigeon flying horizontally at 9.0 m/s. The falcon has a mass of 0.80 kg and the pigeon a mass of 0.36 kg. What are the speed and direction of the falcon (now holding the pigeon) immediately after impact?

PREPARE This is a perfectly inelastic collision because after the collision the falcon and pigeon move at a common velocity. The total momentum of the falcon + pigeon system is conserved. For a two-dimensional collision, this means that the x-component of the total momentum before the collision must equal the x-component of the total momentum after the collision, and similarly for the y-components. **FIGURE 9.24** is a before-and-after visual overview.

FIGURE 9.24 Before-and-after visual overview for a falcon catching a pigeon.

Before:

$(v_{Px})_i$ $\vec{v}_F$ θ $(v_{Fy})_i$ $\vec{v}_P$ $(v_{Fx})_i$

Known
$m_F = 0.80$ kg	$m_P = 0.36$ kg $\theta = 45°$
$v_F = 18$ m/s	$v_P = 9.0$ m/s

The falcon's initial speed   The pigeon's initial speed

After:

$\vec{v}$ α $(v_y)_f$ $(v_x)_f$

Find: α, v ···· The combined final speed

SOLVE We'll start by finding the x- and y-components of the momentum before the collision. For the x-component we have

The x-component of the initial momentum . . .   (Both velocity components are negative, since they point to the left.)

$$(P_x)_i = m_F(v_{Fx})_i + m_P(v_{Px})_i = m_F(-v_F \cos\theta) + m_P(-v_P)$$

. . . equals the x-component of the initial momentum of the falcon . . .   . . . plus the x-component of the initial momentum of the pigeon.

$$= (0.80\text{ kg})(-18\text{ m/s})(\cos 45°) + (0.36\text{ kg})(-9.0\text{ m/s})$$
$$= -13.4\text{ kg}\cdot\text{m/s}$$

Similarly, for the y-component of the initial momentum we have

$$(P_y)_i = m_F(v_{Fy})_i + m_P(v_{Py})_i = m_F(-v_F \sin\theta) + 0$$
$$= (0.80\text{ kg})(-18.0\text{ m/s})(\sin 45°) = -10.2\text{ kg}\cdot\text{m/s}$$

After the collision, the two birds move with a common velocity $\vec{v}$ that is directed at an angle α from the horizontal. The x-component of the final momentum is then

$$(P_x)_f = (m_F + m_P)(v_x)_f$$

Momentum conservation requires $(P_x)_f = (P_x)_i$, so

$$(v_x)_f = \frac{(P_x)_i}{m_F + m_P} = \frac{-13.4\text{ kg}\cdot\text{m/s}}{(0.80\text{ kg}) + (0.36\text{ kg})} = -11.6\text{ m/s}$$

Similarly, $(P_y)_f = (P_y)_i$ gives

$$(v_y)_f = \frac{(P_y)_i}{m_F + m_P} = \frac{-10.2\text{ kg}\cdot\text{m/s}}{(0.80\text{ kg}) + (0.36\text{ kg})} = -8.79\text{ m/s}$$

Continued

From the figure we see that $\tan\alpha = (v_y)_f/(v_x)_f$, so that

$$\alpha = \tan^{-1}\left(\frac{(v_y)_f}{(v_x)_f}\right) = \tan^{-1}\left(\frac{-8.79 \text{ m/s}}{-11.6 \text{ m/s}}\right) = 37°$$

The magnitude of the final velocity (i.e., the speed) can be found from the Pythagorean theorem as

$$v = \sqrt{(v_x)_f^2 + (v_y)_f^2}$$
$$= \sqrt{(-11.6 \text{ m/s})^2 + (-8.79 \text{ m/s})^2} = 15 \text{ m/s}$$

Thus immediately after impact the falcon, with its meal, is moving 37° below horizontal at a speed of 15 m/s.

ASSESS It makes sense that the falcon slows down after catching the slower-moving pigeon. Also, the final angle is closer to the horizontal than the falcon's initial angle. This seems reasonable because the pigeon was initially flying horizontally, making the total momentum vector more horizontal than the direction of the falcon's initial momentum.

FIGURE 9.25 The momentum vectors of the falcon strike.

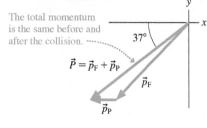

The total momentum is the same before and after the collision.

It's instructive to examine this collision with a picture of the momentum vectors. The vectors $\vec{p}_F$ and $\vec{p}_P$ before the collision, and their sum $\vec{P} = \vec{p}_F + \vec{p}_P$, are shown in FIGURE 9.25. You can see that the total momentum vector makes a 37° angle with the negative x-axis. The individual momenta change in the collision, *but the total momentum does not*.

9.7 Angular Momentum

For a single particle, we can think of the law of conservation of momentum as an alternative way of stating Newton's first law. Rather than saying that a particle will continue to move in a straight line at constant velocity unless acted on by a net force, we can say that the momentum of an isolated particle is conserved. Both express the idea that a particle moving in a straight line tends to "keep going" unless something acts on it to change its motion.

Another important motion you've studied is motion in a circle. The momentum $\vec{p}$ is *not* conserved for a particle undergoing circular motion. Momentum is a vector, and the momentum of a particle in circular motion changes as the direction of motion changes.

Nonetheless, a spinning bicycle wheel would keep turning if it were not for friction, and a ball moving in a circle at the end of a string tends to "keep going" in a circular path. The quantity that expresses this idea for circular motion is called *angular momentum*.

Let's start by looking at an example from everyday life: pushing a merry-go-round, as in FIGURE 9.26. If you push tangentially to the rim, you are applying a *torque* to the merry-go-round. As you learned in ◄ SECTION 7.5, the merry-go-round's angular speed will continue to increase for as long as you apply this torque. If you push *harder* (greater torque) or for a *longer time*, the greater the increase in its angular velocity will be. How can we quantify these observations?

Let's apply a constant torque τ_{net} to the merry-go-round for a time Δt. By how much will the merry-go-round's angular speed increase? In Section 7.5 we found that the angular acceleration α is given by the rotational equivalent of Newton's second law, or

$$\alpha = \frac{\tau_{net}}{I} \tag{9.16}$$

where I is the merry-go-round's moment of inertia.

Now the angular acceleration is the rate of change of the angular velocity, so

$$\alpha = \frac{\Delta\omega}{\Delta t} \tag{9.17}$$

Setting Equations 9.16 and 9.17 equal to each other gives

$$\frac{\Delta\omega}{\Delta t} = \frac{\tau_{net}}{I}$$

or, rearranging,

$$\tau_{net}\,\Delta t = I\,\Delta\omega \tag{9.18}$$

Class Video

FIGURE 9.26 By applying a torque to the merry-go-round, the girl is increasing its angular momentum.

If you recall the impulse-momentum theorem for *linear* motion, which is

$$\vec{F}_{net}\, \Delta t = m\, \Delta \vec{v} = \Delta \vec{p} \qquad (9.19)$$

you can see that Equation 9.18 is an analogous statement about rotational motion. Because the quantity $I\omega$ is evidently the rotational equivalent of $m\vec{v}$, the linear momentum $\vec{p}$, it seems reasonable to define the **angular momentum** L to be

$$L = I\omega \qquad (9.20)$$

Angular momentum of an object with moment
of inertia I rotating at angular velocity ω

The SI units of angular momentum are those of moment of inertia times angular velocity, or $kg \cdot m^2/s$.

Just as an object in linear motion can have a large momentum by having either a large mass or a high speed, a rotating object can have a large angular momentum by having a large moment of inertia or a large angular velocity. The merry-go-round in Figure 9.26 has a larger angular momentum if it's spinning fast than if it's spinning slowly. Also, the merry-go-round (large I) has a much larger angular momentum than a toy top (small I) spinning with the same angular velocity.

Table 9.2 summarizes the analogies between linear and rotational quantities that you learned in Chapter 7 and adds the analogy between linear momentum and angular momentum.

TABLE 9.2 Rotational and linear dynamics

Rotational dynamics	Linear dynamics
Torque τ_{net}	Force $\vec{F}_{net}$
Moment of inertia I	Mass m
Angular velocity ω	Velocity $\vec{v}$
Angular momentum $L = I\omega$	Linear momentum $\vec{p} = m\vec{v}$

Conservation of Angular Momentum

Having now defined angular momentum, we can write Equation 9.18 as

$$\tau_{net}\, \Delta t = \Delta L \qquad (9.21)$$

in exact analogy with its linear dynamics equivalent, Equation 9.19. This equation states that the change in the angular momentum of an object is proportional to the net torque applied to the object. If the net external torque on an object is *zero*, then the change in the angular momentum is zero as well. That is, a rotating object will continue to rotate with *constant* angular momentum—to "keep going"—unless acted upon by an external torque. We can state this conclusion as the *law of conservation of angular momentum*:

Video Tutor
Demo

Video Tutor
Demo

> **Law of conservation of angular momentum** The angular momentum of a rotating object subject to no net external torque ($\tau_{net} = 0$) is a constant. The final angular momentum L_f is equal to the initial angular momentum L_i.

Often we will consider a system that consists of more than one rotating object. In such a case we define the **total angular momentum** as the sum of the angular momenta of all the objects in the system. If no net external torque acts on the system, the mathematical statement of the law of conservation of angular momentum is then

Final (f) and initial (i) moment of inertia and angular velocity of object 1

$$\underbrace{(I_1)_f(\omega_1)_f} + \underbrace{(I_2)_f(\omega_2)_f} + \cdots = \underbrace{(I_1)_i(\omega_1)_i} + \underbrace{(I_2)_i(\omega_2)_i} + \cdots \qquad (9.22)$$

Final (f) and initial (i) moment of inertia and angular velocity of object 2

Recall that the moment of inertia of an object depends not only on its mass but also on how that mass is distributed within the object. If, for example, a rider on a merry-go-round moves closer to its center, her moment of inertia will decrease. In Equation 9.22 we have allowed for the possibility that the initial and final moments of inertia of an object may not be the same.

EXAMPLE 9.10 **Period of a merry-go-round**

Joey, whose mass is 36 kg, stands at the center of a 200 kg merry-go-round that is rotating once every 2.5 s. While it is rotating, Joey walks out to the edge of the merry-go-round, 2.0 m from its center. What is the rotational period of the merry-go-round when Joey gets to the edge?

PREPARE Take the system to be Joey + merry-go-round and assume frictionless bearings. There is no external torque on this system, so the angular momentum of the system will be conserved. As shown in the visual overview of FIGURE 9.27, we model the merry-go-round as a uniform disk of radius $R = 2.0$ m. From Table 7.1, the moment of inertia of a disk is $I_{disk} = \frac{1}{2}MR^2$. If we model Joey as a particle of mass m, his moment of inertia is zero when he is at the center, but it increases to mR^2 when he reaches the edge.

FIGURE 9.27 Visual overview of the merry-go-round.

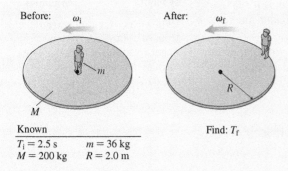

Known	
$T_i = 2.5$ s	$m = 36$ kg
$M = 200$ kg	$R = 2.0$ m

Find: T_f

SOLVE The mathematical statement of the law of conservation of angular momentum is Equation 9.22. The initial angular momentum is

$$L_i = (I_{Joey})_i(\omega_{Joey})_i + (I_{disk})_i(\omega_{disk})_i = 0 \cdot \omega_i + \frac{1}{2}MR^2\omega_i = \frac{1}{2}MR^2\omega_i$$

Here we have used the fact that both Joey and the disk have the same initial angular velocity, which we have called ω_i. Similarly, the final angular momentum is

$$L_f = (I_{Joey})_f\omega_f + (I_{disk})_f\omega_f = mR^2\omega_f + \frac{1}{2}MR^2\omega_f = \left(mR^2 + \frac{1}{2}MR^2\right)\omega_f$$

where ω_f is the common final angular velocity of both Joey and the disk.

The law of conservation of angular momentum states that $L_f = L_i$, so that

$$\left(mR^2 + \frac{1}{2}MR^2\right)\omega_f = \frac{1}{2}MR^2\omega_i$$

Canceling the R^2 terms from both sides and solving for ω_f gives

$$\omega_f = \left(\frac{M}{M + 2m}\right)\omega_i$$

The initial angular velocity is related to the initial period of rotation T_i by

$$\omega_i = \frac{2\pi}{T_i} = \frac{2\pi}{2.5 \text{ s}} = 2.51 \text{ rad/s}$$

Thus the final angular velocity is

$$\omega_f = \left(\frac{200 \text{ kg}}{200 \text{ kg} + 2(36 \text{ kg})}\right)(2.51 \text{ rad/s}) = 1.85 \text{ rad/s}$$

When Joey reaches the edge, the period of the merry-go-round has increased to

$$T_f = \frac{2\pi}{\omega_f} = \frac{2\pi}{1.85 \text{ rad/s}} = 3.4 \text{ s}$$

ASSESS The merry-go-round rotates *more slowly* after Joey moves out to the edge. This makes sense because if the system's moment of inertia increases, as it does when Joey moves out, the angular velocity must decrease to keep the angular momentum constant.

FIGURE 9.28 A spinning figure skater.

Large moment of inertia; slow spin

Small moment of inertia; fast spin

The *linear* momentum of an isolated particle is constant. Because its mass does not change, this implies that its *velocity* is constant as well. In contrast, the moment of inertia of a single rotating object *can* change, if the positions of its parts are moved relative to each other. Thus, a rotating object's angular velocity can change, even if its angular momentum is conserved, if its moment of inertia changes. For example, because no external torques act, the angular momentum of a platform diver is conserved while she's in the air. Just as for Joey and the merry-go-round of Example 9.10, she spins slowly when her moment of inertia is large; by decreasing her moment of inertia, she increases her rate of spin. Divers can thus markedly increase their spin rate by changing their body from an extended posture to a tuck position. Figure skaters also increase their spin rate by decreasing their moment of inertia, as shown in FIGURE 9.28. The following example gives a simplified treatment of this process.

EXAMPLE 9.11 **Analyzing a spinning ice skater**

An ice skater spins around on the tips of his blades while holding a 5.0 kg weight in each hand. He begins with his arms straight out from his body and his hands 140 cm apart. While spinning at 2.0 rev/s, he pulls the weights in and holds them 50 cm apart against his shoulders. If we neglect the mass of the skater, how fast is he spinning after pulling the weights in?

PREPARE There is no external torque acting on the system consisting of the skater and the weights, so their angular momentum is conserved. **FIGURE 9.29** shows a before-and-after visual overview, as seen from above.

FIGURE 9.29 Top-view visual overview of the spinning ice skater.

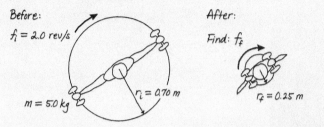

Before:
$f_i = 2.0$ rev/s
$r_i = 0.70$ m
$m = 5.0$ kg

After:
Find: f_f
$r_f = 0.25$ m

SOLVE The two weights have the same mass, move in circles with the same radius, and have the same angular velocity. Thus the total angular momentum is twice that of one weight. The mathematical statement of angular momentum conservation, $I_f \omega_f = I_i \omega_i$, is

There are two weights.

$$(2\, mr_f^2)\omega_f = (2\, mr_i^2)\omega_i$$
$$\underbrace{}_{I_f} \qquad \underbrace{}_{I_i}$$

Because the angular velocity is related to the rotation frequency f by $\omega = 2\pi f$, this equation simplifies to

$$f_f = \left(\frac{r_i}{r_f}\right)^2 f_i$$

When he pulls the weights in, his rotation frequency increases to

$$f_f = \left(\frac{0.70\ \text{m}}{0.25\ \text{m}}\right)^2 \times 2.0\ \text{rev/s} = 16\ \text{rev/s}$$

ASSESS Pulling in the weights increases the skater's spin from 2 rev/s to 16 rev/s. This is somewhat high because we neglected the mass of the skater, but it illustrates how skaters do "spin up" by pulling their mass in toward the rotation axis.

In this example, although the mass of the skater is larger than the mass of the weights, ignoring the skater's mass is not a bad approximation. Moment of inertia depends on the *square* of the distance of the mass from the axis of rotation. The skater's mass is concentrated in his torso, which has an effective radius (i.e., where most of the mass is concentrated) of only 9 or 10 cm. The weights move in much larger circles and have a disproportionate influence on his motion.

Solving either of the two preceding examples using Newton's laws would be quite difficult. We would have to deal with internal forces, such as Joey's feet against the merry-go-round, and other complications. For problems like these, where we're interested in only the before-and-after aspects of the motion, using a conservation law makes the solution much simpler.

▶ **The eye of a hurricane** As air masses from the slowly rotating outer zones are drawn toward the low-pressure center, their moment of inertia decreases. Because the angular momentum of these air masses is conserved, their speed must *increase* as they approach the center, leading to the high wind speeds near the center of the storm.

STOP TO THINK 9.7 The left figure shows two boys of equal mass standing halfway to the edge on a turntable that is freely rotating at angular speed ω_i. They then walk to the positions shown in the right figure. The final angular speed ω_f is

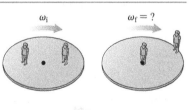

ω_i $\qquad$ $\omega_f = ?$

A. Greater than ω_i. $\qquad$ B. Less than ω_i. $\qquad$ C. Equal to ω_i.

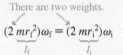

INTEGRATED EXAMPLE 9.12   Aerial firefighting

A forest fire is easiest to attack when it's just getting started. In remote locations, this often means using airplanes to rapidly deliver large quantities of water and fire suppressant to the blaze.

The "Superscooper" is an amphibious aircraft that can pick up a 6000 kg load of water by skimming over the surface of a river or lake and scooping water directly into its storage tanks. As it approaches the water's surface at a speed of 35 m/s, an empty Superscooper has a mass of 13,000 kg.

a. It takes the plane 12 s to pick up a full load of water. If we ignore the force on the plane due to the thrust of its propellers, what is its speed immediately after picking up the water?
b. What is the impulse delivered to the plane by the water?
c. What is the average force of the water on the plane?
d. The plane then flies over the fire zone at 40 m/s. It releases water by opening doors in the belly of the plane, allowing the water to fall straight down with respect to the plane. What is the plane's speed after dropping the water if it takes 5.0 s to do so?

PREPARE We can solve part a, and later part d, using conservation of momentum, following Problem-Solving Strategy 9.1. We'll need to choose the system with care, so that $\vec{F}_{net} = \vec{0}$. The plane alone is not an appropriate system for using conservation of momentum: As the plane scoops up the water, the water exerts a large external drag force on the plane, so $\vec{F}_{net}$ is definitely not zero. Instead, we should choose the plane *and* the water it is going to scoop up as the system. Then there are no external forces in the x-direction, and the net force in the y-direction is zero, since neither plane nor water accelerates appreciably in this direction during the scooping process. The complicated forces between plane and water are now *internal* forces that do not change the total momentum of the plane + water system.

With the system chosen, we follow the steps of Tactics Box 9.1 to prepare the before-and-after visual overview shown in FIGURE 9.30.

Parts b and c are impulse-and-momentum problems, so to solve them we'll use the impulse-momentum theorem, Equation 9.8. The impulse-momentum theorem considers the dynamics of a *single* object—here, the plane—subject to external forces—in this case, from the water.

FIGURE 9.30 Visual overview of the plane and water.

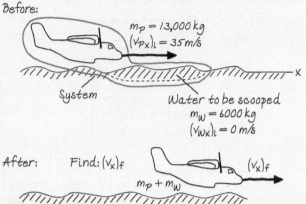

SOLVE a. The x-component of the law of conservation of momentum is

$$(P_x)_f = (P_x)_i$$

or

$$(m_P + m_W)(v_x)_f = m_P(v_{Px})_i + m_W(v_{Wx})_i = m_P(v_{Px})_i + 0$$

Here we've used the facts that the initial velocity of the water is zero and that the final situation, as in an inelastic collision, has the combined mass of the plane and water moving with the same velocity $(v_x)_f$. Solving for $(v_x)_f$, we find

$$(v_x)_f = \frac{m_P(v_{Px})_i}{m_P + m_W} = \frac{(13{,}000 \text{ kg})(35 \text{ m/s})}{(13{,}000 \text{ kg}) + (6000 \text{ kg})} = 24 \text{ m/s}$$

b. The impulse-momentum theorem is $J_x = \Delta p_x$, where $\Delta p_x = m_P \Delta v_x$ is the change in the plane's momentum. Thus

$$J_x = m_P \Delta v_x = m_P[(v_x)_f - (v_{Px})_i]$$
$$= (13{,}000 \text{ kg})(24 \text{ m/s} - 35 \text{ m/s}) = -1.4 \times 10^5 \text{ kg} \cdot \text{m/s}$$

c. From Equation 9.1, the definition of impulse, we have

$$(F_{avg})_x = \frac{J_x}{\Delta t} = \frac{-1.4 \times 10^5 \text{ kg} \cdot \text{m/s}}{12 \text{ s}} = -12{,}000 \text{ N}$$

d. Because the water drops straight down *relative to the plane*, it has the same x-component of velocity immediately after being dropped as before being dropped. That is, simply opening the doors doesn't cause the water to speed up or slow down horizontally, so the water's horizontal momentum doesn't change upon being dropped. Because the total momentum of the plane + water system is conserved, the momentum of the plane doesn't change either. The plane's speed after the drop is still 40 m/s.

ASSESS The mass of the water is nearly half that of the plane, so the significant decrease in the plane's velocity as it scoops up the water is reasonable. The force of the water on the plane is large, but is still only about 10% of the plane's weight, $mg = 130{,}000$ N, so the answer seems to be reasonable.

SUMMARY

Goal: To learn about impulse, momentum, and a new problem-solving strategy based on conservation laws.

GENERAL PRINCIPLES

Conservation Laws

When a quantity *before* an interaction is the same *after* the interaction, we say that the quantity is **conserved.**

Conservation of momentum

The total momentum $\vec{P} = \vec{p}_1 + \vec{p}_2 + \cdots$ of an **isolated system**—one on which no net force acts—is a constant. Thus

$$\vec{P}_\text{f} = \vec{P}_\text{i}$$

Conservation of angular momentum

The angular momentum L of a rotating object or system of objects subject to zero net external torque is a constant. Thus

$$L_\text{f} = L_\text{i}$$

This can be written in terms of the initial and final moments of inertia I and angular velocities ω as

$$(I_1)_\text{f}(\omega_1)_\text{f} + (I_2)_\text{f}(\omega_2)_\text{f} + \cdots = (I_1)_\text{i}(\omega_1)_\text{i} + (I_2)_\text{i}(\omega_2)_\text{i} + \cdots$$

Solving Momentum Conservation Problems

PREPARE Choose an isolated system or a system that is isolated during at least part of the problem. Draw a visual overview of the system before and after the interaction.

SOLVE Write the law of conservation of momentum in terms of vector components:

$$(p_{1x})_\text{f} + (p_{2x})_\text{f} + \cdots = (p_{1x})_\text{i} + (p_{2x})_\text{i} + \cdots$$
$$(p_{1y})_\text{f} + (p_{2y})_\text{f} + \cdots = (p_{1y})_\text{i} + (p_{2y})_\text{i} + \cdots$$

In terms of masses and velocities, this is

$$m_1(v_{1x})_\text{f} + m_2(v_{2x})_\text{f} + \cdots = m_1(v_{1x})_\text{i} + m_2(v_{2x})_\text{i} + \cdots$$
$$m_1(v_{1y})_\text{f} + m_2(v_{2y})_\text{f} + \cdots = m_1(v_{1y})_\text{i} + m_2(v_{2y})_\text{i} + \cdots$$

ASSESS Is the result reasonable?

IMPORTANT CONCEPTS

Momentum $\vec{p} = m\vec{v}$

Impulse J_x = area under force curve

Impulse and momentum are related by the **impulse-momentum theorem**

$$\Delta p_x = J_x$$

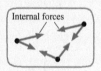

This is an alternative statement of Newton's second law.

Angular momentum $L = I\omega$ is the rotational analog of linear momentum $\vec{p} = m\vec{v}$.

System A group of interacting particles

Isolated system A system on which the net external force is zero

Internal forces

Before-and-after visual overview

- Define the system.
- Use two drawings to show the system *before* and *after* the interaction.
- List known information and identify what you are trying to find.

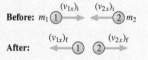

Before: m_1 ① $(v_{1x})_\text{i}$ → ← $(v_{2x})_\text{i}$ ② m_2

After: ← $(v_{1x})_\text{f}$ ① ② $(v_{2x})_\text{f}$ →

APPLICATIONS

Collisions Two or more particles come together. In a perfectly inelastic collision, they stick together and move with a common final velocity.

① → $(v_{1x})_\text{i}$ ← $(v_{2x})_\text{i}$ ②

Explosions Two or more particles move away from each other.

← $(v_{1x})_\text{f}$ ① ② $(v_{2x})_\text{f}$ →

Two dimensions Both the x- and y-components of the total momentum $\vec{P}$ must be conserved, giving two simultaneous equations.

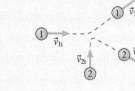

Problem difficulty is labeled as | (straightforward) to |||| (challenging). Problems labeled INT integrate significant material from earlier chapters; BIO are of biological or medical interest.

For assigned homework and other learning materials, go to
MasteringPhysics®

Scan this QR code to launch a Video Tutor Solution that will help you solve problems for this chapter.

QUESTIONS

Conceptual Questions

1. Rank in order, from largest to smallest, the momenta p_{1x} through p_{5x} of the objects presented in Figure Q9.1. Explain.

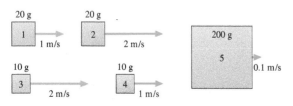

FIGURE Q9.1

2. Starting from rest, object 1 is subject to a 12 N force for 2.0 s. Object 2, with twice the mass, is subject to a 15 N force for 3.0 s. Which object has the greater final speed? Explain.
3. A 0.2 kg plastic cart and a 20 kg lead cart can roll without friction on a horizontal surface. Equal forces are used to push both carts forward for a time of 1 s, starting from rest. After the force is removed at $t = 1$ s, is the momentum of the plastic cart greater than, less than, or equal to the momentum of the lead cart? Explain.
4. Two pucks, of mass m and $4m$, lie on a frictionless table. Equal forces are used to push both pucks forward a distance of 1 m.
   a. Which puck takes longer to travel the distance? Explain.
   b. Which puck has the greater momentum upon completing the distance? Explain.
5. A stationary firecracker explodes into three pieces. One piece travels off to the east; a second travels to the north. Which of the vectors of Figure Q9.5 could be the velocity of the third piece? Explain.

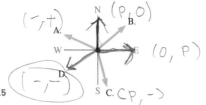

FIGURE Q9.5

6. Two students stand at rest, facing each other on frictionless skates. They then start tossing a heavy ball back and forth between them. Describe their subsequent motion.
7. A 2 kg cart rolling to the right at 3 m/s runs into a 3 kg cart rolling to the left. After the collision, both carts are stationary. What was the original speed of the 3 kg cart?
8. Automobiles are designed with "crumple zones" intended to collapse in a collision. Why would a manufacturer design part of a car so that it collapses in a collision?
9. You probably know that it feels better to catch a baseball if you are wearing a padded glove. Explain why this is so, using the ideas of momentum and impulse.

10. In the early days of rocketry, some people claimed that rockets couldn't fly in outer space as there was no air for the rockets to push against. Suppose you were an early investigator in the field of rocketry and met someone who made this argument. How would you convince the person that rockets could travel in space?

11. Two ice skaters, Megan and Jason, push off from each other on frictionless ice. Jason's mass is twice that of Megan.
   a. Which skater, if either, experiences the greater impulse during the push? Explain.
   b. Which skater, if either, has the greater speed after the push-off? Explain.
12. Suppose a rubber ball and a steel ball collide. Which, if either, receives the larger impulse? Explain.
13. While standing still on a basketball court, you throw the ball to a teammate. Why do you not move backward as a result? Is the law of conservation of momentum violated?
14. To win a prize at the county fair, you're trying to knock down a heavy bowling pin by hitting it with a thrown object. Should you choose to throw a rubber ball or a beanbag of equal size and weight? Explain.
15. Rank in order, from largest to smallest, the angular momenta L_1 through L_5 of the balls shown in Figure Q9.15. Explain.

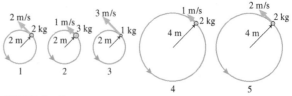

FIGURE Q9.15

16. Monica stands at the edge of a circular platform that is slowly rotating on a frictionless axle. She then walks toward the opposite edge, passing through the platform's center. Describe the motion of the platform as Monica makes her trip.
17. If the earth warms significantly, the polar ice caps will melt. Water will move from the poles, near the earth's rotation axis, and will spread out around the globe. In principle, this will change the length of the day. Why? Will the length of the day increase or decrease?

18. The disks shown in Figure Q9.18 have equal mass. Is the angular momentum of disk 2, on the right, larger than, smaller than, or equal to the angular momentum of disk 1? Explain.

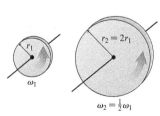

FIGURE Q9.18

Multiple-Choice Questions

19. | Curling is a sport played with 20 kg stones that slide across an ice surface. Suppose a curling stone sliding at 1 m/s strikes another, stationary stone and comes to rest in 2 ms. Approximately how much force is there on the stone during the impact?
A. 200 N   B. 1000 N   C. 2000 N   D. 10,000 N

20. | Two balls are hung from cords. The first ball, of mass 1.0 kg, is pulled to the side and released, reaching a speed of 2.0 m/s at the bottom of its arc. Then, as shown in Figure Q9.20, it hits and sticks to another ball. The speed of the pair just after the collision is 1.2 m/s. What is the mass of the second ball?
A. 0.67 kg
B. 2.0 kg
C. 1.7 kg
D. 1.0 kg

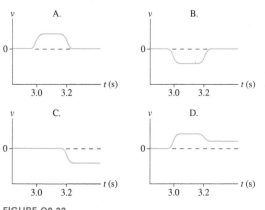

FIGURE Q9.20

21. | Figure Q9.21 shows two blocks sliding on a frictionless surface. Eventually the smaller block catches up with the larger one, collides with it, and sticks. What is the speed of the two blocks after the collision?
A. $v_i/2$ B. $4v_i/5$ C. v_i D. $5v_i/4$ E. $2v_i$

FIGURE Q9.21

22. | Two friends are sitting in a stationary canoe. At $t = 3.0$ s the person at the front tosses a sack to the person in the rear, who catches the sack 0.2 s later. Which plot in Figure Q9.22 shows the velocity of the boat as a function of time? Positive velocity is forward, negative velocity is backward. Neglect any drag force on the canoe from the water.

FIGURE Q9.22

23. ‖ Two blocks, with masses $m_1 = 2.5$ kg and $m_2 = 14$ kg, approach each other along a horizontal, frictionless track. The initial velocities of the blocks are $v_1 = 12.0$ m/s to the right and $v_2 = 3.4$ m/s to the left. The two blocks then collide and stick together. Which of the graphs could represent the force of block 1 on block 2 during the collision?

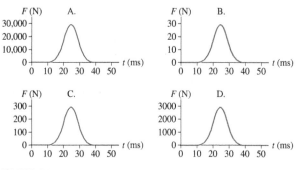

FIGURE Q9.23

24. | A small puck is sliding to the right with momentum $\vec{p}_i$ on a horizontal, frictionless surface, as shown in Figure Q9.24. A force is applied to the puck for a short time and its momentum afterward is $\vec{p}_f$. Which lettered arrow shows the direction of the impulse that was delivered to the puck?

FIGURE Q9.24

25. | A red ball, initially at rest, is simultaneously hit by a blue ball traveling from west to east at 3 m/s and a green ball traveling east to west at 3 m/s. All three balls have equal mass. Afterward, the red ball is traveling south and the green ball is moving to the east. In which direction is the blue ball traveling?
A. West
B. North
C. Between north and west
D. Between north and east
E. Between south and west

26. ‖ A 4.0-m-diameter playground merry-go-round, with a moment of inertia of 400 kg·m², is freely rotating with an angular velocity of 2.0 rad/s. Ryan, whose mass is 80 kg, runs on the ground around the outer edge of the merry-go-round in the opposite direction to its rotation. Still moving, he jumps directly onto the rim of the merry-go-round, bringing it (and himself) to a halt. How fast was Ryan running when he jumped on?
A. 2.0 m/s
B. 4.0 m/s
C. 5.0 m/s
D. 7.5 m/s
E. 10 m/s

27. ‖ A disk rotates freely on a vertical axis with an angular velocity of 30 rpm. An identical disk rotates above it in the same direction about the same axis, but without touching the lower disk, at 20 rpm. The upper disk then drops onto the lower disk. After a short time, because of friction, they rotate together. The final angular velocity of the disks is
A. 50 rpm
B. 40 rpm
C. 25 rpm
D. 20 rpm
E. 10 rpm

PROBLEMS

Section 9.1 Impulse

Section 9.2 Momentum and the Impulse-Momentum Theorem

1. | At what speed do a bicycle and its rider, with a combined mass of 100 kg, have the same momentum as a 1500 kg car traveling at 1.0 m/s?
2. | A 57 g tennis ball is served at 45 m/s. If the ball started from rest, what impulse was applied to the ball by the racket?
3. || A student throws a 120 g snowball at 7.5 m/s at the side of the schoolhouse, where it hits and sticks. What is the magnitude of the average force on the wall if the duration of the collision is 0.15 s?
4. ||| In Figure P9.4, what value of F_{max} gives an impulse of 6.0 N·s?

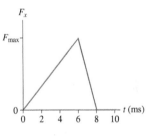

FIGURE P9.4

5. | A sled and rider, gliding over horizontal, frictionless ice at 4.0 m/s, have a combined mass of 80 kg. The sled then slides over a rough spot in the ice, slowing down to 3.0 m/s. What impulse was delivered to the sled by the friction force from the rough spot?

Section 9.3 Solving Impulse and Momentum Problems

6. || Use the impulse-momentum theorem to find how long a stone falling straight down takes to increase its speed from 5.5 m/s to 10.4 m/s.
7. || a. A 2.0 kg object is moving to the right with a speed of 1.0 m/s when it experiences the force shown in Figure P9.7a. What are the object's speed and direction after the force ends?
   b. Answer this question for the force shown in Figure P9.7b.

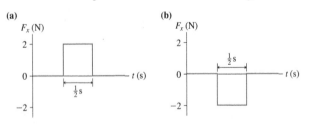

FIGURE P9.7

8. ||| A 60 g tennis ball with an initial speed of 32 m/s hits a wall and rebounds with the same speed. Figure P9.8 shows the force of the wall on the ball during the collision. What is the value of F_{max}, the maximum value of the contact force during the collision?

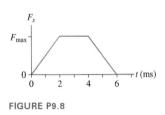

FIGURE P9.8

9. || A child is sliding on a sled at 1.5 m/s to the right. You stop the sled by pushing on it for 0.50 s in a direction opposite to its motion. If the mass of the child and sled is 35 kg, what average force do you need to apply to stop the sled? Use the concepts of impulse and momentum.
10. ||| An ice hockey puck slides along the ice at 12 m/s. A hockey stick delivers an impulse of 4.0 kg·m/s, causing the puck to move off in the opposite direction with the same speed. What is the mass of the puck?
11. | As part of a safety investigation, two 1400 kg cars traveling at 20 m/s are crashed into different barriers. Find the average forces exerted on (a) the car that hits a line of water barrels and takes 1.5 s to stop, and (b) the car that hits a concrete barrier and takes 0.10 s to stop.
12. || In a Little League baseball game, the 145 g ball reaches the batter with a speed of 15.0 m/s. The batter hits the ball, and it leaves his bat with a speed of 20.0 m/s in exactly the opposite direction.
    a. What is the magnitude of the impulse delivered by the bat to the ball?
    b. If the bat is in contact with the ball for 1.5 ms, what is the magnitude of the average force exerted by the bat on the ball?

Section 9.4 Conservation of Momentum

13. |||| A small, 100 g cart is moving at 1.20 m/s on a frictionless track when it collides with a larger, 1.00 kg cart at rest. After the collision, the small cart recoils at 0.850 m/s. What is the speed of the large cart after the collision?
14. || A man standing on very slick ice fires a rifle horizontally. The mass of the man together with the rifle is 70 kg, and the mass of the bullet is 10 g. If the bullet leaves the muzzle at a speed of 500 m/s, what is the final speed of the man?
15. ||| A 2.7 kg block of wood sits on a frictionless table. A 3.0 g bullet, fired horizontally at a speed of 500 m/s, goes completely through the block, emerging at a speed of 220 m/s. What is the speed of the block immediately after the bullet exits?
16. | A strong man is compressing a lightweight spring between two weights. One weight has a mass of 2.3 kg, the other a mass of 5.3 kg. He is holding the weights stationary, but then he loses his grip and the weights fly off in opposite directions. The lighter of the two is shot out at a speed of 6.0 m/s. What is the speed of the heavier weight?
17. || A 10,000 kg railroad car is rolling at 2.00 m/s when a 4000 kg load of gravel is suddenly dropped in. What is the car's speed just after the gravel is loaded?
18. | A 5000 kg train car, with its top open, is rolling on frictionless rails at 22.0 m/s when it starts pouring rain. A few minutes later, the car's speed is 20.0 m/s. What mass of water has collected in the car?
19. | A 55 kg hunter, standing on frictionless ice, shoots a 42 g bullet at a speed of 620 m/s. What is the recoil speed of the hunter?
20. || A 9.5 kg dog takes a nap in a canoe and wakes up to find the canoe has drifted out onto the lake but now is stationary. He walks along the length of the canoe at 0.50 m/s, relative to the water, and the canoe simultaneously moves in the opposite direction at 0.15 m/s. What is the mass of the canoe?

Section 9.5 Inelastic Collisions

21. ‖ A 300 g bird flying along at 6.0 m/s sees a 10 g insect heading straight toward it with a speed of 30 m/s. The bird opens its mouth wide and enjoys a nice lunch. What is the bird's speed immediately after swallowing?

22. ‖ A 71 kg baseball player jumps straight up to catch a hard-hit ball. If the 140 g ball is moving horizontally at 28 m/s, and the catch is made when the ballplayer is at the highest point of his leap, what is his speed immediately after stopping the ball?

23. ‖‖ A kid at the junior high cafeteria wants to propel an empty milk carton along a lunch table by hitting it with a 3.0 g spit ball. If he wants the speed of the 20 g carton just after the spit ball hits it to be 0.30 m/s, at what speed should his spit ball hit the carton?

24. ‖ The parking brake on a 2000 kg Cadillac has failed, and it is rolling slowly, at 1 mph, toward a group of small children. Seeing the situation, you realize you have just enough time to drive your 1000 kg Volkswagen head-on into the Cadillac and save the children. With what speed should you impact the Cadillac to bring it to a halt?

25. ‖ A 2.0 kg block slides along a frictionless surface at 1.0 m/s. A second block, sliding at a faster 4.0 m/s, collides with the first from behind and sticks to it. The final velocity of the combined blocks is 2.0 m/s. What was the mass of the second block?

26. ‖ Erica (36 kg) and Danny (47 kg) are bouncing on a trampoline. Just as Erica reaches the high point of her bounce, Danny is moving upward past her at 4.1 m/s. At that instant he grabs hold of her. What is their speed just after he grabs her?

Section 9.6 Momentum and Collisions in Two Dimensions

27. ‖‖ At a wild-west show, a marksman fires a bullet at a 12 g coin that's thrown straight up into the air. The marksman points his rifle at a 45° angle above the ground, then fires a 15 g bullet at a speed of 550 m/s. Just as the coin reaches its highest point, the bullet hits it and glances off, giving the coin an exactly vertical velocity of 120 m/s. At what angle measured with respect to the horizontal does the bullet ricochet away from this collision?

28. ‖‖‖ A 20 g ball of clay traveling east at 3.0 m/s collides with a 30 g ball of clay traveling north at 2.0 m/s. What are the speed and the direction of the resulting 50 g blob of clay?

29. ‖ Two particles collide and bounce apart. Figure P9.29 shows the initial momenta of both and the final momentum of particle 2. What is the final momentum of particle 1? Show your answer by copying the figure and drawing the final momentum vector on the figure.

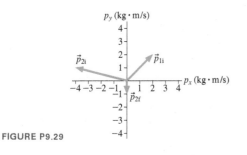

FIGURE P9.29

30. ‖ A 20 g ball of clay traveling east at 2.0 m/s collides with a 30 g ball of clay traveling 30° south of west at 1.0 m/s. What are the speed and direction of the resulting 50 g blob of clay?

31. ‖ A firecracker in a coconut blows the coconut into three pieces. Two pieces of equal mass fly off south and west, perpendicular to each other, at 20 m/s. The third piece has twice the mass of the other two. What are the speed and direction of the third piece?

Section 9.7 Angular Momentum

32. ‖‖ What is the angular momentum of the moon around the earth? The moon's mass is 7.4×10^{22} kg and it orbits 3.8×10^8 m from the earth.

33. ‖‖ A little girl is going on the merry-go-round for the first time, and wants her 47 kg mother to stand next to her on the ride, 2.6 m from the merry-go-round's center. If her mother's speed is 4.2 m/s when the ride is in motion, what is her angular momentum around the center of the merry-go-round?

34. ‖ What is the angular momentum about the axle of the 500 g rotating bar in Figure P9.34?

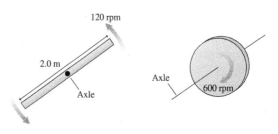

FIGURE P9.34   **FIGURE P9.35**

35. ‖‖‖ What is the angular momentum about the axle of the 2.0 kg, 4.0-cm-diameter rotating disk in Figure P9.35?

36. ‖ Divers change their body position in midair while rotating about their center of mass. In one dive, the diver leaves the board with her body nearly straight, then tucks into a somersault position. If the moment of inertia of the diver in a straight position is $14 \text{ kg} \cdot \text{m}^2$ and in a tucked position is $4.0 \text{ kg} \cdot \text{m}^2$, by what factor is her angular velocity when tucked greater than when straight?

37. ‖ Ice skaters often end their performances with spin turns, where they spin very fast about their center of mass with their arms folded in and legs together. Upon ending, their arms extend outward, proclaiming their finish. Not quite as noticeably, one leg goes out as well. Suppose that the moment of inertia of a skater with arms out and one leg extended is $3.2 \text{ kg} \cdot \text{m}^2$ and for arms and legs in is $0.80 \text{ kg} \cdot \text{m}^2$. If she starts out spinning at 5.0 rev/s, what is her angular speed (in rev/s) when her arms and one leg open outward?

General Problems

38. ‖‖ What is the impulse on a 3.0 kg particle that experiences the force described by the graph in Figure P9.38?

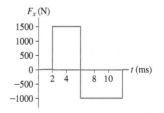

FIGURE P9.38

39. ‖| A 600 g air-track glider collides with a spring at one end of the track. Figure P9.39 shows the glider's velocity and the force exerted on the glider by the spring. How long is the glider in contact with the spring?

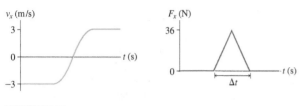

FIGURE P9.39

40. ‖ Far in space, where gravity is negligible, a 425 kg rocket traveling at 75.0 m/s in the positive x-direction fires its engines. Figure P9.40 shows the thrust force as a function of time. The mass lost by the rocket during these 30.0 s is negligible.
    a. What impulse does the engine impart to the rocket?
    b. At what time does the rocket reach its maximum speed? What is the maximum speed?

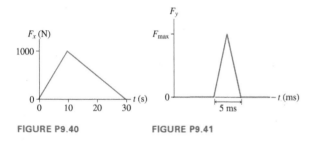

FIGURE P9.40   **FIGURE P9.41**

41. ‖‖ A 200 g ball is dropped from a height of 2.0 m, bounces on a hard floor, and rebounds to a height of 1.5 m. Figure P9.41 shows the impulse received from the floor. What maximum force does the floor exert on the ball?

42. ‖‖ A 200 g ball is dropped from a height of 2.0 m and bounces on a hard floor. The force on the ball from the floor is shown in Figure P9.42. How high does the ball rebound?

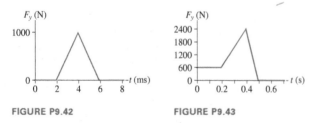

FIGURE P9.42   **FIGURE P9.43**

43. ‖| Figure P9.43 is a graph of the force exerted by the floor on a woman making a vertical jump. At what speed does she leave the ground?
    Hint: The force of the floor is not the only force acting on the woman.

44. ‖ A sled slides along a horizontal surface for which the coefficient of kinetic friction is 0.25. Its velocity at point A is 8.0 m/s and at point B is 5.0 m/s. Use the impulse-momentum theorem to find how long the sled takes to travel from A to B.

45. ‖| A 140 g baseball is moving horizontally to the right at 35 m/s when it is hit by the bat. The ball flies off to the left at 55 m/s, at an angle of 25° above the horizontal. What are the magnitude and direction of the impulse that the bat delivers to the ball?

46. ‖ Squids rely on jet propulsion, a versatile technique to move around in water. A 1.5 kg squid at rest suddenly expels 0.10 kg of water backward to quickly get itself moving forward at 3.0 m/s. If other forces (such as the drag force on the squid) are ignored, what is the speed with which the squid expels the water?

47. ‖‖ The flowers of the bunchberry plant open with astonishing force and speed, causing the pollen grains to be ejected out of the flower in a mere 0.30 ms at an acceleration of 2.5×10^4 m/s^2. If the acceleration is constant, what impulse is delivered to a pollen grain with a mass of 1.0×10^{-7} g?

48. ‖ a. With what speed are pollen grains ejected from a bunchberry flower? See Problem 47 for information.
    b. Suppose that 1000 ejected pollen grains slam into the abdomen of a 5.0 g bee that is hovering just above the flower. If the collision is perfectly inelastic, what is the bee's speed immediately afterward? Is the bee likely to notice?

49. ‖‖ A tennis player swings her 1000 g racket with a speed of 10 m/s. She hits a 60 g tennis ball that was approaching her at a speed of 20 m/s. The ball rebounds at 40 m/s.
    a. How fast is her racket moving immediately after the impact? You can ignore the interaction of the racket with her hand for the brief duration of the collision.
    b. If the tennis ball and racket are in contact for 10 ms, what is the average force that the racket exerts on the ball?

50. ‖ A 20 g ball of clay is thrown horizontally at 30 m/s toward a 1.0 kg block sitting at rest on a frictionless surface. The clay hits and sticks to the block.
    a. What is the speed of the block and clay right after the collision?
    b. Use the block's initial and final speeds to calculate the impulse the clay exerts on the block.
    c. Use the clay's initial and final speeds to calculate the impulse the block exerts on the clay.
    d. Does $\vec{J}_{block\,on\,clay} = -\vec{J}_{clay\,on\,block}$?

51. ‖ Dan is gliding on his skateboard at 4.0 m/s. He suddenly jumps backward off the skateboard, kicking the skateboard forward at 8.0 m/s. How fast is Dan going as his feet hit the ground? Dan's mass is 50 kg and the skateboard's mass is 5.0 kg.

52. ‖ James and Sarah stand on a stationary cart with frictionless wheels. The total mass of the cart and riders is 130 kg. At the same instant, James throws a 1.0 kg ball to Sarah at 4.5 m/s, while Sarah throws a 0.50 kg ball to James at 1.0 m/s. (Both speeds are measured relative to the ground.) James's throw is to the right and Sarah's is to the left.
    a. While the two balls are in the air, what are the speed and direction of the cart and its riders?
    b. After the balls are caught, what are the speed and direction of the cart and riders?

53. ‖‖ Ethan, whose mass is 80 kg, stands at one end of a very long, stationary wheeled cart that has a mass of 500 kg. He then starts sprinting toward the other end of the cart. He soon reaches his top speed of 8.0 m/s, measured relative to the cart. What is the speed of the cart when Ethan has reached his top speed?

54. ‖‖ A small cart rolls freely along the floor. As it rolls, balls of clay, each having one-fourth the mass of the cart, are dropped straight down onto the cart and stick to it. How many balls must be dropped until the cart is moving at less than one-third its initial speed?

55. ‖ Three identical train cars, coupled together, are rolling east at 2.0 m/s. A fourth car traveling east at 4.0 m/s catches up with the three and couples to make a four-car train. A moment later, the train cars hit a fifth car that was at rest on the tracks, and it couples to make a five-car train. What is the speed of the five-car train?

56. | A 110 kg linebacker running at 2.0 m/s and an 82 kg quarterback running at 3.0 m/s have a head-on collision in midair. The linebacker grabs and holds onto the quarterback. Who ends up moving forward after they hit?

57. ‖ Most geologists believe that the dinosaurs became extinct
NT 65 million years ago when a large comet or asteroid struck the earth, throwing up so much dust that the sun was blocked out for a period of many months. Suppose an asteroid with a diameter of 2.0 km and a mass of 1.0×10^{13} kg hits the earth with an impact speed of 4.0×10^4 m/s.
   a. What is the earth's recoil speed after such a collision? (Use a reference frame in which the earth was initially at rest.)
   b. What percentage is this of the earth's speed around the sun? (Use the astronomical data inside the back cover.)

58. ‖‖ Two ice skaters, with masses of 75 kg and 55 kg, stand facing
NT each other on a 15-m-wide frozen river. The skaters push off against each other, glide backward straight toward the river's edges, and reach the edges at exactly the same time. How far did the 75 kg skater glide?

59. ‖ Two ice skaters, with masses of 50 kg and 75 kg, are at the
NT center of a 60-m-diameter circular rink. The skaters push off against each other and glide to opposite edges of the rink. If the heavier skater reaches the edge in 20 s, how long does the lighter skater take to reach the edge?

60. ‖‖ One billiard ball is shot east at 2.00 m/s. A second, identical billiard ball is shot west at 1.00 m/s. The balls have a glancing collision, not a head-on collision, deflecting the second ball by 90° and sending it north at 1.41 m/s. What are the speed and direction of the first ball after the collision?

61. ‖‖‖ A 10 g bullet is fired into a 10 kg wood block that is at rest
NT on a wood table. The block, with the bullet embedded, slides 5.0 cm across the table. What was the speed of the bullet?

62. ‖‖ A typical raindrop is much more massive than a mosquito and
BIO falling much faster than a mosquito flies. How does a mosquito survive the impact? Recent research has found that the collision of a falling raindrop with a mosquito is a perfectly inelastic collision. That is, the mosquito is "swept up" by the raindrop and ends up traveling along with the raindrop. Once the relative speed between the mosquito and the raindrop is zero, the mosquito is able to detach itself from the drop and fly away.
   a. A hovering mosquito is hit by a raindrop that is 40 times as massive and falling at 8.2 m/s, a typical raindrop speed. How fast is the raindrop, with the attached mosquito, falling immediately afterward if the collision is perfectly inelastic?
   b. Because a raindrop is "soft" and deformable, the collision duration is a relatively long 8.0 ms. What is the mosquito's average acceleration, in g's, during the collision? The peak acceleration is roughly twice the value you found, but the mosquito's rigid exoskeleton allows it to survive accelerations of this magnitude. In contrast, humans cannot survive an acceleration of more than about 10g.

63. ‖‖‖ A 15 g bullet is fired at 610 m/s into a 4.0 kg block that sits
NT at the edge of a 75-cm-high table. The bullet embeds itself in the block and carries it off the table. How far from the point directly below the table's edge does the block land?

64. ‖‖‖ A 1500 kg weather rocket accelerates upward at 10.0 m/s².
NT It explodes 2.00 s after liftoff and breaks into two fragments, one twice as massive as the other. Photos reveal that the lighter fragment traveled straight up and reached a maximum height of 530 m. What were the speed and direction of the heavier fragment just after the explosion?

65. ‖ Two 500 g blocks of wood are 2.0 m apart on a frictionless table. A 10 g bullet is fired at 400 m/s toward the blocks. It passes all the way through the first block, then embeds itself in the second block. The speed of the first block immediately afterward is 6.0 m/s. What is the speed of the second block after the bullet stops?

66. | A 495 kg cannon fires a 10.0 kg cannonball with a speed of 211 m/s relative to the muzzle. The cannon is on wheels that roll without friction. When the cannon fires, what is the speed of the cannonball relative to the ground?

67. ‖‖‖ Laura, whose mass is 35 kg, jumps horizontally off a 55 kg canoe at 1.5 m/s relative to the canoe. What is the canoe's speed just after she jumps?

68. ‖ A spaceship of mass 2.0×10^6 kg is cruising at a speed of 5.0×10^6 m/s when the antimatter reactor fails, blowing the ship into three pieces. One section, having a mass of 5.0×10^5 kg, is blown straight backward with a speed of 2.0×10^6 m/s. A second piece, with mass 8.0×10^5 kg, continues forward at 1.0×10^6 m/s. What are the direction and speed of the third piece?

69. ‖‖‖ At the county fair, Chris throws a 0.15 kg baseball at a 2.0 kg wooden milk bottle, hoping to knock it off its stand and win a prize. The ball bounces straight back at 20% of its incoming speed, knocking the bottle straight forward. What is the bottle's speed, as a percentage of the ball's incoming speed?

70. ‖‖ Figure P9.70 shows a collision between three balls of clay. The three hit simultaneously and stick together. What are the speed and direction of the resulting blob of clay?

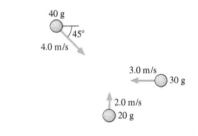

FIGURE P9.70

71. ‖‖‖ The carbon isotope ^{14}C is used for carbon dating of archeological artifacts. ^{14}C (mass 2.34×10^{-26} kg) decays by the process known as *beta decay* in which the nucleus emits an electron (the beta particle) and a subatomic particle called a neutrino. In one such decay, the electron and the neutrino are emitted at right angles to each other. The electron (mass 9.11×10^{-31} kg) has a speed of 5.00×10^7 m/s and the neutrino has a momentum of 8.00×10^{-24} kg · m/s. What is the recoil speed of the nucleus?

72. ‖‖‖ A 1.0-m-long massless rod is pivoted at one end and swings around in a circle on a frictionless table. A block with a hole through the center can slide in and out along the rod. Initially, a small piece of wax holds the block 30 cm from the pivot. The block is spun at 50 rpm, then the temperature of the rod is slowly increased. When the wax melts, the block slides out to the end of the rod. What is the final angular speed? Give your answer in rpm.

73. ‖ A 200 g puck revolves in a circle on a frictionless table at the
NT end of a 50.0-cm-long string. The puck's angular momentum about the center of the circle is 3.00 kg · m²/s. What is the tension in the string?

74. ⦀ Figure P9.74 shows a 100 g puck revolving at 100 rpm on
NT a 20-cm-radius circle on a frictionless table. A string attached
    to the puck passes through a hole in the middle of the table.
    The end of the string below the table is then slowly pulled
    down until the puck is revolving in a 10-cm-radius circle.
    How many revolutions per minute does the puck make at this
    new radius?

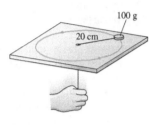

100 g

20 cm

FIGURE P9.74

75. ‖ A 2.0 kg, 20-cm-diameter turntable rotates at 100 rpm on
    frictionless bearings. Two 500 g blocks fall from above, hit the
    turntable simultaneously at opposite ends of a diagonal, and
    stick. What is the turntable's angular speed, in rpm, just after
    this event?

76. ‖ Joey, from Example 9.10, stands at rest at the outer edge of
    the frictionless merry-go-round of Figure 9.27. The merry-go-
    round is also at rest. Joey then begins to run around the perim-
    eter of the merry-go-round, finally reaching a constant speed,
    measured relative to the ground, of 5.0 m/s. What is the final
    angular speed of the merry-go-round?

77. ⦀ A 3.0-m-diameter merry-go-round with a mass of 250 kg is
    spinning at 20 rpm. John runs around the merry-go-round at
    5.0 m/s, in the same direction that it is turning, and jumps onto
    the outer edge. John's mass is 30 kg. What is the merry-go-
    round's angular speed, in rpm, after John jumps on?

78. ⦀ Disk A, with a mass of 2.0 kg and a radius of 40 cm, rotates
    clockwise about a frictionless vertical axle at 30 rev/s. Disk B, also
    2.0 kg but with a radius of 20 cm, rotates counterclockwise about
    that same axle, but at a greater height than disk A, at 30 rev/s. Disk
    B slides down the axle until it lands on top of disk A, after which
    they rotate together. After the collision, what is their common
    angular speed (in rev/s) and in which direction do they rotate?

MCAT-Style Passage Problems

Hitting a Golf Ball

Consider a golf club hitting a golf ball. To a good approximation, we can
model this as a collision between the rapidly moving head of the golf club
and the stationary golf ball, ignoring the shaft of the club and the golfer.

   A golf ball has a mass of 46 g. Suppose a 200 g club head is moving at a speed of 40 m/s just before striking the golf ball. After the
collision, the golf ball's speed is 60 m/s.

79. │ What is the momentum of the club + ball system right before
    the collision?
    A. 1.8 kg · m/s   B. 8.0 kg · m/s
    C. 3220 kg · m/s   D. 8000 kg · m/s

80. │ Immediately after the collision, the momentum of the club +
    ball system will be
    A. Less than before the collision.
    B. The same as before the collision.
    C. Greater than before the collision.

81. │ A manufacturer makes a golf ball that compresses more than
    a traditional golf ball when struck by a club. How will this affect
    the average force during the collision?
    A. The force will decrease.
    B. The force will not be affected.
    C. The force will increase.

82. │ By approximately how much does the club head slow down
    as a result of hitting the ball?
    A. 4 m/s   B. 6 m/s   C. 14 m/s   D. 26 m/s

STOP TO THINK ANSWERS

Chapter Preview Stop to Think: C. The force of the hammer on
the nail and the force of the nail on the hammer are the two members
of an action/reaction pair and thus, according to Newton's third law,
must have the same magnitude.

Stop to Think 9.1: C. Impulse equals the area under the force-
versus-time curve or, for these rectangular graphs, the maximum force
times the duration Δt. The force from the golf club in graph B has half
the maximum force of the hockey stick in graph A, but its duration
is twice as long. Hence both curves have the same area and the same
impulse. The masses of the puck and the ball are not relevant.

Stop to Think 9.2: D. We know that $\vec{p}_f = \vec{p}_i + \vec{J}$. As
shown in the figure at right, the only initial momentum
vector that satisfies this relationship is vector D.

$\vec{J}$

$\vec{p}_i$ $\vec{p}_f$

Stop to Think 9.3: F. The cart is initially moving in the negative
x-direction, so $(p_x)_i = -20$ kg · m/s. After it bounces, $(p_x)_f =$
10 kg · m/s. Thus $\Delta p = (10$ kg · m/s$) - (-20$ kg · m/s$) = 30$ kg · m/s.

Stop to Think 9.4: B. The clay ball goes from $(v_x)_i = v$ to $(v_x)_f = 0$,
so $J_{clay} = \Delta p_x = -mv$. The rubber ball rebounds, going from $(v_x)_i = v$
to $(v_x)_f = -v$ (same speed, opposite direction). Thus $J_{rubber} = \Delta p_x =$
$-2mv$. The rubber ball has a greater momentum change, and this
requires a greater impulse.

Stop to Think 9.5: Right end. The balls started at rest, so the total
momentum of the system is zero. It's an isolated system, so the total
momentum after the explosion is still zero. The 6 g ball has momentum $6v$. The 4 g ball, with velocity $-2v$, has momentum $-8v$. The
combined momentum of these two balls is $-2v$. In order for P to be
zero, the third ball must have a *positive* momentum ($+2v$) and thus
a positive velocity.

Stop to Think 9.6: B. The momentum of particle 1 is
$(0.40$ kg$)(2.5$ m/s$) = 1.0$ kg · m/s, while that of particle 2 is
$(0.80$ kg$)(-1.5$ m/s$) = -1.2$ kg · m/s. The total momentum is then
1.0 kg · m/s $- 1.2$ kg · m/s $= -0.2$ kg · m/s. Because it's negative,
the total momentum, and hence the final velocity of the particles, is
directed to the left.

Stop to Think 9.7: B. Angular momentum $L = I\omega$ is conserved. Both
boys have mass m and initially stand distance $R/2$ from the axis. Thus
the initial moment of inertia is $I_i = I_{disk} + 2 \times m(R/2)^2 = I_{disk} + \frac{1}{2}mR^2$.
The final moment of inertia is $I_f = I_{disk} + 0 + mR^2$, because the boy
standing at the axis contributes nothing to the moment of inertia.
Because $I_f > I_i$ we must have $\omega_f < \omega_i$.

10 Energy and Work

As this bungee jumper falls, he gains kinetic energy, the energy of motion. Where does this energy come from? And where does it go as he slows at the bottom of his fall?

LOOKING AHEAD ▸

Goal: To introduce the concept of energy and to learn a new problem-solving strategy based on conservation of energy.

Forms of Energy

This dolphin has lots of **kinetic energy** as it leaves the water. At its highest point its energy is mostly **potential energy.**

You'll learn about several of the most important forms of energy—kinetic, potential, and thermal.

Work and Energy

As the band is stretched, energy is *transferred* to it as **work.** This energy is then *transformed* into kinetic energy of the rock.

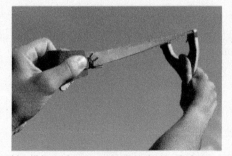

You'll learn how to calculate the work done by a force, and how this work is related to the *change* in a system's energy.

Conservation of Energy

As they slide, their potential energy decreases and their kinetic energy increases, but their total energy is unchanged: It is **conserved.**

How fast will they be moving when they reach the bottom? You'll use a new before-and-after analysis to find out.

LOOKING BACK ◂

Motion with Constant Acceleration

In Chapter 2 you learned how to describe the motion of a particle that has a constant acceleration. In this chapter, you'll use the constant-acceleration equations to connect work and energy.

A particle's final velocity is related to its initial velocity, its acceleration, and its displacement by

$$(v_x)_f^2 = (v_x)_i^2 + 2a_x \Delta x$$

STOP TO THINK

A car pulls away from a stop sign with a constant acceleration. After traveling 10 m, its speed is 5 m/s. What will its speed be after traveling 40 m?

A. 10 m/s   B. 20 m/s
C. 30 m/s   D. 40 m/s

10.1 The Basic Energy Model

Energy. It's a word you hear all the time. We use chemical energy to heat our homes and bodies, electric energy to run our lights and computers, and solar energy to grow our crops and forests. We're told to use energy wisely and not to waste it. Athletes and weary students consume "energy bars" and "energy drinks."

But just what is energy? The concept of energy has grown and changed over time, and it is not easy to define in a general way just what energy is. Rather than starting with a formal definition, we'll let the concept of energy expand slowly over the course of several chapters. In this chapter we introduce several fundamental forms of energy, including kinetic energy, potential energy, and thermal energy. Our goal is to understand the characteristics of energy, how energy is used, and, especially important, how energy is transformed from one form into another. Much of modern technology is concerned with transforming energy, such as changing the chemical energy of oil molecules into electric energy or into the kinetic energy of your car.

We'll also learn how energy can be transferred to or from a system by the application of mechanical forces. By pushing on a sled, you increase its speed, and hence its energy of motion. By lifting a heavy object, you increase its gravitational potential energy.

These observations will lead us to discover a very powerful conservation law for energy. Energy is neither created nor destroyed: If one form of energy in a system decreases, it must appear in an equal amount in another form. Many scientists consider the law of conservation of energy to be the most important of all the laws of nature. This law will have implications throughout the rest of this text.

Systems and Energy

FIGURE 10.1 A system and its energies.

A system can have many different kinds of energy.   System boundary

System

K, U_g, U_s, E_{th}, E_{chem}, $+ \cdots$

$E = K + U + E_{th} + E_{chem} + \cdots$

The total energy E is the sum of the energies present in the system.

In Chapter 9 we introduced the idea of a *system* of interacting objects. A system can be as simple as a falling acorn or as complex as a city. But whether simple or complex, every system in nature has associated with it a quantity we call its **total energy** E. The total energy is the sum of the different kinds of energies present in the system. In the table below, we give a brief overview of some of the more important forms of energy; in the rest of the chapter, we'll look at several of these forms of energy in much greater detail.

A system may have many of these kinds of energy at one time. For instance, a moving car has kinetic energy of motion, chemical energy stored in its gasoline, thermal energy in its hot engine, and many other forms of energy. FIGURE 10.1 illustrates the idea that the total energy of the system, E, is the *sum* of all the different energies present in the system:

$$E = K + U_g + U_s + E_{th} + E_{chem} + \cdots \qquad (10.1)$$

The energies shown in this sum are the forms of energy in which we'll be most interested in this and the next chapter. The ellipses ($\cdots$) stand for other forms of energy, such as nuclear or electric, that also might be present. We'll treat these and others in later chapters.

Some important forms of energy

Kinetic energy K	Gravitational potential energy U_g	Elastic or spring potential energy U_s

Kinetic energy is the energy of *motion*. All moving objects have kinetic energy. The heavier an object and the faster it moves, the more kinetic energy it has. The wrecking ball in this picture is effective in part because of its large kinetic energy.	Gravitational potential energy is *stored* energy associated with an object's *height above the ground*. As this coaster ascends, energy is stored as gravitational potential energy. As it descends, this stored energy is converted into kinetic energy.	Elastic potential energy is energy stored when a spring or other elastic object, such as this archer's bow, is *stretched*. This energy can later be transformed into the kinetic energy of the arrow.

Continued

Thermal energy E_{th}

Hot objects have more *thermal energy* than cold ones because the molecules in a hot object jiggle around more than those in a cold object. Thermal energy is the sum of the microscopic kinetic and potential energies of all the molecules in an object. In boiling water, some molecules have enough energy to escape the water as steam.

Chemical energy E_{chem}

Electric forces cause atoms to bind together to make molecules. Energy can be stored in these bonds, energy that can later be released as the bonds are rearranged during chemical reactions. When we burn fuel to run our car or eat food to power our bodies, we are using *chemical energy*.

Nuclear energy $E_{nuclear}$

An enormous amount of energy is stored in the *nucleus*, the tiny core of an atom. Certain nuclei can be made to break apart, releasing some of this *nuclear energy*, which is transformed into the kinetic energy of the fragments and then into thermal energy. The ghostly blue glow of a nuclear reactor results from high-energy fragments as they travel through water.

Energy Transformations

We've seen that all systems contain energy in many different forms. But if the amounts of each form of energy never changed, the world would be a very dull place. What makes the world interesting is that **energy of one kind can be *transformed* into energy of another kind.** The gravitational potential energy of the roller coaster at the top of the track is rapidly transformed into kinetic energy as the coaster descends; the chemical energy of gasoline is transformed into the kinetic energy of your moving car. The following table illustrates a few common energy transformations. In this table, we use an arrow $\rightarrow$ as a shorthand way of representing an energy transformation.

Some energy transformations

A weightlifter lifts a barbell over her head
The barbell has much more gravitational potential energy when high above her head than when on the floor. To lift the barbell, she transforms chemical energy in her body into gravitational potential energy of the barbell.

$$E_{chem} \rightarrow U_g$$

A base runner slides into the base
When running, he has lots of kinetic energy. After sliding, he has none. His kinetic energy is transformed mainly into thermal energy: The ground and his legs are slightly warmer.

$$K \rightarrow E_{th}$$

A burning campfire
The wood contains considerable chemical energy. When the carbon in the wood combines chemically with oxygen in the air, this chemical energy is transformed largely into thermal energy of the hot gases and embers.

$$E_{chem} \rightarrow E_{th}$$

A springboard diver
Here's a two-step energy transformation. At the instant shown, the board is flexed to its maximum extent, so that elastic potential energy is stored in the board. Soon this energy will begin to be transformed into kinetic energy; then, as the diver rises into the air and slows, this kinetic energy will be transformed into gravitational potential energy.

$$U_s \rightarrow K \rightarrow U_g$$

FIGURE 10.2 Energy transformations occur within the system.

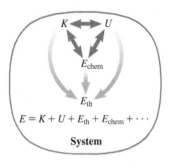

$$E = K + U + E_{\text{th}} + E_{\text{chem}} + \cdots$$

System

Class Video

FIGURE 10.3 Work and heat are energy transfers into and out of the system.

The *environment* is everything that is *not* part of the system.

Environment

Energy is transferred from the environment to the system.

Work, heat

Energy is transferred from the system to the environment.

System

FIGURE **10.2** reinforces the idea that **energy transformations are changes of energy *within* the system from one form to another.** (The *U* in this figure is a generic potential energy; it could be gravitational potential energy U_g, spring potential energy U_s, or some other form of potential energy.) Note that it is easy to convert kinetic, potential, and chemical energies into thermal energy, but converting thermal energy back into these other forms is not so easy. How it can be done, and what possible limitations there might be in doing so, will form a large part of the next chapter.

Energy Transfers and Work

We've just seen that energy *transformations* occur between forms of energy *within* a system. But every physical system also interacts with the world around it—that is, with its *environment.* In the course of these interactions, the system can exchange energy with the environment. **An exchange of energy between system and environment is called an energy *transfer.*** There are two primary energy-transfer processes: **work,** the *mechanical* transfer of energy to or from a system by pushing or pulling on it, and **heat,** the *nonmechanical* transfer of energy from the environment to the system (or vice versa) because of a temperature difference between the two.

FIGURE **10.3,** which we call the **basic energy model,** shows how our energy model is modified to include energy transfers into and out of the system as well as energy transformations within the system. In this chapter we'll consider only energy transfers by means of work; the concept of heat will be developed much further in Chapters 11 and 12.

"Work" is a common word in the English language, with many meanings. When you first think of work, you probably think of physical effort or the job you do to make a living. After all, we talk about "working out," or we say, "I just got home from work." But that is not what work means in physics.

In physics, "work" is the process of *transferring* energy from the environment to a system, or from a system to the environment, by the application of mechanical forces—pushes and pulls—to the system. Once the energy has been transferred to the system, it can appear in many forms. Exactly what form it takes depends on the details of the system and how the forces are applied. The table below gives three examples of energy transfers due to work. We use *W* as the symbol for work.

Energy transfers: work

Putting a shot

The system: The shot

The environment: The athlete

As the athlete pushes on the shot to get it moving, he is doing work on the system; that is, he is transferring energy from himself to the shot. The energy transferred to the system appears as kinetic energy.

The transfer: $W \rightarrow K$

Striking a match

The system: The match and matchbox

The environment: The hand

As the hand quickly pulls the match across the box, the hand does work on the system, increasing its thermal energy. The match head becomes hot enough to ignite.

The transfer: $W \rightarrow E_{\text{th}}$

Firing a slingshot

The system: The slingshot

The environment: The boy

As the boy pulls back on the elastic bands, he does work on the system, increasing its elastic potential energy.

The transfer: $W \rightarrow U_s$

Notice that in each example on the preceding page, the environment applies a force while the system undergoes a *displacement*. Energy is transferred as work only when the system *moves* while the force acts. A force applied to a stationary object, such as when you push against a wall, transfers no energy to the object and thus does no work.

> NOTE ▶ In the table on the preceding page, energy is being transferred *from* the athlete *to* the shot by the force of his hand. We say he "does work" on the shot, or "work is done" by the force of his hand. ◄

The Law of Conservation of Energy

Work done on a system represents energy that is transferred into or out of the system. This transferred energy *changes* the system's energy by exactly the amount of work W that was done. Writing the change in the system's energy as ΔE, we can represent this idea mathematically as

$$\Delta E = W \tag{10.2}$$

Now the total energy E of a system is, according to Equation 10.1, the sum of the different energies present in the system. Thus the change in E is the sum of the *changes* of the different energies present. Then Equation 10.2 gives what is called the *work-energy equation:*

> **The work-energy equation** The total energy of a system changes by the amount of work done on it:
>
> $$\Delta E = \Delta K + \Delta U_g + \Delta U_s + \Delta E_{th} + \Delta E_{chem} + \cdots = W \tag{10.3}$$

> NOTE ▶ Equation 10.3, the work-energy equation, is the mathematical representation of the basic energy model of Figure 10.3. Together, they are the heart of what the subject of energy is all about. ◄

Suppose now we have an **isolated system,** one that is separated from its surrounding environment in such a way that no energy is transferred into or out of the system. This means that *no work is done on the system*. The energy within the system may be transformed from one form into another, but it is a deep and remarkable fact of nature that, during these transformations, the total energy of an isolated system—the *sum* of all the individual kinds of energy—remains *constant*, as shown in FIGURE 10.4. We say that **the total energy of an isolated system is *conserved*.**

For an isolated system, we must set $W = 0$ in Equation 10.3, leading to the following *law of conservation of energy:*

> **Law of conservation of energy** The total energy of an isolated system remains constant:
>
> $$\Delta E = \Delta K + \Delta U_g + \Delta U_s + \Delta E_{th} + \Delta E_{chem} + \cdots = 0 \tag{10.4}$$

The law of conservation of energy is similar to the law of conservation of momentum. A system's momentum changes when an external force acts on it, but the total momentum of an *isolated* system doesn't change. Similarly, a system's energy changes when external forces do work on it, but the total energy of an *isolated* system doesn't change.

In solving momentum problems, we adopted a new before-and-after perspective: The momentum *after* an interaction was the same as the momentum *before* the interaction. We will introduce a similar before-and-after perspective for energy that will lead to an extremely powerful problem-solving strategy.

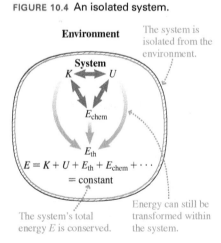

FIGURE 10.4 An isolated system.

Before using energy ideas to solve problems, however, we first need to develop quantitative expressions for work, kinetic energy, potential energy, and thermal energy. This will be our task in the next several sections.

STOP TO THINK 10.1 A child slides down a playground slide at constant speed. The energy transformation is

A. $U_g \rightarrow K$ B. $K \rightarrow U_g$ C. $W \rightarrow K$ D. $U_g \rightarrow E_{th}$ E. $K \rightarrow E_{th}$

10.2 Work

Our first task is to learn how work is calculated. We've just seen that work is the transfer of energy to or from a system by the application of forces exerted on the system by the environment. Thus work is done on a system by forces from *outside* the system; we call such forces *external forces*. Only external forces can change the energy of a system. *Internal forces*—forces between objects *within* the system—cause energy transformations within the system but don't change the system's total energy.

We also learned that in order for energy to be transferred as work, the system must undergo a displacement—it must *move*—during the time that the force is applied. Let's further investigate the relationship among work, force, and displacement.

Consider a system consisting of a windsurfer at rest, as shown on the left in FIGURE 10.5. Let's assume that there is no friction between his board and the water. Initially the system has no kinetic energy. But if a force from outside the system, such as the force due to the wind, begins to act on the system, the surfer will begin to speed up, and his kinetic energy will increase. In terms of energy transfers, we would say that the energy of the system has increased because of the work done on the system by the force of the wind.

What determines how much work is done by the force of the wind? First, we note that the greater the distance over which the wind pushes the surfer, the faster the surfer goes, and the more his kinetic energy increases. This implies a greater transfer of energy. So, **the larger the displacement, the greater the work done.** Second, if the wind pushes with a stronger force, the surfer speeds up more rapidly, and the change in his kinetic energy is greater than with a weaker force. **The stronger the force, the greater the work done.**

This experiment suggests that the amount of energy transferred to a system by a force $\vec{F}$—that is, the amount of work done by $\vec{F}$—depends on both the magnitude F of the force *and* the displacement d of the system. Many experiments of this kind have established that the amount of work done by $\vec{F}$ is *proportional* to both F and d. For the simplest case described above, where the force $\vec{F}$ is constant and points in the direction of the object's displacement, the expression for the work done is found to be

$$W = Fd \tag{10.5}$$

Work done by a constant force $\vec{F}$ in the direction of a displacement $\vec{d}$

The unit of work, that of force multiplied by distance, is N · m. This unit is so important that it has been given its own name, the **joule** (rhymes with *tool*). We define:

$$1 \text{ joule} = 1 \text{ J} = 1 \text{ N} \cdot \text{m}$$

Because work is simply energy being transferred, **the joule is the unit of *all* forms of energy.** Note that work, unlike momentum, is a *scalar* quantity—it has a magnitude but not a direction.

FIGURE 10.5 The force of the wind does work on the system, increasing its kinetic energy K.

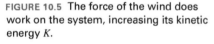

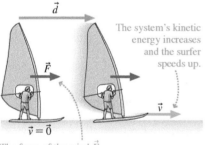

$\vec{d}$

The system's kinetic energy increases and the surfer speeds up.

$\vec{F}$

$\vec{v}$

$\vec{v} = \vec{0}$

The force of the wind $\vec{F}$ does work on the system.

EXAMPLE 10.1   **Work done in pushing a crate**

Sarah pushes a heavy crate 3.0 m along the floor at a constant speed. She pushes with a constant horizontal force of magnitude 70 N. How much work does Sarah do on the crate?

PREPARE We begin with the before-and-after visual overview in FIGURE 10.6. Sarah pushes with a constant force in the direction of the crate's motion, so we can use Equation 10.5 to find the work done.

FIGURE 10.6 Sarah pushing a crate.

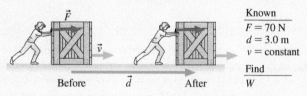

Known
$F = 70$ N
$d = 3.0$ m
$v =$ constant

Find
W

SOLVE The work done by Sarah is

$$W = Fd = (70 \text{ N})(3.0 \text{ m}) = 210 \text{ J}$$

ASSESS Work represents a transfer of energy into a system, so here the energy of the system—the box and the floor—must increase. Unlike the windsurfer, the box doesn't speed up, so its kinetic energy doesn't increase. Instead, the work increases the thermal energy in the crate and the part of the floor along which it slides, increasing the temperature of both. Using the notation of Equation 10.3, we can write this energy transfer as $\Delta E_{th} = W$.

Force at an Angle to the Displacement

A force does the greatest possible amount of work on an object when the force points in the same direction as the object's displacement. Less work is done when the force acts at an angle to the displacement. To see this, consider the kite buggy of FIGURE 10.7a, pulled along a horizontal path by the angled force of the kite string $\vec{F}$. As shown in FIGURE 10.7b, we can divide $\vec{F}$ into a component $F_\perp$ perpendicular to the motion, and a component $F_\parallel$ parallel to the motion. Only the parallel component acts to accelerate the rider and increase her kinetic energy, so only the parallel component does work on the rider. From Figure 10.7b, we see that if the angle between $\vec{F}$ and the displacement is θ, then the parallel component is $F_\parallel = F\cos\theta$. So, when the force acts at an angle θ to the direction of the displacement, we have

$$W = F_\parallel d = Fd\cos\theta \tag{10.6}$$

Work done by a constant force $\vec{F}$ at an angle θ to the displacement $\vec{d}$

Notice that this more general definition of work agrees with Equation 10.5 if $\theta = 0°$.

Tactics Box 10.1 shows how to calculate the work done by a force at any angle to the direction of motion. The system illustrated is a block sliding on a frictionless, horizontal surface, so that only the kinetic energy is changing. However, the same relationships hold for any object undergoing a displacement.

The quantities F and d are always positive, so **the sign of W is determined entirely by the angle θ between the force and the displacement.** Note that Equation 10.6, $W = Fd\cos\theta$, is valid for any angle θ. In three special cases, $\theta = 0°$, $\theta = 90°$, and $\theta = 180°$, however, there are simple versions of Equation 10.6 that you can use. These are noted in Tactics Box 10.1.

FIGURE 10.7 Finding the work done when the force is at an angle to the displacement.

(a)

(b) The rider undergoes a displacement $\vec{d}$.

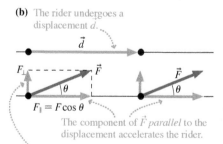

The component of $\vec{F}$ *parallel* to the displacement accelerates the rider.

The component of $\vec{F}$ *perpendicular* to the displacement only pulls up on the rider. It doesn't accelerate her.

TACTICS BOX 10.1 Calculating the work done by a constant force

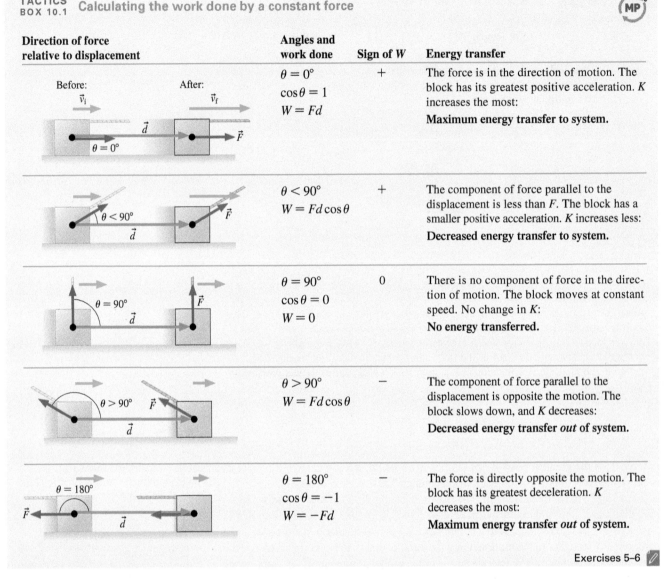

Direction of force relative to displacement	Angles and work done	Sign of W	Energy transfer
Before: $\vec{v}_i$ After: $\vec{v}_f$ $\theta = 0°$	$\theta = 0°$ $\cos\theta = 1$ $W = Fd$	+	The force is in the direction of motion. The block has its greatest positive acceleration. K increases the most: **Maximum energy transfer to system.**
$\theta < 90°$	$\theta < 90°$ $W = Fd\cos\theta$	+	The component of force parallel to the displacement is less than F. The block has a smaller positive acceleration. K increases less: **Decreased energy transfer to system.**
$\theta = 90°$	$\theta = 90°$ $\cos\theta = 0$ $W = 0$	0	There is no component of force in the direction of motion. The block moves at constant speed. No change in K: **No energy transferred.**
$\theta > 90°$	$\theta > 90°$ $W = Fd\cos\theta$	−	The component of force parallel to the displacement is opposite the motion. The block slows down, and K decreases: **Decreased energy transfer *out* of system.**
$\theta = 180°$	$\theta = 180°$ $\cos\theta = -1$ $W = -Fd$	−	The force is directly opposite the motion. The block has its greatest deceleration. K decreases the most: **Maximum energy transfer *out* of system.**

Exercises 5–6

EXAMPLE 10.2 Work done in pulling a suitcase

A strap inclined upward at a 45° angle pulls a suitcase through the airport. The tension in the strap is 20 N. How much work does the tension do if the suitcase is pulled 100 m at a constant speed?

PREPARE FIGURE 10.8 shows a visual overview. Since the suitcase moves at a constant speed, there must be a rolling friction force (not shown) acting to the left.

SOLVE We can use Equation 10.6, with force $F = T$, to find that the tension does work

$$W = Td\cos\theta = (20 \text{ N})(100 \text{ m})\cos 45° = 1400 \text{ J}$$

ASSESS Because a person is pulling on the other end of the strap, causing the tension, we would say informally that the person does 1400 J of work on the suitcase. This work represents

FIGURE 10.8 A suitcase pulled by a strap.

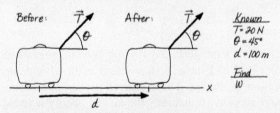

energy transferred into the suitcase + floor system. Since the suitcase moves at a constant speed, the system's kinetic energy doesn't change. Thus, just as for Sarah pushing the crate in Example 10.1, the work done goes entirely into increasing the thermal energy E_{th} of the suitcase and the floor.

CONCEPTUAL EXAMPLE 10.3 Work done by a parachute

A drag racer is slowed by a parachute. What is the sign of the work done?

REASON The drag force on the drag racer is shown in FIGURE 10.9, along with the dragster's displacement as it slows. The force points in the direction opposite the displacement, so the angle θ in

FIGURE 10.9 The force acting on a drag racer.

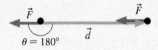

Equation 10.6 is 180°. Then $\cos\theta = \cos(180°) = -1$. Because F and d in Equation 10.6 are magnitudes, and hence positive, the work $W = Fd\cos\theta = -Fd$ done by the drag force is *negative*.

ASSESS Applying Equation 10.3 to this situation, we have

$$\Delta K = W$$

because the only system energy that changes is the racer's kinetic energy K. Because the kinetic energy is decreasing, its change ΔK is negative. This agrees with the sign of W. This example illustrates the general principle that **negative work represents a transfer of energy out of the system.**

If several forces act on an object that undergoes a displacement, each does work on the object. The **total** (or **net**) **work** W_{total} is the sum of the work done by each force. The total work represents the total energy transfer *to* the system from the environment (if $W_{total} > 0$) or *from* the system to the environment (if $W_{total} < 0$).

Forces That Do No Work

The fact that a force acts on an object doesn't mean that the force will do work on the object. The table below shows three common cases where a force does no work.

Forces that do no work

 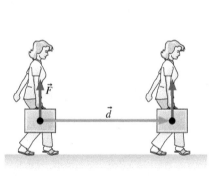 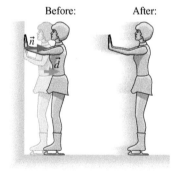

If the object undergoes no displacement while the force acts, no work is done.

This can sometimes seem counterintuitive. The weightlifter struggles mightily to hold the barbell over his head. But during the time the barbell remains stationary, he does no work on it because its displacement is zero. Why then is it so hard for him to hold it there? We'll see in Chapter 11 that it takes a rapid conversion of his internal chemical energy to keep his arms extended under this great load.

A force perpendicular to the displacement does no work.

The woman exerts only a vertical force on the briefcase she's carrying. This force has no component in the direction of the displacement, so the briefcase moves at a constant velocity and its kinetic energy remains constant. Since the energy of the briefcase doesn't change, it must be that no energy is being transferred to it as work. (This is the case where $\theta = 90°$ in Tactics Box 10.1.)

If the part of the object on which the force acts undergoes no displacement, no work is done.

Even though the wall pushes on the skater with a normal force $\vec{n}$ and she undergoes a displacement $\vec{d}$, the wall does no work on her, because the point of her body on which $\vec{n}$ acts—her hands—undergoes no displacement. This makes sense: How could energy be transferred as work from an inert, stationary object? So where does her kinetic energy come from? This will be the subject of much of Chapter 11. Can you guess?

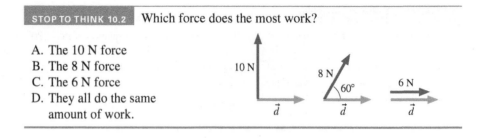

STOP TO THINK 10.2   Which force does the most work?

A. The 10 N force
B. The 8 N force
C. The 6 N force
D. They all do the same
      amount of work.

10.3 Kinetic Energy

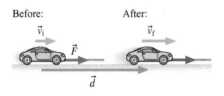

FIGURE 10.10 The work done by the tow rope increases the car's kinetic energy.

Before: After:

We've already qualitatively discussed kinetic energy, an object's energy of motion. Let's now use what we've learned about work, and some simple kinematics, to find a quantitative expression for kinetic energy. Consider a car being pulled by a tow rope, as in FIGURE 10.10. The rope pulls with a constant force $\vec{F}$ while the car undergoes a displacement $\vec{d}$, so the force does work $W = Fd$ on the car. If we ignore friction and drag, the work done by $\vec{F}$ is transferred entirely into the car's energy of motion—its kinetic energy. In this case, the change in the car's kinetic energy is given by the work-energy equation, Equation 10.3, as

$$W = \Delta K = K_f - K_i \qquad (10.7)$$

Using kinematics, we can find another expression for the work done, in terms of the car's initial and final speeds. Recall from ◀ SECTION 2.5 the kinematic equation

$$v_f^2 = v_i^2 + 2a\,\Delta x$$

Applied to the motion of our car, $\Delta x = d$ is the car's displacement and, from Newton's second law, the acceleration is $a = F/m$. Thus we can write

$$v_f^2 = v_i^2 + \frac{2Fd}{m} = v_i^2 + \frac{2W}{m}$$

where we have replaced Fd with the work W. If we now solve for the work, we find

$$W = \frac{1}{2}m\left(v_f^2 - v_i^2\right) = \frac{1}{2}mv_f^2 - \frac{1}{2}mv_i^2$$

If we compare this result with Equation 10.7, we see that

$$K_f = \frac{1}{2}mv_f^2 \qquad \text{and} \qquad K_i = \frac{1}{2}mv_i^2$$

In general, then, an object of mass m moving with speed v has kinetic energy

$$K = \frac{1}{2}mv^2 \qquad (10.8)$$

Kinetic energy of an object of mass m moving with speed v

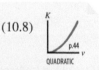

TABLE 10.1 Some approximate kinetic energies

Object	Kinetic energy
Ant walking	1×10^{-8} J
Penny dropped 1 m	2.5×10^{-3} J
Person walking	70 J
Fastball, 100 mph	150 J
Bullet	5000 J
Car, 60 mph	5×10^5 J
Supertanker, 20 mph	2×10^{10} J

From Equation 10.8, the units of kinetic energy are those of mass times speed squared, or $\text{kg} \cdot (\text{m/s})^2$. But

$$1\ \text{kg} \cdot (\text{m/s})^2 = \underbrace{1\ \text{kg} \cdot (\text{m/s}^2)}_{1\,\text{N}} \cdot \text{m} = 1\ \text{N} \cdot \text{m} = 1\ \text{J}$$

We see that the units of kinetic energy are the same as those of work, as they must be. Table 10.1 gives some approximate kinetic energies. Everyday kinetic energies range from a tiny fraction of a fraction of a joule to nearly a million joules for a speeding car.

CONCEPTUAL EXAMPLE 10.4   Kinetic energy changes for a car

Compare the increase in a 1000 kg car's kinetic energy as it speeds up by 5.0 m/s, starting from 5.0 m/s, to its increase in kinetic energy as it speeds up by 5.0 m/s, starting from 10 m/s.

REASON The change in the car's kinetic energy in going from 5.0 m/s to 10 m/s is

$$\Delta K_{5 \to 10} = \frac{1}{2}mv_f^2 - \frac{1}{2}mv_i^2$$

This gives

$$\Delta K_{5 \to 10} = \frac{1}{2}(1000 \text{ kg})(10 \text{ m/s})^2 - \frac{1}{2}(1000 \text{ kg})(5.0 \text{ m/s})^2$$

$$= 3.8 \times 10^4 \text{ J}$$

Similarly, increasing from 10 m/s to 15 m/s requires

$$\Delta K_{10 \to 15} = \frac{1}{2}(1000 \text{ kg})(15 \text{ m/s})^2 - \frac{1}{2}(1000 \text{ kg})(10 \text{ m/s})^2$$

$$= 6.3 \times 10^4 \text{ J}$$

Even though the increase in the car's *speed* is the same in both cases, the increase in kinetic energy is substantially greater in the second case.

ASSESS Kinetic energy depends on the *square* of the speed v. In FIGURE 10.11, which plots kinetic energy versus speed, we see that the energy of the car increases rapidly with speed. We can also see graphically why the change in K for a 5 m/s change in v is greater at high speeds than at low speeds. In part this is why it's harder to accelerate your car at high speeds than at low speeds.

FIGURE 10.11 The kinetic energy increases as the *square* of the speed.

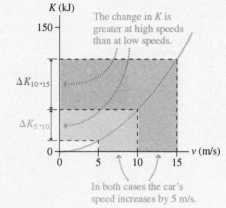

EXAMPLE 10.5   Speed of a bobsled after pushing

A two-man bobsled has a mass of 390 kg. Starting from rest, the two racers push the sled for the first 50 m with a net force of 270 N. Neglecting friction, what is the sled's speed at the end of the 50 m?

PREPARE Because friction is negligible, there is no change in the sled's thermal energy. And, because the sled's height is constant, its gravitational potential energy is unchanged as well. Thus the work-energy equation is simply $\Delta K = W$. We can therefore find the sled's final kinetic energy, and hence its speed, by finding the work done by the racers as they push on the sled. FIGURE 10.12 lists the known quantities and the quantity v_f that we want to find.

FIGURE 10.12 The work done by the pushers increases the sled's kinetic energy.

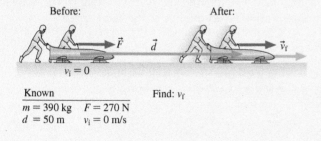

Known	Find: v_f
$m = 390$ kg $F = 270$ N	
$d = 50$ m $v_i = 0$ m/s	

SOLVE From the work-energy equation, Equation 10.3, the change in the sled's kinetic energy is $\Delta K = K_f - K_i = W$. The sled's final kinetic energy is thus

$$K_f = K_i + W$$

Using our expressions for kinetic energy and work, we get

$$\frac{1}{2}mv_f^2 = \frac{1}{2}mv_i^2 + Fd$$

Because $v_i = 0$, the work-energy equation reduces to

$$\frac{1}{2}mv_f^2 = Fd$$

We can solve for the final speed to get

$$v_f = \sqrt{\frac{2Fd}{m}} = \sqrt{\frac{2(270 \text{ N})(50 \text{ m})}{390 \text{ kg}}} = 8.3 \text{ m/s}$$

ASSESS 8.3 m/s, about 18 mph, seems a reasonable speed for two fast pushers to attain.

STOP TO THINK 10.3   Rank in order, from greatest to least, the kinetic energies of the sliding pucks.

FIGURE 10.13 Rotational kinetic energy is due to the circular motion of the particles.

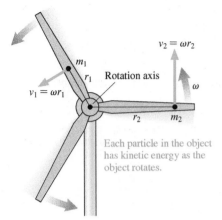

Each particle in the object has kinetic energy as the object rotates.

Rotational Kinetic Energy

We've just found an expression for the kinetic energy of an object moving along a line or some other path. This energy is called **translational kinetic energy.** Consider now an object rotating about a fixed axis, such as a windmill blade. Although the blade has no overall translational motion, each particle in the blade is moving and hence has kinetic energy. Adding up the kinetic energy for all the particles that make up the blade, we find that the blade has **rotational kinetic energy,** the kinetic energy due to rotation.

FIGURE 10.13 shows two of the particles making up a windmill blade that rotates with angular velocity ω. Recall from ◀ SECTION 7.1 that a particle moving with angular velocity ω in a circle of radius r has a speed $v = \omega r$. Thus particle 1, which rotates in a circle of radius r_1, moves with speed $v_1 = r_1\omega$ and so has kinetic energy $\frac{1}{2}m_1v_1^2 = \frac{1}{2}m_1r_1^2\omega^2$. Similarly, particle 2, which rotates in a circle with a larger radius r_2, has kinetic energy $\frac{1}{2}m_2r_2^2\omega^2$. The object's rotational kinetic energy is the sum of the kinetic energies of *all* the particles:

$$K_{rot} = \frac{1}{2}m_1r_1^2\omega^2 + \frac{1}{2}m_2r_2^2\omega^2 + \cdots = \frac{1}{2}\left(\sum mr^2\right)\omega^2$$

You will recognize the term in parentheses as our old friend, the moment of inertia I. Thus the rotational kinetic energy is

$$K_{rot} = \frac{1}{2}I\omega^2 \qquad (10.9)$$

Rotational kinetic energy of an object with moment of inertia I and angular velocity ω

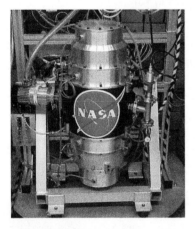

NOTE ▶ Rotational kinetic energy is *not* a new form of energy. It is the ordinary kinetic energy of motion, only now expressed in a form that is especially convenient for rotational motion. Comparison with the familiar $\frac{1}{2}mv^2$ shows again that the moment of inertia I is the rotational equivalent of mass. ◀

A rolling object, such as a wheel, is undergoing both rotational *and* translational motions. Consequently, its total kinetic energy is the sum of its rotational and translational kinetic energies:

$$K = K_{trans} + K_{rot} = \frac{1}{2}mv^2 + \frac{1}{2}I\omega^2 \qquad (10.10)$$

This illustrates an important fact: **The kinetic energy of a rolling object is always greater than that of a nonrotating object moving at the same speed.**

◀ **Rotational recharge** A promising new technology would replace spacecraft batteries that need periodic and costly replacement with a *flywheel*—a cylinder rotating at a very high angular speed. Energy from solar panels is used to speed up the flywheel, which stores energy as rotational kinetic energy that can then be converted back into electric energy as needed.

EXAMPLE 10.6 | **Kinetic energy of a bicycle**

Bike 1 has a 10.0 kg frame and 1.00 kg wheels; bike 2 has a 9.00 kg frame and 1.50 kg wheels. Both bikes thus have the same 12.0 kg total mass. What is the kinetic energy of each bike when they are ridden at 12.0 m/s? Model each wheel as a hoop of radius 35.0 cm.

PREPARE Each bike's frame has only translational kinetic energy $K_{frame} = \frac{1}{2}mv^2$, where m is the mass of the frame. The kinetic energy of each rolling wheel is given by Equation 10.10. From Table 7.1, we find that I for a hoop is MR^2, where M is the mass of one wheel.

SOLVE From Equation 10.10 the kinetic energy of each rolling wheel is

$$K_{wheel} = \frac{1}{2}Mv^2 + \frac{1}{2}I\omega^2 = \frac{1}{2}Mv^2 + \frac{1}{2}\underbrace{(MR^2)}_{I}\underbrace{\left(\frac{v}{R}\right)^2}_{\omega^2} = Mv^2$$

Then the total kinetic energy of a bike is

$$K = K_{frame} + 2K_{wheel} = \frac{1}{2}mv^2 + 2Mv^2$$

The factor of 2 in the second term occurs because each bike has two wheels. Thus the kinetic energies of the two bikes are

$$K_1 = \frac{1}{2}(10.0\ \text{kg})(12.0\ \text{m/s})^2 + 2(1.00\ \text{kg})(12.0\ \text{m/s})^2$$
$$= 1010\ \text{J}$$

$$K_2 = \frac{1}{2}(9.00\ \text{kg})(12.0\ \text{m/s})^2 + 2(1.50\ \text{kg})(12.0\ \text{m/s})^2$$
$$= 1080\ \text{J}$$

The kinetic energy of bike 2 is about 7% higher than that of bike 1. Note that the radius of the wheels was not needed in this calculation.

ASSESS As the cyclists on these bikes accelerate from rest to 12 m/s, they must convert some of their internal chemical energy into the kinetic energy of the bikes. Racing cyclists want to use as little of their own energy as possible. Although both bikes have the same total mass, the one with the lighter wheels will take less energy to get it moving. Shaving a little extra weight off your wheels is more useful than taking that same weight off your frame.

It's important that racing bike wheels are as light as possible.

10.4 Potential Energy

When two or more objects in a system interact, it is sometimes possible to *store* energy in the system in a way that the energy can be easily recovered. For instance, the earth and a ball interact by the gravitational force between them. If the ball is lifted up into the air, energy is stored in the ball + earth system, energy that can later be recovered as kinetic energy when the ball is released and falls. Similarly, a spring is a system made up of countless atoms that interact via their atomic "springs." If we push a box against a spring, energy is stored that can be recovered when the spring later pushes the box across the table. This sort of stored energy is called **potential energy,** since it has the *potential* to be converted into other forms of energy, such as kinetic or thermal energy.

The forces due to gravity and springs are special in that they allow for the storage of energy. Other interaction forces do not. When a crate is pushed across the floor, the crate and the floor interact via the force of friction, and the work done on the system is converted into thermal energy. But this energy is *not* stored up for later recovery—it slowly diffuses into the environment and cannot be recovered.

Interaction forces that can store useful energy are called **conservative forces.** The name comes from the important fact that, as we'll see, the mechanical energy of a system is *conserved* when only conservative forces act. Gravity and elastic forces are conservative forces, and later we'll find that the electric force is a conservative force as well. Friction, on the other hand, is a **nonconservative force.** When two objects interact via a friction force, energy is not stored. It is usually transformed into thermal energy.

Let's look more closely at the potential energies associated with the two conservative forces—gravity and springs—that we'll study in this chapter.

Gravitational Potential Energy

To find an expression for gravitational potential energy, let's consider the system of the book and the earth shown in **FIGURE 10.14a**. The book is lifted at a constant speed from its initial position at y_i to a final height y_f. The lifting force of the hand is external to the system and so does work W on the system, increasing its energy. The book is lifted at a constant speed, so its kinetic energy doesn't change. Because there's no friction, the book's thermal energy doesn't change either. Thus the work done goes entirely into increasing the gravitational potential energy of the system. According to Equation 10.3, the work-energy equation, this can be written as $\Delta U_g = W$. Because $\Delta U_g = (U_g)_f - (U_g)_i$, Equation 10.3 can be written

$$(U_g)_f = (U_g)_i + W \tag{10.11}$$

The work done is $W = Fd$, where $d = \Delta y = y_f - y_i$ is the vertical distance that the book is lifted. From the free-body diagram of **FIGURE 10.14b**, we see that $F = mg$. Thus $W = mg\,\Delta y$, and so

$$(U_g)_f = (U_g)_i + mg\Delta y \tag{10.12}$$

FIGURE 10.14 Lifting a book increases the system's gravitational potential energy.

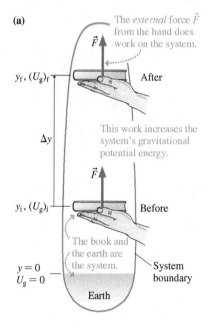

(a)

The *external* force $\vec{F}$ from the hand does work on the system.

$y_f, (U_g)_f$ After

This work increases the system's gravitational potential energy.

Δy

$y_i, (U_g)_i$ Before

The book and the earth are the system.

$y = 0$
$U_g = 0$

System boundary

Earth

(b) Because the book is being lifted at a constant speed, it is in dynamic equilibrium with $\vec{F}_{net} = \vec{0}$. Thus $F = w = mg$.

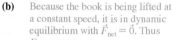

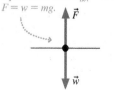

Because our final height was greater than our initial height, Δy is positive and $(U_g)_f > (U_g)_i$. **The higher the object is lifted, the greater the gravitational potential energy in the object + earth system.**

Equation 10.12 gives the final gravitational potential energy $(U_g)_f$ in terms of its initial value $(U_g)_i$. But what is the value of $(U_g)_i$? We can gain some insight by writing Equation 10.12 in terms of energy *changes:*

$$(U_g)_f - (U_g)_i = \Delta U_g = mg\Delta y$$

For example, if we lift a 1.5 kg book up by $\Delta y = 2.0$ m, we increase the system's gravitational potential energy by $\Delta U_g = (1.5 \text{ kg})(9.8 \text{ m/s}^2)(2.0 \text{ m}) = 29.4$ J. This increase is *independent* of the book's starting height: The gravitational potential energy increases by 29.4 J whether we lift the book 2.0 m starting at sea level or starting at the top of the Washington Monument. This illustrates an important general fact about *every* form of potential energy: **Only *changes* in potential energy are significant.**

Because of this fact, we are free to choose a *reference level* where we define U_g to be zero. Our expression for U_g is particularly simple if we choose this reference level to be at $y = 0$. We then have

$$U_g = mgy \qquad (10.13)$$

Gravitational potential energy of an object of mass m at height y
(assuming $U_g = 0$ when the object is at $y = 0$)

EXAMPLE 10.7 **Racing up a skyscraper**

In the Empire State Building Run-Up, competitors race up the 1576 steps of the Empire State Building, climbing a total vertical distance of 320 m. How much gravitational potential energy does a 70 kg racer gain during this race?

Racers head up the staircase in the Empire State Building Run-Up.

PREPARE We choose $y = 0$ m and hence $U_g = 0$ J at the ground floor of the building.

SOLVE At the top, the racer's gravitational potential energy is

$$U_g = mgy = (70 \text{ kg})(9.8 \text{ m/s}^2)(320 \text{ m}) = 2.2 \times 10^5 \text{ J}$$

Because the racer's gravitational potential energy was 0 J at the ground floor, the change in his potential energy is 2.2×10^5 J.

ASSESS This is a large amount of energy. According to Table 10.1, it's comparable to the energy of a speeding car. But if you think how hard it would be to climb the Empire State Building, it seems like a plausible result.

FIGURE 10.15 The hiker's gravitational potential energy depends only on his height above the $y = 0$ m reference level.

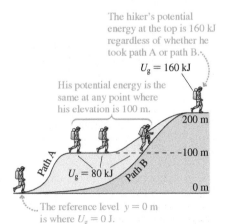

The hiker's potential energy at the top is 160 kJ regardless of whether he took path A or path B.

$U_g = 160$ kJ

His potential energy is the same at any point where his elevation is 100 m.

200 m

100 m

Path A   Path B

$U_g = 80$ kJ

0 m

The reference level $y = 0$ m is where $U_g = 0$ J.

An important conclusion from Equation 10.13 is that gravitational potential energy depends only on the height of the object above the reference level $y = 0$, not on the object's horizontal position. To understand why, consider carrying a briefcase while walking on level ground at a constant speed. As shown in the table on page 291, the vertical force of your hand on the briefcase is *perpendicular* to the displacement. No work is done on the briefcase, so its gravitational potential energy remains constant as long as its height above the ground doesn't change.

This idea can be applied to more complicated cases, such as the 82 kg hiker in **FIGURE 10.15**. His gravitational potential energy depends *only* on his height y above the reference level. Along path A, it's the same value $U_g = mgy = 80$ kJ at any point where he is at height $y = 100$ m above the reference level. If he had instead taken path B, his gravitational potential energy at $y = 100$ m would be the same 80 kJ. It doesn't matter *how* he gets to the 100 m elevation; his potential energy at that height is always the same. **Gravitational potential energy depends only on the *height* of an object and not on the path the object took to get to that position.** This fact will allow us to use the law of conservation of energy to easily solve a variety of problems that would be very difficult to solve using Newton's laws alone.

STOP TO THINK 10.4 Rank in order, from largest to smallest, the gravitational potential energies of identical balls 1 through 4.

Elastic Potential Energy

Energy can also be stored in a compressed or extended spring as **elastic** (or **spring**) **potential energy** U_s. We can find out how much energy is stored in a spring by using an external force to slowly compress the spring. This external force does work on the spring, transferring energy to the spring. Since only the elastic potential energy of the spring is changing, Equation 10.3 reads

$$\Delta U_s = W \qquad (10.14)$$

That is, we can find out how much elastic potential energy is stored in the spring by calculating the amount of work needed to compress the spring.

FIGURE 10.16 shows a spring being compressed by a hand. In ◀ SECTION 8.3 we found that the force the spring exerts on the hand is $F_s = -k\,\Delta x$ (Hooke's law), where Δx is the displacement of the end of the spring from its equilibrium position and k is the spring constant. In Figure 10.16 we have set the origin of our coordinate system at the equilibrium position. The displacement from equilibrium Δx is therefore equal to x, and the spring force is then $-kx$. By Newton's third law, the force that the hand exerts on the spring is thus $F = +kx$.

As the hand pushes the end of the spring from its equilibrium position to a final position x, the applied force increases from 0 to kx. This is not a constant force, so we can't use Equation 10.5, $W = Fd$, to find the work done. However, it seems reasonable to calculate the work by using the *average* force in Equation 10.5. Because the force varies from $F_i = 0$ to $F_f = kx$, the average force used to compress the spring is $F_{avg} = \frac{1}{2}kx$. Thus the work done by the hand is

$$W = F_{avg}d = F_{avg}x = \left(\frac{1}{2}kx\right)x = \frac{1}{2}kx^2$$

This work is stored as potential energy in the spring, so we can use Equation 10.14 to find that as the spring is compressed, the elastic potential energy increases by

$$\Delta U_s = \frac{1}{2}kx^2$$

Just as in the case of gravitational potential energy, we have found an expression for the *change* in U_s, not U_s itself. Again, we are free to set $U_s = 0$ at any convenient spring extension. An obvious choice is to set $U_s = 0$ at the point where the spring is in equilibrium, neither compressed nor stretched—that is, at $x = 0$. With this choice we have

$$U_s = \frac{1}{2}kx^2 \qquad (10.15)$$

Elastic potential energy of a spring displaced a distance x from equilibrium (assuming $U_s = 0$ when the end of the spring is at $x = 0$)

NOTE ▶ Because U_s depends on the *square* of the displacement x, U_s is the same whether x is positive (the spring is compressed as in Figure 10.16) or negative (the spring is stretched). ◀

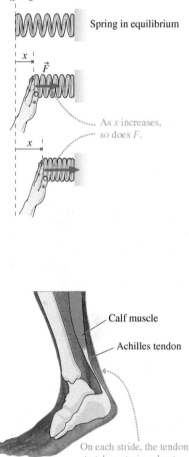

FIGURE 10.16 The force required to compress a spring is not constant.

$x = 0$

Spring in equilibrium

As x increases, so does F.

Calf muscle

Achilles tendon

On each stride, the tendon stretches, storing about 35 J of energy.

Spring in your step BIO As you run, you lose some of your mechanical energy each time your foot strikes the ground; this energy is transformed into unrecoverable thermal energy. Luckily, about 35% of the decrease of your mechanical energy when your foot lands is stored as elastic potential energy in the stretchable Achilles tendon of the lower leg. On each plant of the foot, the tendon is stretched, storing some energy. The tendon springs back as you push off the ground again, helping to propel you forward. This recovered energy reduces the amount of internal chemical energy you use, increasing your efficiency.

EXAMPLE 10.8   **Pulling back on a bow**

An archer pulls back the string on her bow to a distance of 70 cm from its equilibrium position. To hold the string at this position takes a force of 140 N. How much elastic potential energy is stored in the bow?

PREPARE A bow is an elastic material, so we will model it as obeying Hooke's law, $F_s = -kx$, where x is the distance the string is pulled back. We can use the force required to hold the string, and the distance it is pulled back, to find the bow's spring constant k. Then we can use Equation 10.15 to find the elastic potential energy.

SOLVE From Hooke's law, the spring constant is

$$k = \frac{F}{x} = \frac{140 \text{ N}}{0.70 \text{ m}} = 200 \text{ N/m}$$

Then the elastic potential energy of the flexed bow is

$$U_s = \frac{1}{2}kx^2 = \frac{1}{2}(200 \text{ N/m})(0.70 \text{ m})^2 = 49 \text{ J}$$

ASSESS When the arrow is released, this elastic potential energy will be transformed into the kinetic energy of the arrow. According to Table 10.1, the kinetic energy of a 100 mph fastball is about 150 J, so 49 J of kinetic energy for a fast-moving arrow seems reasonable.

FIGURE 10.17 A molecular view of thermal energy.

Hot object: Fast-moving molecules have lots of kinetic and elastic potential energy.

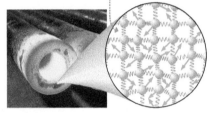

Cold object: Slow-moving molecules have little kinetic and elastic potential energy.

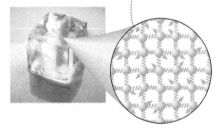

FIGURE 10.18 A thermograph of a box that's been dragged across the floor.

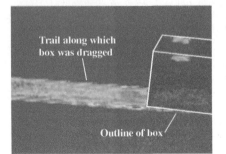

Trail along which box was dragged

Outline of box

STOP TO THINK 10.5   When a spring is stretched by 5 cm, its elastic potential energy is 1 J. What will its elastic potential energy be if it is *compressed* by 10 cm?

A. −4 J   B. −2 J   C. 2 J   D. 4 J

10.5 Thermal Energy

We noted earlier that thermal energy is related to the microscopic motion of the molecules of an object. As **FIGURE 10.17** shows, the molecules in a hot object jiggle around their average positions more than the molecules in a cold object. This has two consequences. First, each atom is on average moving faster in the hot object. This means that each atom has a higher *kinetic energy*. Second, each atom in the hot object tends to stray farther from its equilibrium position, leading to a greater stretching or compressing of the spring-like molecular bonds. This means that each atom has on average a higher *potential energy*. The potential energy stored in any one bond and the kinetic energy of any one atom are both exceedingly small, but there are incredibly many bonds and atoms. The sum of all these microscopic potential and kinetic energies is what we call **thermal energy**. Increasing an object's thermal energy corresponds to increasing its temperature.

Creating Thermal Energy

FIGURE 10.18 shows a thermograph of a heavy box and the floor across which it has just been dragged. In this image, warmer areas appear light blue or green. You can see that the bottom of the box and the region of the floor that the box moved over are noticeably warmer than their surroundings. In the process of dragging the box, thermal energy has appeared in the box and the floor.

We can find a quantitative expression for the change in thermal energy by considering such a box pulled by a rope at a constant speed. As the box is pulled across the floor, the rope exerts a constant forward force $\vec{F}$ on the box, while the friction force $\vec{f}_k$ exerts a constant force on the box that is directed backward. Because the box moves at a constant speed, the magnitudes of these two forces are equal: $F = f_k$.

As the box moves through a displacement $d = \Delta x$, the rope does work $W = F\Delta x$ on the box. This work represents energy transferred into the system, so the system's energy must *increase*. In what form is this increased energy? The box's speed remains constant, so there is no change in its kinetic energy ($\Delta K = 0$). And its height doesn't change, so its gravitational potential energy is unchanged as well ($\Delta U_g = 0$). Instead, the increased energy must be in the form of *thermal* energy E_{th}. As Figure 10.18 shows, this energy appears as an increased temperature of both the box *and* the floor across which it was dragged.

We can write the work-energy equation, Equation 10.3, for the case where only thermal energy changes:

$$\Delta E_{th} = W$$

or, because the work is $W = F\Delta x = f_k \Delta x$,

$$\Delta E_{th} = f_k \Delta x \qquad (10.16)$$

Change in thermal energy for a system consisting of an object and the surface it slides on, as the object undergoes displacement Δx while acted on by friction force f_k

This increase in thermal energy is a general feature of any system where friction between sliding objects is present. An atomic-level explanation is shown in FIGURE 10.19. Although we arrived at Equation 10.16 by considering energy transferred into the system via work done by an external force, the equation is equally valid for the transformation of mechanical energy into thermal energy when, for instance, an object slides to a halt on a rough surface. Equation 10.16 also applies to rolling friction; we need only replace f_k by f_r.

FIGURE 10.19 How friction causes an increase in thermal energy.

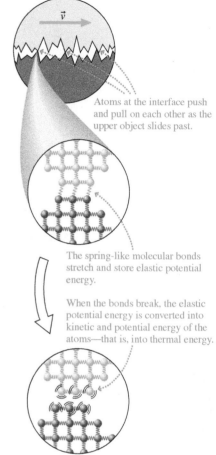

Atoms at the interface push and pull on each other as the upper object slides past.

The spring-like molecular bonds stretch and store elastic potential energy.

When the bonds break, the elastic potential energy is converted into kinetic and potential energy of the atoms—that is, into thermal energy.

STOP TO THINK 10.6 A block with an initial kinetic energy of 4.0 J comes to rest after sliding 1.0 m. How far would the block slide if it had 8.0 J of initial kinetic energy?

A. 1.4 m   B. 2.0 m   C. 3.0 m   D. 4.0 m

EXAMPLE 10.9 Creating thermal energy by rubbing

A 0.30 kg block of wood is rubbed back and forth against a wood table 30 times in each direction. The block is moved 8.0 cm during each stroke and pressed against the table with a force of 22 N. How much thermal energy is created in this process?

PREPARE The hand holding the block does work to push the block back and forth. Work transfers energy into the block + table system, where it appears as thermal energy according to Equation 10.16. The force of friction can be found from the model of kinetic friction introduced in Chapter 5, $f_k = \mu_k n$; from Table 5.2 the coefficient of kinetic friction for wood sliding on wood is $\mu_k = 0.20$. To find the normal force n acting on the block, we draw the free-body diagram of FIGURE 10.20, which shows only the *vertical* forces acting on the block.

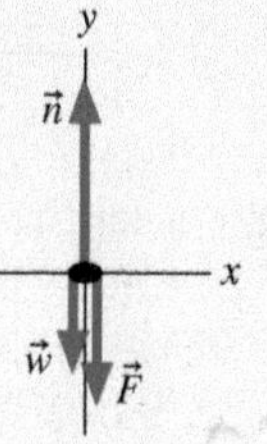

FIGURE 10.20 Free-body diagram (vertical forces only) for a block being rubbed against a table.

SOLVE From Equation 10.16 we have $\Delta E_{th} = f_k \Delta x$, where $f_k = \mu_k n$. The block is not accelerating in the y-direction, so from the free-body diagram Newton's second law gives

$$\sum F_y = n - w - F = ma_y = 0$$

or

$$n = w + F = mg + F = (0.30 \text{ kg})(9.8 \text{ m/s}^2) + 22 \text{ N} = 24.9 \text{ N}$$

The friction force is then $f_k = \mu_k n = (0.20)(24.9 \text{ N}) = 4.98 \text{ N}$. The total displacement of the block is $2 \times 30 \times 8.0 \text{ cm} = 4.8 \text{ m}$. Thus the thermal energy created is

$$\Delta E_{th} = f_k \Delta x = (4.98 \text{ N})(4.8 \text{ m}) = 24 \text{ J}$$

ASSESS This modest amount of thermal energy seems reasonable for a person to create by rubbing.

10.6 Using the Law of Conservation of Energy

Class Video

The work-energy equation, Equation 10.3, states that the change in the total energy of a system equals the energy transferred to or from the system as work. If we consider only those forms of energy that are typically transformed during the motion of ordinary objects—kinetic energy K, gravitational and elastic potential energies U_g and U_s, and thermal energy E_{th}—then the work-energy equation can be written as

$$\Delta K + \Delta U_g + \Delta U_s + \Delta E_{th} = W \qquad (10.17)$$

Just as for momentum conservation, we wish to develop a before-and-after perspective for energy conservation. We can do so by noting that $\Delta K = K_f - K_i$, $\Delta U_g = (U_g)_f - (U_g)_i$, and so on. Then Equation 10.17 can be written as

$$K_f + (U_g)_f + (U_s)_f + \Delta E_{th} = K_i + (U_g)_i + (U_s)_i + W \qquad (10.18)$$

Video Tutor
Demo

Equation 10.18 is the before-and-after version of the work-energy equation. It states that a system's final energy, including any change in the system's thermal energy, equals its initial energy *plus* any energy added to the system as work. This equation will be the basis for a powerful problem-solving strategy, presented on the next page.

NOTE ▶ We don't write ΔE_{th} as $(E_{th})_f - (E_{th})_i$ in Equation 10.18 because the initial and final values of the thermal energy are typically unknown; only their *difference* ΔE_{th} can be measured. ◀

Conservation of Energy

In Section 10.1, we introduced the idea of an *isolated system*—one in which no work is done on the system so that no energy is transferred into or out of the system. In that case, $W = 0$ in Equation 10.18, so that

$$K_f + (U_g)_f + (U_s)_f + \Delta E_{th} = K_i + (U_g)_i + (U_s)_i \qquad (10.18a)$$

Equation 10.18a states that **for an isolated system, energy is conserved**—the final energy, including any change in thermal energy, equals the initial energy. This is the law of conservation of energy, Equation 10.4, but restricted to those forms of energy typical for mechanical motion. Table 10.2 shows how to choose an isolated system for four common situations.

TABLE 10.2 Choosing an isolated system

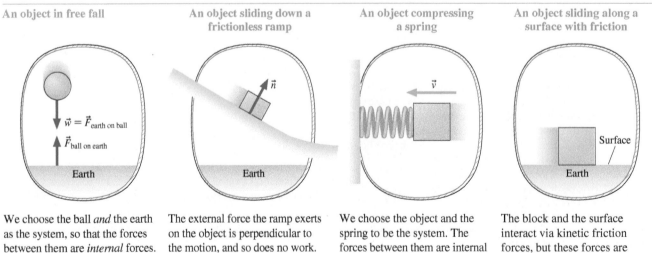

An object in free fall	An object sliding down a frictionless ramp	An object compressing a spring	An object sliding along a surface with friction
We choose the ball *and* the earth as the system, so that the forces between them are *internal* forces. There are no external forces to do work, so the system is isolated.	The external force the ramp exerts on the object is perpendicular to the motion, and so does no work. The object and the earth together form an isolated system.	We choose the object and the spring to be the system. The forces between them are internal forces, so no work is done.	The block and the surface interact via kinetic friction forces, but these forces are internal to the system. There are no external forces to do work, so the system is isolated.

If we further restrict ourselves to cases where friction can be neglected, so that $\Delta E_{th} = 0$, the law of conservation of energy, Equation 10.18a, becomes

$$K_f + (U_g)_f + (U_s)_f = K_i + (U_g)_i + (U_s)_i \qquad (10.19)$$

The sum of the kinetic and potential energies, $K + U_g + U_s$, is called the **mechanical energy** of the system, so Equation 10.19 says that **the mechanical energy is conserved for an isolated system without friction.**

PROBLEM-SOLVING
STRATEGY 10.1   **Conservation of energy problems**   (MP)

The work-energy equation and the law of conservation of energy relate a system's *final* energy to its *initial* energy. We can solve for initial and final heights, speeds, and displacements from these energies.

PREPARE Draw a before-and-after visual overview, as was outlined in Tactics Box 9.1. Note the known quantities, and identify what you're trying to find.

SOLVE Apply the before-and-after version of the work-energy equation, Equation 10.18:

$$K_f + (U_g)_f + (U_s)_f + \Delta E_{th} = K_i + (U_g)_i + (U_s)_i + W$$

There are three common situations:

- If work is done on the system, then use the full version of Equation 10.18.
- If the system is isolated, no work is done. Use Equation 10.18 with $W = 0$ (the law of conservation of energy, Equation 10.18a).
- If the system is isolated *and* there's no friction, the mechanical energy is conserved. Use Equation 10.18 with both $W = 0$ and $\Delta E_{th} = 0$—that is, Equation 10.19.

Depending on the problem, you'll need to calculate the initial and/or final values of these energies. You can then solve for the unknown energies, and from these any unknown speeds (from K), heights and distances (from U_g and U_s), or displacements or friction forces (from $\Delta E_{th} = f_k \Delta x$).

ASSESS Check the signs of your energies. Kinetic energy is always positive, as is the change in thermal energy. Check that your result has the correct units, is reasonable, and answers the question.

Exercise 23 🖉

Spring into action BIO A locust can jump as far as 1 meter, an impressive distance for such a small animal. To make such a jump, its legs must extend much more rapidly than muscles can ordinarily contract. Thus, instead of using its muscles to make the jump directly, the locust uses them to more slowly stretch an internal "spring" near its knee joint. This stores elastic potential energy in the spring. When the muscles relax, the spring is suddenly released, and its energy is rapidly converted into kinetic energy of the insect.

EXAMPLE 10.10   **Hitting the bell**

At the county fair, Katie tries her hand at the ring-the-bell attraction, as shown in FIGURE 10.21. She swings the mallet hard enough to give the ball an initial upward speed of 8.0 m/s. Will the ball ring the bell, 3.0 m from the bottom?

PREPARE We'll follow the steps of Problem-Solving Strategy 10.1. From Table 10.2, we see that once the ball is in the air, the system consisting of the ball and the earth is isolated. If we assume that the track along which the ball moves is frictionless, then the system's mechanical energy is conserved. Figure 10.21 shows a before-and-after visual overview in which we've chosen $y = 0$ m to be at the ball's starting point. We can then use conservation of mechanical energy, Equation 10.19.

FIGURE 10.21 Visual overview of the ring-the-bell attraction.

We'll calculate how high the ball would go if the bell weren't there. Then we'll see if that height is enough to have reached the bell.

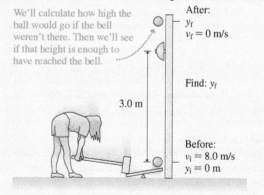

After:
y_f
$v_f = 0$ m/s

Find: y_f

3.0 m

Before:
$v_i = 8.0$ m/s
$y_i = 0$ m

Continued

SOLVE Equation 10.19 tells us that $K_f + (U_g)_f = K_i + (U_g)_i$. We can use our expressions for kinetic and potential energy to write this as

$$\frac{1}{2}mv_f^2 + mgy_f = \frac{1}{2}mv_i^2 + mgy_i$$

Let's ignore the bell for the moment and figure out how far the ball would rise if there were nothing in its way. We know that the ball starts at $y_i = 0$ m and that its speed v_f at the highest point is 0 m/s. Thus the energy equation simplifies to

$$mgy_f = \frac{1}{2}mv_i^2$$

This is easily solved for the height y_f:

$$y_f = \frac{v_i^2}{2g} = \frac{(8.0 \text{ m/s})^2}{2(9.8 \text{ m/s}^2)} = 3.3 \text{ m}$$

This is higher than the point where the bell sits, so the ball would actually hit it on the way up.

ASSESS It seems reasonable that Katie could swing the mallet hard enough to make the ball rise by about 3 m.

EXAMPLE 10.11 **Speed at the bottom of a water slide**

While at the county fair, Katie tries the water slide, whose shape is shown in FIGURE 10.22. The starting point is 9.0 m above the ground. She pushes off with an initial speed of 2.0 m/s. If the slide is frictionless, how fast will Katie be traveling at the bottom?

PREPARE Table 10.2 showed that the system consisting of Katie and the earth is isolated because the normal force of the slide is perpendicular to Katie's motion and does no work. If we assume the slide is frictionless, we can use the conservation of mechanical energy equation. Figure 10.22 is a visual overview of the problem.

FIGURE 10.22 Before-and-after visual overview of Katie on the water slide.

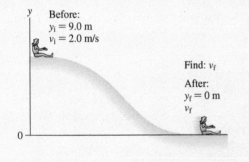

SOLVE Conservation of mechanical energy gives

$$K_f + (U_g)_f = K_i + (U_g)_i$$

or

$$\frac{1}{2}mv_f^2 + mgy_f = \frac{1}{2}mv_i^2 + mgy_i$$

Taking $y_f = 0$ m, we have

$$\frac{1}{2}mv_f^2 = \frac{1}{2}mv_i^2 + mgy_i$$

which we can solve to get

$$v_f = \sqrt{v_i^2 + 2gy_i}$$
$$= \sqrt{(2.0 \text{ m/s})^2 + 2(9.8 \text{ m/s}^2)(9.0 \text{ m})} = 13 \text{ m/s}$$

ASSESS This speed is about 30 mph. This is probably faster than you really would go on a water slide but, because we have ignored friction, our answer is reasonable. It is important to realize that the *shape* of the slide does not matter because gravitational potential energy depends only on the *height* above a reference level. **If you slide down any (frictionless) slide of the same height, your speed at the bottom is the same.**

EXAMPLE 10.12 **Speed of a spring-launched ball**

A spring-loaded toy gun is used to launch a 10 g plastic ball. The spring, which has a spring constant of 10 N/m, is compressed by 10 cm as the ball is pushed into the barrel. When the trigger is pulled, the spring is released and shoots the ball back out. What is the ball's speed as it leaves the barrel? Assume that friction is negligible.

PREPARE Assume the spring obeys Hooke's law, $F_s = -kx$, and is massless so that it has no kinetic energy of its own. Using Table 10.2, we choose the isolated system to be the spring and the ball. There's no friction; hence the system's mechanical energy $K + U_s$ is conserved.

FIGURE 10.23 shows a before-and-after visual overview. The compressed spring will push on the ball until the spring has returned to its equilibrium length. We have chosen the origin of the coordinate system at the equilibrium position of the free end of the spring, making $x_i = -10$ cm and $x_f = 0$ cm.

FIGURE 10.23 Before-and-after visual overview of a ball being shot out of a spring-loaded toy gun.

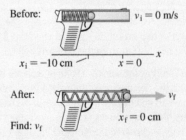

SOLVE The energy conservation equation is $K_f + (U_s)_f = K_i + (U_s)_i$. We can use the elastic potential energy of the spring, Equation 10.15, to write this as

$$\frac{1}{2}mv_f^2 + \frac{1}{2}kx_f^2 = \frac{1}{2}mv_i^2 + \frac{1}{2}kx_i^2$$

We know that $x_f = 0$ m and $v_i = 0$ m/s, so this simplifies to

$$\tfrac{1}{2}mv_f^2 = \tfrac{1}{2}kx_i^2$$

It is now straightforward to solve for the ball's speed:

$$v_f = \sqrt{\frac{kx_i^2}{m}} = \sqrt{\frac{(10 \text{ N/m})(-0.10 \text{ m})^2}{0.010 \text{ kg}}} = 3.2 \text{ m/s}$$

ASSESS This is *not* a problem that we could have easily solved with Newton's laws. The acceleration is not constant, and we have not learned how to handle the kinematics of nonconstant acceleration. But with conservation of energy—it's easy!

EXAMPLE 10.13   **Pulling a bike trailer**

Monica pulls her daughter Jessie in a bike trailer. The trailer and Jessie together have a mass of 25 kg. Monica starts up a 100-m-long slope that's 4.0 m high. On the slope, Monica's bike pulls on the trailer with a constant force of 8.0 N. They start out at the bottom of the slope with a speed of 5.3 m/s. What is their speed at the top of the slope?

PREPARE We'll again follow the steps of Problem-Solving Strategy 10.1. Taking Jessie and the trailer as the system, we see that Monica's bike is applying a force to the system as it moves through a displacement; that is, Monica's bike is doing work on the system. Thus we'll need to use the full version of Equation 10.18, including the work term W. **FIGURE 10.24** shows a before-and-after visual overview in which we've chosen $y = 0$ m to be the trailer's starting height.

FIGURE 10.24 Before-and-after visual overview of a bike trailer being pulled uphill.

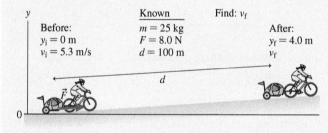

SOLVE If we assume there's no friction, so that $\Delta E_{th} = 0$, then Equation 10.18 is

$$K_f + (U_g)_f = K_i + (U_g)_i + W$$

or

$$\frac{1}{2}mv_f^2 + mgy_f = \frac{1}{2}mv_i^2 + mgy_i + W$$

Taking $y_i = 0$ m and writing $W = Fd$, we can solve for the final speed:

$$v_f^2 = v_i^2 - 2gy_f + \frac{2Fd}{m}$$

$$= (5.3 \text{ m/s})^2 - 2(9.8 \text{ m/s}^2)(4.0 \text{ m}) + \frac{2(8.0 \text{ N})(100 \text{ m})}{25 \text{ kg}}$$

$$= 13.7 \text{ m}^2/\text{s}^2$$

from which we find that $v_f = 3.7$ m/s. Note that we took the work to be a positive quantity because the force is in the same direction as the displacement.

ASSESS A speed of 3.7 m/s—about 8 mph—seems reasonable for a bicycle's speed. Jessie's final speed is less than her initial speed, indicating that the uphill force of Monica's bike on the trailer is less than the downhill component of gravity.

Friction and Thermal Energy

Thermal energy is always created when kinetic friction is present, so we must use the more general conservation of energy equation, Equation 10.18a, which includes thermal-energy changes ΔE_{th}. Furthermore, we know from Section 10.5 that the change in the thermal energy when an object slides a distance Δx while subject to a friction force f_k is $\Delta E_{th} = f_k \Delta x$.

EXAMPLE 10.14   **Where will the sled stop?**

A sledder, starting from rest, slides down a 10-m-high hill. At the bottom of the hill is a long horizontal patch of rough snow. The hill is nearly frictionless, but the coefficient of friction between the sled and the rough snow at the bottom is $\mu_k = 0.30$. How far will the sled slide along the rough patch?

PREPARE In order to be isolated, the system must include the sled, the earth, *and* the rough snow. As Table 10.2 shows, this makes the friction force an internal force so that no work is done on the system. We can use conservation of energy, but we will need to include thermal energy. A visual overview of the problem is shown in **FIGURE 10.25**.

FIGURE 10.25 Visual overview of a sledder sliding downhill.

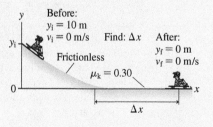

Continued

SOLVE At the top of the hill the sled has only gravitational potential energy $(U_g)_i = mgy_i$. It has no kinetic or potential energy after stopping at the bottom of the hill, so $K_f = (U_g)_f = 0$. However, friction in the rough patch causes an increase in thermal energy. Thus our conservation of energy equation $K_f + (U_g)_f + \Delta E_{th} = K_i + (U_g)_i$ is

$$\Delta E_{th} = (U_g)_i = mgy_i$$

The change in thermal energy is $\Delta E_{th} = f_k \Delta x = \mu_k n \Delta x$. The normal force $\vec{n}$ balances the sled's weight $\vec{w}$ as it crosses the rough patch, so $n = w = mg$. Thus

$$\Delta E_{th} = \mu_k n \Delta x = \mu_k (mg) \Delta x = mgy_i$$

from which we find

$$\Delta x = \frac{y_i}{\mu_k} = \frac{10 \text{ m}}{0.30} = 33 \text{ m}$$

ASSESS It seems reasonable that the sledder would slide a distance that is greater than the height of the hill he started down.

SYNTHESIS 10.1 Energy and its conservation

In this chapter we have considered four basic kinds of energy. The energies present in an isolated system can transform from one kind into another, but the total energy is *conserved*. The unit of all types of energy is the **joule** (J).

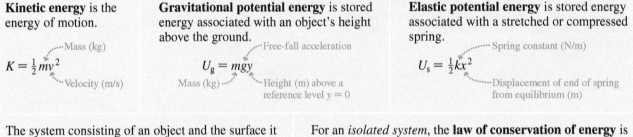

Kinetic energy is the energy of motion.

Mass (kg)

$$K = \tfrac{1}{2}mv^2$$

Velocity (m/s)

Gravitational potential energy is stored energy associated with an object's height above the ground.

Free-fall acceleration

$$U_g = mgy$$

Mass (kg) ⋯ Height (m) above a reference level $y = 0$

Elastic potential energy is stored energy associated with a stretched or compressed spring.

Spring constant (N/m)

$$U_s = \tfrac{1}{2}kx^2$$

Displacement of end of spring from equilibrium (m)

The system consisting of an object and the surface it slides on gains thermal energy if friction is present.

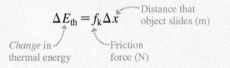

Distance that object slides (m)

$$\Delta E_{th} = f_k \Delta x$$

Change in thermal energy ⋯ Friction force (N)

For an *isolated system*, the **law of conservation of energy** is

$$\underbrace{K_f + (U_g)_f + (U_s)_f + \Delta E_{th}}_{\text{Final total energy}} = \underbrace{K_i + (U_g)_i + (U_s)_i}_{\text{Initial total energy}}$$

A system's final total energy, including any increase in its thermal energy, is equal to its initial energy.

STOP TO THINK 10.7 At the water park, Katie slides down each of the frictionless slides shown. At the top, she is given a push so that she has the same initial speed each time. At the bottom of which slide is she moving the fastest?

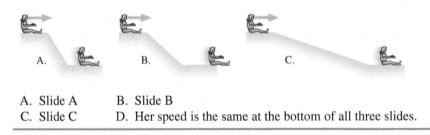

A. B. C.

A. Slide A   B. Slide B
C. Slide C   D. Her speed is the same at the bottom of all three slides.

10.7 Energy in Collisions

In Chapter 9 we studied collisions between two objects. We found that if no external forces are acting on the objects, the total *momentum* of the objects will be conserved. Now we wish to study what happens to *energy* in collisions. The energetics of collisions are important in many applications in biokinetics, such as designing safer automobiles and bicycle helmets.

Let's first re-examine a perfectly inelastic collision. We studied just such a collision in Example 9.8. Recall that in such a collision the two objects stick together and then move with a common final velocity. What happens to the energy?

EXAMPLE 10.15 **Energy transformations in a perfectly inelastic collision**

FIGURE 10.26 shows two train cars that move toward each other, collide, and couple together. In Example 9.8, we used conservation of momentum to find the final velocity shown in Figure 10.26 from the given initial velocities. How much thermal energy is created in this collision?

FIGURE 10.26 Before-and-after visual overview of a completely inelastic collision.

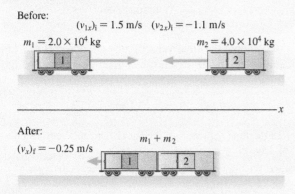

PREPARE We'll choose our system to be the two cars. Because the track is horizontal, there is no change in potential energy. Thus the law of conservation of energy, Equation 10.18a, is $K_f + \Delta E_{th} = K_i$. The total energy before the collision must equal the total energy afterward, but the *mechanical* energies need not be equal.

SOLVE The initial kinetic energy is

$$K_i = \frac{1}{2}m_1(v_{1x})_i^2 + \frac{1}{2}m_2(v_{2x})_i^2$$

$$= \frac{1}{2}(2.0 \times 10^4 \text{ kg})(1.5 \text{ m/s})^2 + \frac{1}{2}(4.0 \times 10^4 \text{ kg})(-1.1 \text{ m/s})^2$$

$$= 4.7 \times 10^4 \text{ J}$$

Because the cars stick together and move as a single object with mass $m_1 + m_2$, the final kinetic energy is

$$K_f = \frac{1}{2}(m_1 + m_2)(v_x)_f^2$$

$$= \frac{1}{2}(6.0 \times 10^4 \text{ kg})(-0.25 \text{ m/s})^2 = 1900 \text{ J}$$

From the conservation of energy equation above, we find that the thermal energy increases by

$$\Delta E_{th} = K_i - K_f = 4.7 \times 10^4 \text{ J} - 1900 \text{ J} = 4.5 \times 10^4 \text{ J}$$

This amount of the initial kinetic energy is transformed into thermal energy during the impact of the collision.

ASSESS About 96% of the initial kinetic energy is transformed into thermal energy. This is typical of many real-world collisions.

Elastic Collisions

FIGURE 10.27 shows a collision of a tennis ball with a racket. The ball is compressed and the racket strings stretch as the two collide, then the ball expands and the strings relax as the two are pushed apart. In the language of energy, the kinetic energy of the objects is transformed into the elastic potential energy of the ball and strings, then back into kinetic energy as the two objects spring apart. If *all* of the kinetic energy is stored as elastic potential energy, and *all* of the elastic potential energy is transformed back into the post-collision kinetic energy of the objects, then mechanical energy is conserved. A collision in which mechanical energy is conserved is called a **perfectly elastic collision**.

Needless to say, most real collisions fall somewhere between perfectly elastic and perfectly inelastic. A rubber ball bouncing on the floor might "lose" 20% of its kinetic energy on each bounce and return to only 80% of the height of the preceding bounce. But collisions between two very hard objects, such as two pool balls or two steel balls, come close to being perfectly elastic. And collisions between microscopic particles, such as atoms or electrons, can be perfectly elastic.

FIGURE 10.28 shows a head-on, perfectly elastic collision of a ball of mass m_1, having initial velocity $(v_{1x})_i$, with a ball of mass m_2 that is initially at rest. The balls' velocities after the collision are $(v_{1x})_f$ and $(v_{2x})_f$. These are velocities, not speeds, and have signs. Ball 1, in particular, might bounce backward and have a negative value for $(v_{1x})_f$.

The collision must obey two conservation laws: conservation of momentum (obeyed in any collision) and conservation of mechanical energy (because the collision is perfectly elastic). Although the energy is transformed into potential energy during the collision, the mechanical energy before and after the collision is purely kinetic energy. Thus,

momentum conservation: $m_1(v_{1x})_i = m_1(v_{1x})_f + m_2(v_{2x})_f$

energy conservation: $\frac{1}{2}m_1(v_{1x})_i^2 = \frac{1}{2}m_1(v_{1x})_f^2 + \frac{1}{2}m_2(v_{2x})_f^2$

FIGURE 10.27 A tennis ball collides with a racket. Notice that the ball is compressed and the strings are stretched.

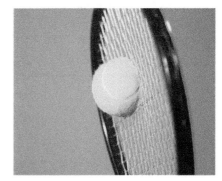

FIGURE 10.28 A perfectly elastic collision.

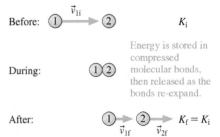

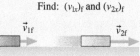

Video Tutor
Demo

Momentum conservation alone is not sufficient to analyze the collision because there are two unknowns: the two final velocities. That is why we did not consider perfectly elastic collisions in Chapter 9. Energy conservation gives us another condition. The complete solution of these two equations involves straightforward but rather lengthy algebra. We'll just give the solution here:

$$(v_{1x})_f = \frac{m_1 - m_2}{m_1 + m_2}(v_{1x})_i \qquad (v_{2x})_f = \frac{2m_1}{m_1 + m_2}(v_{1x})_i \qquad (10.20)$$

Perfectly elastic collision with object 2 initially at rest

Equations 10.20 allow us to compute the final velocity of each object. Let's look at a common and important example: a perfectly elastic collision between two objects of equal mass.

EXAMPLE 10.16 Velocities in an air hockey collision

On an air hockey table, a moving puck, traveling to the right at 2.3 m/s, makes a head-on collision with an identical puck at rest. What is the final velocity of each puck?

PREPARE The before-and-after visual overview is shown in FIGURE 10.29. We've shown the final velocities in the picture, but we don't really know yet which way the pucks will move. Because one puck was initially at rest, we can use Equations 10.20

FIGURE 10.29 A moving puck collides with a stationary puck.

Before: $(v_{1x})_i = 2.3$ m/s $(v_{2x})_i = 0$ m/s

$\vec{v}_{1i}$ $\vec{v}_{2i} = \vec{0}$

After: Find: $(v_{1x})_f$ and $(v_{2x})_f$

$\vec{v}_{1f}$ $\vec{v}_{2f}$

to find the final velocities of the pucks. The pucks are identical, so we have $m_1 = m_2 = m$.

SOLVE We use Equations 10.20 with $m_1 = m_2 = m$ to get

$$(v_{1x})_f = \frac{m - m}{m + m}(v_{1x})_i = 0 \text{ m/s}$$

$$(v_{2x})_f = \frac{2m}{m + m}(v_{1x})_i = (v_{1x})_i = 2.3 \text{ m/s}$$

The incoming puck stops dead, and the initially stationary puck goes off with the same velocity that the incoming one had.

ASSESS You can see that momentum and energy are conserved: The incoming puck's momentum and energy are completely transferred to the outgoing puck. If you've ever played pool, you've probably seen this sort of collision when you hit a ball head-on with the cue ball. The cue ball stops and the other ball picks up the cue ball's velocity.

Other cases where the colliding objects have unequal masses will be treated in the end-of-chapter problems.

Forces in Collisions

The collision between two pool balls occurs very quickly, and the forces are typically very large and difficult to calculate. Fortunately, by using the concepts of momentum and energy conservation, we can often calculate the final velocities of the balls without having to know the forces between them. There are collisions, however, where knowing the forces involved is of critical importance. The next example shows how a helmet helps protect the head from the large forces involved in a bicycle accident.

EXAMPLE 10.17 Protecting your head BIO

A bike helmet—basically a shell of hard, crushable foam—is tested by being strapped onto a 5.0 kg headform and dropped from a height of 2.0 m onto a hard anvil. What force is encountered by the headform if the impact crushes the foam by 3.0 cm?

◄ The foam inside a bike helmet is designed to crush upon impact.

PREPARE A before-and-after visual overview of the test is shown in FIGURE 10.30. We've chosen the endpoint of the problem to be when the headform comes to rest with the foam crushed. We can use the work-energy equation, Equation 10.18, to calculate the force on the headform. We'll choose the headform and the earth to be the system; the foam in the helmet is part of the environment. We make this choice so that the force on the headform due to the foam is an *external* force that does work W on the headform.

SOLVE The work-energy equation, Equation 10.18, tells us that the work done by external forces—in this case, the force of the foam on the headform—changes the energy of the system. The headform starts at rest, speeds up as it falls, then returns to rest during the impact. Overall, then, $K_f = K_i$. Furthermore, $\Delta E_{th} = 0$ because there's no friction to increase the thermal energy. Only the gravitational potential energy changes, so the work-energy equation is

$$(U_g)_f - (U_g)_i = W$$

The upward force of the foam on the headform is opposite the downward displacement of the headform. Referring to Tactics Box 10.1, we see that the work done is negative: $W = -Fd$, where we've assumed that the force is constant. Using this result in the work-energy equation and solving for F, we find

$$F = -\frac{(U_g)_f - (U_g)_i}{d} = \frac{(U_g)_i - (U_g)_f}{d}$$

Taking our reference height to be $y = 0$ m at the anvil, we have $(U_g)_f = 0$. We're left with $(U_g)_i = mgy_i$, so

$$F = \frac{mgy_i}{d} = \frac{(5.0 \text{ kg})(9.8 \text{ m/s}^2)(2.0 \text{ m})}{0.030 \text{ m}} = 3300 \text{ N}$$

This is the force that acts on the head to bring it to a halt in 3.0 cm. More important from the perspective of possible brain injury is the head's *acceleration:*

FIGURE 10.30 Before-and-after visual overview of the bike helmet test.

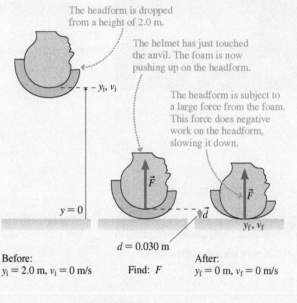

Before:
$y_i = 2.0$ m, $v_i = 0$ m/s   Find: F

After:
$y_f = 0$ m, $v_f = 0$ m/s

$$a = \frac{F}{m} = \frac{3300 \text{ N}}{5.0 \text{ kg}} = 660 \text{ m/s}^2 = 67g$$

ASSESS The accepted threshold for serious brain injury is around $300g$, so this helmet would protect the rider in all but the most serious accidents. Without the helmet, the rider's head would come to a stop in a much shorter distance and thus be subjected to a much larger acceleration.

STOP TO THINK 10.8 A small ball with mass M is at rest. It is then struck by a ball with twice the mass, moving at speed v_0. The situation after the collision is shown in the figure. Is this possible?

A. Yes
B. No, because momentum is not conserved
C. No, because energy is not conserved
D. No, because neither momentum nor energy is conserved

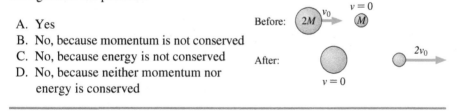

10.8 Power

We've now studied how energy can be transformed from one kind into another and how it can be transferred between the environment and the system as work. In many situations we would like to know *how quickly* the energy is transformed or transferred. Is a transfer of energy very rapid, or does it take place over a long time? In passing a truck, your car needs to transform a certain amount of the chemical energy in its fuel into kinetic energy. It makes a *big* difference whether your engine can do this in 20 s or 60 s!

The question How quickly? implies that we are talking about a *rate*. For example, the velocity of an object—how fast it is going—is the *rate of change* of position. So, when we raise the issue of how fast the energy is transformed, we are talking about the *rate of transformation* of energy. Suppose in a time interval Δt an amount of energy ΔE is transformed from one form to another. The rate at which this energy is transformed is called the **power** P and is defined as

Both these cars take about the same energy to reach 60 mph, but the race car gets there in a much shorter time, so its *power* is much greater.

The English unit of power is the *horsepower.* The conversion factor to watts is

$$1 \text{ horsepower} = 1 \text{ hp} = 746 \text{ W}$$

Many common appliances, such as motors, are rated in hp.

$$P = \frac{\Delta E}{\Delta t} \qquad (10.21)$$

Power when an amount of energy ΔE is transformed in a time interval Δt

The unit of power is the **watt**, which is defined as 1 watt = 1 W = 1 J/s.

Power also measures the rate at which energy is transferred into or out of a system as work W. If work W is done in time interval Δt, the rate of energy *transfer* is

$$P = \frac{W}{\Delta t} \qquad (10.22)$$

Power when an amount of work W is done in a time interval Δt

A force that is doing work (i.e., transferring energy) at a rate of 3 J/s has an "output power" of 3 W. A system that is gaining energy at the rate of 3 J/s is said to "consume" 3 W of power. Common prefixes used for power are mW (milliwatts), kW (kilowatts), and MW (megawatts).

We can express Equation 10.22 in a different form. If in the time interval Δt an object undergoes a displacement Δx, the work done by a force acting on the object is $W = F\Delta x$. Then Equation 10.22 can be written as

$$P = \frac{W}{\Delta t} = \frac{F\Delta x}{\Delta t} = F\frac{\Delta x}{\Delta t} = Fv$$

The rate at which energy is transferred to an object as work—the power—is the product of the force that does the work and the velocity of the object:

$$P = Fv \qquad (10.23)$$

Rate of energy transfer due to a force F acting on an object moving at velocity v

---

EXAMPLE 10.18   **Power to pass a truck**

Your 1500 kg car is behind a truck traveling at 60 mph (27 m/s). To pass the truck, you speed up to 75 mph (34 m/s) in 6.0 s. What engine power is required to do this?

PREPARE Your engine is transforming the chemical energy of its fuel into the kinetic energy of the car. We can calculate the rate of transformation by finding the change ΔK in the kinetic energy and using the known time interval.

SOLVE We have

$$K_i = \frac{1}{2}mv_i^2 = \frac{1}{2}(1500 \text{ kg})(27 \text{ m/s})^2 = 5.47 \times 10^5 \text{ J}$$

$$K_f = \frac{1}{2}mv_f^2 = \frac{1}{2}(1500 \text{ kg})(34 \text{ m/s})^2 = 8.67 \times 10^5 \text{ J}$$

so that

$$\Delta K = K_f - K_i$$
$$= (8.67 \times 10^5 \text{ J}) - (5.47 \times 10^5 \text{ J}) = 3.20 \times 10^5 \text{ J}$$

To transform this amount of energy in 6 s, the power required is

$$P = \frac{\Delta K}{\Delta t} = \frac{3.20 \times 10^5 \text{ J}}{6.0 \text{ s}} = 53,000 \text{ W} = 53 \text{ kW}$$

This is about 71 hp. This power is in addition to the power needed to overcome drag and friction and cruise at 60 mph, so the total power required from the engine will be even greater than this.

ASSESS You use a large amount of energy to perform a simple driving maneuver such as this. 3.20×10^5 J is enough energy to lift an 80 kg person 410 m in the air—the height of a tall skyscraper. And 53 kW would lift him there in only 6 s!

STOP TO THINK 10.9   Four students run up the stairs in the times shown. Rank in order, from largest to smallest, their power outputs P_A through P_D.

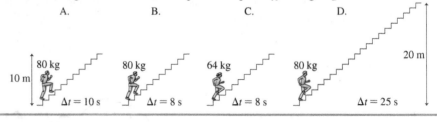

Stopping a runaway truck

A truck's brakes can overheat and fail while descending mountain highways, leading to an extremely dangerous runaway truck. Some highways have *runaway-truck ramps* to safely bring out-of-control trucks to a stop. These uphill ramps are covered with a deep bed of gravel. The uphill slope and the large coefficient of rolling friction as the tires sink into the gravel bring the truck to a safe halt.

A runaway-truck ramp along Interstate 70 in Colorado.

A 22,000 kg truck heading down a 3.5° slope at 20 m/s ($\approx$ 45 mph) suddenly has its brakes fail. Fortunately, there's a runaway-truck ramp 600 m ahead. The ramp slopes upward at an angle of 10°, and the coefficient of rolling friction between the truck's tires and the loose gravel is $\mu_r = 0.40$. Ignore air resistance and rolling friction as the truck rolls down the highway.

a. Use conservation of energy to find how far along the ramp the truck travels before stopping.
b. By how much does the thermal energy of the truck and ramp increase as the truck stops?

PREPARE Parts a and b can be solved using energy conservation by following Problem-Solving Strategy 10.1. **FIGURE 10.31** shows a before-and-after visual overview. Because we're going to need to determine friction forces to calculate the increase in thermal energy, we've also drawn a free-body diagram for the truck as it moves up the ramp. One slight complication is that the y-axis of free-body diagrams is drawn perpendicular to the slope, whereas the calculation of gravitational potential energy needs a vertical y-axis to measure height. We've dealt with this by labeling the free-body diagram axis the y'-axis.

FIGURE 10.31 Visual overview of the runaway truck.

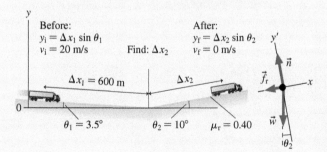

SOLVE a. The law of conservation of energy for the motion of the truck, from the moment its brakes fail to when it finally stops, is

$$K_f + (U_g)_f + \Delta E_{th} = K_i + (U_g)_i$$

Because friction is present only along the ramp, thermal energy will be created only as the truck moves up the ramp. This thermal energy is then given by $\Delta E_{th} = f_r \Delta x_2$, because Δx_2 is the length of the ramp. The conservation of energy equation then is

$$\frac{1}{2}mv_f^2 + mgy_f + f_r \Delta x_2 = \frac{1}{2}mv_i^2 + mgy_i$$

From Figure 10.31 we have $y_i = \Delta x_1 \sin\theta_1$, $y_f = \Delta x_2 \sin\theta_2$, and $v_f = 0$, so the equation becomes

$$mg \Delta x_2 \sin\theta_2 + f_r \Delta x_2 = \frac{1}{2}mv_i^2 + mg \Delta x_1 \sin\theta_1$$

To find $f_r = \mu_r n$ we need to find the normal force n. The free-body diagram shows that

$$\sum F_{y'} = n - mg\cos\theta_2 = a_{y'} = 0$$

from which $f_r = \mu_r n = \mu_r mg \cos\theta_2$. With this result for f_r, our conservation of energy equation is

$$mg \Delta x_2 \sin\theta_2 + \mu_r mg\cos\theta_2 \Delta x_2 = \frac{1}{2}mv_i^2 + mg \Delta x_1 \sin\theta_1$$

which, after we divide both sides by mg, simplifies to

$$\Delta x_2 \sin\theta_2 + \mu_r \cos\theta_2 \Delta x_2 = \frac{v_i^2}{2g} + \Delta x_1 \sin\theta_1$$

Solving this for Δx_2 gives

$$\Delta x_2 = \frac{\dfrac{v_i^2}{2g} + \Delta x_1 \sin\theta_1}{\sin\theta_2 + \mu_r \cos\theta_2}$$

$$= \frac{\dfrac{(20 \text{ m/s})^2}{2(9.8 \text{ m/s}^2)} + (600 \text{ m})(\sin 3.5°)}{\sin 10° + 0.40(\cos 10°)} = 100 \text{ m}$$

b. We know that $\Delta E_{th} = f_r \Delta x_2 = (\mu_r mg \cos\theta_2)\Delta x_2$, so that

$$\Delta E_{th} = (0.40)(22,000 \text{ kg})(9.8 \text{ m/s}^2)(\cos 10°)(100 \text{ m})$$

$$= 8.5 \times 10^6 \text{ J}$$

ASSESS It seems reasonable that a truck that speeds up as it rolls 600 m downhill takes only 100 m to stop on a steeper, high-friction ramp. We also expect the thermal energy to be roughly comparable to the kinetic energy of the truck, since it's largely the kinetic energy that is transformed into thermal energy. At the top of the hill the truck's kinetic energy is $K_i = \frac{1}{2}mv_i^2 = \frac{1}{2}(22,000 \text{ kg})(20 \text{ m/s})^2 = 4.4 \times 10^6 \text{ J}$, which is of the same order of magnitude as ΔE_{th}. Our answer is reasonable.

SUMMARY

Goal: To introduce the concept of energy and to learn a new problem-solving strategy based on conservation of energy.

GENERAL PRINCIPLES

Basic Energy Model

Within a system, energy can be **transformed** between various forms.

Energy can be **transferred** into or out of a system in two basic ways:

- **Work:** The transfer of energy by mechanical forces

- **Heat:** The nonmechanical transfer of energy from a hotter to a colder object

Energy is *transformed* within the system.

Environment
System
$K \longleftrightarrow U$

E_{chem}

E_{th}

Work, heat

Energy is *transferred* to or from the system from or to the environment.

Conservation of Energy

When work W is done on a system, the system's total energy changes by the amount of work done. In mathematical form, this is the **work-energy equation:**

$$\Delta E = \Delta K + \Delta U_g + \Delta U_s + \Delta E_{th} + \Delta E_{chem} + \cdots = W$$

A system is **isolated** when no energy is transferred into or out of the system. This means the work is zero, giving the **law of conservation of energy:**

$$\Delta K + \Delta U_g + \Delta U_s + \Delta E_{th} + \Delta E_{chem} + \cdots = 0$$

Solving Energy Transfer and Energy Conservation Problems

PREPARE Draw a before-and-after visual overview.

SOLVE

- If work is done on the system, then use the before-and-after version of the work-energy equation:

$$K_f + (U_g)_f + (U_s)_f + \Delta E_{th} = K_i + (U_g)_i + (U_s)_i + W$$

- If the system is isolated but there's friction present, then the total energy is conserved:

$$K_f + (U_g)_f + (U_s)_f + \Delta E_{th} = K_i + (U_g)_i + (U_s)_i$$

- If the system is isolated and there's no friction, then mechanical energy is conserved:

$$K_f + (U_g)_f + (U_s)_f = K_i + (U_g)_i + (U_s)_i$$

ASSESS Kinetic energy is always positive, as is the change in thermal energy.

IMPORTANT CONCEPTS

Kinetic energy is an energy of motion:

$$K = \frac{1}{2}mv^2 + \frac{1}{2}I\omega^2$$

Translational ⸱⸱⸱ ⤴   ⤴ ⸱⸱⸱ Rotational

Potential energy is energy stored in a system of interacting objects.

- **Gravitational potential energy:** $U_g = mgy$

- **Elastic potential energy:** $U_s = \frac{1}{2}kx^2$

Mechanical energy is the sum of a system's kinetic and potential energies:

$$\text{Mechanical energy} = K + U = K + U_g + U_s$$

Thermal energy is the sum of the microscopic kinetic and potential energies of all the molecules in an object. The hotter an object, the more thermal energy it has. When kinetic (sliding) friction is present, the increase in the thermal energy is $\Delta E_{th} = f_k \Delta x$.

Work is the process by which energy is transferred to or from a system by the application of mechanical forces.

If a particle moves through a displacement $\vec{d}$ while acted upon by a constant force $\vec{F}$, the force does work

$$W = F_{\parallel}d = Fd\cos\theta$$

Only the component of the force parallel to the displacement does work.

APPLICATIONS

Perfectly elastic collisions
Both mechanical energy and momentum are conserved.

$$(v_{1x})_f = \frac{m_1 - m_2}{m_1 + m_2}(v_{1x})_i$$

$$(v_{2x})_f = \frac{2m_1}{m_1 + m_2}(v_{1x})_i$$

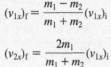

Object 2 initially at rest

$(v_{1x})_i$

Before: ① ⟶ ② K_i

After: $K_f = K_i$ ① ⟶ ② ⟶
$(v_{1x})_f$ $(v_{2x})_f$

Power is the rate at which energy is transformed . . .

$$P = \frac{\Delta E}{\Delta t}$$ ⸱⸱⸱ Amount of energy transformed
⸱⸱⸱ Time required to transform it

. . . or at which work is done.

$$P = \frac{W}{\Delta t}$$ ⸱⸱⸱ Amount of work done
⸱⸱⸱ Time required to do work

Problem difficulty is labeled as | (straightforward) to |||| (challenging). Problems labeled INT integrate significant material from earlier chapters; BIO are of biological or medical interest.

 For assigned homework and other learning materials, go to MasteringPhysics®

 Scan this QR code to launch a Video Tutor Solution that will help you solve problems for this chapter.

QUESTIONS

Conceptual Questions

1. The brake shoes of your car are made of a material that can tolerate very high temperatures without being damaged. Why is this so?
2. When you pound a nail with a hammer, the nail gets quite warm. Describe the energy transformations that lead to the addition of thermal energy in the nail.

For Questions 3 through 10, give a specific example of a system with the energy transformation shown. In these questions, W is the work done on the system, and K, U, and E_{th} are the kinetic, potential, and thermal energies of the system, respectively. Any energy not mentioned in the transformation is assumed to remain constant; if work is not mentioned, it is assumed to be zero.

3. $W \rightarrow K$
4. $W \rightarrow U$
5. $K \rightarrow U$
6. $K \rightarrow W$
7. $U \rightarrow K$
8. $W \rightarrow \Delta E_{th}$
9. $U \rightarrow \Delta E_{th}$
10. $K \rightarrow \Delta E_{th}$

11. A ball of putty is dropped from a height of 2 m onto a hard floor, where it sticks. What object or objects need to be included within the system if the system is to be isolated during this process?
12. A 0.5 kg mass on a 1-m-long string swings in a circle on a horizontal, frictionless table at a steady speed of 2 m/s. How much work does the tension in the string do on the mass during one revolution? Explain.
13. Particle A has less mass than particle B. Both are pushed forward across a frictionless surface by equal forces for 1 s. Both start from rest.
    a. Compare the amount of work done on each particle. That is, is the work done on A greater than, less than, or equal to the work done on B? Explain.
    b. Compare the impulses delivered to particles A and B. Explain.
    c. Compare the final speeds of particles A and B. Explain.
14. Puck B has twice the mass of puck A. Starting from rest, both pucks are pulled the same distance across frictionless ice by strings with the same tension.
    a. Compare the final kinetic energies of pucks A and B.
    b. Compare the final speeds of pucks A and B.
15. To change a tire, you need to use a jack to raise one corner of your car. While doing so, you happen to notice that pushing the jack handle down 20 cm raises the car only 0.2 cm. Use energy concepts to explain why the handle must be moved so far to raise the car by such a small amount.
16. You drop two balls from a tower, one of mass m and the other of mass $2m$. Just before they hit the ground, which ball, if either, has the larger kinetic energy? Explain.

17. A roller coaster car rolls down a frictionless track, reaching speed v at the bottom.
    a. If you want the car to go twice as fast at the bottom, by what factor must you increase the height of the track?
    b. Does your answer to part a depend on whether the track is straight or not? Explain.
18. A spring gun shoots out a plastic ball at speed v. The spring is then compressed twice the distance it was on the first shot.
    a. By what factor is the spring's potential energy increased?
    b. By what factor is the ball's speed increased? Explain.
19. Sandy and Chris stand on the edge of a cliff and throw identical mass rocks at the same speed. Sandy throws her rock horizontally while Chris throws his upward at an angle of 45° to the horizontal. Are the rocks moving at the same speed when they hit the ground, or is one moving faster than the other? If one is moving faster, which one? Explain.
20. A solid cylinder and a hollow cylinder have the same mass, same radius, and turn on frictionless, horizontal axles. (The hollow cylinder has lightweight spokes connecting it to the axle.) A rope is wrapped around each cylinder and tied to a block. The FIGURE Q10.20 blocks have the same mass and are held the same height above the ground as shown in Figure Q10.20. Both blocks are released simultaneously. The ropes do not slip. Which block hits the ground first? Or is it a tie? Explain.
21. BIO You are much more likely to be injured if you fall and your head strikes the ground than if your head strikes a gymnastics pad. Use energy and work concepts to explain why this is so.
22. Jason slides a large crate down a ramp from a truck to the ground. To control the crate and keep it sliding at constant speed, Jason backs down the ramp in front of the crate while pushing upward on it. During its trip to the bottom, is the thermal energy created in the ramp and the crate greater than, less than, or equal to the crate's loss of gravitational potential energy? Explain.

Multiple-Choice Questions

23. || A roller coaster starts from rest at its highest point and then descends on its (frictionless) track. Its speed is 30 m/s when it reaches ground level. What was its speed when its height was half that of its starting point?
    A. 11 m/s   B. 15 m/s   C. 21 m/s   D. 25 m/s
24. | You and a friend each carry a 15 kg suitcase up two flights of stairs, walking at a constant speed. Take each suitcase to be the system. Suppose you carry your suitcase up the stairs in 30 s while your friend takes 60 s. Which of the following is true?
    A. You did more work, but both of you expended the same power.
    B. You did more work and expended more power.
    C. Both of you did equal work, but you expended more power.
    D. Both of you did equal work, but you expended less power.

25. | A woman uses a pulley and a rope to raise a 20 kg weight to a height of 2 m. If it takes 4 s to do this, about how much power is she supplying?
    A. 100 W   B. 200 W   C. 300 W   D. 400 W

26. | A hockey puck sliding along frictionless ice with speed v to the right collides with a horizontal spring and compresses it by 2.0 cm before coming to a momentary stop. What will be the spring's maximum compression if the same puck hits it at a speed of $2v$?
    A. 2.0 cm   B. 2.8 cm   C. 4.0 cm
    D. 5.6 cm   E. 8.0 cm

27. ‖ A block slides down a smooth ramp, starting from rest at a height h. When it reaches the bottom it's moving at speed v. It then continues to slide up a second smooth ramp. At what height is its speed equal to $v/2$?
    A. $h/4$ B. $h/2$ C. $3h/4$ D. $2h$

28. | A wrecking ball is suspended from a 5.0-m-long cable that makes a 30° angle with the vertical. The ball is released and swings down. What is the ball's speed at the lowest point?
    A. 7.7 m/s   B. 4.4 m/s   C. 3.6 m/s   D. 3.1 m/s

PROBLEMS

Section 10.2 Work

1. ‖ A 2.0 kg book is lying on a 0.75-m-high table. You pick it up and place it on a bookshelf 2.3 m above the floor. During this process,
   a. How much work does gravity do on the book?
   b. How much work does your hand do on the book?

2. ‖ The two ropes seen in Figure P10.2 are used to lower a 255 kg piano exactly 5 m from a second-story window to the ground. How much work is done by each of the three forces?

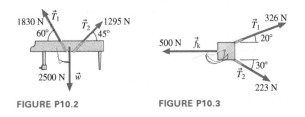

FIGURE P10.2   **FIGURE P10.3**

3. | The two ropes shown in the bird's-eye view of Figure P10.3 are used to drag a crate exactly 3 m across the floor. How much work is done by each of the ropes on the crate?

4. ‖ a. An escalator carries you from one level to the next in an airport terminal. The upper level is 4.5 m above the lower level, and the length of the escalator is 7.0 m. How much work does the up escalator do on you when you ride it from the lower level to the upper level?
   b. How much work does the down escalator do on you when you ride it from the upper level to the lower level?

5. | A boy flies a kite with the string at a 30° angle to the horizontal. The tension in the string is 4.5 N. How much work does the string do on the boy if the boy
   a. Stands still?
   b. Walks a horizontal distance of 11 m away from the kite?
   c. Walks a horizontal distance of 11 m toward the kite?

6. ‖ A crate slides down a ramp that makes a 20° angle with the ground. To keep the crate from sliding too fast, Paige pushes back on it with a 68 N horizontal force. How much work does Paige do on the crate as it slides 3.5 m down the ramp?

Section 10.3 Kinetic Energy

7. | Which has the larger kinetic energy, a 10 g bullet fired at 500 m/s or a 10 kg bowling ball sliding at 10 m/s?

8. ‖ At what speed does a 1000 kg compact car have the same kinetic energy as a 20,000 kg truck going 25 km/h?

9. | A car is traveling at 10 m/s.
   a. How fast would the car need to go to double its kinetic energy?
   b. By what factor does the car's kinetic energy increase if its speed is doubled to 20 m/s?

10. | The cheetah is the fastest land animal, reaching speeds as high as 33 m/s, or roughly 75 mph. What is the kinetic energy of a 60 kg cheetah running at top speed?

11. | How fast would an 80 kg man need to run in order to have the same kinetic energy as an 8.0 g bullet fired at 400 m/s?

12. ‖ Sam's job at the amusement park is to slow down and bring to a stop the boats in the log ride. If a boat and its riders have a mass of 1200 kg and the boat drifts in at 1.2 m/s, how much work does Sam do to stop it?

13. ‖ A 20 g plastic ball is moving to the left at 30 m/s. How much work must be done on the ball to cause it to move to the right at 30 m/s?

14. ‖ The turntable in a microwave oven has a moment of inertia of 0.040 kg·m² and rotates continuously, making a complete revolution every 4.0 s. What is its kinetic energy?

15. ‖ An energy storage system based on a flywheel (a rotating disk) can store a maximum of 4.0 MJ when the flywheel is rotating at 20,000 revolutions per minute. What is the moment of inertia of the flywheel?

Section 10.4 Potential Energy

16. ‖ The lowest point in Death Valley is 85.0 m below sea level. The summit of nearby Mt. Whitney has an elevation of 4420 m. What is the change in gravitational potential energy of an energetic 65.0 kg hiker who makes it from the floor of Death Valley to the top of Mt. Whitney?

17. | a. What is the kinetic energy of a 1500 kg car traveling at a speed of 30 m/s ($\approx$ 65 mph)?
    b. From what height should the car be dropped to have this same amount of kinetic energy just before impact?
    c. Does your answer to part b depend on the car's mass?

18. | The world's fastest humans can reach speeds of about 11 m/s. In order to increase his gravitational potential energy by an amount equal to his kinetic energy at full speed, how high would such a sprinter need to climb?

19. | A 72 kg bike racer climbs a 1200-m-long section of road that has a slope of 4.3°. By how much does his gravitational potential energy change during this climb?

20. ‖ A 1000 kg wrecking ball hangs from a 15-m-long cable. The ball is pulled back until the cable makes an angle of 25° with the vertical. By how much has the gravitational potential energy of the ball changed?

21. ‖ How far must you stretch a spring with $k = 1000$ N/m to store 200 J of energy?

22. ‖ How much energy can be stored in a spring with a spring constant of 500 N/m if its maximum possible stretch is 20 cm?

23. ‖‖‖ The elastic energy stored in your tendons can contribute up BIO to 35% of your energy needs when running. Sports scientists have studied the change in length of the knee extensor tendon in sprinters and nonathletes. They find (on average) that the sprinters' tendons stretch 41 mm, while nonathletes' stretch only 33 mm. The spring constant for the tendon is the same for both groups, 33 N/mm. What is the difference in maximum stored energy between the sprinters and the nonathletes?

Section 10.5 Thermal Energy

24. ‖ Marissa drags a 23 kg duffel bag 14 m across the gym floor. If the coefficient of kinetic friction between the floor and bag is 0.15, how much thermal energy does Marissa create?

25. ‖ Mark pushes his broken car 150 m down the block to his friend's house. He has to exert a 110 N horizontal force to push the car at a constant speed. How much thermal energy is created in the tires and road during this short trip?

26. ‖‖‖ A 900 N crate slides 12 m down a ramp that makes an angle of 35° with the horizontal. If the crate slides at a constant speed, how much thermal energy is created?

27. ‖‖‖ A 25 kg child slides down a playground slide at a *constant speed*. The slide has a height of 3.0 m and is 7.0 m long. Using the law of conservation of energy, find the magnitude of the kinetic friction force acting on the child.

Section 10.6 Using the Law of Conservation of Energy

28. ‖ A boy reaches out of a window and tosses a ball straight up with a speed of 10 m/s. The ball is 20 m above the ground as he releases it. Use conservation of energy to find
   a. The ball's maximum height above the ground.
   b. The ball's speed as it passes the window on its way down.
   c. The speed of impact on the ground.

29. ‖ a. With what minimum speed must you toss a 100 g ball straight up to just barely hit the 10-m-high ceiling of the gymnasium if you release the ball 1.5 m above the floor? Solve this problem using energy.
   b. With what speed does the ball hit the floor?

30. ‖‖‖ What minimum speed does a 100 g puck need to make it to the top of a frictionless ramp that is 3.0 m long and inclined at 20°?

31. ‖ A car is parked at the top of a 50-m-high hill. Its brakes fail and it rolls down the hill. How fast will it be going at the bottom? (Ignore friction.)

32. ‖‖‖ A 1500 kg car is approaching the hill shown in Figure P10.32 at 10 m/s when it suddenly runs out of gas.
   a. Can the car make it to the top of the hill by coasting?
   b. If your answer to part a is yes, what is the car's speed after coasting down the other side?

FIGURE P10.32

33. ‖ A 10 kg runaway grocery cart runs into a spring, attached wall, with spring constant 250 N/m and compresses it by 60 What was the speed of the cart just before it hit the spring?

34. ‖ As a 15,000 kg jet lands on an aircraft carrier, its tail hoo snags a cable to slow it down. The cable is attached to a spring with spring constant 60,000 N/m. If the spring stretches 30 m to stop the plane, what was the plane's landing speed?

35. ‖ Your friend's Frisbee has become stuck 16 m above the ground in a tree. You want to dislodge the Frisbee by throwing a rock at it. The Frisbee is stuck pretty tight, so you figure the rock needs to be traveling at least 5.0 m/s when it hits the Frisbee. If you release the rock 2.0 m above the ground, with what minimum speed must you throw it?

36. ‖ A fireman of mass 80 kg slides down a pole. When he reaches the bottom, 4.2 m below his starting point, his speed is 2.2 m/s. By how much has thermal energy increased during his slide?

37. ‖ A 20 kg child slides down a 3.0-m-high playground slide. She starts from rest, and her speed at the bottom is 2.0 m/s.
   a. What energy transfers and transformations occur during the slide?
   b. What is the total change in the thermal energy of the slide and the seat of her pants?

38. ‖ A hockey puck is given an initial speed of 5.0 m/s. If the coefficient of kinetic friction between the puck and the ice is 0.05, how far does the puck slide before coming to rest? Solve this problem using conservation of energy.

39. ‖ In the winter activity of tubing, riders slide down snow-covered slopes while sitting on large inflated rubber tubes. To get to the top of the slope, a rider and his tube, with a total mass of 80 kg, are pulled at a constant speed by a tow rope that maintains a constant tension of 340 N. How much thermal energy is created in the slope and the tube during the ascent of a 30-m-high, 120-m-long slope?

40. ‖ A cyclist is coasting at 12 m/s when she starts down a 450-m-long slope that is 30 m high. The cyclist and her bicycle have a combined mass of 70 kg. A steady 12 N drag force due to air resistance acts on her as she coasts all the way to the bottom. What is her speed at the bottom of the slope?

Section 10.7 Energy in Collisions

41. ‖ A 50 g marble moving at 2.0 m/s strikes a 20 g marble at rest. What is the speed of each marble immediately after the collision? Assume the collision is perfectly elastic and the marbles collide head-on.

42. ‖ Ball 1, with a mass of 100 g and traveling at 10 m/s, collides head-on with ball 2, which has a mass of 300 g and is initially at rest. What are the final velocities of each ball if the collision is (a) perfectly elastic? (b) perfectly inelastic?

43. ‖ An air-track glider undergoes a perfectly inelastic collision with an identical glider that is initially at rest. What fraction of the first glider's initial kinetic energy is transformed into thermal energy in this collision?

44. ‖ Two balls undergo a perfectly elastic head-on collision, with one ball initially at rest. If the incoming ball has a speed of 200 m/s, what are the final speed and direction of each ball if
   a. The incoming ball is *much* more massive than the stationary ball?
   b. The stationary ball is *much* more massive than the incoming ball?

_ch work must you do to push a 10 kg block of
_across a steel table at a steady speed of 1.0 m/s for
_0 s? The coefficient of kinetic friction for steel on steel
is 0.60.

b. What is your power output while doing so?

a. How much work does an elevator motor do to lift a
1000 kg elevator a height of 100 m at a constant speed?

b. How much power must the motor supply to do this in 50 s
at constant speed?

47. ||| A 1000 kg sports car accelerates from 0 to 30 m/s in 10 s.
What is the average power of the engine?

48. ||| In just 0.30 s, you compress a spring (spring constant 5000 N/m),
which is initially at its equilibrium length, by 4.0 cm. What is
your average power output?

49. || An elite Tour de France cyclist can maintain an output power
of 450 W during a sustained climb. At this output power, how
long would it take an 85 kg cyclist (including the mass of his
bike) to climb the famed 1100-m-high Alpe d'Huez mountain
stage?

50. || A 710 kg car drives at a constant speed of 23 m/s. It is subject
to a drag force of 500 N. What power is required from the car's
engine to drive the car
a. On level ground?
b. Up a hill with a slope of 2.0°?

51. ||| An elevator weighing 2500 N ascends at a constant speed of
8.0 m/s. How much power must the motor supply to do this?

General Problems

52. | How much work does Scott do to push a 80 kg sofa 2.0 m
INT across the floor at a constant speed? The coefficient of kinetic
friction between the sofa and the floor is 0.23.

53. || A 550 kg elevator accelerates upward at 1.2 m/s² for the first
INT 15 m of its motion. How much work is done during this part of
its motion by the cable that lifts the elevator?

54. || A 2.3 kg box, starting from rest, is pushed up a ramp by a
10 N force parallel to the ramp. The ramp is 2.0 m long and
tilted at 17°. The speed of the box at the top of the ramp is
0.80 m/s. Consider the system to be the box + ramp + earth.
a. How much work W does the force do on the system?
b. What is the change ΔK in the kinetic energy of the
system?
c. What is the change ΔU_g in the gravitational potential energy
of the system?
d. What is the change ΔE_{th} in the thermal energy of the system?

55. | A 55 kg skateboarder wants
to just make it to the upper
edge of a "half-pipe" with a
radius of 3.0 m, as shown in
Figure P10.55. What speed v_i
does he need at the bottom if
he is to coast all the way up?

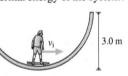

FIGURE P10.55

a. First do the calculation treating the skateboarder and board
as a point particle, with the entire mass nearly in contact with
the half-pipe.
b. More realistically, the mass of the skateboarder in a deep
crouch might be thought of as concentrated 0.75 m from the
half-pipe. Assuming he remains in that position all the way
up, what v_i is needed to reach the upper edge?

56. ||| Fleas have remarkable jumping ability. A 0.50 mg flea, jump-
BIO ing straight up, would reach a height of 40 cm if there were no
air resistance. In reality, air resistance limits the height to 20 cm.
a. What is the flea's kinetic energy as it leaves the ground?
b. At its highest point, what fraction of the initial kinetic energy
has been converted to potential energy?

57. ||| You are driving your 1500 kg car at 20 m/s down a hill with
a 5.0° slope when a deer suddenly jumps out onto the roadway.
You slam on your brakes, skidding to a stop. How far do you skid
before stopping if the kinetic friction force between your tires and
the road is 1.2×10^4 N? Solve this problem using conservation
of energy.

58. || A 20 kg child is on a swing that hangs from 3.0-m-long
chains, as shown in Figure P10.58. What is her speed v_i at
the bottom of the arc if she swings out to a 45° angle before
reversing direction?

FIGURE P10.58   FIGURE P10.59

59. || Suppose you lift a 20 kg box by a height of 1.0 m.
a. How much work do you do in lifting the box?
Instead of lifting the box straight up, suppose you push it
up a 1.0-m-high ramp that makes a 30° degree angle with
the horizontal, as shown in Figure P10.59. Being clever, you
choose a ramp with no friction.
b. How much force F is required to push the box straight up the
slope at a constant speed?
c. How long is the ramp?
d. Use your force and distance results to calculate the work you
do in pushing the box up the ramp. How does this compare
to your answer to part a?

60. || A cannon tilted up at a 30° angle fires a cannon ball at 80 m/s
from atop a 10-m-high fortress wall. What is the ball's impact
speed on the ground below? Ignore air resistance.

61. | The sledder shown in Figure P10.61 starts from the top
of a frictionless hill and slides down into the valley. What
initial speed v_i does the sledder need to just make it over the
next hill?

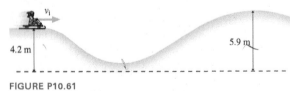

FIGURE P10.61

62. ||||| In a physics lab experiment, a spring clamped to the table shoots
INT a 20 g ball horizontally. When the spring is compressed 20 cm, the
ball travels horizontally 5.0 m and lands on the floor 1.5 m below
the point at which it left the spring. What is the spring constant?

63. |||| A 50 g ice cube can slide without friction up and down a 30°
slope. The ice cube is pressed against a spring at the bottom of
the slope, compressing the spring 10 cm. The spring constant
is 25 N/m. When the ice cube is released, what distance will it
travel up the slope before reversing direction?

64. IIIII The maximum energy a bone can absorb without breaking is
BIO surprisingly small. For a healthy human of mass 60 kg, experimental data show that the leg bones of both legs can absorb about 200 J.
    a. From what maximum height could a person jump and land rigidly upright on both feet without breaking his legs? Assume that all the energy is absorbed in the leg bones in a rigid landing.
    b. People jump from much greater heights than this; explain how this is possible.
    Hint: Think about how people land when they jump from greater heights.

65. ‖ In an amusement park water slide, people slide down an
NT essentially frictionless tube. The top of the slide is 3.0 m above the bottom where they exit the slide, moving horizontally, 1.2 m above a swimming pool. What horizontal distance do they travel from the exit point before hitting the water? Does the mass of the person make any difference?

66. ‖ The 5.0-m-long rope in Figure P10.66 hangs vertically from a tree right at the edge of a ravine. A woman wants to use the rope to swing to the other side of the ravine. She runs as fast as she can, grabs the rope, and swings out over the ravine.

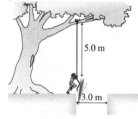

FIGURE P10.66

    a. As she swings, what energy conversion is taking place?
    b. When she's directly over the far edge of the ravine, how much higher is she than when she started?
    c. Given your answers to parts a and b, how fast must she be running when she grabs the rope in order to swing all the way across the ravine?

67. IIII You have been asked to design a "ballistic spring system" to
NT measure the speed of bullets. A bullet of mass m is fired into a block of mass M. The block, with the embedded bullet, then slides across a frictionless table and collides with a horizontal spring whose spring constant is k. The opposite end of the spring is anchored to a wall. The spring's maximum compression d is measured.
    a. Find an expression for the bullet's initial speed v_B in terms of m, M, k, and d.
    Hint: This is a two-part problem. The bullet's collision with the block is an inelastic collision. What quantity is conserved in an inelastic collision? Subsequently the block hits a spring on a frictionless surface. What quantity is conserved in this collision?
    b. What was the speed of a 5.0 g bullet if the block's mass is 2.0 kg and if the spring, with $k = 50$ N/m, was compressed by 10 cm?
    c. What fraction of the bullet's initial kinetic energy is "lost"? Where did it go?

68. IIII A new event, shown in
NT Figure P10.68, has been proposed for the Winter Olympics. An athlete will sprint 100 m, starting from rest, then leap onto a 20 kg bobsled. The person and bobsled will then slide down a 50-m-long

FIGURE P10.68

ice-covered ramp, sloped at 20°, and into a spring with a carefully calibrated spring constant of 2000 N/m. The athlete who compresses the spring the farthest wins the gold medal. Lisa, whose mass is

40 kg, has been training for this event. She can reach a maximum speed of 12 m/s in the 100 m dash.
    a. How far will Lisa compress the spring?
    b. The Olympic committee has very exact specifications about the shape and angle of the ramp. Is this necessary? If the committee asks your opinion, what factors about the ramp will you tell them are important?

69. ‖ Boxes A and B in Figure P10.69 have masses of 12.0 kg and 4.0 kg, respectively. The two boxes are released from rest. Use conservation of energy to find the boxes' speed when box B has fallen a distance of 0.50 m. Assume a frictionless upper surface.

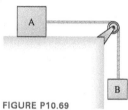

FIGURE P10.69

70. IIII What would be the speed of the boxes in Problem 69 if the coefficient of kinetic friction between box A and the surface it slides on were 0.20? Use conservation of energy.

71. IIIII A 20 g ball is fired horizontally with initial speed v_i toward a 100 g ball that is hanging motionless from a 1.0-m-long string. The balls undergo a head-on, perfectly elastic collision, after which the 100 g ball swings out to a maximum angle $\theta_{max} = 50°$. What was v_i?

72. ‖ Two coupled boxcars are rolling along at 2.5 m/s when they
NT collide with and couple to a third, stationary boxcar.
    a. What is the final speed of the three coupled boxcars?
    b. What fraction of the cars' initial kinetic energy is transformed into thermal energy?

73. ‖ A fish scale, consisting of a spring with spring constant $k = 200$ N/m, is hung vertically from the ceiling. A 5.0 kg fish is attached to the end of the unstretched spring and then released. The fish moves downward until the spring is fully stretched, then starts to move back up as the spring begins to contract. What is the maximum distance through which the fish falls?

74. ‖ A 70 kg human sprinter can accelerate from rest to 10 m/s in
BIO 3.0 s. During the same time interval, a 30 kg greyhound can accelerate from rest to 20 m/s. Compute (a) the change in kinetic energy and (b) the average power output for each.

75. IIII A 50 g ball of clay traveling at 6.5 m/s hits and sticks to a
NT 1.0 kg block sitting at rest on a frictionless surface.
    a. What is the speed of the block after the collision?
    b. Show that the mechanical energy is *not* conserved in this collision. What percentage of the ball's initial kinetic energy is "lost"? Where did this kinetic energy go?

76. ‖ A package of mass m is released from rest at a warehouse
NT loading dock and slides down a 3.0-m-high frictionless chute to a waiting truck. Unfortunately, the truck driver went on a break without having removed the previous package, of mass $2m$, from the bottom of the chute as shown in Figure P10.76.
    a. Suppose the packages stick together. What is their common speed after the collision?
    b. Suppose the collision between the packages is perfectly elastic. To what height does the package of mass m rebound?

FIGURE P10.76

77. ▥ A 50 kg sprinter, starting from rest, runs 50 m in 7.0 s at constant acceleration.
    a. What is the magnitude of the horizontal force acting on the sprinter?
    b. What is the sprinter's average power output during the first 2.0 s of his run?
    c. What is the sprinter's average power output during the final 2.0 s?

78. ‖ Bob can throw a 500 g rock with a speed of 30 m/s. During the time the rock is in his hand, his hand moves forward by 1.0 m.
    a. How much force, assumed to be constant, does Bob apply to the rock?
    b. How much work does Bob do on the rock?

79. ‖ The mass of an elevator and its occupants is 1200 kg. The electric motor that lifts the elevator can provide a maximum power of 15 kW. What is the maximum constant speed at which this motor can lift the elevator?

80. ‖ The human heart has to pump the average adult's 6.0 L of blood through the body every minute. The heart must do work to overcome frictional forces that resist the blood flow. The average blood pressure is $1.3 \times 10^4 \, \text{N/m}^2$.
    a. Compute the work done moving the 6.0 L of blood completely through the body, assuming the blood pressure always takes its average value.
    b. What power output must the heart have to do this task once a minute?
    Hint: When the heart contracts, it applies force to the blood. Pressure is just force/area, so we can write work = pressure × area × distance. But area × distance is just the blood volume passing through the heart.

MCAT-Style Passage Problems

Tennis Ball Testing

A tennis ball bouncing on a hard surface compresses and then rebounds. The details of the rebound are specified in tennis regulations. Tennis balls, to be acceptable for tournament play, must have a mass of 57.5 g. When dropped from a height of 2.5 m onto a concrete surface, a ball must rebound to a height of 1.4 m. During impact, the ball compresses by approximately 6 mm.

81. ▎ How fast is the ball moving when it hits the concrete surface? (Ignore air resistance.)
    A. 5 m/s   B. 7 m/s   C. 25 m/s   D. 50 m/s

82. ▎ If the ball accelerates uniformly when it hits the floor, what is its approximate acceleration as it comes to rest before rebounding?
    A. 1000 m/s²   B. 2000 m/s²   C. 3000 m/s²   D. 4000 m/s²

83. ▎ The ball's kinetic energy just after the bounce is less than just before the bounce. In what form does this lost energy end up?
    A. Elastic potential energy
    B. Gravitational potential energy
    C. Thermal energy
    D. Rotational kinetic energy

84. ▎ By approximately what percent does the kinetic energy decrease?
    A. 35%   B. 45%   C. 55%   D. 65%

85. ▎ When a tennis ball bounces from a racket, the ball loses approximately 30% of its kinetic energy to thermal energy. A ball that hits a racket at a speed of 10 m/s will rebound with approximately what speed?
    A. 8.5 m/s   B. 7.0 m/s   C. 4.5 m/s   D. 3.0 m/s

Work and Power in Cycling

When you ride a bicycle at constant speed, almost all of the energy you expend goes into the work you do against the drag force of the air. In this problem, assume that *all* of the energy expended goes into working against drag. As we saw in Section 5.6, the drag force on an object is approximately proportional to the square of its speed with respect to the air. For this problem, assume that $F \propto v^2$ exactly and that the air is motionless with respect to the ground unless noted otherwise. Suppose a cyclist and her bicycle have a combined mass of 60 kg and she is cycling along at a speed of 5 m/s.

86. ▎ If the drag force on the cyclist is 10 N, how much energy does she use in cycling 1 km?
    A. 6 kJ   B. 10 kJ   C. 50 kJ   D. 100 kJ

87. ▎ Under these conditions, how much power does she expend as she cycles?
    A. 10 W   B. 50 W   C. 100 W   D. 200 W

88. ▎ If she doubles her speed to 10 m/s, how much energy does she use in cycling 1 km?
    A. 20 kJ   B. 40 kJ   C. 200 kJ   D. 400 kJ

89. ▎ How much power does she expend when cycling at that speed?
    A. 100 W   B. 200 W   C. 400 W   D. 1000 W

90. ▎ Upon reducing her speed back down to 5 m/s, she hits a headwind of 5 m/s. How much power is she expending now?
    A. 100 W   B. 200 W   C. 500 W   D. 1000 W

STOP TO THINK ANSWERS

Chapter Preview Stop to Think: A. Because the car starts from rest, $v_i = 0$ and the kinematic equation is $(v_x)_f^2 = 2a_x \Delta x$, so that $(v_x)_f = \sqrt{2a_x \Delta x}$. Thus the speed is proportional to the square root of the displacement. If, as in this question, the displacement increases by a factor of 4, the speed only doubles. So the speed will increase from 5 m/s to 10 m/s.

Stop to Think 10.1: D. Since the child slides at a constant speed, his kinetic energy doesn't change. But his gravitational potential energy decreases as he descends. It is transformed into thermal energy in the slide and his bottom.

Stop to Think 10.2: C. $W = Fd\cos\theta$. The 10 N force at 90° does no work at all. $\cos 60° = \frac{1}{2}$, so the 8 N force does less work than the 6 N force.

Stop to Think 10.3: B > D > A = C. $K = (1/2)mv^2$. Using the given masses and velocities, we find $K_A = 2.0 \, \text{J}$, $K_B = 4.5 \, \text{J}$, $K_C = 2.0 \, \text{J}$, $K_D = 4.0 \, \text{J}$.

Stop to Think 10.4: $(U_g)_3 > (U_g)_2 = (U_g)_4 > (U_g)_1$. Gravitational potential energy depends only on height, not speed.

Stop to Think 10.5: D. The potential energy of a spring depends on the *square* of the displacement x, so the energy is positive whether the spring is compressed or extended. Furthermore, if the spring is compressed by twice the amount it had been stretched, the energy will increase by a factor of $2^2 = 4$. So the energy will be $4 \times 1\ \text{J} = 4\ \text{J}$.

Stop to Think 10.6: B. We can use conservation of energy to write $\Delta K + \Delta E_{th} = 0$. Now if the initial kinetic energy doubles, so does ΔK, so ΔE_{th} must double as well. But $\Delta E_{th} = f_k \Delta x$, so if ΔE_{th} doubles, then Δx doubles to 2.0 m.

Stop to Think 10.7: D. In all three cases, Katie has the same initial kinetic energy and potential energy. Thus her energy must be the same at the bottom of the slide in all three cases. Because she has only kinetic energy at the bottom, her speed there must be the same in all three cases as well.

Stop to Think 10.8: C. The initial momentum is $(2M)v_0 + 0$, and the final momentum is $0 + M(2v_0)$. These are equal, so momentum is conserved. The initial kinetic energy is $\frac{1}{2}(2M)v_0^2 = Mv_0^2$, and the final kinetic energy is $\frac{1}{2}M(2v_0)^2 = 2Mv_0^2$. The final kinetic energy is *greater* than the initial kinetic energy, so this collision is not possible. (If the final kinetic energy had been less than the initial kinetic energy, the collision could be possible because the difference in energy could be converted into thermal energy.)

Stop to Think 10.9: $P_B > P_A = P_C > P_D$. The power here is the rate at which each runner's internal chemical energy is converted into gravitational potential energy. The change in gravitational potential energy is $mg\Delta y$, so the power is $mg\Delta y/\Delta t$. For runner A, the ratio $m\Delta y/\Delta t$ equals $(80\ \text{kg})(10\ \text{m})/(10\ \text{s}) = 80\ \text{kg} \cdot \text{m/s}$. For C, the ratio is also $80\ \text{kg} \cdot \text{m/s}$. For B, it's $100\ \text{kg} \cdot \text{m/s}$, while for D the ratio is $64\ \text{kg} \cdot \text{m/s}$.

Mathematics Review

Algebra

Using exponents: $a^{-x} = \dfrac{1}{a^x}$ $a^x a^y = a^{(x+y)}$ $\dfrac{a^x}{a^y} = a^{(x-y)}$ $(a^x)^y = a^{xy}$

$a^0 = 1$ $a^1 = a$ $a^{1/n} = \sqrt[n]{a}$

Fractions: $\left(\dfrac{a}{b}\right)\left(\dfrac{c}{d}\right) = \dfrac{ac}{bd}$ $\dfrac{a/b}{c/d} = \dfrac{ad}{bc}$ $\dfrac{1}{1/a} = a$

Logarithms: Natural (base e) logarithms: If $a = e^x$, then $\ln(a) = x$ $\ln(e^x) = x$ $e^{\ln(x)} = x$

Base 10 logarithms: If $a = 10^x$, then $\log_{10}(a) = x$ $\log_{10}(10^x) = x$ $10^{\log_{10}(x)} = x$

The following rules hold for both natural and base 10 algorithms:

$$\ln(ab) = \ln(a) + \ln(b) \qquad \ln\left(\dfrac{a}{b}\right) = \ln(a) - \ln(b) \qquad \ln(a^n) = n\ln(a)$$

The expression $\ln(a + b)$ cannot be simplified.

Linear equations: The graph of the equation $y = ax + b$ is a straight line. a is the slope of the graph. b is the y-intercept.

Proportionality: To say that y is proportional to x, written $y \propto x$, means that $y = ax$, where a is a constant. Proportionality is a special case of linearity. A graph of a proportional relationship is a straight line that passes through the origin. If $y \propto x$, then

$$\dfrac{y_1}{y_2} = \dfrac{x_1}{x_2}$$

Slope $a = \dfrac{\text{rise}}{\text{run}} = \dfrac{\Delta y}{\Delta x}$

Δy

Δx

y-intercept $= b$

Quadratic equation: The quadratic equation $ax^2 + bx + c = 0$ has the two solutions $x = \dfrac{-b \pm \sqrt{b^2 - 4ac}}{2a}$.

Geometry and Trigonometry

Area and volume:

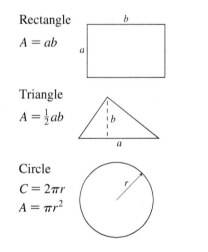

Rectangle
$A = ab$

Triangle
$A = \frac{1}{2}ab$

Circle
$C = 2\pi r$
$A = \pi r^2$

Rectangular box
$V = abc$

Right circular cylinder
$V = \pi r^2 l$

Sphere
$A = 4\pi r^2$
$V = \frac{4}{3}\pi r^3$

Arc length and angle:

The angle θ in radians is defined as $\theta = s/r$.

The arc length that spans angle θ is $s = r\theta$.

$2\pi \text{ rad} = 360°$

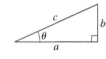

Right triangle:

Pythagorean theorem $c = \sqrt{a^2 + b^2}$ or $a^2 + b^2 = c^2$

$$\sin\theta = \frac{b}{c} = \frac{\text{far side}}{\text{hypotenuse}} \qquad \theta = \sin^{-1}\left(\frac{b}{c}\right)$$

$$\cos\theta = \frac{a}{c} = \frac{\text{adjacent side}}{\text{hypotenuse}} \qquad \theta = \cos^{-1}\left(\frac{a}{c}\right)$$

$$\tan\theta = \frac{b}{a} = \frac{\text{far side}}{\text{adjacent side}} \qquad \theta = \tan^{-1}\left(\frac{b}{a}\right)$$

In general, if it is known that sine of an angle θ is x, so $x = \sin\theta$, then we can find θ by taking the *inverse sine* of x, denoted $\sin^{-1}x$. Thus $\theta = \sin^{-1}x$. Similar relations apply for cosines and tangents.

General triangle:

$\alpha + \beta + \gamma = 180° = \pi \text{ rad}$

Identities:

$$\tan\alpha = \frac{\sin\alpha}{\cos\alpha} \qquad\qquad \sin^2\alpha + \cos^2\alpha = 1$$

$$\sin(-\alpha) = -\sin\alpha \qquad\qquad \cos(-\alpha) = \cos\alpha$$

$$\sin(2\alpha) = 2\sin\alpha\cos\alpha \qquad\qquad \cos(2\alpha) = \cos^2\alpha - \sin^2\alpha$$

Expansions and Approximations

Binomial approximation: $(1 + x)^n \approx 1 + nx$ if $x \ll 1$

Small-angle approximation: If $\alpha \ll 1$ rad, then $\sin\alpha \approx \tan\alpha \approx \alpha$ and $\cos\alpha \approx 1$.

The small-angle approximation is excellent for $\alpha < 5°$ (≈ 0.1 rad) and generally acceptable up to $\alpha \approx 10°$.

Periodic Table of Elements

1																		2
H 1.0																		He 4.0
3 Li 6.9	4 Be 9.0											5 B 10.8	6 C 12.0	7 N 14.0	8 O 16.0	9 F 19.0	10 Ne 20.2
11 Na 23.0	12 Mg 24.3											13 Al 27.0	14 Si 28.1	15 P 31.0	16 S 32.1	17 Cl 35.5	18 Ar 39.9
19 K 39.1	20 Ca 40.1	21 Sc 45.0	22 Ti 47.9	23 V 50.9	24 Cr 52.0	25 Mn 54.9	26 Fe 55.8	27 Co 58.9	28 Ni 58.7	29 Cu 63.5	30 Zn 65.4	31 Ga 69.7	32 Ge 72.6	33 As 74.9	34 Se 79.0	35 Br 79.9	36 Kr 83.8
37 Rb 85.5	38 Sr 87.6	39 Y 88.9	40 Zr 91.2	41 Nb 92.9	42 Mo 95.9	43 Tc [98]	44 Ru 101.1	45 Rh 102.9	46 Pd 106.4	47 Ag 107.9	48 Cd 112.4	49 In 114.8	50 Sn 118.7	51 Sb 121.8	52 Te 127.6	53 I 126.9	54 Xe 131.3
55 Cs 132.9	56 Ba 137.3	71 Lu 175.0	72 Hf 178.5	73 Ta 180.9	74 W 183.9	75 Re 186.2	76 Os 190.2	77 Ir 192.2	78 Pt 195.1	79 Au 197.0	80 Hg 200.6	81 Tl 204.4	82 Pb 207.2	83 Bi 209.0	84 Po [209]	85 At [210]	86 Rn [222]
87 Fr [223]	88 Ra [226]	103 Lr [262]	104 Rf [265]	105 Db [268]	106 Sg [271]	107 Bh [272]	108 Hs [270]	109 Mt [276]	110 Ds [281]	111 Rg [280]	112 Cn [285]	113	114 Fl [289]	115	116 Lv [293]	117	118

Period 1 2 3 4 5 6 7

Atomic number — 27
Co — Symbol
Atomic mass — 58.9

Transition elements

Lanthanides 6

57 La 138.9	58 Ce 140.1	59 Pr 140.9	60 Nd 144.2	61 Pm 144.9	62 Sm 150.4	63 Eu 152.0	64 Gd 157.3	65 Tb 158.9	66 Dy 162.5	67 Ho 164.9	68 Er 167.3	69 Tm 168.9	70 Yb 173.0

Actinides 7

89 Ac [227]	90 Th 232.0	91 Pa 231.0	92 U 238.0	93 Np [237]	94 Pu [244]	95 Am [243]	96 Cm [247]	97 Bk [247]	98 Cf [251]	99 Es [252]	100 Fm [257]	101 Md [258]	102 No [259]

Inner transition elements

An atomic mass in brackets is that of the longest-lived isotope of an element with no stable isotopes.

Video Resources (MP)

The following lists describe the videos available in the Pearson eText and in the Study Area of MasteringPhysics, with the corresponding textbook section and page references.

Class Videos

Video Tutor Solutions

Video Tutor Demonstrations

Atomic and Nuclear Data

Atomic Number (Z)	Element	Symbol	Mass Number (A)	Atomic Mass (u)	Percent Abundance	Decay Mode	Half-Life $t_{1/2}$
0	(Neutron)	n	1	1.008 665		β^-	10.4 min
1	Hydrogen	H	1	1.007 825	99.985	stable	
	Deuterium	D	2	2.014 102	0.015	stable	
	Tritium	T	3	3.016 049		β^-	12.33 yr
2	Helium	He	3	3.016 029	0.000 1	stable	
			4	4.002 602	99.999 9	stable	
			6	6.018 886		β^-	0.81 s
3	Lithium	Li	6	6.015 121	7.50	stable	
			7	7.016 003	92.50	stable	
			8	8.022 486		β^-	0.84 s
4	Beryllium	Be	9	9.012 174	100	stable	
			10	10.013 534		β^-	1.5×10^6 yr
5	Boron	B	10	10.012 936	19.90	stable	
			11	11.009 305	80.10	stable	
			12	12.014 352		β^-	0.020 2 s
6	Carbon	C	10	10.016 854		β^+	19.3 s
			11	11.011 433		β^+	20.4 min
			12	12.000 000	98.90	stable	
			13	13.003 355	1.10	stable	
			14	14.003 242		β^-	5 730 yr
			15	15.010 599		β^-	2.45 s
7	Nitrogen	N	12	12.018 613		β^+	0.011 0 s
			13	13.005 738		β^+	9.96 min
			14	14.003 074	99.63	stable	
			15	15.000 108	0.37	stable	
			16	16.006 100		β^-	7.13 s
			17	17.008 450		β^-	4.17 s
8	Oxygen	O	15	15.003 065		β^+	122 s
			16	15.994 915	99.76	stable	
			17	16.999 132	0.04	stable	
			18	17.999 160	0.20	stable	
			19	19.003 577		β^-	26.9 s
9	Fluorine	F	18	18.000 937		β^+	109.8 min
			19	18.998 404	100	stable	
			20	19.999 982		β^-	11.0 s
10	Neon	Ne	19	19.001 880		β^+	17.2 s
			20	19.992 435	90.48	stable	
			21	20.993 841	0.27	stable	
			22	21.991 383	9.25	stable	
17	Chlorine	Cl	35	34.968 853	75.77	stable	
			36	35.968 307		β^-	3.0×10^5 yr
			37	36.965 903	24.23	stable	

Atomic Number (Z)	Element	Symbol	Mass Number (A)	Atomic Mass (u)	Percent Abundance	Decay Mode	Half-Life $t_{1/2}$
18	Argon	Ar	36	35.967 547	0.34	stable	
			38	37.962 732	0.06	stable	
			39	38.964 314		β^-	269 yr
			40	39.962 384	99.60	stable	
			42	41.963 049		β^-	33 yr
19	Potassium	K	39	38.963 708	93.26	stable	
			40	39.964 000	0.01	β^-	1.28×10^9 yr
			41	40.961 827	6.73	stable	
26	Iron	Fe	54	54.939 613	5.9	stable	
			56	55.934 940	91.72	stable	
			57	56.935 396	2.1	stable	
			58	57.933 278	0.28	stable	
			60	59.934 072		β^-	1.5×10^6 yr
27	Cobalt	Co	59	58.933 198	100	stable	
			60	59.933 820		β^-	5.27 yr
38	Strontium	Sr	84	83.913 425	0.56%	stable	
			86	85.909 262	9.86%	stable	
			87	86.908 879	7.00%	stable	
			88	87.905 614	82.58%	stable	
			89	88.907 450		β^-	50.53 days
			90	89.907 738		β^-	27.78 yr
53	Iodine	I	127	126.904 474	100	stable	
			129	128.904 984		β^-	1.6×10^7 yr
			131	130.906 124		β^-	8 days
54	Xenon	Xe	128	127.903 531	1.9	stable	
			129	128.904 779	26.4	stable	
			130	129.903 509	4.1	stable	
			131	130.905 069	21.2	stable	
			132	131.904 141	26.9	stable	
			133	132.905 906		β^-	5.4 days
			134	133.905 394	10.4	stable	
			136	135.907 215	8.9	stable	
55	Cesium	Cs	133	132.905 436	100	stable	
			137	136.907 078		β^-	30 yr
82	Lead	Pb	204	203.973 020	1.4	stable	
			206	205.974 440	24.1	stable	
			207	206.975 871	22.1	stable	
			208	207.976 627	52.4	stable	
			210	209.984 163		α, β^-	22.3 yr
			211	210.988 734		β^-	36.1 min
83	Bismuth	Bi	209	208.980 374	100	stable	
			211	210.987 254		α	2.14 min
			215	215.001 836		β^-	7.4 min
86	Radon	Rn	219	219.009 477		α	3.96 s
			220	220.011 369		α	55.6 s
			222	222.017 571		α, β^-	3.823 days

Atomic Number (Z)	Element	Symbol	Mass Number (A)	Atomic Mass (u)	Percent Abundance	Decay Mode	Half-Life $t_{1/2}$
88	Radium	Ra	223	223.018 499		α	11.43 days
			224	224.020 187		α	3.66 days
			226	226.025 402		α	1 600 yr
			228	228.031 064		β^-	5.75 yr
90	Thorium	Th	227	227.027 701		α	18.72 days
			228	228.028 716		α	1.913 yr
			229	229.031 757		α	7 340 yr
			230	230.033 127		α	7.54×10^4 yr
			231	231.036 299		α, β^-	25.52 h
			232	232.038 051	100	α	1.40×10^{10} yr
			234	234.043 593		β^-	24.1 days
92	Uranium	U	233	233.039 630		α	1.59×10^5 yr
			234	234.040 946		α	2.45×10^5 yr
			235	235.043 924	0.72	α	7.04×10^8 yr
			236	236.045 562		α	2.34×10^7 yr
			238	238.050 784	99.28	α	4.47×10^9 yr
93	Neptunium	Np	237	237.048 168		α	2.14×10^6 yr
			238	238.050 946		β^-	2.12 days
			239	239.052 939		β^-	2.36 days
94	Plutonium	Pu	238	238.049 555		α	87.7 yr
			239	239.052 157		α	2.412×10^4 yr
			240	240.053 808		α	6560 yr
			242	242.058 737		α	3.73×10^6 yr
			244	244.064 200		α	8.1×10^7 yr
95	Americium	Am	241	241.056 823		α	432.21 yr
			243	243.061 375		α	7 370 yr

Answers

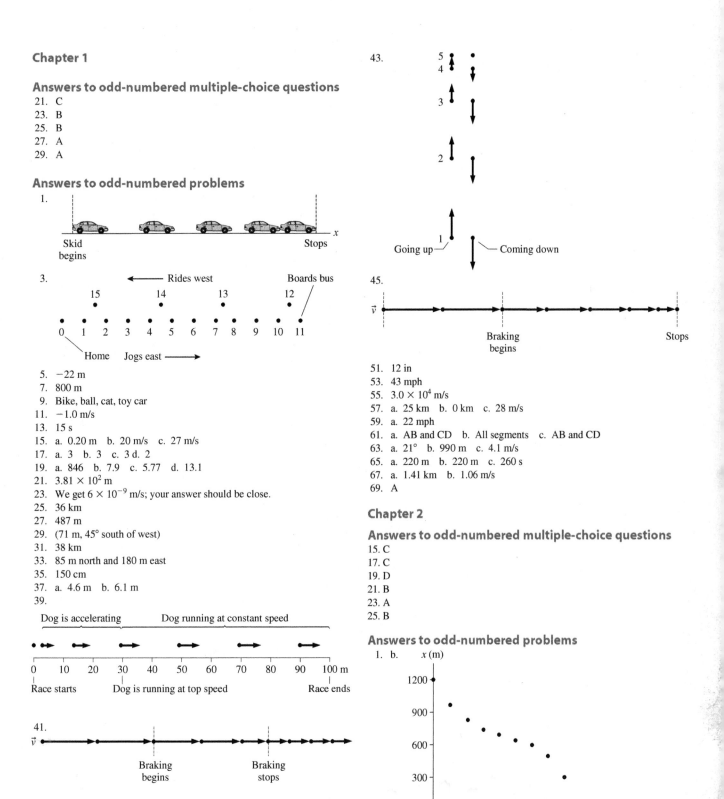

Chapter 1

Answers to odd-numbered multiple-choice questions

21. C
23. B
25. B
27. A
29. A

Answers to odd-numbered problems

1.

Skid begins   Stops

3.

← Rides west   Boards bus

15   14   13   12

0 1 2 3 4 5 6 7 8 9 10 11

Home   Jogs east →

5. −22 m
7. 800 m
9. Bike, ball, cat, toy car
11. −1.0 m/s
13. 15 s
15. a. 0.20 m   b. 20 m/s   c. 27 m/s
17. a. 3   b. 3   c. 3 d. 2
19. a. 846   b. 7.9   c. 5.77   d. 13.1
21. 3.81×10^2 m
23. We get 6×10^{-9} m/s; your answer should be close.
25. 36 km
27. 487 m
29. (71 m, 45° south of west)
31. 38 km
33. 85 m north and 180 m east
35. 150 cm
37. a. 4.6 m   b. 6.1 m
39.

Dog is accelerating   Dog running at constant speed

0   10   20   30   40   50   60   70   80   90   100 m
Race starts   Dog is running at top speed   Race ends

41.

$\vec{v}$

Braking begins   Braking stops

43.

5
4

3

2

1
Going up   Coming down

45.

$\vec{v}$

Braking begins   Stops

51. 12 in
53. 43 mph
55. 3.0×10^4 m/s
57. a. 25 km   b. 0 km   c. 28 m/s
59. a. 22 mph
61. a. AB and CD   b. All segments   c. AB and CD
63. a. 21°   b. 990 m   c. 4.1 m/s
65. a. 220 m   b. 220 m   c. 260 s
67. a. 1.41 km   b. 1.06 m/s
69. A

Chapter 2

Answers to odd-numbered multiple-choice questions

15. C
17. C
19. D
21. B
23. A
25. B

Answers to odd-numbered problems

1. b. x (m)

1200

900

600

300

0

0   2   4   6   8   10
t (s)

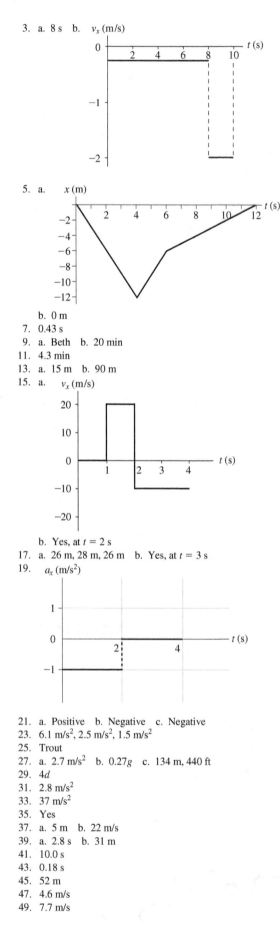

3. a. 8 s   b. v_x (m/s)

b. 0 m
7. 0.43 s
9. a. Beth   b. 20 min
11. 4.3 min
13. a. 15 m   b. 90 m
15. a. v_x (m/s)

b. Yes, at $t = 2$ s
17. a. 26 m, 28 m, 26 m   b. Yes, at $t = 3$ s
19. a_x (m/s^2)

21. a. Positive   b. Negative   c. Negative
23. 6.1 m/s^2, 2.5 m/s^2, 1.5 m/s^2
25. Trout
27. a. 2.7 m/s^2 b. 0.27g c. 134 m, 440 ft
29. 4d
31. 2.8 m/s^2
33. 37 m/s^2
35. Yes
37. a. 5 m   b. 22 m/s
39. a. 2.8 s   b. 31 m
41. 10.0 s
43. 0.18 s
45. 52 m
47. 4.6 m/s
49. 7.7 m/s

51. a. 3.0 s   b. 15 m/s   c. -31 m/s, -35 m/s
53. 57 mph
55. b. x (m)

c. 4 m   d. 4 m   e. 4 m/s   f. 2 m/s   g. -2 m/s^2
57. a. 2.3 m/s^2, 0.23g b. 35 s   c. 4.2 km
59. 8.7 y
61. a. ≈ 1.0 cm   b. 35 m/s^2 c. 0.84 m/s
63. a. 1000 m/s^2 b. 1.0 ms   c. 5.1 cm
65. 3.2 s
67. a. 54.8 km   b. 230 s
c. v_y (m/s)

69. 110 m
71. a. 4.1 s   b. Equal speeds
73. Man: 10.2 s; horse: 10.7 s; man wins
75. Honda wins by 1.0 s
77. 5.5 m/s^2
79. B

Chapter 3

Answers to odd-numbered multiple-choice questions
19. C
21. a. B   b. E   c. C   d. D   e. B   f. A
23. A
25. a. C   b. C
27. C
29. E

Answers to odd-numbered problems
5. 8.0 m
7. a. 34°   b. 1.7 m/s
9. 87 m/s
11. a. $d_x = 71$ m, $d_y = -71$ m   b. $v_x = 280$ m/s, $v_y = 100$ m/s
c. $a_x = 0.0$ m/s^2, $a_y = -5.0$ m/s^2
13. a. 45 m/s, 63°   b. 6.3 m/s^2, $-72°$
15. 7.1 km, 7.5 km
17. 530 m
19. 1.3 s
21. Ball 1: 5 m/s; ball 2: 15 m/s
23. Ball 1: 15 m/s; ball 2: 5 m/s
25. 2.0 km/h

27. a. $\vec{v}_i = (5, 0 \text{ m/s, horizontal})$ d.   10 m

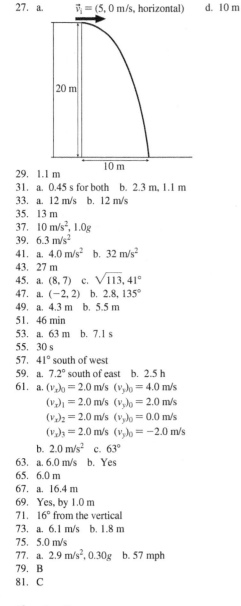

29. 1.1 m
31. a. 0.45 s for both   b. 2.3 m, 1.1 m
33. a. 12 m/s   b. 12 m/s
35. 13 m
37. $10 \text{ m/s}^2, 1.0g$
39. 6.3 m/s^2
41. a. 4.0 m/s^2 b. 32 m/s^2
43. 27 m
45. a. (8, 7)   c. $\sqrt{113}, 41°$
47. a. (−2, 2)   b. 2.8, 135°
49. a. 4.3 m   b. 5.5 m
51. 46 min
53. a. 63 m   b. 7.1 s
55. 30 s
57. 41° south of west
59. a. 7.2° south of east   b. 2.5 h
61. a. $(v_x)_0 = 2.0 \text{ m/s}$ $(v_y)_0 = 4.0 \text{ m/s}$
       $(v_x)_1 = 2.0 \text{ m/s}$ $(v_y)_0 = 2.0 \text{ m/s}$
       $(v_x)_2 = 2.0 \text{ m/s}$ $(v_y)_0 = 0.0 \text{ m/s}$
       $(v_x)_3 = 2.0 \text{ m/s}$ $(v_y)_0 = -2.0 \text{ m/s}$
     b. 2.0 m/s^2 c. 63°
63. a. 6.0 m/s   b. Yes
65. 6.0 m
67. a. 16.4 m
69. Yes, by 1.0 m
71. 16° from the vertical
73. a. 6.1 m/s   b. 1.8 m
75. 5.0 m/s
77. a. $2.9 \text{ m/s}^2, 0.30g$ b. 57 mph
79. B
81. C

Chapter 4

Answers to odd-numbered multiple-choice questions
21. C
23. D
25. D
27. C
29. B

Answers to odd-numbered problems
1. First is rear-end; second is head on
5.

7. Weight, tension force by rope

9. Weight, normal force by ground, kinetic friction force by ground
11. Weight, normal force by slope, kinetic friction force by slope
13. $m_1 = 0.080$ kg and $m_3 = 0.50$ kg
15. 3.0 m/s^2
17. a. 16 m/s^2 b. 4.0 m/s^2 c. 8.0 m/s^2 d. 32 m/s^2
19. 2.5 m/s^2
21. 0.25 kg
23. a. No   b. 10 m/s^2 to the left
25. 5.0 h
31.

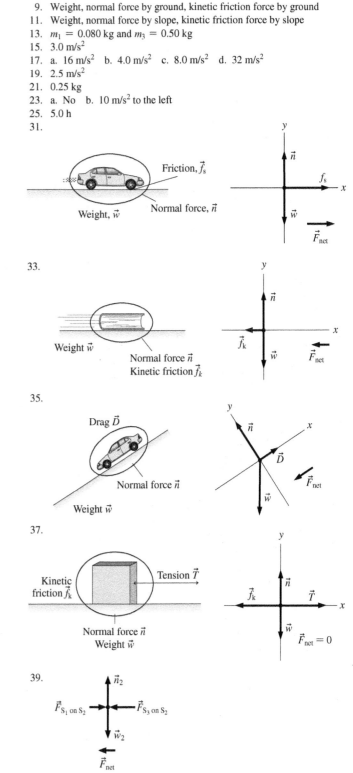

33.

35.

37.

39.

41. Normal force of road on car and of car on road; friction force of road on car and of car on road.

43. **Motion diagram**   45. **Motion diagram**

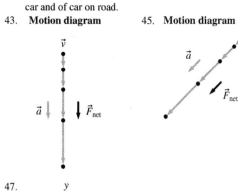

47.

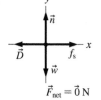

$\vec{F}_{net} = \vec{0}$ N

49.

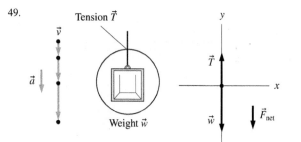

53.

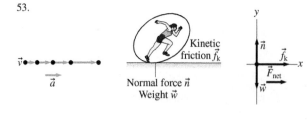

55.

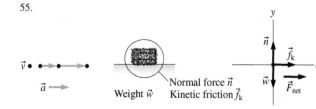

57.

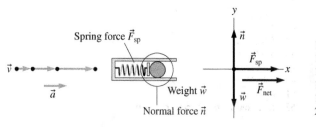

59.

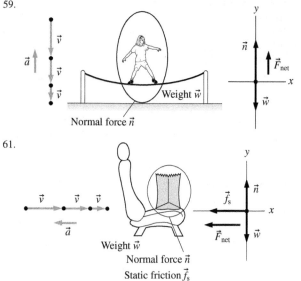

61.

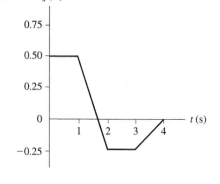

63. a. Up   b. Normal force and weight   c. Upward   d. Larger
65. b. 150 N   c. 1/5
67. Greater on the way up
69. C
71. C

Chapter 5

Answers to odd-numbered multiple-choice questions
23. D
25. a. B   b. D
27. C
29. C
31. D

Answers to odd-numbered problems
1. $T_1 = 87$ N, $T_2 = 50$ N
3. 110 N
5. 49 N
7. 170 kg
9.

11. $a_x = 1.0$ m/s^2, $a_y = 0.0$ m/s^2
13. a. d b. $4d$
15. 310 N
17. 0 N
19. a. 590 N   b. 740 N   c. 590 N
21. a. 780 N   b. 1100 N

23. 1000 N, 740 N, 590 N
25. b. 180 N
27. 4800 N
29. No
31. a.

c. 4.9 m/s^2 d. 2.9 m/s^2
33. 63 N
35. 7.0 kg
37. 170 m/s
39. a. 3000 N   b. 3000 N
41. a. 6.0 N   b. 10 N
43. a. 20 N   b. 21 N
45. a. 1.0 N   b. 50 N
47. a. 530 N   b. 5300 N
49. F_{net}(1 s) = 8 N, F_{net}(4 s) = 0 N, F_{net}(7 s) = −12 N
51. a. 490 N   b. 740 N
53. a. 490 N   b. 240 N
55. a. 6800 N   b. 1.4 × 10^6 N   c. 11 times, 2300 times
57. a. 5.2 m/s^2 b. 1000 kg
59. 160 N
61. 60 m
63. 11°
65. 0.98 kg
67. Stay at rest
69. T_1 = 17 N, T_2 = 27 N
71. 14 N
73. a. −49 m/s^2 b. 29 m/s^2
75. Down at 0.93 m/s^2
77. C
79. B

Chapter 6

Answers to odd-numbered multiple-choice questions
21. D
23. E
25. B
27. A
29. A
31. D

Answers to odd-numbered problems
1. 3.9 m/s
3. a. 0.56 rev/s   b. 1.8 s
5. a. 6.0 ms, 170 rev/s   b. 63 m/s   c. 6700g
7. a. 3.0 × 10^4 m/s   b. 6.0 × 10^{-3} m/s^2
9. 5.7 m/s, 110 m/s^2
11. 34 m/s
13. $T_3 > T_1 = T_4 > T_2$
15. a. 3.9 m/s   b. 6.2 N
17. 9400 N, toward center, static friction
19. a. 1700 m/s^2 b. 240 N
21. 14 m/s
23. 270 N
25. 3.3 m/s
27. 20 m/s

29. a. 1.8 × 10^4 m/s^2 b. 4.4 × 10^3 m/s^2
31. 99 min
33. 1/2
35. 6.0 × 10^{-4}
37. 3.9 m/s^2
39. a. 3.53 × 10^{22} N   b. 1.99 × 10^{29} N   c. 0.564%
41. a. 3.77 m/s^2 b. 25.9 m/s^2
43. 7000 m/s
45. 4.37 × 10^{11} m, 1.7 × 10^4 m/s
47. 0.039 au
49. 11 h
53. North pole, by 2.5 N
55. 5.5 m/s
57. a. 5.0 N   b. 30 rpm   c. 3.0 s
59. 5.4 m/s
61. 22 m/s
63. $T_1 = 12\pi^2 mf^2 l$, $T_2 = 8\pi^2 mf^2 l$
65. 2400 m
67. a. 3.0 × 10^{24} kg   b. 0.89 m/s^2
69. a. 3.8 m/s^2 b. 1.6 m/s
71. (12 cm, 0 cm)
73. 6.5 × 10^{23} kg
75. 0.48 m/s
77. C
79. A

Chapter 7

Answers to odd-numbered multiple-choice questions
17. C
19. D
21. D
23. A
25. D
27. D

Answers to odd-numbered problems
1. a. $\theta = \pi/2$ b. $\theta = 0$ c. $\theta = 4\pi/3$
3. 1.7 × 10^{-3} rad/s
5. a. 1.3 rad, 72°   b. 3.9 × 10^{-5} rad/s
7. 3.0 rad
9. a. 25 rad   b. −5 rad/s
11. a. 0.105 rad/s   b. 0.00105 m/s
13. 11 m/s
15. a. 160 rad/s^2 b. 50 rev
17. $\tau_1 < \tau_2 = \tau_3 < \tau_4$
19. −0.20 N·m
21. 5.7 N
23. a. 13.2 N·m   b. 117 N·m
25. 5.5 N·m
27. a. 0.62 m   b. 0.65 m
29. 12 N·m
31. −4.9 N·m
33. a. 34 N·m   b. 24 N·m
35. a. 1.70 m   b. −833 N·m
37. 7.2 × 10^{-7} kg·m^2
39. 120 kg·m^2
41. 1.8 kg
43. 6.0 × 10^{-3} N·m
45. 8.0 N·m
47. −44 rad/s^2
49. 0.11 N·m
51. 0.50 s
53. a. 14 rad/s   b. 11 m/s   c. 7.9 m/s

55. b.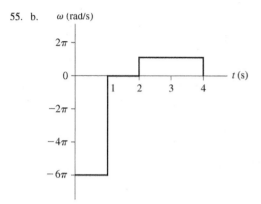

57. 55 rotations
59. $-0.94 \, \text{N} \cdot \text{m}$
61. 7.5 cm
63. a. $1.4 \, \text{kg} \cdot \text{m}^2$ b. Increase
65. $3.5 \, \text{rad/s}^2$
67. $-0.28 \, \text{N} \cdot \text{m}$
69. 1.1 s
71. 2.2 N
73. B
75. B
77. B

Chapter 8

Answers to odd-numbered multiple-choice questions
15. C
17. A
19. B
21. B
23. B
25. D

Answers to odd-numbered problems
1. Right 470 N; left 160 N
3. 15 cm
5. 140 N
7. $12 \, \text{N} \cdot \text{m}$
9. 1.0
11. 1.4 m
13. 590 N
15. 98 N
17. 16°
19. 40 cm
21. 0.93 m
23. 830 N/m
25. 0.30 N/m
27. a. 390 N/m b. 18 cm
29. 1.2 cm
31. a. 2 mm b. 0.25 mm
33. 7900 N
35. a. 980 N b. 4.0 m
37. 16 cm
39. 0.0078%
41. a. 3.8 mN b. 8.6 mm
43. $F_1 = 750 \, \text{N}, F_2 = 1000 \, \text{N}$
45. 860 N
47. a. 2000 N b. 2200 N
49. 1.0 m
51. a. 49 N b. 1500 N/m c. 3.4 cm
53. 17.4 N

55. 31 cm
57. a. 2300 N b. 1.0 m
59. $0.25 = 25\%$
61. a. $4.8 \times 10^{-2} \, \text{m}^2$ b. 70,000 kg c. 0.14
63. B
65. B

Part I Problems

Answers to odd-numbered problems
1. A
3. A
5. C
7. B
9. D
11. A
13. B
15. D
17. B
19. A
21. b. 2.8 N

Chapter 9

Answers to odd-numbered multiple-choice questions
19. D
21. D
23. D
25. D
27. C

Answers to odd-numbered problems
1. 15 m/s
3. 6.0 N
5. $80 \, \text{kg} \cdot \text{m/s}$
7. a. 1.5 m/s to the right b. 0.5 m/s to the right
9. $-110 \, \text{N}$
11. a. $-19 \, \text{kN}$ b. $-280 \, \text{kN}$
13. 0.205 m/s
15. 0.31 m/s
17. 1.4 m/s
19. 0.47 m/s
21. 4.8 m/s
23. 2.3 m/s
25. 1.0 kg
27. 37°
29. $(-2 \, \text{kg} \cdot \text{m/s}, 4 \, \text{kg} \cdot \text{m/s})$
31. 14 m/s at 45° north of east
33. $510 \, \text{kg} \cdot \text{m}^2/\text{s}$
35. $0.025 \, \text{kg} \cdot \text{m}^2/\text{s}$, into the page
37. 1.3 rev/s
39. 0.20 s
41. 930 N
43. 3.8 m/s
45. $12 \, \text{kg} \cdot \text{m/s}$, left; 15° above the horizontal
47. $7.5 \times 10^{-10} \, \text{kg} \cdot \text{m/s}$
49. a. 6.4 m/s b. 360 N
51. 3.6 m/s
53. 1.1 m/s
55. 2.0 m/s
57. a. $6.7 \times 10^{-8} \, \text{m/s}$ b. $2.2 \times 10^{-10}\%$
59. 13 s
61. 440 m/s
63. 0.89 m

65. 2.0 m/s
67. −0.58 m/s
69. 9.0%
71. 1.97×10^3 m/s
73. 360 N
75. 50 rpm
77. 22 rpm
79. B
81. A

Chapter 10

Answers to odd-numbered multiple-choice questions

23. C
25. A
27. C

Answers to odd-numbered problems

1. a. −30 J   b. 30 J
3. Rope 1: 0.919 kJ; rope 2: 0.579 kJ
5. a. 0 J   b. −43 J   c. 43 J
7. The bullet
9. a. 14 m/s   b. Factor of 4
11. 4.0 m/s
13. 0.0 J
15. $1.8 \text{ kg} \cdot \text{m}^2$
17. a. 6.8×10^5 J   b. 46 m   c. No
19. 63 kJ
21. 0.63 m
23. 9.7 J

25. 16.5 kJ
27. 110 N
29. a. 13 m/s   b. 14 m/s
31. 31 m/s
33. 3.0 m/s
35. 17 m/s
37. a. The child's gravitational potential energy will be changing into kinetic energy and thermal energy.   b. 550 J
39. 17 kJ
41. 0.86 m/s and 2.9 m/s
43. 1/2
45. a. 1.8×10^2 J   b. 59 W
47. 45 kW
49. 34 min
51. 2.0×10^4 W
53. 91 kJ
55. a. 7.7 m/s   b. 6.6 m/s
57. 28 m
59. a. 200 J   b. 98 N   c. 2.0 m   d. 200 J
61. 5.8 m
63. 51 cm
65. 3.8 m, not dependent on mass
67. a. $\sqrt{\dfrac{(m + M)kd^2}{m^2}}$ b. 2.0×10^2 m/s   c. 99.8%
69. 1.6 m/s
71. 7.9 m/s
73. 49 cm
75. a. 0.31 m/s   b. 95%
77. a. 100 N   b. 0.21 kW   c. 1.2 kW
79. 1.3 m/s
81. B
83. C
85. A
87. B
89. C

Credits

Index

For users of the two-volume edition, pages 1–533 are in Volume 1 and pages 534–1011 are in Volume 2.